The Correspondence of John Tyndall

VOLUME 6

THE CORRESPONDENCE OF JOHN TYNDALL

VOLUME 6

The Correspondence, November 1856–February 1859

Edited by
Michael D. Barton, Janet Browne, Ken Corbett,
and Norman McMillan

General Editors
James Elwick, Roland Jackson, Bernard Lightman,
and Michael S. Reidy

UNIVERSITY OF PITTSBURGH PRESS

Published by the University of Pittsburgh Press, Pittsburgh, Pa., 15260
Copyright © 2018, University of Pittsburgh Press
All rights reserved
Manufactured in the United States of America
Printed on acid-free paper
10 9 8 7 6 5 4 3 2 1

Cataloging-in-Publication data is available from the Library of Congress

ISBN 13: 978-0-8229-4533-8

CONTENTS

ACKNOWLEDGMENTS

This volume of *The Correspondence of John Tyndall* owes its creation to many colleagues whose willingness to share their time and expertise is deeply appreciated. We especially acknowledge the work of many graduate students and other colleagues in providing the original transcriptions. We greatly valued the work of James Braund and Juliette Reboul in checking the transcriptions and translations of German and French letters, respectively. Kostas Gavroglu of the National and Kapodistrian University of Athens assisted in looking over the transcription of Greek words in one letter. We sometimes required new scans of letters, and for this we warmly thank Jane Harrison of the Royal Institution of Great Britain and Rosemary Clarkson of the Darwin Correspondence Project. For help in creating high-quality image files of the drawings within letters in this volume, we thank Erin Grosjean. Finally, for invaluable help in answering a variety of questions about letters in this volume, we would like to acknowledge the general editors of *The Correspondence of John Tyndall*, James Elwick, Roland Jackson, Bernard Lightman, and Michael Reidy, as well as Ruth Barton, William Brock, Richard Carter, Thony Christie, Jim Endersby, Diarmid Finnegan, Frank A. J. L. James, Cam Sharp Jones (Joseph Hooker Correspondence Project), Elizabeth Neswald, Efram Sera-Shriar, James Spedding, and Kurt Tobler (Secretary of the Naturforschende Gesellschaft in Zürich).

LIST OF ABBREVIATIONS

ADB	*Allgemeine Deutsche Biographie*
ANB	*American National Biography*
Annal. Chim. et Phys.	*Annales de Chimie et de Physique*
BAAS	British Association for the Advancement of Science
BL	British Library, London
Brit. Assoc. Rep.	*Reports of the British Association for the Advancement of Science*
Brock & MacLeod, *Hirst Journals*	William H. Brock and Roy M. MacLeod, eds., *Natural Knowledge in Social Context: The Journals of Thomas Archer Hirst FRS* (London: Mansell microfiche, 1980)
Brock, 'Queenwood'	W. H. Brock, 'Queenwood College Revisited,' article XVII in *Science for All: Studies in the History of Victorian Science and Education* (Aldershot: Ashgate Variorum, 1996)
CDSB	*Complete Dictionary of Scientific Biography*, 27 vols. (Detroit: Charles Scribner's Sons, 2008)
CUL SC	Cambridge University Library, Stokes Correspondence
Faraday Correspondence	*The Correspondence of Michael Faraday*, ed. F. A. J. L. James, 6 vols. (London: Institution of Electrical Engineers/Institution of Engineering and Technology, 1991–2012)
FRS	Fellow of the Royal Society
Glaciers of the Alps	John Tyndall, *Glaciers of the Alps* (London: John Murray, 1860)
HLS	*Historisches Lexikon der Schweiz*
IC HP	Imperial College of Science, Technology and Medicine, College Archives, Huxley Papers
Journal	John Tyndall's Journal, 1848–55, typescript (RI MS JT/2/13b); and 1855–72, typescript (RI MS JT/2/13c)
Kew JDH	Royal Botanical Gardens, Kew, Joseph Dalton Hooker Correspondence
LI	London Institution
LT	Louisa Tyndall, John Tyndall's widow
NDB	*Neue Deutsche Biographie*
NLS	National Library of Scotland

NRCC Plücker — National Research Council of Canada Archive, Plücker letters fonds: 1814–1963

ODNB — *Oxford Dictionary of National Biography*

OED — *Oxford English Dictionary*

Paris, Comptes Rendus — *Comptes rendus hebdomadaires des séances de l'Académie des Sciences*

Phil. Mag. — *Philosophical Magazine*

Phil. Trans. — *Philosophical Transactions of the Royal Society of London*

Poggend. Annal. — *Poggendorff's Annalen der Physik und Chemie*

RGO — Cambridge University Library, Royal Greenwich Observatory Archives

RGO MS.RGO 6 — Royal Greenwich Observatory Archives, Papers of George Airy

RI — Royal Institution of Great Britain

RI MS JT — Royal Institution, London, John Tyndall Papers

Roy. Inst. Proc. — *Proceedings of the Royal Institution of Great Britain*

Roy. Soc. Proc. — *Proceedings of the Royal Society of London*

RS — Royal Society of London

RS HS — Royal Society of London, Herschel Papers

RS RR — Royal Society of London, Referees Reports

Russell, *Edward Frankland* — Colin A. Russell, *Edward Frankland: Chemistry, Controversy and Conspiracy in Victorian England* (Cambridge: Cambridge University Press, 1996)

StA JDF — Saint Andrews, Papers of James David Forbes

Tyndall Correspondence, vol. 2 — Melinda Baldwin and Janet Browne, eds., *The Correspondence of John Tyndall, Volume 2: The Correspondence, September 1843–December 1849* (Pittsburgh: University of Pittsburgh Press, 2016)

Tyndall Correspondence, vol. 4 — Ian Hesketh and Efram Sera-Shriar, eds., *The Correspondence of John Tyndall, Volume 4: The Correspondence, January 1853–December 1854* (Pittsburgh: University of Pittsburgh Press, 2018)

Tyndall Correspondence, vol. 5 — William H. Brock and Geoffrey Cantor, eds., *The Correspondence of John Tyndall, Volume 5: The Correspondence, January 1855–October 1856* (Pittsburgh: University of Pittsburgh Press, 2018)

UB Mbg — University Library of Marburg

Frontispiece: John Tyndall, photographed by Maull & Polyblank, 21 September 1857, in London, dated by an entry in Tyndall's journal (Journal, RI MS JT/2/13c/1034). Tyndall's correspondents refer to this photograph in several letters in this volume. Credit: Royal Institution of Great Britain, Science Photo Library.

INTRODUCTION TO VOLUME 6

This sixth volume of *The Correspondence of John Tyndall* contains 302 letters covering a period of twenty-eight months, from 6 November 1856 to 26 February 1859. The volume begins shortly after Tyndall returned from his first glacier researches in the Alps and follows him as he experimented and lectured on physics in central London at the Royal Institution of Great Britain (RI), visited friends, joined London's fashionable social circles, published and reviewed scientific articles, corresponded with fellow men of science on a wide range of topics, and developed his theories about the structure and movement of glaciers. At this point Tyndall was in his mid-thirties (the year of his birth is not certain).[1] Importantly, this volume includes Tyndall's expeditions to the Alps to deepen his researches on glaciers—and also documents some of his most dangerous mountaineering exploits and appreciation of the natural wonders of these regions. In letters to his closest friends—who included his colleague and mentor Michael Faraday and personal friends Thomas Archer Hirst and Thomas Henry Huxley—Tyndall captured the excitement and achievement of his expeditions. This period also saw changes in his role as an examiner in science and mathematics for students at two British military schools, as well as changes in his own role as professor of natural philosophy at the RI. Perhaps unusually for the period, Tyndall wrote with relaxed familiarity to a few female friends showing his delight at being drawn into their family circles. The letters printed here engage with every aspect of this energetic and sociable life and include incidental contemporary comment on events such as the Second Anglo-Chinese War of 1856–60, the Indian Rebellion of 1857, the launching of the SS *Great Eastern*, David Livingstone's first African expedition, the sighting and naming of the famous Donati's comet, and the "Great Stink" of London. These references place Tyndall and his correspondents into the larger context of British society and its ever-expanding empire. By the end of the period covered by this volume, Tyndall was celebrated as a charismatic lecturer at the RI, was a seasoned Alpine traveler known for his controversial theories about the movement of glaciers, was a beloved friend to colleagues, and was increasingly respected as a scientist in the wider academic world.

Glacier Research in the Alps

In a June 1857 letter to his literary friend Juliet Pollock, the wife of William Frederick Pollock, Master of the Court of Exchequer, Tyndall wrote, "I hope this glorious weather has an invigorating effect on you. It will on me as soon as I escape from London, but I believe not before."[2] The "escape from London" was to the Alps. Tyndall spent that summer hiking and traveling in the Alpine region of continental Europe, setting up a pattern of regular expeditions that served two essential purposes for Tyndall: first, to restore his strength and relax his overworked brain, and second, to allow him to investigate in person the movements of the ice in Alpine glaciers. Judging from his letters, these trips were highly invigorating despite the physical toil and risks involved. Tyndall had taken his first trip to the Swiss and Austrian Alps to study glaciers with Huxley in the summer of 1856, just before this volume opens.[3] Subsequently, he eagerly looked forward to the end of his lecturing duties at the RI so that he would be free to visit again. In this volume, Tyndall's trips to the Alps in the summers of 1857 and 1858 established his lifelong love of the region (a map that shows the principal Alpine locales mentioned in this correspondence is included in this volume).[4] He would continue to study the structure and movement of glaciers, and find other avenues of research in the mountains, many of which not only depended on but were related to the physics of an increase in elevation, or "verticality."[5] He became a notable and competitive Alpine mountaineer, and, much later, from 1877, spent his summers with his wife, Louisa, at their home in the Alpine village of Bel Alp.

Tyndall's interest in glaciers began in 1856 when he gave a Friday Evening Discourse at the RI on 6 June, on the "Comparative view of the Cleavage of Crystals and Slate Rocks."[6] Although this was a discourse about how the formation of the cleavage plane was not due to crystalline forces but to pressure alone, it seems that following the lecture, Huxley suggested to him that the same explanation might also apply to the structure of glaciers, which are formed from layers of compacted snow and ice. That summer they traveled together to the Tyrol region to examine the structure of glaciers.[7] Their mutual friend, the botanist Joseph Dalton Hooker, the assistant director of the botanic gardens at Kew who had formerly traveled in the Himalayas, accompanied them part of the way (some time was also spent in the region with chemist Edward Frankland). All that autumn, as this volume of correspondence opens, Tyndall was enthusiastically thinking through the problem in mechanical and physical terms. Tyndall read his and Huxley's resulting scientific paper offering a new interpretation of the topic on 15 January 1857 at the Royal Society of London (RS), the most prestigious scientific venue in the United Kingdom, as well as delivering a more accessible Friday Evening

Discourse at the RI a week later. The paper was published as "On the Structure and Motion of Glaciers" in the *Phil. Trans.*,[8] and generated much attention. Tyndall undertook some effective self-promotion, and this volume shows him circulating reports of his conclusions, and copies of the public lecture at the RI, long before the RS published the full paper.[9]

Several descriptions of glacier motion had been proposed before, mostly suggesting that the ice slid over the landscape under the force of gravity. A common metaphor in use was the phrase "rivers of ice" (although, technically, sliding and river flow are not the same). The subject became current in 1840 when Louis Agassiz, the great Swiss naturalist, published his *Études sur les glaciers.* Agassiz demonstrated that the velocity is greater in the center than at the sides, and that a glacier moves more slowly at its head than at the snout or terminus. James David Forbes then proposed that even though glacial movement may include some sliding due to the weight of the ice and the downward pull of gravity, most movement was actually an imperceptible flow, as if the ice were a very viscous fluid.[10] This appeared to be supported by some experiments conducted by James Thomson.[11] In contrast, Tyndall expressed the view that the internal pressure and weight of a glacier was so great that the ice constantly cracked, melted, and congealed in new configurations, hence leading to downward movement. He wrote to his friend Rudolf Clausius in 1856, "There is in my mind no doubt whatever that the structure is produced by the united action of pressure and of tension."[12] There are some letters around this time in which Tyndall sought an appropriate word for this process, eventually settling on "fracture and regelation" (the term *regelation* was suggested by Hooker).[13] Toward the end of this volume of letters, Tyndall's views had developed to the point that he argued that while tension caused constant fracturing, and pressure aided in the regelation, together they also squeezed the glacier down the mountain, and began to think that the crystalline structure of ice played a role, since it would deform just like any other crystalline solid at temperatures near its melting point. This modified view was seemingly based on Thomson's work that he had initially disparaged.[14]

In all this, Tyndall believed he was diametrically at odds with the view adopted by James David Forbes.[15] He expressed a certain amount of glee at the prospect of an intellectual battle, and he vigorously argued against Forbes's view in papers, lectures, and letters. In a letter published here, to his friend Thomas Archer Hirst in December 1856, he wrote, "I think I have so smashed up this theory; so utterly annihilated it, that its author will hardly acknowledge it. Of course this may be all self delusion—others must judge of the force of my arguments, but unless I labour under a hallucination altogether new, [Forbes's] theory is killed without the hope of recovery."[16] The controversy drew in other scientists of the period and this volume of correspondence includes queries and suggestions from notable figures such as William Hopkins and George Biddell Airy.[17] Many years later, as it turned out, both Forbes and Tyndall were shown to be partly right. The letters reveal Tyndall's use of his position as an increasingly celebrated public lecturer to gather popular opinion behind him. In one instance, in his summer expedition of 1858 he deliberately employed the same mountain guide, Auguste Balmat, as had Forbes in previous years, to accompany him on Mont Blanc: to upstage Forbes was a great satisfaction to Tyndall. However, Balmat's time at the summit during that trip proved harrowing, as described in this volume. A year earlier Balmat had suggested to Tyndall that they might bury a self-registering thermometer in the ice at the summit. Through Tyndall's efforts the RS Council granted funds to purchase the thermometer. In burying the thermometer on 12 September 1858 Balmat suffered severe frostbite on his hands and ended up losing six of his fingernails. In another instance, difficulties that Tyndall encountered that summer with respect to the local regulations of the Commission of Guides of Chamonix led him to speak out about access at the British Association for the Advancement of Science (BAAS) meeting in Leeds that September. Garnering support from fellow attendees, a resolution was drafted that "application would be made to the Sardinian authorities for increased facilities for making scientific observations in the Alps."[18] At that point in time, the Duchy of Savoy (including the Chamonix valley and surrounding mountains) was ruled by the Kingdom of Sardinia. While Tyndall made sure his summit of Mont Blanc was known to fellow men of science, Faraday ensured that Tyndall's daring—and the very first—solo summit of Monte Rosa would be known to the public by forwarding Tyndall's vivid letter to Sarah Faraday to be published in the *Times*.[19] Faraday's pleasure in seeing the scientific development of his protégé seems palpable, even in the few letters that have survived.

The scientific aim of most of Tyndall's expeditions to the Alps was to find out for himself how movement varied from place to place on one glacier, and from glacier to glacier. In this volume of correspondence, it is clear that he mostly used simple field observations as best he could, noted the veining, and took meticulous barometric and temperature readings in different places and at different heights. Some of Tyndall's surveying and drawing instruments are held at the Carlow County Museum in Carlow, Ireland. Back at the RI during the winter and spring lecturing seasons he was able to carry out laboratory experiments on the regelation of ice. Perhaps one of the attractions of research into glaciers for Tyndall was the opportunity of exploring—and vindicating—Faraday's method of scientific induction and experimentation. Coupled with this, skills that Tyndall first developed while he was a surveyor for the Ordnance Survey could be brought into use along with his remarkable powers of physical endurance.[20] And he also cherished the aesthetic appeal of the landscapes he encountered. There are some remarkable letters published here that show Tyndall's descriptive powers. On 4 August 1858 he was at the Hotel Jungfrau, under the Eggishorn, resting after climbing the Finster-aarhorn. There he wrote to his friend Heinrich Debus in terms that reveal a marked appreciation of the sublime:

> Five valleys, each flanked by noble mountains send down their frozen contents to the same place of convergence and forms there the trunk of the great Aletsch glacier. This scene is most wonderful and most beautiful. The mountain masses are grand: the Jungfrau, the Monk, the Eiger the Aletsch horn and others of similar magnitude send down the snows from their shining shoulders. In the upper regions the snow appears as smooth as polished marble and clothing as it does mountain forms in themselves beautiful, renders the scene lovely as well as majestic. . . . A world of indescribable magnificence lay before & around us. Water boils there at 187° Faht.[21]

During these excursions Tyndall and his friends relied heavily on the services of mountain guides. The Alps were already well known as a summer tourist location, for the European tourist market was dramatically expanding at this time with the development of rail transport and hotels catering to prosperous British visitors, to the extent that parties of tourists would frequently ascend to scenic viewpoints, and more experienced climbers set out to summit challenging mountains, accompanied by knowledgeable local experts. A carefully regulated system of climbing guides operating under centralized rules for accessing the mountains was in existence several years before Tyndall began to visit. Yet Tyndall's scientific aims were different from most and his own sense of his personal climbing ability generated impatience with the regulations. Several of the letters in this volume mention his guides, in terms both admirable and less so, the restrictions they at times placed on him, and his delight in circumventing the system.[22] His preferred mode of ascent was to go with only one guide to accompany him. In the summer of 1858 Tyndall lodged a formal complaint to the superintendent of the Commission of Guides of Chamonix, based in Bonneville, Savoy, that men of science needed special access and more leeway in pursuing their investigations.[23] As noted above, his case was taken up by the BAAS.[24] Tyndall was indeed very skilled (and seemingly fearless) at climbing. On 27 November 1858 he was elected to the newly established Alpine Club in Britain, after having declined to be a founding member in December 1857.[25]

Equally, here in this volume, many letters reveal the emerging cult in Britain of heroic activities and the public interest in geographical exploration in general. These trends are notable for the ways in which they enhanced the image of scientific researchers during the middle years of the nineteenth century. Through the adventurous and frequently hazardous exploits of Tyndall, Forbes, and others in the Alps, David Livingstone in Africa, John Franklin in the Arctic, and more, it increasingly became possible for scientists to call on popular public imagery of the heroic traveler, the researcher with his lonely and selfless dedication to increasing the public's knowledge of nature, always willing to push onward in a quest for discovery despite great personal risk. A large element of this imagery called on the Victorian notion of manliness, expressed here through Tyndall's mountain-climbing expeditions, and elsewhere in Victorian literature relating to empire and imperial expansion, in popular ideas about military and public virtue, in the educational ethos of well-known boys' schools, in sporting competitions, and the new interest in "muscular Christianity." Alpinists such as Tyndall and Forbes earned great scientific authority when they returned from their expeditions, especially among audiences who understood that each had individually overcome dangerous situations in order to gather information to further the advancement of science. Authority such as this was hard won. Tyndall's exploits generated respect in much the same way as traveling naturalists such as Alexander von Humboldt, Charles Darwin, or Joseph Dalton Hooker had accomplished in their perilous travels by land and sea in the decades before. The actual experience of such masculine adventures, coupled with the rhetoric of scientific commitment and the importance of firsthand experience, proved to be a heady combination in Victorian England and Tyndall found himself increasingly "lionized" as his work became more widely known among the fashionable public.[26]

Scientific Professional

During these years Tyndall consolidated his scientific prestige. He was already known in Europe's scientific circles, having been elected a fellow of the RS in 1852. From 1853 he was, after being recruited by his friend and mentor Michael Faraday, professor of natural philosophy at the RI,[27] a position that facilitated original laboratory research and required his participation in the regular series of lectures delivered at the RI to members and their guests.[28] Tyndall usually lectured on Thursday afternoons, delivering themed courses on topics like heat during the autumn and spring. He also delivered occasional Friday Evening Discourses, public events for which the RI was—and still is— noted. These discourses introduced scientific topics in an accessible way to London's elite social circles, sometimes including royalty. Prince Albert, the eldest son of Queen Victoria, then aged seventeen, attended Tyndall's series in winter 1858. Tyndall said of this:

> The Prince comes 10 minutes before the time & it sometimes falls to my lot to
> amuse him during this interval. Some days ago, I half choked him by the fumes
> of explosive antimony—Yesterday I showed him Chladni's figures, and caused
> him to use the fiddle bow himself. He is a mild amicable boy.[29]

Tyndall was talented at producing lively, visible demonstrations of physical
concepts, often with apparatus borrowed or constructed for the purpose. In
January 1859, for example, he borrowed unspecified display apparatus from
the Astronomer Royal, George Biddell Airy.[30] He worked hard at producing a
spirited performance every time. In a letter to Hirst, written in February 1858
about two weeks after he delivered a Friday Evening Discourse, he thought
that insomnia was jeopardizing his lectures. This insomnia plagued him his
entire life. Nevertheless, he pulled a performance out of his pocket:

> I have not written to you since I gave my Friday evening. On that evening I was
> a weary broken down man, still the gods gave me power & utterance. I thought
> the lecture was not far from being a failure, but I received a flight of notes from
> my friends next day congratulating me upon its wonderful success. My great
> enemy is want of sleep, but I still hope to discover the conditions which shall
> secure it to me.[31]

Increasingly through this volume Tyndall, in fact, found his lecturing sched-
ule and high level of public performance completely draining. He complained about ill health brought on by incessant work and the lack of time
to conduct his own research. Revealing letters to Faraday, written by Tyndall
in April 1857 and February 1858, describe his situation at the RI very clearly.
In these, Tyndall inquired about the possibility of reducing his lecturing load
or perhaps increasing his salary, at that point £200 per annum.[32] His contract
with the RI had been drawn up in 1853 by a board of trustees, known as the
"Managers," to whom Faraday formally submitted Tyndall's request. Following this, Tyndall felt it necessary to explain his financial situation in detail to
Faraday and the various options he had once considered before accepting the
RI, including positions at the Government School of Mines, the Royal Military Academy at Woolwich, and the London Institution. In spring of 1855
he had also been offered a position at the University of Toronto in Canada,
but declined.[33] The situation was resolved in March 1858 by allowing Tyndall
to offer fewer lectures for the same pay.[34] This exchange provides fascinating
primary evidence about the emergence of scientific professional employment
in the Victorian era and the mutual negotiations—and misunderstandings—
involved as the conventions of a scientific career began to shift away from
gentlemanly practitioners with private income to those, like Tyndall, who

began with few social advantages, were not educated at a socially prestigious university, and had little or no financial security.

Still, Tyndall continued to fret over his salary and position. In 1856 he took on a paid role as an examiner at the Royal Military Academy at Woolwich for entrants into the Royal Artillery and Engineers, and the significance of these fees for his income is displayed by the shock he expressed when the pay was reduced in 1857.[35] He promptly resigned from this position, saying, "They thought proper to cut down the rates of remuneration, and on asking me whether I was prepared to undertake the examinations at the reduced rate I answered 'No!'"[36] From 1858, under the newly formed Military Council on Education, Tyndall transitioned to examine officer candidates for the Royal Military College, Sandhurst on "Experimental sciences." He also delivered lectures and directed laboratory training in science at the London Institution, an educational organization with buildings in central London that made scientific instruction more widely available. This role and his previous time spent as a mathematics teacher at Queenwood College (1847–48 and 1851–53), gave him the expertise to respond to several inquiries about enhancing science teaching in British private schools.[37] In 1857 he additionally gave six lectures to boys at Eton College. An interesting exchange between Tyndall and the head of the school discusses the misbehavior of some students during his lectures.[38] That same year, the English educationalist Lyon Playfair discussed with him the role of the state in providing science instruction to its citizens, and a proposed educational museum.[39] These discussions mark the start of Tyndall's lifelong interest in the reform of science education.[40] As he took on these time-consuming paid positions, his scientific reputation continued to soar. In 1857, for example, he was elected to the prestigious Academia Caesarea Leopoldino Carolina Naturae Curiosorum.[41]

This volume also reveals Tyndall's relish for professional matters. He had experienced a meteoric rise in scientific influence since his arrival in London in 1851, reflected in being elected a fellow of the RS in 1852 at a younger age than was typical (he was in his early thirties),[42] and the following year nominated for the Royal Medal of the RS, which he declined to accept.[43] Before then, he had one of the best possible European educations in science, having as professors at the University of Marburg the chemist Robert Bunsen, the mathematician Friedrich Ludwig Stegmann, and two physicists, Christian Gerling and Hermann Knoblauch. He followed his Marburg studies with time in the Berlin laboratory of the great experimental scientist Heinrich Gustav Magnus. By 1856 everything was coming together to give him an unprecedented platform at the RI from which to engage in wider scientific affairs and push Victorian science in directions that he favored. He actively took part in the emerging structure of scientific publishing, such as peer reviewing (refereeing) scientific papers for the *Philosophical Transactions of the Royal Society*, the premier scientific periodical in the United Kingdom. There are several such referral letters to George Gabriel Stokes, the editor of the *Phil. Trans.*[44] In 1857 he read the proofs of the English translation of the third volume of a work by his friend Auguste de la Rive, occasioning some rueful correspondence between them about the poor style of the translator.[45] There is also a letter between Tyndall and Huxley about the possibility of writing science articles for the *Saturday Review*, in order to broadcast a modernizing influence among the Victorian public. Tyndall did indeed write articles for this publication.[46] Tyndall participated (and maneuvered) in the annual cycle of nominations and elections for the RS, having been elected to its council in November 1856. For example, he worked behind the scenes to get his former teacher Robert Bunsen elected as a foreign member of the RS, reflected in a draft nomination prepared for him by Edward Frankland.[47] Tyndall resigned from the council in 1857 because of the pressure of work at the RI.[48] Ultimately, the proposers for Bunsen's foreign membership of the RS were John Wrottesley, William Sharpey, George Gabriel Stokes, Richard Owen, John Percy, and Charles Wheatsone, and Bunsen was elected in 1858. For Tyndall, there was dining and socializing associated with the RS and much professional gatekeeping that clearly favored those with connections to established fellows and officers of the society. Only fellows could communicate research to a meeting, and Tyndall did so on behalf of others on several occasions.[49] Many of these dealings foreshadow the professional influence over the field that Tyndall would later wield as a member of the self-selecting "X Club" that began in 1864 under the stimulus of close friends such as Huxley and Hooker.[50]

Public and Private Life

Letters published here include several to and from female friends. One of these was Frances Hooker, the wife of Tyndall's close friend Joseph Dalton Hooker, who evidently attended Tyndall's lectures at the RI and sometimes wrote to him afterward to ask for more information, congratulate him, and express enthusiasm for science delivered in this personalized format with demonstrations. The first letter in this volume is a poem sent to Tyndall written by Frances Hooker in Alfred, Lord Tennyson's style, now held in John Tyndall's archive at the RI, although unfortunately the covering letter is lost.[51] Frances Hooker invited Tyndall several times to join her husband and the family for weekend parties or holiday celebrations at their home in nearby Kew Gardens.[52] While this volume does not contain any letters from Tyndall to Frances Hooker, her letters to him show that they shared an interest in books and that he described his summer Alpine tours to her. She also occasionally assisted Tyndall in his scientific work by looking over manuscripts for him.[53] There are also friendly letters to one of the Moore sisters, either Julia or Harriet, that indicate something of Tyndall's charm and ability to enter London's social circles.[54] It is not possible categorically to identify which sister it was. Harriet Jane Moore made watercolor drawings of the scientific laboratory of the RI in the 1850s. Tyndall visited the Moores' family home in Petersham a number of times. On one occasion, in June 1858, at a weekend party that included Tyndall and Faraday, the group was photographed in the garden, now a very rare record of the social interaction of scientific people beyond the laboratory.[55]

Tyndall's acquaintance with Juliet Pollock (later Lady Pollock), wife of William Frederick Pollock, Master of the Court of Exchequer, is exceptional in this volume for the easy friendship they both express and Tyndall's clear appreciation of Juliet Pollock's intellect. Tyndall's personal friendship with both William and Juliet Pollock is well reflected in the correspondence printed here, in which they discuss literature, essays, and poetry, exchange books as gifts, allude to theatrical performances, and swap information about the three young Pollock sons. In April 1857 Tyndall wrote to Hirst, "Last night I had a chop with my friend Pollock the son of the Chief Baron, a man of fine cultivated mind, and whose wife overflows, by her excellence, the measure of those beautiful ideals that we used to admire together long ago in the earlier poems of Tennyson."[56] The Pollocks knew almost every notable literary individual in London and had a great love of the theater. Tyndall visited the Pollocks' homes in London and the country a number of times and was a favorite with their son Walter, even to the point of being personally invited in a handwritten note by Walter to his seventh birthday party in February 1857.[57] Walter got his older brother Frederick to write a letter to Tyndall on his behalf in November 1857, testifying to Tyndall's ability to delight a child.[58] The Pollock children attended Faraday's famed Christmas lectures in 1857.[59] Juliet Pollock, in due course, published several volumes of children's stories, novels (serialized in *Fraser's Magazine*), and literary memoirs, the most noted of which was her recollections of the life of the Shakespearean actor Charles Macready, *Macready as I Knew Him* (London, 1884). Her lively correspondence with Tyndall allows a glimpse into Tyndall's movement into cultivated literary circles during this period and his increasingly wide-ranging social life. On one occasion when he was on holiday in the Isle of Wight, Juliet Pollock asked him to visit Alfred and Emily Tennyson, with a book she wished to give Emily Tennyson. This he did, and spent some time with Tennyson discussing poetry, all recounted in a breathless letter to his friend Hirst.[60] Juliet Pollock also had much to say to Tyndall about the education of her sons, on which he provided reassuring advice. Tyndall shared with the Pollocks his Alpine adventures, writing detailed letters to Juliet.[61] She felt she knew him well enough to admonish him after hearing about his daring exploit in climbing Mount Rosa alone:

> Here comes one of my maternal lectures for you: I am dismayed at your solitary ascent of Monte Rosa, and I call upon you to remember that however muscularly strong you may be, you must, in common with the rest of humanity, be subject to casualties, and that such a casualty as a sprained ankle a swelled knee, a sudden faintness, or a sudden fog, would place you (alone on a steep glacier) in considerable jeopardy; and to my thinking it is really not right to risk a valuable life in such a way.[62]

One of the letters from William Frederick Pollock in this volume describes in detail the many Greek words he thought might replace Tyndall's adopted word *regelation* for the process behind glacier movement, for "Greek is the great source of all scientific nomenclature, and to it accordingly one turns in the first instance, when in want of a name."[63]

These married female friends evidently felt warmly toward Tyndall. As an unmarried and increasingly celebrated figure, he was also invited to a number of homes as an attractive bachelor guest. Judging from the extant correspondence printed here, he appears to have fallen in love with at least one, possibly two, of the daughters of Maria Drummond, the wealthy and influential widow of Thomas Drummond, a civil engineer and Irish undersecretary from 1835 until his death in 1840. Tyndall said of Mrs. Drummond, "[S]he is so kind towards me; and her fair daughters are very pleasant blossoms, both physically & intellectually."[64] The Drummonds' social circle was extensive and socially of much higher status than Tyndall's own. In his journal in May 1857, he wrote when discussing the Drummonds, "After all there are great advantages attached to a residence in London: you can pick from the mass the spirits which harmonise with your own, and thus surround yourself with social influences of the highest order."[65] Tyndall had met the Drummonds by 8 June 1855, when he noted in his journal that he helped one of the daughters, probably the eldest, Mary Elizabeth Drummond (known as May), repeat one of her father's experiments.[66] By 1857 the Drummond daughters appear regularly in Tyndall's journal entries, and he appears to have revealed his feelings to one of them in a letter before departing for the Alps in 1857.[67] Notes in Tyndall's journal are unclear about whether this declaration was addressed to May, aged twenty-one, or to her younger sister Emily (known as Pussey), aged nineteen. Later, after a whirlwind weekend in the Drummonds' company in 1858, he recounted his emotions in a letter to his friend Hirst that included deep anxieties about being neither wealthy enough nor sufficiently well born to be an appropriate suitor:

> These girls have been brought up in opulence and among fashionable people, still there is not a trace of artificiality among them. A noble truthfulness of demeanour, a direct reliance upon high natural qualities, and a lovingness of disposition which warms and beautifies all, are the characteristics which I have ever observed among them. I sometimes ask myself, is it manly to permit them to take such a hold of me? would it not be manlier to say sternly I will see them no more, and get on with my work?[68]

It may be that Tyndall made some form of overture to Mary around this time. He certainly expressed perplexity about the family's attitude toward him and could not tell if he was being allowed to court Mary or not. Whatever the case, in October 1858 he was clearly thinking of marriage, telling his closest confidant, Hirst, "I would not drag her down, I am unable to maintain her at her present level, and the thought of her dragging me up is so terribly repugnant to me that even supposing what is far too daring to suppose, I could not bear it. Thus on all sides this question is surrounded with difficulties."[69] In December 1858 Tyndall related to Hirst that he had "received a peremptory order from her mother not to write to May—[. . .] I do not care a straw for the order."[70] Tyndall must have been visibly affected by the failed relationship, as he noted in his journal how Faraday "hesitatingly" brought him an extract of one of Charles Mackay's popular poems:

> He appeared a little affected and left me in the lecturer's room to read the scrap alone. I read it twice over—it was indeed wonderfully true to my feelings, and my position. I came out and said 'ye quite true.' He squeezed my hand and pointing to the last line, 'that is the best of it[.]' 'Yes thank God' I replied 'I have resources which they are not aware of.' 'I know it['] said he 'and on that I congratulate you.'[71]

As it turned out, Mary Drummond would marry Joseph Kay, a political economist and judge on the Northern Circuit, in 1863.

One of the most striking correspondences in this volume is that between Tyndall and his closest personal friend, Thomas Archer Hirst, whom he had met some ten years previously as a surveyor working in Yorkshire.[72] It is rare to find such an exchange of letters, full of personal insight and affection, between Victorian men. Tyndall was about eight years older than Hirst and had always taken a protective role in their relationship as well as enjoying a genuine meeting of minds. As a talented mathematician, Hirst had many of the same scientific aspirations as Tyndall. He was to be Tyndall's lifelong friend and confidant, sharing his achievements and setbacks, encouraging his scientific work, vacationing together, climbing together in the Alps, and exchanging family news. In 1857 Hirst published his first mathematical paper. In January that year, Tyndall wrote to him, "I have strong hope that your mathematical paper when ready will as you say be worth something. To me it is impossible that you could continue looking for a length of time steadfastly at such a subject without making it worth something."[73] While earlier in their friendship Hirst had helped Tyndall financially, several letters in this volume attest to Tyndall often assisting Hirst in the same fashion.[74] He also supported Hirst through the emotional trauma of his wife, Anna's illness and death in July 1857. Many of Hirst's letters in this volume were written from Pau, in the southern Pyrenees, where he had taken his wife for her health.[75]

Tyndall attended Anna Hirst's funeral in Paris on 4 July and then took Hirst with him to the Alps. In mid-July he shared the unfortunate news with Faraday and Huxley in letters written at Chamonix, north of Mont Blanc.[76] Hirst did not return to England for several years following Anna's death but remained on the Continent to pursue research and make a small tour of classical sites such as Rome. He went to Paris, then Italy for a year, then visited Marburg, where he had once stayed with Tyndall, before going back to London in 1860.

Tyndall's letters to Hirst are among the most revealing and intimate letters he ever wrote. In April 1857, weary with his lectures yet flushed with success, he wrote to Hirst, drawing on an essay on friendship by the seventeenth-century scholar and philosopher Francis Bacon:

> With what treble comfort can I read the Essays of Bacon feeling as I do that I possess in you all that a man ever possessed in a friend. There is a certain wildness in my heart when I think of our relationship to each other; I mean that wildness which would make self sacrifice cheap if offered in the cause of that holy friendship. If you ever felt that 'fine excess' you will understand me: if not it must be an Irish feeling which words would not explain. Ever & always John.[77]

Conclusion

The letters published in this volume reveal Tyndall as an increasingly imposing figure on the professional scientific stage, intent on building his reputation and yet deeply committed to the importance of knowledge—not just knowledge for the elite but knowledge available for everyone. His intellect, ability for warm personal friendships, and sheer performative skills had brought him to a point in his career where he could regard his achievements with pleasure.

Achievements such as these were often commemorated in Victorian times by photography. Were there any photographs of Tyndall at this early point in his career? Tyndall indeed had his portrait taken twice by the London-based photographers Maull & Polyblank in 1857. The first of these generates perplexing and unresolved historical questions. The first photographic sitting occurred in London on 26 August 1857, a day after his return from the Alps.

Tyndall noted in his journal that he was "rough."[78] He may have meant that he still sported a beard from his summer field trip, since he usually let it grow during his expeditions: "I have never once shaved since I came to Switzerland. my beard is quite furious looking!" he told Heinrich Debus in August 1858.[79] Or he may have meant that he felt and looked tired. While an actual copy of this photograph has not been located, there is an 1859 lithograph by Rudolph Hoffmann of Austria that states in the inscription that it is based on a photograph by Maull & Polyblank. This is probably the portrait that Tyndall referred to here. The lithograph was published in *Gallerie ausgezeichneter Naturforscher nach Photographien* (Vienna, 1860).

Maull & Polyblank were noted for their interest in photographing notable persons, and in 1855 had produced a series of fifty-four portraits called *Literary and Scientific Men: Photographs by Maull & Polyblank*. Tyndall was not included in this set, although Faraday was. On 21 September 1857, just one month after the previous sitting, Tyndall returned to Maull & Polyblank with his friend Eduard Magnus to have their individual portraits taken.[80] Eduard's brother, Gustav Magnus, deemed this portrait "excellent" and requested a copy from Tyndall.[81] However, Tyndall wrote in his journal on 31 October that "Mrs. Coxe, and all other ladies that have seen it speak against my portrait";[82] and on 19 March 1858 he wrote: "Joined, after a time, by Miss Drummond, when they all united in abusing my portrait, calling it the most horrid thing in the world, and affirming that they had never seen the Mr. Tyndall therein represented."[83] Despite the contemporary criticism, this photograph is the frontispiece to this volume, reproduced from the original in the RI archives.

In spring 1858 Tyndall wrote in his journal about an additional photography session:

> On Friday went over the Westminster Bridge works with Mr Faraday, and afterwards sought out a sixpenny photographers, where according to an old agreement he presented me with a sixpenny photograph of himself, and I him with a similar one of me. He proposed to have us both taken together, and this was done, and for each of us. It is now before me, I with a pair of compasses in my hand measuring off a distance on a card which I hold in the other, and Faraday looking intently at what I am doing. I must keep this picture from the action of the light, and thus protect it as a precious relic. We stayed nearly two hours at the photographers.[84]

Sadly, none of the images mentioned have ever been located.

At much the same time, June 1858, Tyndall posed with Michael Faraday, Sarah Faraday, and two unidentified women, probably Faraday's niece Jane Barnard, and either Harriet Jane Moore or Julia Moore, for an open-air photographic portrait around a garden bench. Tyndall referred to the taking of this photograph in a June 1858 letter to Juliet Pollock.[85] This photograph is reproduced as the frontispiece to volume 5 of Tyndall's correspondence. It appears from the letter published in this volume that "Miss Moore" took the photograph, indicating that perhaps it was Julia (the younger sister) in the image.

Tyndall appears as a well-dressed, respectably whiskered, and pensive figure. There is little in his pose to suggest the active mountaineer, the dynamic public lecturer, the lively thinker and man of affairs, or innovative scientific experimenter, all documented so exuberantly in his correspondence, but much to suggest the thoughtful cast of mind and resolute air that were fundamental to the success he began to enjoy during these years.

Following his now-customary Alpine travels in the summer months of 1858, Tyndall began to write a popular book about his climbing excursions and study of glaciers. While he mentioned the new project only a handful of times in his correspondence, the noticeable decline in letters toward the end of this volume perhaps indicates the time and effort that he was expending in writing. The book, Tyndall's first, was published in London by John Murray in 1860 (see volume 7). He titled it, simply, *Glaciers of the Alps.* In a way, the authorship and publication of this book on glaciers confirm Tyndall's transformation into a public figure and respected man of science. He had come a long way from his first position as a surveyor on the Irish Ordnance Survey, some twenty years previously, and as his letters indicate, his mental abilities, energy, and grasp on scientific problems were not only incisive and insightful but also dramatically expanding in scope. He lectured widely, actively participated in contemporary scientific education, and moved freely around London's fashionable social circles, all the while conducting research experiments in physics at the Royal Institution of Great Britain and developing his theories about the structure and movement of glaciers. The themes arising in this sixth volume of Tyndall's correspondence were to develop into central features of his professional and personal life.

Notes

1. The general introduction in volume 1 of *The Correspondence of John Tyndall* notes that while older scholarship, and Louisa Tyndall, give Tyndall's birth year as 1820, this is doubtful and 1822 is more likely. Either way, this would make him thirty-four or thirty-six at the beginning of this volume and thirty-six or thirty-eight when this volume ends. In April 1857 Tyndall remarked to Joseph Hooker, "I was born at Leighlin Bridge Ireland and am 36 years old. It may be a year more or less for aught I know" (see letter 1366).
2. Letter 1401.
3. See *Tyndall Correspondence*, vol. 5.
4. Tyndall would publish his first book about his travels in the Alps and glacier research in 1860, *Glaciers of the Alps* (London: John Murray). Two years later he published

Mountaineering in 1861: A Vacation Tour (London: Longman, Green, Longman, and Roberts, 1862), and his third book appeared a decade later, *Hours of Exercise in the Alps* (London: Longmans, Green, and Co., 1871). In the biography of Tyndall using the material Louisa Tyndall had accumulated before her death in 1940 (A. S. Eve and C. H. Creasey, *Life and Work of John Tyndall* [London: Macmillan & Co., 1945]), there is a section on Tyndall's work on "The Structure and Movement of Glaciers" (297–308) as well as a chapter devoted to "Tyndall as a Mountaineer," written by Lord (Claud) Schuster of the Alpine Club (340–92). Other works on Tyndall's mountaineering and glacier research include Ronald W. Clark, "Tyndall as a Mountaineer," in *John Tyndall: Essays on a Natural Philosopher*, ed. W. H. Brock, N. D. McMillan, and R. C. Mollan (Dublin: Royal Dublin Society, 1981), 61–68; John Hollier and Anita Hollier, "The Glacier Theory of Louis Rendu (1789–1859) and the Forbes–Tyndall Controversy," *Earth Sciences History* 35, no. 2 (2016): 346–53; Michael S. Reidy, "Evolutionary Naturalism on High: The Victorians Sequester the Alps," in *Victorian Scientific Naturalism: Community, Identity, Continuity*, ed. Gowan Dawson and Bernard Lightman (Chicago: University of Chicago Press, 2014), 55–78; J. S. Rowlinson, "Tyndall's Work on Glaciology and Geology," in Brock, McMillan, and Mollan, *John Tyndall: Essays on a Natural Philosopher*, 113–28; Werner Sackmann, "John Tyndall (1820–93) and His Relationship to the Alps and Switzerland," *Gesnerus* 50, no. 1/2 (1993): 66–78; and E. J. Wiseman, "John Tyndall: Scientific Work and Social Life in the Alps," in Brock, McMillan, and Mollan, *John Tyndall: Essays on a Natural Philosopher*, 69–79.

5. On Tyndall and "verticality," see Michael S. Reidy, "John Tyndall's Vertical Physics: From Rock Quarries to Icy Peaks," *Physics in Perspective* 12, no. 2 (2010): 122–45; and Michael S. Reidy, "From Oceans to Mountains: Spatial Science in an Age of Empire," in *Knowing Global Environments: New Historical Perspectives on the Field Sciences*, ed. Jeremy Vetter (New Brunswick, NJ: Rutgers University Press, 2010), 17–38.

6. This lecture was published as John Tyndall, "Comparative View of the Cleavage of Crystals and Slate Rocks," *Phil. Mag.* 12, no. 76 (1856): 35–48.

7. See *Tyndall Correspondence*, vol. 5.

8. John Tyndall and Thomas H. Huxley, "On the Structure and Motion of Glaciers," *Phil. Trans.* 147 (1857): 327–46. This was also published the following year as John Tyndall and Thomas H. Huxley, "On the Structure and Motion of Glaciers," *Phil. Mag.* 15 (1858): 365–88.

9. See Aileen Fyfe and Noah Moxham, "Making Public Ahead of Print: Meetings and Publications at the Royal Society, 1752–1892," *Notes and Records of the Royal Society of London* 70, no. 4 (2016): 361–79.

10. Forbes published his theory of viscous glacial motion in *Travels through the Alps of Savoy and Other Parts of the Pennine Chain with Observations on the Phenomena of Glaciers* (Edinburgh: Adam and Charles Black; London: Longman, Brown, Green, and Longmans, 1843); and in three papers for the RS: James D. Forbes, "Illustrations of the Viscous Theory of Glacier Motion Part I. Containing Experiments on the Flow of Plastic Bodies, and Observations on the Phenomena of Lava," *Phil. Trans.* 136 (1846): 143–55; "Illustrations of the Viscous Theory of Glacier Motion Part II. An Attempt to Establish by Observation the Plasticity of Glacier Ice," *Phil. Trans.* 136 (1846): 157–75; and "Illustrations of the Viscous Theory of Glacier Motion Part III," *Phil. Trans.* 136 (1846): 177–210.

11. James Thomson, "On the Plasticity of Ice, as Manifested in Glaciers," *Roy. Soc. Proc.* 8 (1857): 455–58. Read to the RS by his brother William Thomson on 7 May 1857. Thomson further discussed his theory of glacial motion in a second paper, "On Recent Theories and Experiments Regarding Ice at or near Its Melting-Point," *Roy. Soc. Proc.* 10 (1859): 151–60. Tyndall discussed Thomson's papers in *Glaciers of the Alps*, 340–45.

12. Letter 1302.
13. Letters 1322, 1323, 1328, and 1358.
14. Letter 1435. Also see *Glaciers of the Alps*, 340–45.
15. See Nanna Katrine Lüders Kaalund, "A Frosty Disagreement: John Tyndall, James David Forbes, and the Early Formation of the X-Club," *Annals of Science* 74 (2017): 282–98.
16. Letter 1306.
17. For Hopkins, see letters 1359 and 1468, and for Airy, see letters 1517, 1518, and 1541.
18. Tyndall, *Glaciers of the Alps*, 162.
19. Letter 1548.
20. For Tyndall and the Ordnance Survey, see Melinda Baldwin and Janet Browne, eds., *The Correspondence of John Tyndall*, vol. 2, *The Correspondence, September 1843–December 1849* (Pittsburgh: University of Pittsburgh Press, 2016).
21. Letter 1536.
22. For example, letters 1419, 1453, 1473, 1527, 1536, 1547, and 1557.
23. Letters 1551 and 1556. On Tyndall's "difficulties at Chamouni," see Tyndall, *Glaciers of the Alps*, 169–72, 192–94.
24. See Tyndall, *Glaciers of the Alps*, 192–94. For a summary of Tyndall's testimony for the BAAS at Leeds, see "Particulars of an Ascent of Mont Blanc," *Brit. Assoc. Rep. 1858*, 39–40.
25. Letter 1466.
26. On Tyndall, risk, and masculinity, see R. D. Eaton, "In the 'World of Death and Beauty': Risk, Control, and John Tyndall as Alpinist," *Victorian Literature and Culture* 41, no. 1 (2013): 55–73; Francis O'Gorman, "'The Mightiest Evangel of the Alpine Club': Masculinity and Agnosticism in the Alpine Writing of John Tyndall," in *Masculinity and Spirituality in Victorian Culture*, ed. Andrew Bradstock et al. (New York: Palgrave, 2000), 134–48; Bruce Hevly, "The Heroic Science of Glacier Motion," *Osiris* 11 (1996): 66–86; and Michael S. Reidy, "Mountaineering, Masculinity, and the Male Body in Victorian Britain," *Osiris* 30 (2015): 158–81.
27. On Tyndall at the RI, see J. D. Burchfield, "John Tyndall at the Royal Institution," in *The Common Purposes of Life: Science and Society at the Royal Institution of Great Britain*, ed. Frank A. J. L. James (Burlington, VT: Ashgate, 2002), 147–68; Sophie Forgan, "Tyndall at the Royal Institution," in Brock, McMillan, and Mollan, *John Tyndall: Essays on a Natural Philosopher*, 49–60; and D. Thompson, "John Tyndall and the Royal Institution," *Annals of Science* 13, no. 1 (1957): 9–21.
28. On Tyndall's lecturing, see Jill Howard, "'Physics and Fashion': John Tyndall and His Audiences in Mid-Victorian Britain," *Studies in the History and Philosophy of Science* 35, no. 4 (2004): 729–58; and Charles A. Taylor, "Tyndall as Lecture Demonstrator," in Brock, McMillan, and Mollan, *John Tyndall: Essays on a Natural Philosopher*, 205–16.
29. Letter 1478.
30. Letters 1582, 1585, and 1586.
31. Letter 1490.
32. Letters 1377 and 1493.
33. See letter 1493, n. 13.
34. Letter 1500.
35. Letters 1375 and 1376. On Tyndall and others as examiners, see James Elwick, "Economies of Scales: Evolutionary Naturalists and the Victorian Examination System," in Dawson and Lightman, *Victorian Scientific Naturalism: Community, Identity, Continuity*, 131–56.
36. Letter 1398.
37. On Tyndall and education, see *Tyndall Correspondence*, vol. 2.
38. Letters 1441 and 1442.

39. Letter 1369.

40. N. D. McMillan, "Ireland and The Reform of the Politics and Government of British Science and Education," in *Prometheus's Fire: A History of Scientific and Technological Education in Ireland*, ed. N. D. McMillan, J. Cooke, and D. D. G. McMillan (Carlow: Tyndall Publications, 2000), 481–582.

41. Letters 1365, 1367, and 1427.

42. In 1852 Tyndall was probably around thirty or thirty-two years old. See n. 1.

43. See letter 1360, n. 1, and Roland Jackson, "John Tyndall and the Royal Medal that Was Never Struck," *Notes and Records of the Royal Society of London* 68, no. 2 (2014): 151–64.

44. Letters 1495, 1498, 1523, 1564, 1571, 1579, and 1583. On Tyndall as referee for Stokes, see Melinda Baldwin, "Tyndall and Stokes: Correspondence, Referee Reports and the Physical Sciences in Victorian Britain," in *The Age of Scientific Naturalism: Tyndall and His Contemporaries*, ed. Bernard Lightman and Michael S. Reidy (London: Pickering & Chatto, 2014), 171–86.

45. Letter 1378. On Tyndall as translator of foreign language works, see Bernard Lightman, "Scientific Naturalists and Their Language Games," *History of Science* 53, no. 4 (2015): 395–416.

46. Letter 1509. Two letters in this volume are from the *Saturday Review*'s editor John Cook, 1528 and 1574.

47. Letter 1321. Also see a similar letter to Tyndall from Heinrich Debus, 1326.

48. Tyndall wrote in his Journal on 17 December 1857: "I have left the Council of the Royal Society—partly at my own desire, as my lectures interfere with my attendance" (Journal, RI MS JT/2/13c/1041).

49. See Fyfe and Moxham, "Making Public Ahead of Print." For example, Tyndall read Samuel Haughton's paper "On the Physical Structure of the Old Red Sandstone of the County of Waterford, Considered with Relation to Cleavage, Joint Surfaces, and Faults" for the RS on 21 August 1858 (see letter 1484).

50. On the "X Club," see Ruth Barton, "'Huxley, Lubbock, and Half a Dozen Others': Professionals and Gentlemen in the Formation of the X Club, 1851–1864," *Isis* 89, no. 3 (1998): 410–44; Ruth Barton, "'An Influential Set of Chaps': The X-Club and Royal Society Politics 1864–85," *British Journal for the History of Science* 23, no. 1 (1990): 53–81; Ruth Barton, *The X Club: Power and Authority in Victorian Science* (Chicago: University of Chicago Press, 2018); Adrian Desmond, "Redefining the X Axis: 'Professionals,' 'Amateurs' and the Making of Mid-Victorian Biology—A Progress Report," *Journal of the History of Biology* 34, no. 1 (2001): 3–50; J. Vernon Jensen, "The X Club: Fraternity of Victorian Scientists," *British Journal for the History of Science* 5, no. 1 (1970): 63–72; and Roy M. MacLeod, "The X-Club: A Social Network of Science in Late-Victorian England," *Notes and Records of the Royal Society of London* 24, no. 2 (1970): 305–22.

51. Letter 1292.

52. Letters 1368, 1395, 1432, 1464, and 1482.

53. Letter 1516, n. 8.

54. Letters 1515 and 1539.

55. The taking of this photograph is referred to in letter 1527, and it is reproduced as the frontispiece to *Tyndall Correspondence*, vol. 5.

56. Letter 1367.

57. Letters 1336, 1337, and 1339.

58. Letter 1450.

59. Letter 1474.

60. Letter 1507. Also see letter 1510.

61. Letters 1420, 1540, and 1547.

62. Letter 1550.

63. Letter 1358.
64. Letter 1367.
65. Journal, 2 May 1857, RI MS JT/2/13c/931.
66. Journal, 8 June 1855, RI MS JT/2/13c/752.
67. Letter 1406.
68. Letter 1566.
69. Letter 1566.
70. Letter 1578.
71. Journal, 4 March 1859, RI MS JT/2/13c/1155. It has not been possible to identify the poem.
72. See *Tyndall Correspondence*, vol. 2.
73. Letter 1319.
74. Letters 1351, 1355, 1361, 1404, 1458, 1460, and 1472.
75. Letters 1298, 1316, 1355, and 1387.
76. Letters 1415 and 1416.
77. Letter 1367.
78. Journal, 26 August 1857, RI MS JT/2/13c/1032.
79. Letter 1536.
80. Journal, 21 September 1857, RI MS JT/2/13c/1034.
81. See letter 1505.
82. Journal, 31 October 1857, RI MS JT/2/13c/1038. In letter 1445, Mary Anne Coxe had requested a copy of the portrait from Tyndall.
83. Journal, 19 March 1858, RI MS JT/2/13c/1061.
84. Journal, 18 April 1858, RI MS JT/2/13c/1070. These photographs have not been located.
85. Letter 1527.

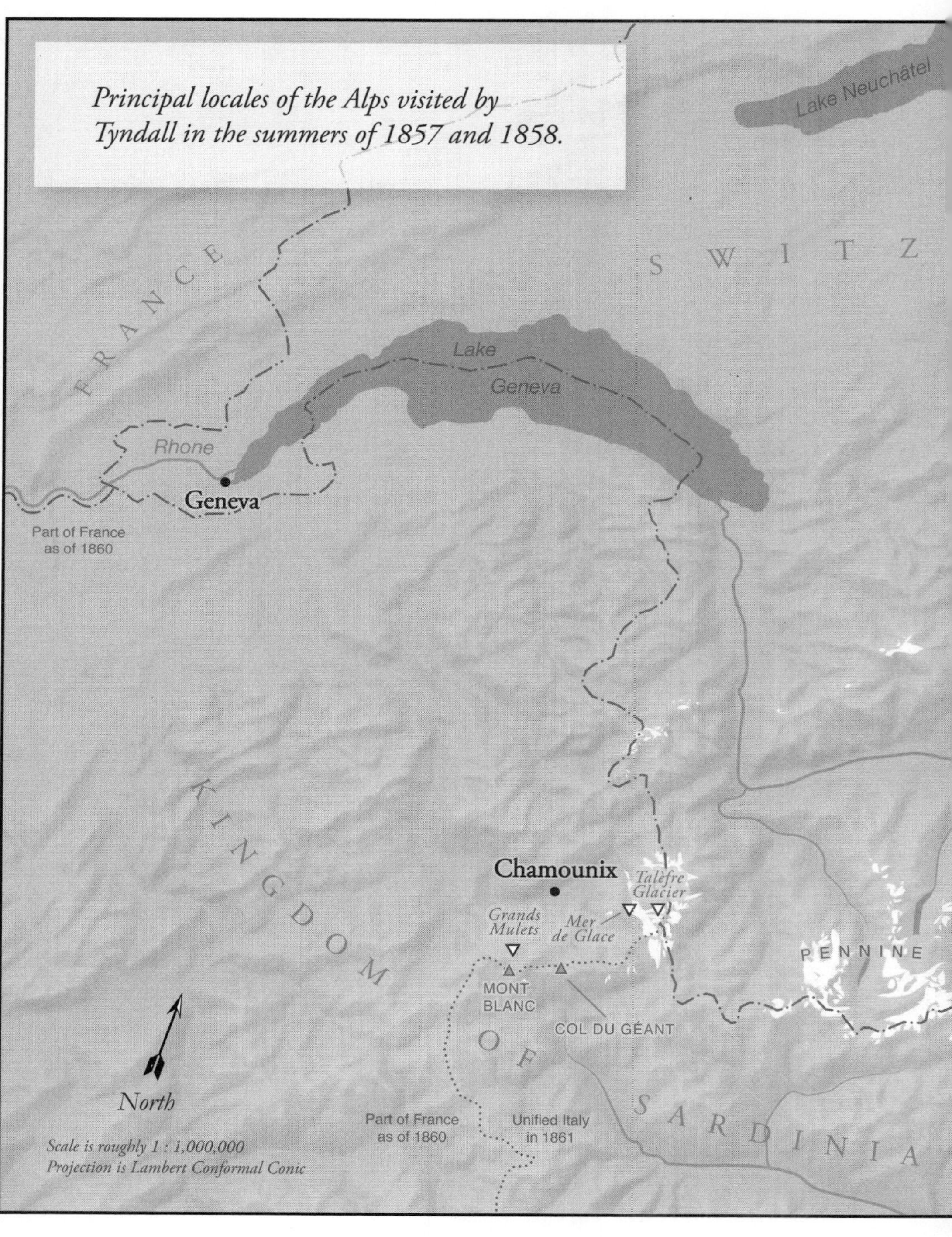

Principal locales of the Alps visited by
Tyndall in the summers of 1857 and 1858.
Lake Neuchâtel
FRANCE
SWITZ
Lake
Geneva
Rhone
Geneva
Part of France
as of 1860
KINGDOM
Chamounix
Talèfre
Glacier
Grands
Mulets
Mer
de Glace
PENNINE
MONT
BLANC
COL DU GÉANT
OF
North
Part of France
as of 1860
Unified Italy
in 1861
SARDINIA
Scale is roughly 1 : 1,000,000
Projection is Lambert Conformal Conic

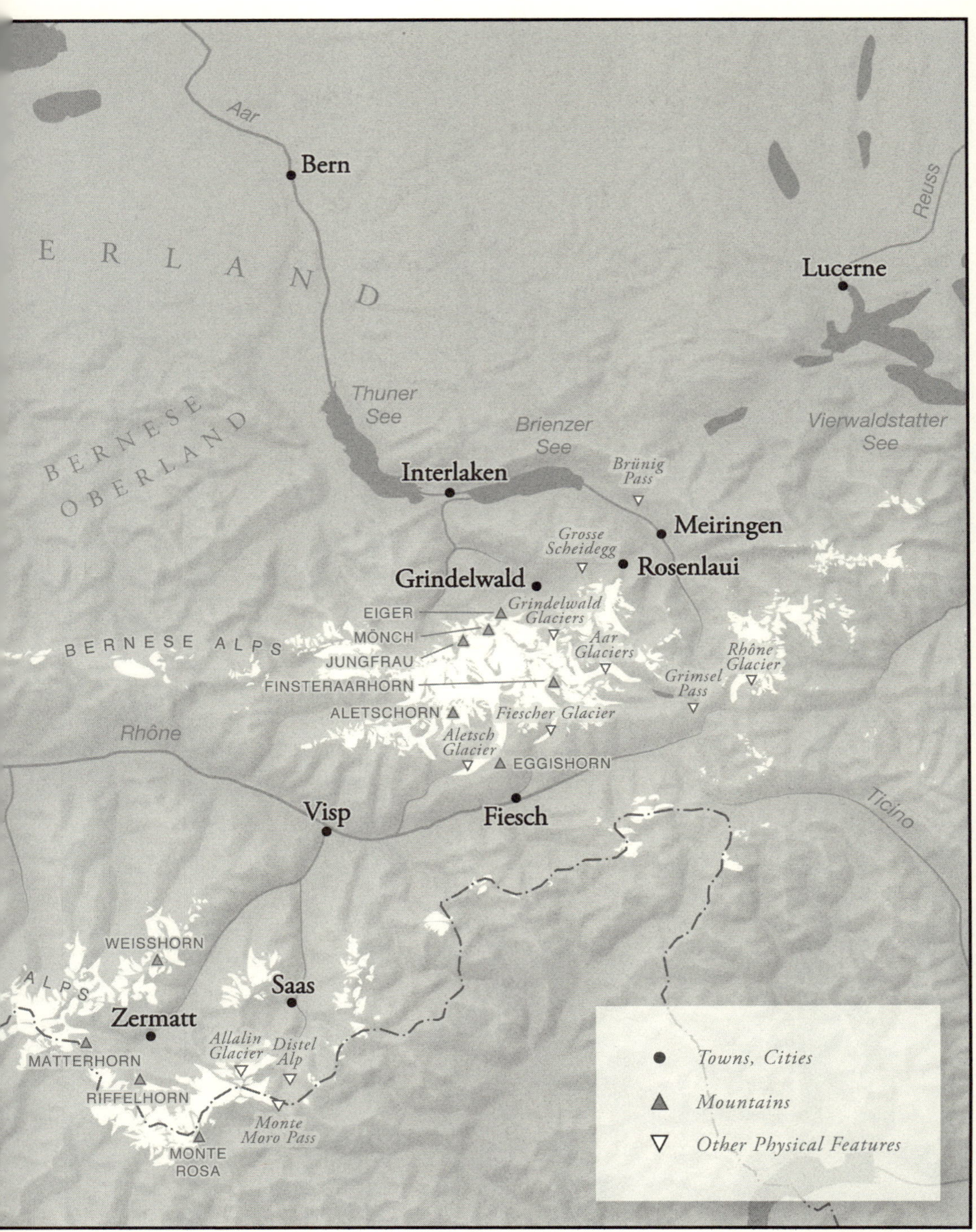

Credit: *Benjamin Meader*

EDITORIAL PRINCIPLES

Our aim is to include all letters to and from John Tyndall that are currently extant. Some of the letters are from published sources, such as letters written to newspapers or journals, but the majority of them are either the originals or typescripts produced by Louisa Tyndall from originals. One central editorial principle has been to transcribe, wherever possible, from original letters rather than Louisa's typescripts. This is because of Louisa's conscious and unconscious editorial interventions, a point discussed in more detail below. We have also aimed to reproduce, as accurately as possible, the text of the letters as they were written. Spelling mistakes that have appeared in the handwritten letters have therefore been preserved; mistakes appearing where only the typescript letter has survived are silently corrected as in our judgment they are almost always typographical errors. But, in general, non-standard spellings (e.g., the use of dialect) have been retained even when in a typescript-only letter (though there is the possibility that the spelling is Louisa's, not her husband's). Where the transcription is based on a typescript transcription by Louisa (or one of her assistants), whether for typescript letters or in her typescript of Tyndall's journal, this will be indicated by "LT Transcript Only." For those cases in which our transcription comes from typescript letters not done by Louisa, or it is unclear if she was involved, this will be indicated by "Typed Transcript Only." For the rare case when only a handwritten transcript of a letter exists, this will be indicated by "Transcript Only."

We have included illustrations that are part of the letter, reprinting the original handwritten image. While quotations within some handwritten and typescript letters use double quotation marks, for consistency we have used single marks in the transcriptions. However, if our source for a letter was a publication such as a journal, newspaper, or book, we have retained the quote marks as they appear in that publication.

The letters are presented in chronological order of writing. Where more than one letter has the same date, letters from Tyndall (in alphabetical order of addressee) come first, followed by letters to Tyndall (in alphabetical order of writer), with the exception that if the order of writing can be determined, that order takes precedence. Where letters cannot be dated precisely they are placed at the latest likely date of writing. To indicate the beginning of a new year, the date of that year has been inserted in a large font before the first letter from that year.

Both the transcription of non-English language letters and the English translations are presented. The transcription of the non-English letter will appear first, followed by the English translation in a smaller font. Editorial notes are inserted in the English translation. Some foreign letters are in an original hand while others exist only as transcriptions by Louisa or others who often did not have a full command of the language. In the latter case these letters contain frequent spelling, grammatical, and interpretation errors that do not seem to have been committed by the original author. In these typescript-only cases, and only in these cases, spelling and grammatical errors have been corrected, as have poor interpretations where a better word choice was obvious from the context.

There is a standard format for each letter. The first line lists the author, if it is to Tyndall, or the recipient, if it is from Tyndall; the date; and the letter number. The second line lists the place the letter was written (if known), and any information supplied by the writer in their salutation, such as the date or time of day when the letter was written. (In some cases, though, where the writer put the date at the foot of the letter we have left it there in order to reproduce the text accurately.) Information from stationery—for instance from institutional letterhead—is also supplied in the second line if extant. The opening salutation, the text of the letter, and closing then follow. Then come the source and editorial notes, the latter indicated in the text by arabic numerals.

Editorial notes are intended to provide the contextualization necessary to understand the contents of each letter. There are basically six types of editorial notes. Many of them are informative notes on persons, real or mythical, mentioned in the letters. If a person referred to in a letter is not described in a note, they have been mentioned more than twice in the volume and they will have an entry in the biographical register at the back of the volume. For the rest—those mentioned once or twice—biographical information will appear in the editorial note appended to the letter in which he or she is first mentioned. When there is a second reference to that person a note will refer back to the letter in which he or she was first mentioned. Biographical notes on well-known figures such as Faraday or Huxley stress what that person was doing at the time when the letter was written rather than providing a full overview of the person's life.

A second type of editorial note identifies allusions and quotations. In the case of quotations, where possible, the first edition of the work is cited unless a different edition is mentioned in the letter. If a quote is not in English a translation is given. Glosses on obscure places, abstruse words, and words used in an unfamiliar sense constitute the third type of editorial note.

The fourth type provides information about the context of the letter, drawing on such sources as contemporary publications, scholarly sources, Tyndall's personal journals and field notes, and other letters in the Tyndall Correspondence. The fifth type of editorial note is cross-references to other letters in the Correspondence, referring to the letter number. The last type is information on significant textual changes that Tyndall made to the original that is thought to be relevant to the meaning of the letter.

There are a series of conventions for indicating when editors have given additional information about the text of a letter:

When Louisa Tyndall has made an annotation or marginal insertion that the editors have deemed important enough to retain, we have included her annotation in an editorial footnote.

When editors other than she have made insertions they are indicated by unitalicized square brackets, for example, [word].

When certain words are ambiguous and we have had to conjecture their meaning, they are indicated by italicized square brackets, for example, *[words]*; when we have considered such words illegible the number of illegible words is indicated in full italicization and square brackets, for example, *[3 words illeg]*.

If a word or series of words are not present or have been destroyed—by inkblots, holes, or mold, for example—they are indicated as missing, italicized and in angular brackets, for example, *<3 words missing>*.

In extremely rare cases where it is clear that Louisa herself has intentionally destroyed a few words, they are indicated as excised italicized and in angular brackets, for example, *<3 words excised>*.

Finally, a brief word about Louisa Tyndall. A Tyndall correspondence project would be very difficult, perhaps even impossible, without the work done by Louisa, much of which seems to have been done in order to write a biography of her late husband. This biography was not completed. We have therefore treated Louisa as a de facto member of the editorial team. Some of her handwritten comments on the typewritten letters are extremely helpful and thus have been retained as footnotes to the relevant passage. Conversely, many of these handwritten annotations seem to have been inserted to help write her biography—for instance, summaries of passages or interesting quotes. When the editors have deemed these notes of Louisa's to add no new information, they have not been included for publication.

It is also important to note that Louisa was not always reliable as a transcriber. She wasn't always accurate and may have gone back and edited portions of Tyndall's writing to make his letters seem more consistent or grammatically correct. For example, she made punctuation more formal, corrected grammar, and, occasionally, corrected quotations. She also left out details that she may have considered too private or too salacious for public consumption. She equated postmark and date of writing and overlooked the time taken to write some letters. When she inferred dates from internal evidence her reasoning was sometimes faulty. In cases where there is an original, handwritten letter, therefore, we have considered that as taking precedence over a typed transcription. Lastly, the patterns of punctuation differ between LT sources and manuscript sources. Tyndall himself used dashes more often than commas, semicolons, and stops, and he used dashes of many different lengths that are all standardized here to em-dashes. It is often difficult to distinguish between very short dashes and commas or very short dashes and stops. Similarly, capitalization is often ambiguous; distinguishing S/s for "science," for example, is a particular problem. Therefore, scholars for whom punctuation and capitalization are significant will need to consult the manuscripts. Users should also be aware that throughout this project the transcribers and editors have worked from scans and photographs rather than the original manuscripts.

NOTE ON MONEY

In the nineteenth century the currency used in England was the pound sterling, usually abbreviated to £ (or, occasionally, to L). A pound consisted of 20 shillings (abbreviated *s*), and each shilling contained 12 pennies (*d*). A guinea was 21 shillings; that is, £1 1*s*. A crown was 5 shillings (5*s*), and half a crown equaled two shillings and sixpence (2*s* 6*d*). The smallest coin was a farthing, one-quarter of a penny (¼*d*).

Several forms of representation were in contemporary use. Sometimes the numbers for pounds, shillings, and pence were separated by dots, forward slashes, or spaces; thus, for example, 3 pounds, 4 shillings, and 5 pennies could be written as £3.4.5, £3/4/5, or £3 4 5. A zero could be represented by a dash; for example, 7 shillings was often written as 7/- or just 7/. A half crown was often written as 2/6. A florin per diem was a daily allowance of one florin, a British coin worth two shillings.

French currency is also occasionally mentioned in this volume. The franc was the national currency in France from 1795 until 1999 when France adopted the Euro. In the mid-nineteenth century the exchange rate between British and French currencies was approximately 25 francs per £. A centime was a coin worth one-hundredth of a franc.

Additionally, a Reichsthaler was a silver coin issued by Prussia from 1750 that became a common currency throughout much of Germany in the mid-nineteenth century before being replaced by the Vereinsthaler in 1857. When used in letters, a standard symbol from Tyndall's time was used, but given the lack of such a symbol as a typing character today, we have inserted the standard abbreviation "Rthl" in the text of letters.

TIMELINE OF JOHN TYNDALL'S LIFE

Year	Event
c. 1822	John Tyndall born at Leighlin Bridge, County Carlow, Ireland
c. 1836	Attends John Conwill's National School in Ballinabranna
1839	Begins employment as civil assistant with Ordnance Survey of Ireland in Carlow
1840	Becomes civil assistant in the Ordnance Survey Office at Youghal, Cork County
1841	First appearance in print (poem in *Carlow Sentinel* under pseudonym "W[alter] S[nooks]")
1842	Joins English Ordnance Survey in Preston
1843	Leads written criticism of Ordnance Survey in *Liverpool Mercury*; dismissed November
1844	Home in Ireland, mostly unemployed, until obtaining a position with the firm of Nevins and Lawton, Surveyors, of Manchester
1845–47	Works for Richard Carter, Surveyor, of Halifax
1845	Meets Thomas Archer Hirst in Halifax
1847	Begins teaching mathematics and surveying at Queenwood College, Hampshire
1848–50	At University of Marburg, working for PhD with Robert Bunsen and Friedrich Stegmann
1850	Returns to England; meets Michael Faraday and William Francis and attends his first British Association meeting, before returning to Marburg
1851	Moves from Marburg to Berlin (April–June); then returns to Queenwood
1852	Elected fellow of the Royal Society
1852	Friendship developed with Thomas Henry Huxley
1853	First lecture at Royal Institution (RI)
1853	Meets Herbert Spencer
1853	Appointed professor of natural philosophy at RI
1853	Awarded Royal Medal for magnetic work; declines when controversy arises over award
1856	Meets Thomas Carlyle
1857	Climbs Mont Blanc for first time with Hirst
1859	Demonstrates existence of greenhouse gases and climatic implications
1859	Joins Government School of Mines as professor of natural philosophy
1860	*The Glaciers of the Alps* published

Year	Event
1860	Meets John Lubbock and his wife, Ellen Lubbock
1861	First ascent of the Weisshorn
1861	Delivers his first Christmas Lectures at RI
1862	*Mountaineering in 1861* published
1862	Begins to champion Mayer's priority in discovery of conservation of energy
1863	*Heat Considered as a Mode of Motion* published
1864	Awarded Royal Society's Rumford Medal
1864	First meeting of X Club
1865	*On Radiation* published
1865	Engages in public controversy over the efficacy of prayer
1866	Succeeds Michael Faraday as scientific adviser to Trinity House
1867	Faraday dies; Tyndall becomes superintendent of the RI
1867	*Sound: A Course of Eight Lectures* published
1868	Successfully climbs Matterhorn
1868	Discovers the cause of light scattering, to be known as "Tyndall Effect"
1868	*Faraday as a Discoverer* published
1868	Lecture on "Scientific Materialism" at the British Association
1868	Resigns from Royal School of Mines
1869	Joins Metaphysical Society
1870	"On the Scientific Use of the Imagination" lecture at British Association
1870	*Three Scientific Addresses* published
1870	*Researches on Diamagnetism and Magne-Crystallic Action* published
1871	Meets Louis Pasteur for first time while in Paris
1871	*Hours of Exercise in the Alps* published
1871	*Light and Electricity* published
1871	*Fragments of Science* published
1872	Debates over how to measure the efficacy of prayer, aka the "Prayer-Gauge Debate"
1872	*Contributions to Molecular Physics in the Domain of Radiant Heat* published
1872	*The Forms of Water in Clouds and Rivers, Ice and Glaciers* published
1872–73	USA lecture tour
1873	*Six Lectures on Light* published
1874	As president of the BAAS, delivers the "Belfast Address"
1876	Marries Louisa Charlotte Hamilton
1877	Develops "Tyndallization" (lengthy sterilization process to destroy heat-resistant spores)
1877	Tyndall and Louisa build summer cottage, Alp Lusgen, at Bel Alp, northern side of Valais, above Brieg
1877	*Fermentation and its Bearings on Phenomena of Disease* published
1879	*Fragments of Science*, which had gradually expanded, first published in two volumes
1881	*Essays on the Floating Matter of the Air* published
1885	Tyndall and Louisa build Hindhead retreat, Surrey Downs
1887	Resigns from RI

Year	Event
1890	Fight with Gladstone in the *Times* over Irish Home Rule
1892	Hirst dies
1892	*New Fragments* published
1893	Dies at Haslemere, Surrey, accidentally poisoned

TIMELINE OF EVENTS IN JOHN TYNDALL'S LIFE SPECIFIC TO VOLUME 6

Month and year	Biographical details	Notable social, political, and cultural events
November 1856	Gives Friday Evening Discourse at the RI, "Comparative View of the Cleavage of Crystals and Slate Rock" (6 June); published in *Roy. Inst. Proc.* Elected to the council of the RS (27th) Hirst in Pau (France) with wife, Anna, for her illness	Second Anglo-Chinese War between Britain and China begins (8 October) when Chinese officials board and arrest men from the British trade vessel *Arrow* in Canton; official British and French declaration on 3 March 1857 Anglo-Persian War between Britain and Persia begins (1st) James Buchanan wins US presidential election (4th) Henry Bessemer receives a patent in the United States for his method of the mass production of steel (11th) RS awards Henri Milne-Edwards the Copley Medal, "for his researches in comparative anatomy and zoology" (30th) RS awards William Thomson the Royal Medal, "for his various chemical researches relating to electricity, to the motive power of heat, and to other subjects" (30th) RS awards John Richardson the Royal Medal, "for his contributions to natural history and physical geography" (30th) RS awards Louis Pasteur the Rumford Medal, "for his discovery of the nature of racemic acid and its relations to polarized light, and for the researches to which he was led by that discovery" (30th)

Month and year	Biographical details	Notable social, political, and cultural events
November 1856		Eunice Foote publishes "Circumstances Affecting the Heat of the Sun's Rays" in *American Journal of Science and Arts*, in which she describes how carbon dioxide could warm the earth's atmosphere; her paper had been read at the American Association for the Advancement of Science's meeting on 23 August by Joseph Henry
		Elizabeth Barrett Browning publishes *Aurora Leigh*
December 1856		National Portrait Gallery, London established (2nd); opens to the public in January 1859
		Bushehr (Persia) surrenders to Britain (9th)
		Faraday gives annual Christmas Lectures for youth at the RI, on "attractive forces"
1857		Richard Dawes publishes *Effective Primary Instruction: The Only Sure Road to Success in the Reading-Room, Library, and Institutes for Secondary Education*
		Robert Bunsen publishes *Gasometrische Methoden* (published in English the same year as *Gasometry*)
		German astronomer Peter Andreas Hansen publishes his lunar tables, at the expense of the British government
January 1857	Picks up writing entries in his regular journal (2nd); last entry was 10 August 1856 (he had filled a separate journal with his summer travels in the Tyrol region of the eastern Alps, where he first began to examine glaciers)	Huxley begins lecture course at the RI, "Twelve Lectures on Physiology and Comparative Anatomy" (20th)
	Reads "Observations on Glaciers" at the RS (15th); published with Huxley in *Phil. Trans.* as "On the Structure and Motion of Glaciers"	

Month and year	Biographical details	Notable social, political, and cultural events
January 1857	Lectures at the London Institution, "On the Nature and Phenomena of Light" (19th) Begins "Eleven Lectures on Sound" at the RI (22nd); through 2 April Gives Friday Evening Discourse at the RI, "Observations on Glaciers" (23rd) Attends the funeral of Alice Maud Wright in Bossington (26th)	
February 1857	Darwin writes to Tyndall, "It is most beautiful your having given cleavage to ice" (4th) Recounts his experiments on the structure of glacial ice to the Royal Society Philosophical Club (5th)	Faraday gives Bakerian Lecture at the RS, "Experimental Relations of Gold (and other Metals) to Light" (5th) Faraday lectures at the RI, "On the Conservation of Force" (27th)
March 1857		US Supreme Court ruling in *Dred Scott v. Sandford* states that persons of African descent (slaves or descendants of slaves) are not US citizens and cannot file lawsuits in federal courts (6th)
April 1857	Takes a trip to the Isle of Wight with Heinrich Debus and his son Harry (5th–15th) Receives letter from War Office asking if he would continue being an examiner for commissioned officers of the Royal Artillery and Royal Engineers for lower pay (17th) Asks Faraday if he could request of the RI managers that his annual lecture load be decreased (18th) Replies to the War Office stating he will not continue being an examiner for lower pay (18th) Council of the RS grants Tyndall £40 for glacier research (20th) Begins another series of lectures on sound (eight) at the RI (23rd); through 11 June	Treaty of Paris ends Anglo-Persian War (4th) Edward Frankland begins lecture course (seven) at the RI, "Relations of Chemistry to Graphic and Plastic Art" (25th) Andrew Ramsay gives Friday Evening Discourse at the RI, "On Certain Peculiarities of Climate During Part of the Permian Epoch" (27th) RS holds their first meeting in Burlington House in London after moving from Somerset House Gustave Flaubert publishes *Madame Bovary* as a single volume

Month and year	Biographical details	Notable social, political, and cultural events
May 1857	Visits the spiritualist Camilla Dufour Crosland with Latimer Clark (8th) Publishes "Remarks on Foam and Hail" in the *Phil. Mag.*	James Thomson's paper "On the Plasticity of Ice, as Manifested in Glaciers" read to the RS by his brother William (7th); Thomson further discussed his theory of glacial motion in an 1859 paper in *Roy. Soc. Proc.*, "On Recent Theories and Experiments Regarding Ice at or near Its Melting-Point" Indian Rebellion begins when Indian soldiers (*sepoys*) of the British East India Company, then ruling India on behalf of Britain, mutiny in the town of Meerut northeast of Delhi (10th); Dehli captured (11th) Huxley gives Friday Evening Discourse at the RI, "On the Present State of Knowledge as to the Structure and Function of Nerve" (15th) The British North American Exploring Expedition to Canada begins (16th); ends in 1860 James Paget gives Croonian Lecture for the RS, "On the Cause of the Rhythmic Motion of the Heart" (28th) Hirst publishes first part of his paper "On Equally Attracting Bodies" in the *Phil. Mag.* (second part in September 1858)
June 1857	Gives Friday Evening Discourse at the RI, "On Lissajous' Acoustic Experiments"; Lissajous present to perform the experiments (5th)	Faraday gives Friday Evening Discourse at the RI, "On the Relations of Gold to Light" (12th) South Kensington Museum in London opens (22nd); later separated into the Victoria and Albert Museum and the Science Museum Queen Victoria grants her husband Albert the title Prince Consort (25th) Queen Victoria awards the first Victoria Crosses, for actions during the Crimean War, at a ceremony in Hyde Park, London (26th)

Month and year	Biographical details	Notable social, political, and cultural events
June 1857		Charles Dickens reads from *A Christmas Carol* at Saint Martin's Hall in London, his first public reading (30th)
		Henry Buckle publishes first volume of *History of Civilization in England*
July 1857	Anna Hirst dies in Paris of tuberculosis (1st)	Tensions between Catholics and Protestants in Belfast lead to ten days of rioting (12th)
	Leaves London to visit Hirst in Paris, attend Anna's funeral, and travel to the Alps for glacier research (2nd); arrives in Paris on 3 July	
	Attends Anna's funeral in Paris (4th)	
	Leaves Paris with Hirst to travel to the Alps (7th)	
	Visits Auguste de la Rive in Geneva, without Hirst (11th)	
	Sees Mont Blanc for the first time; arrives in the valley of Chamonix (12th)	
	Hirst rejoins Tyndall (13th)	
	Examines the Montanvert, Talèfre, Géant, Aiguille, and other glaciers situated on the north side of Mont Blanc (13 July–11 August)	
August 1857	Huxley joins Tyndall and Hirst in the Alps (10th–22nd)	First ascent of the Mönch in the Swiss Alps by Christian Almer, Christian Kaufmann, Ulrich Kaufmann, and Sigismund Porges (15th)
	Summits Mont Blanc with Hirst (13th); Huxley climbs part of the route but stops due to exhaustion	Failure of the Ohio Life Insurance and Trust Company is first sign of economic decline in United States, leading to the Panic of 1857, which also affected Europe (24th)
	Begins journey home with Hirst and Huxley (21st); arrives in Paris (24th), and London (25th)	Matrimonial Causes Act removes requirement for parliamentary approval in granting divorces (28th)
	Has photographic portrait taken by Maull & Polyblank (later Maull & Fox) in London (26th)	BAAS holds its twenty-seventh meeting in Dublin, Ireland; into September
	In Queenwood (21 August–8 September)	Rudolf Clausius publishes "On the Nature of the Motion which We Call Heat" in the *Phil. Mag.*

Month and year	Biographical details	Notable social, political, and cultural events
September 1857	Has portrait taken by Maull & Polyblank (later Maull & Fox) in London (21st) (see frontispiece to this volume)	French philosopher Auguste Comte dies (5th) Charles Darwin writes a letter to American botanist Asa Gray outlining his theory of evolution (5th) Mountain Meadows Massacre in Utah Territory results in more than one hundred members of an emigrant wagon train party killed by Mormon militia (7th–11th) Britain recaptures Delhi from Indian rebels (20th); this leads to the surrender of Bahadur Shah II, the last Mughal emperor Obscene Publications Act defines the sale of obscene material as a statutory offense
October 1857	Begins series of six lectures on physics for students at Eton College (Eton, Berkshire), refuses pay (8th) Huxley's letter to Tyndall of 14 September 1857 published as "Observations on the Structure of Glacial Ice" in the *Phil. Mag.*	Panic of 1857: New York banks close (13th) Philip Gosse publishes *Omphalos: An Attempt to Untie the Geological Knot*
November 1857	Recounts his August 1857 ascent of Mont Blanc to the Philosophical Club (19th)	The *Atlantic Monthly*, a magazine of literature, art, and politics in the United States, is established (1st) Louis Pasteur publishes "Mémoire sur la fermentation appelée lactique" in *Comptes Rendus Chimie* (30th) RS awards Michel Eugène Chevreul the Copley Medal, "for his researches in organic chemistry, particularly on the composition of the fats, and for his researches on the contrast of colours" (30th) RS awards Edward Frankland the Royal Medal, "for the isolation of the organic radicals of the alcohols, and for his researches on the metallic derivatives of alcohol" (30th)

Month and year	Biographical details	Notable social, political, and cultural events
November 1857		RS awards John Lindley the Royal Medal, "for his numerous researches and works on all branches of scientific botany, and especially for his vegetable kingdom, and his genera & species of Orchideae" (30th)
December 1857	Begins "Six Lectures on the Phenomena of Light" at the London Institution (14th); through 18 January 1858 Reads "On Some Physical Properties of Ice" at the RS (17th); published in the *Phil. Trans.* in 1858 Declines invitation to be a founding member of the Alpine Club Publishes "On the Sounds produced by the Combustion of Gases in Tubes" in the *Phil. Mag.* Leaves the council of the RS due to heavy lecturing responsibilities	Hydrographer and Royal Navy officer Francis Beaufort dies (17th) The Alpine Club in London is founded (22nd); Tyndall would join in November 1858 Second Anglo-Chinese War: British seize Canton with French assistance (28th–31st) Queen Victoria names Ottawa, Ontario, as capital of Canada (31st) John Ball publishes "Observations upon the Structure of Glaciers" in the *Phil. Mag.*
1858		Rudolf Virchow publishes *Cellular Pathology* (*Die Cellularpathologie*), which held his well-known aphorism from a few years before: "Every cell stems from another cell"
January 1858	Begins "Ten Lectures on Heat, Considered as a Mode of Motion" at the RI (21st); through 25 March Reads Samuel Haughton's paper "On the Physical Structure of the Old Red Sandstone of the County of Waterford, Considered with Relation to Cleavage, Joint Surfaces, and Faults" at the RS (21st) Gives Friday Evening Discourse at the RI, "On Some Physical Properties of Ice" (22nd) Becomes examiner for the newly established Council of Military Education Converses with Huxley and Frankland about establishing a scientific review publication	William Grove gives Friday Evening Discourse at the RI, "On Molecular Impressions by Light and Electricity" (29th) The SS *Great Eastern* launches from the Thames, at the time the largest ship ever built (31st)

Month and year	Biographical details	Notable social, political, and cultural events
February 1858	Meets explorer David Livingstone at a dinner party (5th)	William Thomson receives a patent in Britain for his design of a mirror galvanometer (9th); it would be used for the first transatlantic telegraph cable
		Faraday gives Friday Evening Discourse at the RI, "Remarks on Static Induction" (12th)
		Richard Burton and John Speke are the first Europeans to reach Lake Tanganyika in East Africa (13th)
		Following an assassination attempt on the French emperor by the Italian republican Felice Orsini with a bomb made in Britain (14 January), the diplomatic crisis forces Prime Minister Palmerston to resign (19th); Derby assumes position for a short duration; Palmerston returns in June
		Rudolf Clausius publishes "On the Conduction of Electricity in Electrolytes" in the *Phil. Mag.*
March 1858	The managers of the RI approve a change to Tyndall's position that decreases his annual lecture load from nineteen to twelve, without a decrease in pay (1st) Observes an annular solar eclipse in England (15th)	John Peter Gassiot gives Bakerian Lecture at the RS, "On the Stratifications and Dark Band in Electrical Discharges as Observed in Torricellian Vacua" (4th)
		Henry Buckle gives Friday Evening Discourse at the RI, "On the Influence of Women on the Progress of Knowledge," his only public lecture (19th)
		John Barlow gives Friday Evening Discourse at the RI, "On Mineral Candles and other Products Manufactured at Belmont and Sherwood" (26th)
April 1858	Takes a trip to the Isle of Wight with Henry Bence Jones; visits Alfred, Tennyson (3rd–8th)	J. D. Forbes reads "On Some Properties of Ice near Its Melting Point" at the Royal Society of Edinburgh (19th)

Month and year	Biographical details	Notable social, political, and cultural events
April 1858	Begins an additional three lectures added to his course on heat at the RI (15th); through 29 April Has photographic portrait taken with Faraday by a sixpenny photographer in London (16th)	Arthur Ramsay gives Friday Evening Discourse at the RI, "On the Geological Causes That Have Influenced the Scenery of Canada and the North-East Provinces of the United States" (30th)
May 1858	Reads "On the Physical Phenomena of Glaciers.—Part I. Observations on the Mer de Glace" at the RS (20th); published in the *Phil. Trans.* in 1859 as "On the Physical Phenomena of Glaciers – Part I. Observations on the Mer de Glace" Publishes "On the Structure and Motion of Glaciers" in the *Phil. Mag.*	Eastern region of the Minnesota territory in United States admitted as a state (11th) Royal Opera House, London, opens (15th) Second Anglo-Chinese War: following the takeover of Taku forts near Tiensin (20th), hostilities between Britain and France against China cease Huxley gives Friday Evening Discourse at the RI, "On the Phenomena of Gemmation" (21st) Edward Frankland gives Friday Evening Discourse at the RI, "On the Production of Organic Bodies Without the Agency of Vitality" (28th)
June 1858	Gives Friday Evening Discourse at the RI, "On the Mer-de-Glace" (4th) Has open-air photographic portrait taken with Michael and Sarah Faraday, Harriet Moore, and Jane Barnard, positioned around a garden bench, at Petersham (see frontispiece to *Tyndall Correspondence*, vol. 5)	Giovanni Battista Donati discovers comet (Donati's comet), which is visible in European skies for several months (2nd) Faraday gives Friday Evening Discourse at the RI, "On Wheatstone's Electric Telegraph in Relation to Science (Being an Argument in Favour of the Full Recognition of Science as a Branch of Education)" (11th) Abraham Lincoln accepts Republican nomination for a seat in US Senate, delivers "House Divided" speech (16th) British physician John Snow, a founder in the field of epidemiology, dies (16th) Thomas Huxley gives Croonian Lecture for the RS, "On the Theory of the Vertebrate Skull" (17th)

Month and year	Biographical details	Notable social, political, and cultural events
June 1858		Charles Darwin receives a letter from naturalist Alfred Russel Wallace describing natural selection (18th)
		Indian Rebellion ends when Indian soldiers (*sepoys*) of the British East India Company are defeated in Gwalior, south of Delhi (20th)
		Second Anglo-Chinese War: through the Treaty of Tientsin, France, Russia, the United States, and Britain force China to open ports to Western trade (25th)
July 1858	Asks John Murray if he would be interested in publishing his glacier book (12th); published in 1860 as *Glaciers of the Alps*	A joint presentation of papers by Charles Darwin and Alfred Russel Wallace given at the Linnean Society (1st)
	Leaves London with Ramsay to travel to the Alps for continued glacier research; arrives in Paris (16th)	The Plombières Agreement between France and Italy creates an alliance against Austria (21st)
	Visits Fick and Clausius in Zurich (20th)	Lionel de Rothschild takes his seat as the first Jewish Member of Parliament (26th)
	Examines the Rosenlaui and Upper and Lower Grindelwald glaciers, crosses the Strahlegg and Grimsel passes, examines the Rhone glacier, summits the Eggishorn, examines the Aletsch glacier, and summits the Finsteraarhorn (23 July–3 August)	British officer William Herschel conducts first contract finger-print signature in Bengal (28th)
		The "Great Stink" occurs in London, in which hot summer weather intensifies the stench of untreated sewage on and around the Thames (through August); public reaction leads to authorities accelerating upgrades to the sewer system
		Charles Darwin begins writing an "abstract" of his planned work on natural selection; published 24 November 1859 as *On the Origin of Species*
		Auguste de la Rive publishes third volume of the English translation of his *Traité de l'électricité théorique et appliquée* (*A Treatise on Electricity in Theory and Practice*), translated by Charles Walker and proofed by Tyndall

Month and year	Biographical details	Notable social, political, and cultural events
July 1858		Rudolf Clausius publishes "On the Nature of Ozone" in the *Phil. Mag.*
		George Bentham publishes *Handbook of the British Flora*
August 1858	Arrives in Zermatt (8th–15th)	Government of India Act 1858 passed, describes the transfer of India from the British East India Company to Britain (2nd)
	Summits Monte Rosa (9th)	
	Summits the Riffelhorn and examines the Gorner glacier (14th–15th)	Medical Act 1858 passes in Britain, regulates "the Qualifications of Practitioners in Medicine and Surgery" (2nd)
	Summits Monte Rosa solo (17th)	
	Summits the Distel Alp (18th)	British Columbia in North America established as a British colony by Richard Clement Moody (2nd)
	Crosses Monte Moro pass, examines the Allalin glacier, arrives in Saas to find Hirst at the hotel, and they examine the Fee glacier (23rd–27th)	
		John Speke is the first European to reach Lake Victoria, source of the Nile (3rd)
		First transatlantic telegraph cable completed, from Britain to the United States via Newfoundland (5th); stops working on 1 September
		First ascent of the Eiger in the Swiss Alps by Swiss guides Christian Almer and Peter Bohren, and Irishman Charles Barrington (11th)
		US President James Buchanan exchanges greetings with Queen Victoria through transatlantic telegraph cable (16th)
		Seven debates between Abraham Lincoln and Stephen Douglas during US Senate campaigns begin (21st); run through 15 October
		Henry Gray and illustrator Henry Vandyke Carter publish *Gray's Anatomy*
September 1858	Writes to the president of the Chamonix Guide Company about difficulties with plans to summit Mont Blanc, regarding the use of porters and guides (1st)	First ascent of the Dom in the Alps by British theologian John Llewelyn-Davies and three guides (11th)
		BAAS holds its twenty-eighth meeting in Leeds (22nd–29th); Tyndall attends

Month and year	Biographical details	Notable social, political, and cultural events
September 1858	Faraday notifies Tyndall that he had sent his 23 August letter to Sarah Faraday, which describes his solo ascent of Monte Rosa a week before, to the *Times* (2nd); published 3 September Describes first seeing Donati's comet (11th) Summits Mont Blanc with Auguste Balmat and Alfred Wills (12th) Begins travel back toward London, from Geneva (17th) Obtains a small grant from the RS in order for Auguste Balmat to place thermometers on the summit of Mont Blanc	Hirst publishes second part of his paper "On Equally Attracting Bodies" in the *Phil. Mag.* (first part in May 1857) Thomas Carlyle publishes first two volumes (of six) of *History of Friedrich II of Prussia*
October 1858	Describes his examination of veined structure of Alpine glaciers (comparable to slaty cleavage in rocks) at the Philosophical Club (28th) Travels to Queenwood to work on his glacier book	Florence Nightingale sends a copy of her privately published *Notes on Matters Affecting the Health, Efficiency and Hospital Administration of the British Army* to Queen Victoria (11th) William (I) assumes regency in Prussia, in lieu of his brother Frederick William IV, who became incapable following a stroke (26th) William Parker Foulke discovers the first dinosaur bones (*Hadrosaurus*) in the United States (Haddonfield, New Jersey)
November 1858	David Brewster offers Tyndall the Chair of Natural Philosophy at Saint Andrews University in Scotland (25th) Elected a member of the Alpine Club (27th) Lectures at the London Institution, "On Light" (29th)	British government takes power in India from the British East India Company following the unsuccessful Indian Rebellion (1st) RS awards Charles Lyell the Copley Medal, "for his various researches and writings by which he has contributed to the advance of geology" (30th) RS awards Albany Hancock the Royal Medal, "for his various researches on the anatomy of the Mollusca" (30th)

Month and year	Biographical details	Notable social, political, and cultural events
November 1858		RS awards William Lassell the Royal Medal, "for his various astronomical discoveries and researches" (30th)
		RS awards Jules Jamin the Rumford Medal in 1858, "for his various experimental researches on light" (30th)
		Henry Sorby publishes "On the Microscopical Structure of Crystals, Indicating the Origin of Minerals and Rocks" in the *Quarterly Journal of the Geological Society*
December 1858		South Foreland Lighthouse in Saint Margaret's Bay (Dover, Kent, England) becomes first to be lit with electricity (8th); this is not sustained, but renewed on 28 March 1859
		Faraday gives annual Christmas Lectures for youth at the RI, on "the metallic properties"
		William Thomson publishes "On the Stratification of Vesicular Ice by Pressure" in the *Phil. Mag.*
January 1859	Begins "Twelve Lectures on the Force of Gravity" at the RI (27th); through 14 April	Elizabeth Blackwell becomes the first woman to be entered on the General Medical Council's Medical Register (1st)
		National Portrait Gallery, London opens to the public (15th)
		William Grove gives Friday Evening Discourse at the RI, "On the Electrical Discharge and its Stratified Appearance in Rarefied Media" (28th)
February 1859	Paper "On the Veined Structures of Glaciers; with Observation upon White Ice-Seams, Air-bubbles and Dirt-bands, and Remarks upon Glacier Theories" read at the RS (24th); published later in the *Phil. Trans.*	George Eliot publishes *Adam Bede* (1st)
		Richard Owen gives Friday Evening Discourse at the RI, "On the Gorilla" (4th)
		Western region of the Oregon territory in United States admitted as a state (11th)

Month and year	Biographical details	Notable social, political, and cultural events
February 1859	Publishes "Remarks on Ice and Glaciers" in the *Phil. Mag.* (Forbes published a reply to this in the March 1859 issue)	French colonial forces capture Saigon (now Ho Chi Minh City) in Vietnam, becoming the center of French Indochina (which includes Laos and Cambodia) (18th)
		Swede Svante Arrhenius born (19th); would later study the effects of increased atmospheric carbon dioxide on the Earth's surface temperature
		Faraday gives Friday Evening Discourse at the RI, "On Schšnbein's Ozone and Antozone" (25th)
		John Stuart Mill publishes *On Liberty*
March 1859	Gives Friday Evening Discourse at the RI, "On the Veined Structure of Glaciers" (4th)	

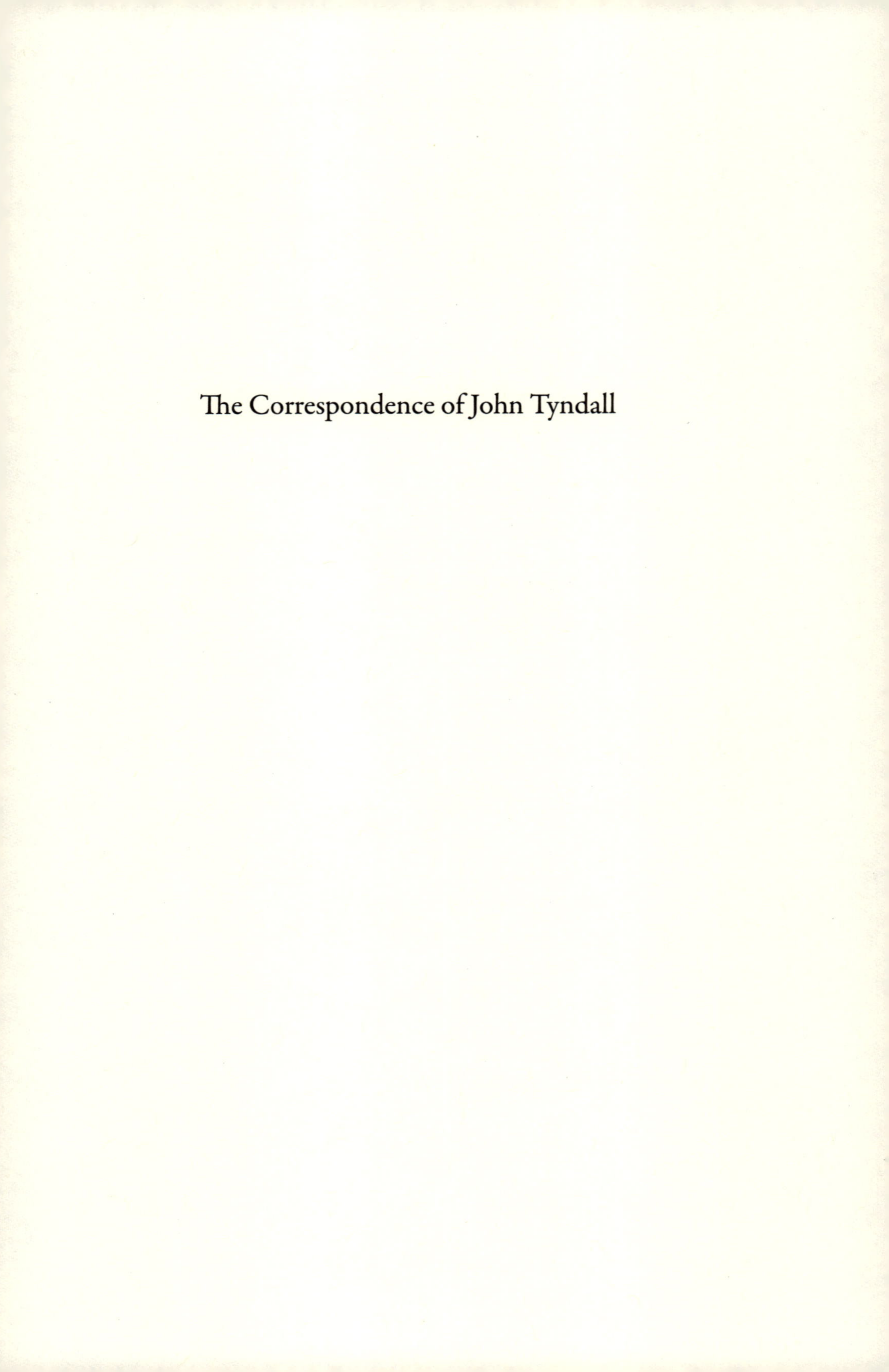

The Correspondence of John Tyndall

1856

From Frances Hooker   6 November 1856   1292

> Cold was the air; the wind blew keen around;
> Autumn's brief day was drawing to its close:—
> For fallen leaves you scarce might see the ground,
> And giant, naked trunks around me rose.
> The glorious; setting sun athwart the stems
> Poured forth a golden flood of slanting rays,
> Which decked the trees with such resplendent gems,
> I could not choose but stand awhile, and gaze.
> The red, red sun! a ball of liquid fire:
> Tinging the murky haze with rose light;
> Touching with gold the unseen village spire,—
> Bidding the blushing trees a warm good night.
> Slowly his glories paled: his reign was past;
> And the white haze proclaimed the frost-king's power:—
> I turned, and through the pine-woods hurried fast,
> Filled with the beauties of that fleeting hour.[1]

Nov. 6th. 1856.

RI MS JT/1/TYP/8/2799
LT Typescript Only

––––––––––

1. *Cold was the air . . . of that fleeting hour*: a poem by Frances Hooker written in the style of
   Tennyson. An annotation from LT reads: 'In Mrs Hooker's writing'. The covering letter for
   this enclosure is missing.

From Auguste de la Rive   7 November 1856   1293

Venise le 7 novembre 1856,

Mon cher Monsieur,
    C'est de Venise, où je viens de recevoir votre lettre qu'on m'a transmise de

Genève, que je viens vous répondre. J'ai quitté Genève il y a dix jours & je me dirige *[sur]* Florence & *[vers]* Rome où je vais passer l'hiver avec une partie de ma famille. Vous seriez bien aimable, si vous en trouvez le temps, de m'y adresser une fois quelques lignes pour me donner de vos nouvelles, de celles de nos amis & en particulier de Faraday, & aussi quelques nouvelles scientifiques dont je serai bien privé cet hiver. Je serai à <u>Rome</u> du 15 décembre à la fin de Mars et vous n'aurez qu'à m'y écrire <u>poste restante</u> ; j'espère que vous me ferez ce plaisir.

Venant à l'objet de votre lettre, je vous dirai qu'il est parfaitement exact que la couleur de l'eau du Rhône à sa sortie du lac est d'un beau bleu tout à fait semblable (peut-être un peu plus foncé) à la couleur fluorescente de la solution de sulfate de quinine. Cette couleur si remarquable présente cependant d'autres nuances ; près des piliers des ponts qui traversent le Rhône, l'eau présente une nuance violette très prononcée ; elle a aussi des teintes verdâtres en certaines places. Arago, dans son séjour à Genève, était très occupé de ces diverses apparences, & je l'ai trouvé souvent en contemplation devant ces effets de couleur de l'eau du Rhône qu'il admirait sans pouvoir s'en rendre compte. Cependant, il me paraissait disposé à y voir des phénomènes de contraste. Il me semble qu'il a publié une fois quelque chose sur ce sujet dans une de ses notices, mais je n'en suis pas bien sûr ! Davy s'en était occupé & il croyait que la couleur bleue de l'eau tenait à la présence d'un peu d'iode, ce qui n'est pas admissible. Un Colonel <u>Jackson</u>, établi à Saint Petersburg, a publié un mémoire sur la question, il a été inséré dans la <u>Bibl.[iothèque] Universelle de Genève</u>, je ne me souviens pas dans quelle année ; ce doit être entre 1826 et 1836 ; au moyen des tables analytiques vous trouveriez facilement ce mémoire. Je crois me rappeler qu'il y en a également un à peu près à la même époque de M. <u>Xavier de Maistre</u>. Enfin vous trouverez dans la <u>Bibl.[iothèque] Universelle</u> (N° de <u>Juillet, d'Août ou de Septembre</u> 1847), un travail de Melloni sur la grotte d'azur de l'île de Capri, qui peut jeter quelque lumière sur le sujet. Le fait est que, ainsi que j'ai pu en juger par expérience, la couleur bleue de l'eau dans l'intérieur de cette grotte est tout à fait semblable à celle de l'eau du Rhône dans les beaux jours d'été.

Quant à moi, j'étais personnellement disposé à regarder le <u>bleu</u> comme la couleur naturelle de l'eau parfaitement pure, couleur qui ne peut se manifester que lorsque l'eau est en grande masse, comme cela arrive à beaucoup de liquides très légèrement colorés. La glace en gros blocs est également bleue. Les autres couleurs que l'eau affecte seraient un effet de la présence de quelque substance en solution, ou à la réflexion sur le fond du bassin qui contient l'eau, de la lumière du jour. Aussi, l'eau n'a son apparence bleue que lorsqu'elle est très pure et que le fond est blanc ou grisâtre (gravier & sable) ; dans le lac de Neuchâtel, on remarque des places où l'eau est constamment verte & il se trouve que ces places correspondent à un fond de sable jaune. Quant à la couleur violette qu'on observe près des piliers en bois des ponts du Rhône à Genève, elle tiendrait à la réflexion contre ces piliers brunâtres de la lumière du jour. Je ne nie point que cette manière de voir présente bien des objections & je me souviens l'avoir présentée, je ne me rappelle ni où, ni quand, avec quelque défiance, en reconnaissant ce qu'elle laissait à désirer. Je serais bien charmé d'apprendre que vous avez réussi à *[éclaircir]* la question, en la rattachant peut-être aux phénomènes de la <u>fluorescence</u> comme j'en ai eu quelque fois l'idée. Mais il vous faut avant tout venir <u>vous-même</u> à Genève voir le phénomène ; aussi je vous attends l'été prochain; vous êtes sûr, si Dieu me prête vie, de m'y trouver de la fin de <u>Mai</u> jusqu'au mois de <u>Novembre</u>.

Voilà une lettre bien longue pour vous dire malheureusement bien peu de choses. Mais il me vient l'idée de profiter de ce que je vous écris pour vous demander un très grand service, mais à une condition, c'est que vous me le refuserez s'il en résulte pour vous un trop grand dérangement. Vous le pourrez d'autant mieux que je pourrai m'adresser à quelqu'un d'autre ; il est vrai que ce sera très différent pour moi. Voici ce dont il s'agit. Monsieur Walker est occupé à faire la traduction de mon 3ᵉᵐᵉ volume : il la fait sur le texte français parfaitement revu & corrigé ; néanmoins, j'ai toujours reçu les dernières épreuves de chaque feuille pour les deux premiers volumes ; il est vrai que ces épreuves étaient tellement correctes que le plus souvent il n'y avait rien à corriger. Maintenant, il est impossible qu'il m'envoie ces épreuves en Italie comme il me les envoyait à Genève. Serait-ce bien indiscret de vous prier de les recevoir & de les lire ? Je le répète ; il est probable que vous n'auriez que bien peu de corrections à faire ; vous n'auriez donc que l'ennui de lire, ce qui est bien assez. La personne à qui je pourrais demander ce service, si vous ne pouvez me le rendre, n'est point scientifique, c'est ce qui fait que j'hésitais à le lui demander quand, en vous écrivant, l'idée m'est venue de m'adresser à vous. Comme j'ai besoin sur ce point d'une réponse un peu prompte, auriez-vous l'extrême bonté de m'écrire aussitôt que vous pouvez à <u>Florence (poste restante)</u> ; j'y serai jusqu'au commencement de décembre.

Votre dévoué et affectionné | Aug. de la Rive.

Venice 7 November 1856,

My dear Sir,

It is from Venice,[1] where I just received your letter[2] that was delivered to me from Geneva,[3] that I will respond. I left Geneva ten days ago & I am heading for Florence[4] and Rome where I will spend the winter with some of my family. You would be very kind if you were to find the time to send me a few lines about yourself, our friends & in particular Faraday, as well as a few scientific updates which I will be missing this winter. I will be in <u>Rome</u> from the 15th of December to the end of March, and you can simply address your letters to <u>POSTE RESTANTE</u>;[5] I hope you will give me that pleasure.

On the subject of your letter, I would tell you that it is perfectly clear that the colour of the Rhône[6] as it leaves the lake is a beautiful blue absolutely similar (perhaps slightly darker) to the fluorescent solution of quinine sulfate. Nevertheless, this remarkable colour displays other nuances; close to the pillars of the bridges crossing the Rhône, the water shows a very accentuated purple nuance; it also has greenish shades in some places. During his stay in Geneva, Arago[7] was really taken up by these diverse appearances, & I often found him gazing at the colour effects of the Rhône waters that he admired without realizing it. Nevertheless, he seemed prone to explain them by a phenomenon of contrast. I think he published once something on this matter in one of his notes, but I am not that sure! Davy was occupied by it & believed that the blue colour of the water resulted from the presence of a little iodine, which is not acceptable. A Colonel <u>Jackson</u>,[8] from St. Petersburg, published an essay[9] on the question; it was *[printed]* in the <u>Bibl. Universelle de Genève</u> but I do not remember which year; it is probably 1826 and 1836; you could easily find the essay using the analytical indexes. I seem to recall there was another one by <u>Mr Xavier de Maistre</u>[10] from a similar date. Finally, you will find in the <u>Bibl. Universelle</u> (No from <u>July, August or September</u> 1847), a work[11] by Melloni[12] on the blue grotto[13] of the island of Capri[14] that may enlighten you on the matter. The fact is that, as I know from my own experience, the blue colour from inside the grotto is absolutely similar to that of the Rhône in the sunny days of summer.

As for me, I was personally willing to think of the <u>blue</u> as that natural colour of the perfectly pure water, a colour that only reveals itself when water is in large quantity, as it happens to many lightly coloured solutions. Large blocks of ice are also blue. The other colours exhibited by water would be due to the presence of some liquid substance, or to the reflection of daylight on the bottom of the pool. Hence, water appears blue only when it is really pure and the bottom is white or greyish (gravel & sand); in the Neuchâtel lake,[15] there are places where the water is constantly green, and, in fact, the bottom in these places is usually made of yellow sand. As for the purple colour we observe next to the wooden pillars of the bridges on the Rhône in Geneva, it would be due to the reflection of the daylight against these brownish columns. I am not denying that this approach presents many objections and I remember introducing it, I cannot recollect <u>where or when</u>, with some *[hesitation]*, recognising it left much to be desired. I would delighted to learn that you successfully shed a light on the question, perhaps by linking it to the phenomenon of <u>fluorescence</u> as I sometimes thought. But first, you must come to Geneva to witness <u>with your own eyes</u> the phenomenon;

therefore, I shall wait for you next summer; if God lets me live, you can be sure to find me there from late <u>May</u> to <u>November</u>.

This is such a long letter to unfortunately tell you very few things. But it just occurred to me that I could take advantage of this letter to ask you for a very big favour, on the condition however, that you would turn it down if it was too much trouble. All the more so that I could ask somebody else; it is true that it would be very different for me. Here it is. Mr Walker is employed to translate my 3rd volume:[16] he is doing it from a perfectly revised and corrected French text; nonetheless, I have always received the last proofs of each page for the first two volumes; admittedly, the proofs were so exact that in most cases nothing had to be changed. Now, he won't be able to send these proofs to Italy the way he used to send them to Geneva. Would it be too much to ask you to receive & read them?[17] Once again, you probably would only have a few corrections to make; you would only have the trouble of reading them, which is more than enough. The person to whom I could ask this favor, if you could not do it, is not scientific; which makes me hesitate to ask him when, while writing to you, it occurred to me to ask you. Since I need a rather prompt answer on this matter, would you have the extreme kindness to write to me as soon as you can to <u>Florence (POSTE RESTANTE)</u>; I will be there until the beginning of December.

Your very devoted and affectionate | Aug. de la Rive.

RI MS JT/1/TYP/1/340–42
LT Typescript Only

---

1. *Venice*: a coastal city in northeastern Italy, built on numerous small islands and notable for its system of canals.

2. *your letter*: see letter 1289, John Tyndall to Auguste de la Rive, 26 October 1856, *Tyndall Correspondence*, vol. 5.

3. *Geneva*: a city in western Switzerland at the southern end of Lake Geneva, capital of the canton of Geneva.

4. *Florence*: a city in central Italy to the south of Bologna.

5. <u>*POSTE RESTANTE*</u>: general delivery (French); when a post office holds mail until the recipient requests it.

6. *Rhône*: significant river starting in Switzerland and passing through southeastern France.

7. *Arago*: François Arago (1786–1853), a staunch French Republican and eminent physicist who made a number of contributions especially to the science of optics and who championed Fresnel's wave theory of light (*CDSB*).

8. *Colonel Jackson*: Julian Jackson (1790–1853), a geographer and army officer in the Russian service, who from 1844 was the editor of the *Journal of the Royal Geographical Society* (*ODNB*).

9. *an essay*: see G. R. Jackson, 'Sur la Couleur des Eaux', *Bibliothèque universelle des sciences,*

belles-lettres et arts, 43 (1830), pp. 420–5. The initial 'G' was probably a typographical error, as Julian Jackson's (see n. 8) obituary discussed an 1830 article on the colour of water published in the *Bibliothèque universelle de Genève.* See 'Obituary', *Journal of the Royal Geographical Society,* 23 (1853), pp. lxxi–lxxiii.

10. <u>*Mr Xavier de Maistre*</u>: Xavier de Maistre (1763–1852), a French writer. See *Encyclopædia Britannica,* 11th edn, 29 vols (Cambridge: Cambridge University Press, 1911), vol. 17, p. 446.

11. *a work*: see Xavier de Maistre, 'Sur la couleur de l'air et des eaux profondes, et sur quelques autres couleurs fugitives analogues', *Bibliothèque universelle des sciences, belles-lettres et arts,* 51 (1832), pp. 259–78.

12. *Melloni*: Macedonio Melloni (1798–1854), an Italian physicist and 1834 recipient of the RS's Rumford Medal, 'For his discoveries relevant to radiant heat' (*CDSB*).

13. *blue grotto*: the Grotta Azzurra, a sea cave on the northwestern shore of Capri, which is illuminated in blue by sunlight that shines through the water via an underwater cavity.

14. *island of Capri*: an island in the Tyrrhenian Sea to the west of southern Italy.

15. *Neuchâtel lake*: a large lake in Switzerland, mostly in the canton of Neuchâtel but shared with those of Vaud, Fribourg, and Bern.

16. *my 3rd volume*: Tyndall looked over the English translation proofs from Charles Walker for de la Rive's third volume of *Traité de l'Électricité Théorique et Appliquée* (Paris: Chez J.–B. Bailliére et Fils, 1858). The full series was published as A. de la Rive, *A Treatise on Electricity: In Theory and Practice,* 3 vols (London: Longman, Brown, Green, Longmans, & Roberts, 1853–8).

17. *Would it be too much to ask you to receive & read them?*: for Tyndall's reply, see letter 1299.

From William Pridie   8 November 1856   1294

Comm Bank[1] | Halifax 8th Nov 1856

Dear Sir

I have received through our London Bankers £14.13.7 and have placed the sums to the credit of Hirst's[2] account, which a/c,[3] for money furnished by me to Mrs Booth[4] is now closed.[5]

I will advise our Friend of the payment when I know his address—and on his return home, will send him his Pass Book, containing a statement of sums received and paid.

Thus terminate a series of Charitable acts such as are rarely to be met with, and which will confer lasting honor on Hirst and those of his Friends who took an interest in Frank Booth,[6] and his Mother, two of the worthiest objects of Charity that ever lived.

I remain | Yours Ever | W^m R Pridie

RI MS JT/1/P/258

1. *Comm Bank*: Halifax Commercial Banking Company, Yorkshire.
2. *Hirst's*: Thomas Hirst.
3. *a/c*: account.
4. *Mrs. Booth*: Sarah Booth.
5. *is now closed*: Tyndall provided these funds to cover those furnished by Thomas Hirst to Sarah Booth, the mother of Hirst's friend Francis Booth. Tyndall had notified Hirst that he would do this (see letter 1283, John Tyndall to Thomas Hirst, 9 October 1856, *Tyndall Correspondence*, vol. 5). Tyndall often assisted Hirst in times of financial need, as Hirst had done for him previously.
6. *Frank Booth*: Francis Booth (d. 1853), son of Sarah Booth and a mutual friend of Tyndall and Hirst. In 1851, he became ill with tuberculosis and his health deteriorated quickly. See R. Barton, J. Rankin, and M. Reidy (eds), *The Correspondence of John Tyndall, Volume 3: The Correspondence, January 1850–December 1852* (Pittsburgh: University of Pittsburgh Press, 2017).

To Henry Clifton Sorby   14 November 1856   1295

14th <u>Nov. 1856</u>

Dear Mr Sorby

You have every reason to think hardly of me for not replying to your last letter,[1] but the fact is I mislaid it and after hunting for it many a time—hoping that chance also might assist in causing it to 'turn up'—I have come to the resolution of saying to you that if you repeat your question I shall at once see Faraday upon the subject[2] and obtain for you the information you require.

Believe me | most sinc[ere]ly yours | <u>John Tyndall</u>

With regard to the structure of the ice it was the hope that it was for me a case of cleavage that took me upon the glaciers.[3] It is a difficult question,[4] but I hope within the ensuing six weeks to let you know something of my notions (which are now quite unripe) upon the subject.[5]

Sheffield Archives SLPS/51/77–14 Nov 1856

1. *your last letter*: see letter 1275, Henry Sorby to John Tyndall, 1 September 1856, *Tyndall Correspondence*, vol. 5.
2. *upon the subject*: not identified.
3. *a case of cleavage that took me upon the glaciers*: see J. Tyndall, 'Comparative View of the Cleavage of Crystals and Slate Rocks', *Phil. Mag.*, 12 (1856), pp. 35–48. Tyndall explained cleavage in rocks as occurring by the exertion of compression perpendicular to the cleavage planes (a mechanical model), and not because of crystalline (polar) forces.
4. *a difficult question*: beginning in the summer of 1856 Tyndall investigated the structure and movement of glaciers in the Alps and his explanation for glacial motion brought him

into conflict with James Forbes and others. See letter 1306, n. 8 and n. 10; and J. Tyndall and T. H. Huxley, 'On the Structure and Motion of Glaciers', *Phil. Trans.*, 147 (1857), pp. 327–46.

5.   *to let you know something of my notions . . . upon the subject*: see letter 1344, n. 2.

To Thomas Archer Hirst   16 November 1856   1296

London 16th Nov. 1856

My dear Tom.

I had hoped for a line from you before this; but as none appears and as I have a communication to make to you on business matters I seize the pen this raw, foggy Sunday morning, and with cold fingers, for I have no fire, make in deformed letters my statement of affairs. Firstly, as I told you I had written to Pridie,[1] and soon afterwards received from him a statement of your affairs with Mrs Booth. I sent him duly a post office order for £14–13–7, which was the amount by which your account was overdrawn. Nearly at the same time, I received a note[2] from Pridie, dated 4th November, informing me that on the previous Saturday poor Mrs Booth who had been so <u>long</u> a pensioner on God's bounty in this world had settled her final account with her maker. I enclose the note which contains the acknowledgement of the sum sent to Pridie, and also an allusion to the death of your old pensioner.

Were my journal beside me I should make some extracts from it for Anna's[3] amusement; but it is in the hands of my friend Mrs Pollock[4] and I therefore must be silent on the subject. I met a lady at Pollocks a few evenings ago who spent with her friends the last winter in Pau,[5] and gave a description of the place which quite harmonizes with your own. She said the winter was merely a long drawn Autumn. I hope sincerely that Anna finds the mildness comfortable; and that she is able to enjoy a drive without associating with it such perils as those described in your last letter.[6] John Martin I am informed is with you. This will make your life the pleasanter. Francis[7] expected him to call, and I thought he might be good enough to look in upon me, but he is quite right to go direct to those objects which more immediately interest him. I have a letter from the Mathematical Master of Eton,[8] asking my opinion as to the introduction of a course of experimental science into the school,[9] which should extend over three or four years. He was introduced to me by Mr Moseley; who regards the movement as possessing a very important bearing upon the question of education in this country.[10] I have thought of recommending you to him, but I do not know whether it would meet your wishes, nor do I know indeed precisely what he wants. He would want I imagine lectures on experimental Physics, and perhaps some on elementary Chemistry, but I should recommend to him the combining of a <u>physical class</u> with the lectures. I shall probably write to him before you receive this and will mention you to him—it can do no harm even if neither of you should find the proposal suitable.

I have done very little real earnest work since I wrote to you. I wanted to measure some infinitesimal effects of Magnetism, and for this purpose had to devise an instrument of surpassing delicacy. I have had many trials with it but each trial has suggested some essential change. This week I hope to have it in working order, and I also expect to be able to pursue my enquiries even when the lectures are going on—It is too bad to devote a whole week to one or two lectures which only endure an hour each. Faraday has been working very successfully. You will probably know more about Queenwood[11] than I do, so I will not touch upon that subject.

Pridie has sent me a mourning card—on which stands the following:—

In memory of
Sarah, widow of the late John Booth[12]
of Skircoat[13] near Halifax, Yorkshire.
Died at Bebington[14] November 1st. 1856
Aged 65 years.

Goodbye my dear Tom—give my love to Anna and believe me as ever
Your affectionate | <u>John</u>

RI MS JT/1/T/633
RI MS JT/1/HTYP/482–82a

---

1. *written to Pridie*: William Pridie; letter missing, but see letter 1294 of this volume and letter 1283, John Tyndall to Thomas Hirst, 9 October 1856, *Tyndall Correspondence,* vol. 5.

2. *note from Pridie*: letter missing.

3. *Anna*: Anna Hirst.

4. *Mrs Pollock*: Juliet Pollock.

5. *Pau*: see letter 1298, n. 4.

6. *your last letter*: see letter 1281, Thomas Hirst to John Tyndall, 5 October 1856, *Tyndall Correspondence,* vol. 5.

7. *Francis*: William Francis.

8. *a letter from the Mathematical Master of Eton*: letter missing. Probably from Stephen Hawtrey, who served as Eton's Mathematical Master from 1851–71. Tyndall did receive a letter from Henry Moseley stating that Hawtrey desired to start a course of experimental science at Eton (see letter 1291, 28 October 1856, *Tyndall Correspondence,* vol. 5). Eton College, a boarding school for boys in Windsor, UK.

9. *the school*: Eton College.

10. *a very important bearing upon the question of education in this country*: see letter 1291, Henry Moseley to John Tyndall, 28 October 1856, *Tyndall Correspondence*, vol. 5. Science teaching in schools was pioneered by William Sharp (1805–96) at Rugby School, Warwickshire in 1849. Edward Rupert Humphreys (1820–93) similarly established science education at Cheltenham Grammar School, Gloucestershire in 1852. For a discussion of science education at Queenwood College, see D. Thompson, 'Queenwood College, Hampshire', *Annals of Science,* 11 (1855), 246–54. For Tyndall's work in science education, see N. D. McMillan, 'British physics—the Irish role in the origin, differentiation and organisation of a profession', *Physics Education,* 23 (1988), pp. 272–78; and B. Pippard, 'Schoolmaster-Fellows and the Campaign for Science Education', *Notes and Records of the Royal Society of London,* 56 (2002), pp. 63–81.

11. *Queenwood*: see letter 1298, n. 13.

12. *John Booth*: husband to Sarah Booth and father of Francis Booth.

13. *Skircoat*: a town south of Halifax, Yorkshire.

14. *Bebington*: a town in northeast England, south of Liverpool.

From Rudolf Clausius   16 November 1856   1297

Zürich, d. 16 ten Nov., 1856.

Hochgeehrter Herr Professor,

Nehmen Sie zunächst herzlichen Dank für die schönen Uebersetzungen meiner beiden Abhandlungen, welche Sie in Ihr Phil. Mag. aufgenommen haben. Ich weiss zwar nicht, ob Sie sich selbst die Mühe gemacht haben, die Uebersetzungen anzufertigen; jedenfalls aber glaube ich nicht zu irren, wenn ich annehme, dass Sie die Anregung dazu gegeben haben. Die mechanische Wärmetheorie findet nirgends so allgemeine Theilnahme, wie in England, und es muss daher für mich von besonderem Interesse sein, dass meine Arbeiten in England bekannt werden.

Die Note des H. Joule über meine Abhandlung setzte mich anfangs etwas in Erstaunen, da ich nicht entfernt die Absicht gehabt habe, ihm einen Irrthum zuzuschreiben. Als ich aber den betreffenden Para. meiner Abhandlung nachlas, fand ich, dass allerdings eine Stelle darin vorkommt, welche er möglicher Weise auf sich beziehen konnte. Ich habe es daher für meine Schuldigkeit gehalten, einige Worte zur Aufklärung zu schreiben, und ich möchte Sie bitten, auch diese in ihr Phil. Mag. zu nehmen. Was H. Joule über Prof. Thomson sagt, ist allerdings richtig, aber es widerspricht durchaus nicht dem, was ich gesagt habe. Ich würde dieses leicht nachweisen können; da ich aber Sie und das Publicum nicht gern mit einer Sache belästigen möchte, die doch nur ein persönliches Interesse haben kann, so habe ich diesen Theil der Note ganz unberücksichtigt gelassen.

Ich habe in diesem Herbste das Vergnügen gehabt, durch einen merkwürdigen Zufall die Bekanntschaft eines Ihrer Freunde, des Herrn Huxley zu machen, mit welchem Sie am Tage vorher auf dem Aeggischhorn gewesen waren. Ich habe dabei nur bedauert, dass ich ihn nicht einen Tag früher getroffen habe, als er noch mit Ihnen zusammen war, was mir eine ausserordentlich angenehme Ueberraschung gewesen sein würde. Am folgenden Tage habe ich ihn eben so unerwartet wieder verloren, wie ich ihn gefunden hatte. Als wir mit noch einigen zufälligen Reisegefährten auf dem Wege nach Zermatt waren, begegneten wir einer Gesellschaft, die er kannte. Er blieb im Gespräch mit derselben stehen, und wir übrigen setzten, um ihn nicht zu stören, unseren Weg etwas langsamer fort. Als wir ihn aus dem Gesichte verloren hatten, hielten wir an, und haben wohl eine halbe Stunde auf ihn gewartet. Wahrscheinlich haben seine Freunde ihn beredet, mit ihnen umzukehren. Wenn Sie ihn sehen, seien Sie so gut ihn freundlich von mir zu grüssen, und ihm zu sagen, ich hätte sehr bedauert, dass er nicht mit in Zermatt gewesen wäre. Das Wetter war herrlich, und die Aussicht vom Gorner-Grat ist die grossartigste, die ich gesehen habe.

Es war bei dieser Sache noch ein anderes eigenthümliches Zusammentreffen. Ich hatte Ihre schöne Abhandlung über die Spaltung des Schiefers, welche Sie so gut gewesen sind, mir zu schicken, mit dem grössten Interesse gelesen, und es war mir dabei eingefallen, ob nicht die blauen Bänder der Gletscher, welche ich zwar noch nicht gesehen hatte, von denen mir aber Prof. Studer aus Bern und Escher von der Linth in Zürich mehrfach gesprochen hatten, sich auf dieselbe Weise erklären liessen. Als ich daher auf meiner Reise zum ersten Male an einen Gletscher gekommen war, (den Rhone Gletscher) war ich ziemlich weit hinauf gegangen, und hatte meine Aufmerksamkeit besonders auf diesen Gegenstand gerichtet, und ebenso bei anderen Gletschern die ich auf meiner Reise nach einander erreicht hatte. Ich hatte die Sache freilich nicht so gründlich verfolgt, um in einzelnen Fällen die Streifen mit Sicherheit aus dem vorhandenen Drucke zu erklären, aber die Richtigkeit der Erklärung im Allgemeinen war mir durch den Augenschein noch wahrscheinlicher geworden, besonders am Rhonegletscher, wo ich die Bänder am deutlichsten gesehen habe, und wo mir auch ihre Lage mit der Richtung des Druckes, welchen der Gletscher im Thale erleidet, wo er seine Bewegungsrichtung aendern muss, übereinzustimmen schien. Sie können sich daher denken, wie sehr ich erstaunte als ich erfuhr, dass Sie /in der Schweiz und zum Theile an denselben Gletschern/ dieselben Untersuchungen gemacht haben. Ihre Untersuchungen sind natürlich viel gründlicher gewesen als die meinigen, da ich gar nicht die Absicht hatte eine wissenschaftliche Arbeit über den Gegenstand zu machen, und ich sehe den Resultaten Ihrer Untersuchungen, welche Sie hoffentlich veröffentlichen werden, mit grossem Interesse entgegen.

Mit freundlichem Grusse | der Ihrige | R. Clausius.

Zürich, 16th Nov., 1856.

Esteemed Herr Professor,

First of all, accept my sincere thanks for the fine translations of both of my papers,[1] which you included in your Phil. Mag.[2] Admittedly I do not know whether you went to the trouble of preparing these translations yourself, but at any rate I believe I am not mistaken in assuming that that you came up with idea for this. Nowhere does the mechanical theory of heat find such general interest as in England, and it must therefore be of particular interest for me that my works are becoming well-known in England.

Herr Joule's note[3] about my paper astonished me somewhat at first, as I did not remotely intend ascribing an error to him. However, when I re-read the para. of my paper in question, I found that a passage does indeed occur in it which he was possibly able to interpret as referring to himself. I have therefore considered it my duty to write a few words in explanation,[4] and I would like to ask you to include these in your Phil. Mag. as well. What Herr Joule says about Prof. Thomson[5] is indeed correct, but it by no means contradicts what I have said. I would easily be able to prove this, but as I would not willingly like to bother you and the audience with a matter that can have only a personal interest, I have thus entirely disregarded this part of the note.

I have had the pleasure this autumn, through a curious coincidence, of making the acquaintance of one of your friends, Herr Huxley, with whom you had been on the Eggishorn[6] the day before. The only regret I had about this was that I did not meet him a day earlier, when he was still together with you, which would have been an extraordinarily pleasant surprise for me. The following day, I lost him again just as unexpectedly as I had found him. When we were on the way to Zermatt[7] with a few other travelling companions we had chanced upon, we encountered a party that he knew. He stopped and conversed with them, and, in order not to disturb him, the rest of us continued somewhat more slowly on our way. When we lost sight of him, we stopped and waited for about half an hour for him. His friends probably talked him into turning back with them. When you see him, be so good as to give him kind regards from me, and to tell him that I was very sorry that he was not with us in Zermatt. The weather was magnificent, and the view from the Gorner Ridge[8] is the most wonderful that I have seen.

There was one other peculiar coincidence to do with this matter. I had been reading your fine paper about the cleavage of slate,[9] which you had been so good as to send me, with the greatest interest, and it had occurred to me while I was doing so whether the blue veins of glaciers, which I had still not yet seen but about which Prof. Studer from Berne[10] and Escher von der Linth in Zürich[11] had spoken to me many times, could not be explained in the same way. When

I therefore came to a glacier for the first time on my trip (the Rhône glacier[12]), I went quite far up it and directed my attention especially to this subject, and likewise with other glaciers that I came to in turn on my trip. Admittedly, I did not pursue the matter so thoroughly[13] as to explain the stripes with certainty in individual cases from the existing pressure, but the accuracy of the explanation in general became even more probable to me from what I had seen, especially from the Rhône Glacier, where I saw the veins most clearly, and where, to me, their position also seemed to concur with the direction of the pressure which the glacier experiences in the valley, where it has to to alter the direction of its movement. You can therefore imagine how astonished I was when I learned that you had done the same investigations *[*in Switzerland and, in part, on the same glaciers*]*. Your investigations have, of course, been much more thorough than mine, as I did not have any intention at all of doing a scientific paper on the subject, and I am looking forward to the results of your investigations, which you will hopefully publish, with great interest.

With kind regards | Yours | R. Clausius.

RI MS JT/1/TYP/7/2223, 2225
LT Typescript Only

---

1. *fine translations of both of my papers*: R. Clausius, 'On a modified Form of the second Fundamental Theorem in the Mechanical Theory of Heat', *Phil. Mag.*, 12 (1856), pp. 81–98; and R. Clausius, 'On the Application of the Mechanical Theory of Heat to the Steam-engine', *Phil. Mag.*, 12 (1856), pp. 241–65, 338–54, 426–43.
2. *your Phil. Mag.*: Tyndall served as an editor for the *Phil. Mag.* from 1854–63.
3. *Herr Joule's note*: James Prescott Joule (1818–89), an English physicist and brewer who studied the relationship between work and heat (*ODNB*). See J. P. Joule, 'Note on Prof. Clausius's Application of the Mechanical Theory of Heat to the Steam-engine', *Phil. Mag.*, 12 (1856), pp. 385–86.
4. *a few words in explanation*: R. Clausius, 'Reply to a Note of Mr. Joule, contained in the November issue of the Philosophical Magazine', *Phil. Mag.*, 12 (1856), p. 463.
5. *Prof. Thomson*: William Thomson.
6. *Eggishorn*: a mountain of the Bernese Alps in the Swiss canton of Valais.
7. *Zermatt*: an alpine town in Switzerland surrounded by many of the range's high peaks, including Monte Rosa, the Weisshorn, and the Matterhorn.
8. *Gorner Ridge*: the Gornergrat is a rocky ridge, located southeast of Zermatt, Switzerland, and which overlooks the Gorner glacier.
9. *your fine paper about the cleavage of slate*: J. Tyndall, 'Comparative View of the Cleavage of Crystals and Slate Rocks', *Phil. Mag.*, 12 (1856), pp. 35–48.
10. *Berne*: Bern, the capital city of Switzerland, in the canton of Bern.

11. *Zürich*: a city in north central Switzerland.
12. *the Rhône glacier*: a glacier on the north side of the Alps, situated in Germany, Switzerland, and Austria, that forms the river Rhône.
13. *I did not pursue the matter so thoroughly*: Clausius did write two papers on the colour of the sky, see R. Clausius, 'Ueber die Natur derjenigen Bestandtheile der Erdatmosphäre, durch welche die Lichtreflexion in derselben bewirkt wird', *Poggend. Annal.*, 76 (1849), pp. 161–88; and 'Uber die blaue Farbe des Himmels und die Morgen- und Abendröthe', *Poggend. Annal.*, 76 (1849), pp. 188–95.

From Thomas Archer Hirst   16 November 1856   1298

Pau. | Nov. 16th 1856

My dear John,

It is quite time I acknowledged the receipt of your two last letters,[1] one of which was the testimonial which has already been of service to me and the other was another page or two of the description of your journey which interested both myself, my wife[2] and John Martin greatly. When you have leisure another page or two from the journal will be very acceptable. For the last fortnight or more Anna's <u>average</u> state of health has been much better than it was before that time. The weather has changed, we have had slight frosts at night but bright warm sunny days (generally). Of her worst symptoms, the night perspirations have almost disappeared and the cough and expectoration[3] have both decidedly diminished. On the whole therefore I think I may say that Pau[4] has been of some service to her already and I live in the hope that it will continue to be so and that at the end of the winter she may be better than she was at the beginning. As yet I have but one pupil[5] to whom I give three lessons a week, of an hour each, and for which I get 5 francs a lesson. He is the son of Lord Francis Gordon[6] who is an invalid here. The lad is about 16 years old and has received a bad education at the Charter House.[7] They intend him to enter the army or some of the public offices, and to fit him for passing the necessary examination in mathematics is my task. From several conversations I have had with Lord and Lady Gordon[8] I gather that they by no means approve of the late changes that have been made in the appointments for public offices.[9] One day her Ladyship told me that amongst the candidates for some public office of £60 a year she noticed a Wrangler of Cambridge,[10] and then in the tone of an injured parent she exclaimed 'what chance would my boy have against a wrangler I should like to know?' Indeed, the boy would have but a poor chance amongst far inferior competitors than wranglers. He is not naturally stupid, but he has been allowed to acquire loose, thoughtless habits. His memory and his fingers have been educated but his reason has been sadly neglected. He can convert a circulating decimal into a vulgar fraction with the greatest ease and rapidity, but when I threw 12 franc pieces on the table and asked him to return me five sixths of them he was completely puzzled. He has learnt 3 books of Euclid[11] by heart, but I am convinced he <u>never understood</u> a proposition. I confess I was somewhat surprised to find that the son of a Lord has been educated thus, especially since the boy himself is naturally sharp enough. I have told both Lord and Lady Gordon that it <u>will be necessary</u> for me to go over much that he has already done, in short to unteach him a good deal before we can advance to anything new. They seemed to approve of my plain-speaking, as well as of my intentions but I am not sure that by and bye they will think his progress worth the money, for I can easily perceive that in their sight the value of education is chiefly a marketable one. However we shall see. As to myself I am taking advantage of John Martin's assistance in order to improve my latin. At Berlin[12] I was getting on pretty well, but my going to Queenwood[13] interrupted me, and for three years I scarcely read a sentence. Now I have taken it up again more thoroughly than before, and have not only recovered what I had forgotten in the three years, but have advanced much further. At present I devote a great deal of time to this as I wish this time to get tolerably well grounded in it, and that before John Martin leaves us again, which he will probably do in January. For translating from Latin into English I choose mathematical books, of course, for it is with mathematical latin that I which to become familiar. I have moreover chosen some works of Jacobi on Determinants and Elliptical functions[14] inasmuch as these bear upon the research to which I have long ago decided to devote all my spare time and thought. The most unsatisfactory thing about our residence in France is that I am making little or no progress in speaking French. Except our servant and occasionally a shopman, we have no one with whom to speak French. This is very annoying to me and I have not yet devised any way of improving it. But now I have said enough of myself and of my occupations. I hear you were at Queenwood since you last wrote to me.[15] Tell me what you saw there. I noticed with pleasure Sir John Herschel letter to you in the Phil. Mag.[16] It is gratifying to me to see that he has appreciated and acknowledged the real and intrinsic merits of your late little research on cleavage,[17] viz its philosophic ingenuity, simplicity and elegance. Deal with me better than I deserve and write me soon a long letter. Tell me have you heard from Pridie? Remember me kindly to Francis[18] and Bevington[19] and receive Anna's love with mine.

Yours affectionately | <u>T. A. Hirst</u>

P.S. Postage is so cheap that I hope if you meet with anything likely to interest me in newspapers or scientific journals you will send me it. Here it would be a great boon.

RI MS JT/1/H/229
RI MS JT/1/HTYP/480–81

---

1. *your last two letters*: see letters 1283 and 1284, John Tyndall to Thomas Hirst, both dated 9 October 1856, *Tyndall Correspondence*, vol. 5. The 'testimonial' is letter 1284.

2. *my wife*: Anna Hirst.

3. *expectoration*: the action of coughing up and spitting out mucus or other material from the lungs and respiratory passages (*OED*).

4. *Pau*: in the Pyrenees, southwestern France, which became a popular holiday resort for the British in the second half on the nineteenth century. Thomas and Anna Hirst were visiting Pau for Anna's health.

5. *one pupil*: probably Francis Frederick Gordon (1839–1925), son of Francis Gordon and Isabel Grant. See *Debrett's Peerage, Baronetage, Knightage, and Companionage, 1903* (London: Dean & Son, 1903), p. 448.

6. *Lord Francis Gordon*: Francis Arthur Gordon (1808–57), son of the ninth Marquis of Huntly, a notable amateur cricketer and lieutenant colonel in the 1st Life Guards. Gordon died in Paris while Hirst was tutoring his son. See *Debrett's Peerage*, p. 448.

7. *Charter House*: a boarding school in London founded in 1611 and moved to Surrey in 1872.

8. *Lady Gordon*: Isabel Gordon (née Grant, d. 1892), wife of Francis Arthur Gordon (married in 1835), and daughter of General Sir William Keir Grant (1771–1852). See *Debrett's Peerage*, p. 448.

9. *late changes . . . public offices*: possibly a reference to the establishment of the Civil Service Commission in 1855 in a bid to end patronage and base appointments on open examinations. The commission was influenced by the publication of the Northcote-Trevelyan Report (1854).

10. *Wrangler of Cambridge*: a student who gained first-class honours in their undergraduate degree in mathematics at the University of Cambridge was called a 'Wrangler'.

11. *Euclid*: Euclid of Alexandria (fl. *c.* 295 BC), a Greek mathematician, especially of geometry. One of the three books Hirst's pupil would have read was surely Euclid's *Elements*, a classic mathematical text.

12. *At Berlin*: referring to Hirst's time studying mathematics in Berlin in 1852–53, after he finished his doctorate at the University of Marburg in July 1852.

13. *my going to Queenwood*: referring to Hirst's appointment to a teaching post at Queenwood College, a boy's school in Hampshire, England, in mid-1853 (for mathematics for surveying, until 1856). Tyndall kept an interest in Queenwood College, where he had also taught mathematics for surveying, from 1847–48 and again from 1851–53. He spent the time in between his two periods of employment at Queenwood studying in Marburg, Germany. For letters during Tyndall's time at Queenwood, see *Tyndall Correspondence* vol. 2; R. Barton, J. Rankin, and M. Reidy (eds), *The Correspondence of John Tyndall, Volume 3:*

The Correspondence, January 1850–December 1852 (Pittsburgh: University of Pittsburgh Press, 2017); and *Tyndall Correspondence*, vol. 4.

14. *works of Jacobi on Determinants and Elliptical functions*: Carolo Gustavo Jacob Jacobi, *Fundamenta Nova Theoriae Functionum Ellipticarum* (Regiomonti: Sumptibus Fratrum Borntraeger, 1829).

15. *last wrote to me*: see n. 1.

16. *Sir John Herschel letter to you in the Phil. Mag.*: J. F. W. Herschel, 'Remarks on Slaty Cleavage, and the Contortions of Rocks', *Phil. Mag.*, 12 (1856), pp. 197–98.

17. *late little research on cleavage*: see letter 1295, n. 3.

18. *Francis*: William Francis.

19. *Bevington*: James Buckingham Bevington (1804–92), a wealthy and philanthropic leather merchant, Quaker, and father of one of Tyndall's Queenwood pupils. Tyndall and Bevington were good friends in the 1850s and remained in contact throughout their lives. See *Tyndall Correspondence*, vol. 4, and 'Obituary', *Saddlery and Harness*, 2 (1892), p. 58.

To Auguste de la Rive   17 November 1856   1299

17th, Nov, 1856.

My dear Sir

I am much indebted to you for your friendly letter[1] and shall find the information which it contains of considerable use to me. It is a matter of surprise that a phenomenon so interesting, and so beautiful as the colour of the Rhone[2] should not have received a greater share of attention, and been made the subject of more exact and extensive experiment than has been the case hitherto. Surely the subject comes within the range of experimental examination, and it certainly is one likely to repay the man who takes it up. With regard to my going to Geneva[3] it is within the range of possibility that I may go there next year or the year following, but my attention would be more directed to the glaciers than to the rivers, as what I have seen of these great ice masses has merely augmented their interest, and made me desirous of seeing more of them.

With regard to the '<u>nouvelles</u>'[4] I am in the first place happy to say that Faraday's health is excellent, and he has been working through the autumn with great success. His subject has been the transparency of metallic films, and the modifications which light undergoes in its passage through them. He has either sent, or intends to send, a long paper[5] to the Royal Society upon the subject, which I think will be read with interest by men of science generally

Grove has just returned from Honfleur[6] where he has been staying with his family; his health is good; I do not know what he is now engaged upon—he described to us at our club[7] meeting a few days ago the appearance of Jupiter[8] as he observed it during the recent occultation.

Gassiot is quite well and the same straightforward true-hearted man as usual. Sabine has had a severe cold, but is I believe now better. He caught it in Burlington House,[9] which is now being prepared for the Royal Society. You know the house I suppose; it is near the Burlington Arcade[10] in Piccadilly,[11] and the Royal Society is going to move from Somerset House[12] and establish itself in new quarters: there will be more room, and the position will be more convenient for the majority of the Fellows.

Wheatstone is also well; I spent the whole of last Saturday with him at Hammersmith.[13] He has had a quarrel[14] with Sir David Brewster lately, and a long correspondence on the subject of the stereoscope has been published in the Times.[15] It is a great pity that those personal feelings should be allowed to mix so largely with scientific questions. A man occupying the high position of Sir David Brewster, ought to shew an example of forbearance and magnanimity to younger workers in science. How simply grand Faraday stands out in comparison with these disputants—he has done more than they all, and if he died tomorrow he would not leave a foe behind him. His example is valuable in many respects, but not the least by shewing the possibility of forty years successful work in science, and of living at the same time in peace with his fellow-labourers.

With regard to your book[16] I will write to you in all frankness. I have a great deal of work to do, and therefore could not promise to read your proofs <u>speedily</u>. But if Mr Walker sends them to me <u>gradually</u>, I shall be glad to devote to them every hour I can spare. I do not expect to go through them without making corrections, for there are many points in the style of the two first volumes which I should deem capable of improvement, and which I should hardly pass over in reading the proofs of the third.

I had a very pleasant letter some time ago from M. Matteucci.[17] He and I, as you are aware, have differed upon some scientific points, but this is not likely to make us the worse friends. The greatest blessing of my connexion with Faraday is that he allows me to think for myself: he knows that if I differ from him I do so conscientiously, and not as a partisan, and I feel perfectly sure that there are scientific men in Europe from whom if I dissented so widely and so publicly as I have done from Faraday I should make them my foes for life. But Faraday knows that my reverence for him is far more high and independent, than if I compromised my opinion to make it agree with his. I found my visit to Vienna[18] very pleasant—I met M. Plücker there, but he was not friendly. He has written both to Mr Faraday and to Wheatstone in a manner which I think was hardly justifiable. But I was prepared to talk over our differences in Vienna if he wished to do so. His demeanour however was cold, and seeing this I became indifferent. Time will settle it all. I beg of you to excuse this egotistical letter, but want of other matter compelled me to write about myself.

Believe me dear Sir | Ever Yours Sincerely | John Tyndall

RI MS JT/1/TYP/1/343–44
LT Typescript Only

1. *your friendly letter*: letter 1293.
2. *Rhone*: see letter 1293, n. 6.
3. *Geneva*: see letter 1293, n. 3.
4. *nouvelles*: news (French).
5. *a long paper*: M. Faraday, 'Experimental Relations of Gold (and other Metals) to Light', *Phil. Trans.*, 147 (1857), pp. 145–81.
6. *Honfleur*: a coastal town in northwestern France.
7. *our club*: probably the Philosophical Club of the RS, founded in 1847. Tyndall was elected a member on 10 May 1855 (Journal, 12 May 1855, RI MS JT/2/13c/745).
8. *appearance of Jupiter*: Grove's notes about his observations of the moon passing in front of Jupiter were published as W. R. Grove, 'Note on the Occultation of Jupiter, November 8, 1856', *Monthly Notices of the Royal Astronomical Society*, 17 (1856), p. 3.
9. *Burlington House*: a private mansion purchased by the British government and made home to the Royal Society, Linnean Society, and Chemical Society in 1857. Three other learned societies and the Royal Academy of Arts joined them later.
10. *Burlington Arcade*: a covered shopping arcade on Piccadilly.
11. *Piccadilly*: a main road in London on which Burlington House is located.
12. *Somerset House*: in the Strand, London, the RS's headquarters from 1780 until its move to Burlington House in 1857.
13. *Hammersmith*: Charles Wheatstone's house in Lower Mall, Hammersmith (a district in west London). See B. Bowers, *Sir Charles Wheatstone FRS 1802–1875, IEE History of Technology Series 29* (London: Institution of Electrical Engineers, 2001), p. 188.
14. *a quarrel*: Brewster claimed that Wheatstone did not invent the stereoscope in 1838, but rather James Elliot did in 1834. See Bowers, p. 51.
15. *published in the Times*: Brewster and Wheatstone exchanged letters arguing their perspectives in the *Times,* the dates as follows: 17, 20, 25, and 31 October, and 5 and 15 November 1856. See Bowers, pp. 50–54.
16. *your book*: see letter 1293, n. 16.
17. *I had a very pleasant letter some time ago from M. Matteucci*: see letter 1286, Carlo Matteucci to John Tyndall, 10 October 1856, *Tyndall Correspondence*, vol. 5.
18. *my visit to Vienna*: Tyndall attended the thirty-second meeting of the Gesellschaft Deutscher Naturforscher und Ärzte held in Vienna from 16–22 September 1856. On 19 September he presented a paper, 'Über die Spalten im Gletschereise', to the physical section. See *Tageblatt der 32 Versammlung deutscher Naturforscher und Aerzte in Wien im Jahre 1856* (1856), p. 78.

From Carlo Matteucci   [17 November 1856][1]   1300

Mon cher Monsieur,

Le but de cette lettre serait de vous prier à vouloir bien me donner l'idée principale d'un travail qui parait être le sujet de la <u>Bakerian Lecture</u> cette année de Thomson. Je n'en ai qu'une idée imparfaite ou aucune, le titre seulement, par des journaux non scientifiques et j'en suis très curieux.

J'espère qu'à cette heure-ci vous aurez déjà reçu le petit appareil et les tiges de bismuth cristallisé du Général Sabine.

Ayez la bonté de dire à M. Sabine que j'ai reçu régulièrement le <u>Phil Mag</u> et que j'espère qu'il aura aussi reçu régulièrement le <u>Nuovo Cimento.</u> J'ai donné moi même dans le dernier n° l'extrait de votre Lecture sur les clivages. Si jamais vous avez la bonté de me répondre, ayez la complaisance de m'informer si mon mémoire manuscrit envoyé pour le Phil. Transactions a été imprimé ou quand il le sera. Présentez mes respects à M. Faraday dont j'espère que la santé est bonne et veuillez lui dire qu'il a oublié depuis longtemps ses amis de Pise.

Croyez moi tout dévoué | C Matteucci

My dear Sir,

The aim of this letter would be to ask you to give me the key idea of a research that seems to be the subject of the <u>Bakerian Lecture</u> this year by Thomson.[2] I only have an imperfect idea of it, or none, except for the title, through non-scientific journals and I am very curious about it.

I hope that by now you have already received the small device and the rods of crystallised bismuth from General Sabine.

Have the kindness to tell M[r] Sabine that I have regularly received the <u>Phil Mag</u> and that I hope he would also have regularly received the <u>Nuovo Cimento</u>.[3] I personally gave the summary of your lectures on cleavage in the last issue.[4] If you ever have the kindness to respond, be so good as to let me know if my manuscript essay sent for the Phil. Transactions has been printed or when it will be. Give my respects to Faraday, who I hope is in good health, and please, tell him that he has long-forgotten his friends from Pisa.[5]

Yours faithfully | C Matteucci

RI MS JT/1/M/63

1. *[17 November 1856]*: the reverse side of this letter has on it a postmark for '17 Nov. 56'.
2. *Bakerian Lecture this year by Thomson*: W. Thomson, 'The Bakerian Lecture.—On the Electro-dynamic Qualities of Metals', *Phil. Trans.*, 146 (1856), pp. 649–751. Read before the RS on 28 February 1856.
3. <u>*Nuovo Cimento*</u>: *Il Nuovo Cimento*, a physics journal established in 1855 by Matteucci and Raffaele Piria (1814–65), an Italian chemist, as a continuation of the journal *Il Cimento* (est. 1844).
4. *your lectures on cleavage in the last issue*: see letter 1295, n. 3. Delivered at the RI on 6 June 1856.
5. *Pisa*: a coastal city in central Italy to the west of Florence. Matteucci was professor of physics at the University of Pisa from 1840.

From [Heinrich Gustav Magnus]¹   19 November 1856   1301

Berlin 19 Nov 56.

Mein theurer Tyndall

Ihren lieben Brief vom 14ᵗⁿ d. mit der Einlage von Prf. G. erhielt ich gestern. Sie kennen unsere Einrichtungen hier um zu wissen daß Mʳ. R. Scott sowohl Rose's als meiner Vorlesung ohne Schwierigkeit beiwohnen kann. Wir haben zwar beide schon seit etwa 4 Wochen begonnen, allein wenn der junge Mann bereits einige chemische und physicalische Kenntniße besitzt, so wird es ihm nicht schwer sein, die Vorträge auch jetzt noch benutzen zu können.—Sie schreiben daß Mʳ. Scott gern eine physicalische Untersuchung hier machen möchte, und deßhalb vermuthe ich daß Ihre Anfrage zum Zweck hat zu erfahren, ob Mʳ. S. bei mir eine solche würde ausführen können. Sie erinnern sich daß ich Ihnen öfter gesagt habe, daß ich gern jedem talentvollen jungen Mann mein Privatlaboratorium öffne; und da Mʳ. S. als ein talentvoller junger Mann bezeichnet wird, so werde ich auch ihn sehr gern aufnehmen; allein es scheint mir nothwendig einige Bemerkungen vorauszuschicken, die von der Art sind, daß sie vielleicht eine Aendrung in dem Entschluße des Mʳ. S. herbeiführen könnten. Sie wissen daß ich hier vielfach beschäftigt bin. Ich kann daher dem, der bei mir arbeitet, nicht viel Zeit widmen. Ist derselbe so weit in der Physik zu Hause, daß er nur eines Rathes bedarf, so wird er vielleicht mit Nutzen bei mir arbeiten. Kann er sich aber, nachdem ihm dieser Rath ertheilt ist, nicht selbst helfen, dann verliert er seine Zeit. Ich denke daß ich mit Ihnen so kurz reden kann, da Sie die Verhältnisse meines Laboratoriums genau kennen. Bitte rathen Sie hiernach dem Mʳ. Scott. Es würde mir sehr leid sein wenn er hierher käme und nicht fände, was er erwartete, und doch fürchte ich daß dies der Fall sein wird. Ich muß mir für meine eignen Arbeiten die Zeit stehlen, ich kann daher auf Anderer Untersuchungen nicht viel Zeit verwenden.

Für mich brauchen Sie M[r]. S., falls er kommt, keine Empfehlung zu geben; statt dessen rathe ich ihn bei Rose's und Poggend. zu introduciren, vielleicht auch bei Rieß und Dove. Aber thun Sie mir den Gefallen ihm das mitzutheilen was mich bedenklich macht. In einem chemischen Laborat[.] kann man jemand leicht mit Analysen beschäftigen, aber physicalische Arbeiten die jemand machen könnte ohne viele Vorbereitungen ohne vielleicht ein Jahr lang sich zu quälen bis er eine sichere Methode gefunden, giebt es wenig oder gar nicht.

Ob ich glaube daß Wasser in den Spheroidalen Zustand übergehn kann indem man das Gefäß erhitzt in dem es sich befindet?—Denken Sie sich ein flaches Gefäß von Metall

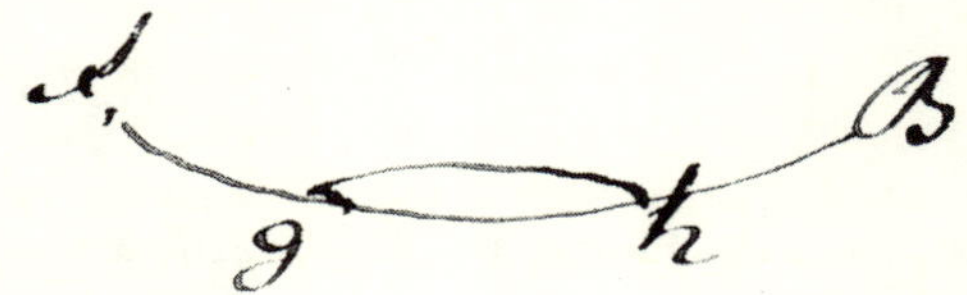

AB und in der Mitte, bei gh Wasser. Wenn man das Stück AgBh erwärmt, und gh kalt erhält, wird man nicht im Stande sein es dahin zu bringen daß das Wasser bei gh nicht mehr das Metall berührt? Ob die ganze Fläche gh nicht mehr berührt, wird von verschiedenen Umständen abhängen, aber der Rand wird nicht mehr berühren. So verhält es sich bei den Dampfkesseln. Das Wasser sinkt unter den normalen Stand, der vom Feuer bespühlte Theil des Kessels wird glühend, und wenn kein Abzug für den Dampf vorhanden, so kocht das Wasser nicht. Dies geschieht erst wenn die Maschine in Gang gesetzt wird; dann

Berlin 19 Nov. 56.

My dear Tyndall

Yesterday I received your kind letter[2] of the 14th of this month with the enclosure from Prof. G.[3] You are familiar enough with our facilities here to know that Mr. R. Scott[4] will be able to attend both Rose's[5] and my lecture without difficulty. We both did start some 4 weeks ago, but if the young man already has some knowledge in chemistry and physics, then it will not be difficult for him to still be able to take advantage of the lectures even now.—You write that Mr. Scott would like to do an investigation in physics here, and I therefore assume that your inquiry has the aim of finding out whether Mr. S. would be able to carry out such a one under my supervision. You will remember that I have frequently told you that I shall gladly open my private laboratory[6] to any talented young man; and as

Mr. S. is described as a talented young man, then I shall gladly accept him too; however, it seems necessary to me to make some remarks in advance, that are of a kind that they could perhaps bring about a change in the decision of Mr. S. You know that I am busy on many different fronts here. I therefore cannot devote much time to anyone who is working under me. If he is sufficiently at home in physics that he only requires a bit of advice, then he will perhaps benefit from working under me. If, however, he cannot help himself after this advice has been given to him, then he will be wasting his time. I think that I can speak to you so directly, as you know the conditions of my laboratory precisely. Please advise Mr. Scott accordingly. I should be very sorry if he came here and did not find what he expected, and yet I fear that this will be the case. I have to steal time for my own work, I therefore cannot spend much time on the investigations of others.

You do not need to give Mr. S. a recommendation for me in the event he does come; instead, I suggest that you introduce him to the Roses[7] and Poggend.[8], perhaps also to Riess[9] and Dove. But do me the favour of informing him about what makes me apprehensive. In a chemistry laboratory, one can easily keep someone busy with analyses, but there are few tasks in physics, or none at all, that someone could do without much preparation, without toiling for perhaps <u>one year</u> until he has found a sure method.

Do I believe that water can make the transition to the spheroidal state by having someone heat the vessel in which it is being held?—Imagine a flat vessel made from metal

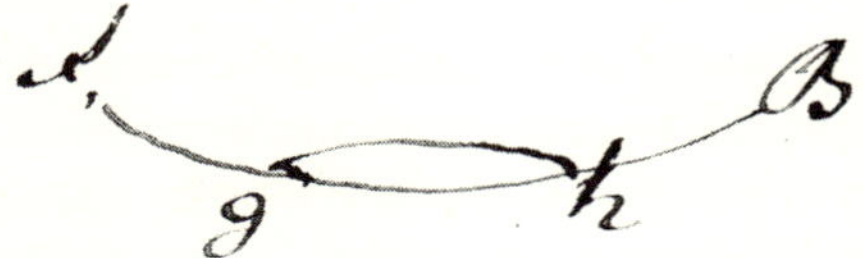

AB, and water in the middle, at <u>gh</u>. If one heats the piece AgBh and keeps gh cool, will one not be in a position to bring it to pass that the water at gh no longer touches the metal? Whether the whole area gh is no longer touched will depend on different circumstances, but the edge will no longer touch. This is what happens with steam boilers. The water falls below its normal level, the part of the boiler lapped by the fire will become red-hot, and if no outlet is present for the steam, then the water will not boil. This only happens when the machine is started; then . . . [10]

RI MS JT/1/UN/4

---

1. *[Heinrich Gustav Magnus]*: the identity of the correspondent is based on the mention of Robert Scott in letter 1309 (see n. 4), from Magnus to Tyndall. Also, the use of 'theurer' in the opening salutation is typical of other Magnus letters around this time, as well as the handwriting.

2. *your kind letter*: letter missing.

3. *Prof. G.*: not identified (perhaps William Grove).

4. *Mr. R. Scott*: Robert Henry Scott (1833–1916), a meteorologist born in Dublin, and edu-
    cated at Rugby School, and Trinity College, Dublin, before departing for Berlin in 1856
    (*ODNB*). While *ODNB* states that Scott worked under Heinrich Dove in Berlin, letter
    1309 also from Magnus states that Scott 'will therefore work under Dr Schneider'.

5. *Rose's*: Heinrich Rose.

6. *my private laboratory*: Magnus's laboratory, located at his home, was largely funded by his
    inherited wealth. See C. Jumgnickel and R. McCormmach, *The Second Physicist: On the
    History of Theoretical Physics in Germany* (Dordrecht: Springer, 2017), p. 134.

7. *the Roses*: Heinrich Rose and his brother Gustav Rose (1798–1873), a German mineralo-
    gist who in 1856 became director of the Mineralogy Museum in Berlin (*CDSB*).

8. *Poggend.*: Johann Poggendorff.

9. *Riess*: Peter Theophil Riess (1805–83), a German physicist who studied electricity, and
    in 1842 became the first Jewish member of the Akademie der Wissenschaften in Berlin
    (*ADB*).

10. *then . . .* : fragment only, letter is incomplete.

To Rudolf Clausius   27 November 1856   1302

Nov. <u>27</u>[th] 1856

My dear Clausius.

My time is so much occupied at present with original work and with my
lectures that I hardly ever find myself able to devote any time to translations,
but I often <u>recommend</u> articles to be translated for the Philosophical Magazine, and in this way I recommended yours.[1] I am happy to say they excite
a great deal of interest, and are considered very important in this country—
they were translated by Hirst and I think on the whole are well done.

It gave me great pleasure to receive your last letter,[2] and I sat down immediately and translated the note for the Philosophical Magazine. I was greatly
amused with your account of meeting and <u>losing</u> M[r] Huxley: he told me he
had met a friend of mine but could not recollect the name, so until your letter reached me I was ignorant that it was you. The party he met was that of
my friend Dr Hooker, the botanist and Himalayan traveller, Huxley and I
had left them behind us at Interlaken[3], and I met them again by accident[4]
in the Hofkirche at Innsbrück,[5] and accompanied them to Vienna,[6] where
we remained during the meeting of the Naturforscher.[7] It would have given
me great pleasure to have met you, but next to this it is a pleasure to me to
know that you are well and able to enjoy the pure air of the mountains. I
found my journey very instructive and it is gratifying to me to learn that the

same thought struck you regarding the structure of the ice as struck Huxley & myself. There is in my mind no doubt whatever that the structure is produced by the united action of pressure and of tension.

I am at the present moment engaged upon this subject, and hope very soon to send a paper[8] to the Royal Society upon it. I have felt far greater pleasure in following it up than I anticipated when I commenced it. I can hardly hope that Prof. Forbes, who owes his chief reputation in England to his writings upon glaciers, will be satisfied with what I have to say upon the subject, for I am compelled to differ essentially from almost all his views. But this cannot be helped—out of the difference of opinion truth will be sifted in the end.

I have recently read a paper[9] in the Bibliotheque Universelle for the year 1843. In Vol. 45. page 131 M. Desor refers to a map constructed by M. Wild[10] for M. Agassiz—Can you tell me whether this map has ever been published?[11]

I have taken the liberty of addressing you without your title,[12] because I think such formalities are not necessary between such old friends as you and I. I have given your message to Huxley and he in return sends you many greetings, and desires me to express his regret for having left you so unceremoniously. He was very sorry not to have had the pleasure at least of saying goodbye.

Believe me dear Clausius | most truly yours | John Tyndall

RI MS JT/1/T/171

---

1. *I recommended yours*: R. Clausius, 'On a modified Form of the second Fundamental Theorem in the Mechanical Theory of Heat', *Phil. Mag.*, 12 (1856), pp. 81–98; and R. Clausius, 'On the Application of the Mechanical Theory of Heat to the Steam-engine', *Phil. Mag.*, 12 (1856), pp. 241–65, 338–54, 426–43. See letter 1297.

2. *your last letter*: letter 1297.

3. *Interlaken*: a town in Switzerland that serves as a gateway to the mountains and lakes of the Swiss Alps.

4. *met them again by accident*: see Tyndall's journal entry of 11 September 1856 (Journal, RI MS JT/2/13c/907).

5. *Hofkirche at Innsbrück*: a gothic church in Innsbrück, the capital city of the Austrian state of Tyrol.

6. *Vienna*: Austrian capital.

7. *meeting of the Naturforscher*: Tyndall attended the thirty-second meeting of the Gesellschaft Deutscher Naturforscher und Ärzte held in Vienna from 16–22 September 1856. On 19 September he presented a paper, 'Über die Spalten im Gletschereise', to the physical section. See *Tageblatt der 32 Versammlung deutscher Naturforscher und Aerzte in Wien im Jahre 1856* (1856), p. 78.

8. *send a paper*: J. Tyndall and T. H. Huxley, 'On the Structure and Motion of Glaciers', *Phil. Trans.*, 147 (1857), pp. 327–46.

9. *recently read a paper*: published separately as E. Desor, *Compte rendu des recherches de M. Agassiz pendant ses deux derniers séjours a l'Hotel des Neuchatelois, sur la glacier inférieur de l'Aar* (Genève: Bibliotheque universelle, Mars 1843), p. 70.

10. *M. Wild*: Johannes Wild (1814–94), a Swiss engineer, cartographer, and professor of topography (*HLS*).

11. *ever been published?*: Wild's map for Agassiz, 'Carte du glacier inférieur de l'Aar', was published in Louis Agassiz, *Nouvelles études et expériences sur les glaciers actuels* (Paris: Victor Masson, 1847).

12. *your title*: Professor.

To Thomas Henry Huxley [28 November 1856][1] 1303

Royal Institution. | Friday night

My dear Huxley,

I want very much to see Agassiz's 'Etudes sur les Glaciers'[2] do you know where I can get a look at it? We have not got it.

Do you know whether Agassiz wrote anything upon the Aar[3] after 1843? A map[4] was constructed for him by an engineer named Wild,[5] but whether it was ever published I know not; do you know anything about it?

Could you lend me those sketches of the crevasses that you took underneath the '<u>Pavillion</u>'[6] I should like to set one of my lady hearers to copy them upon a grand scale.

I am working very hard. And this rascally brain of mine often murmurs and simmers and boils, and almost swears it will work no longer.

Kind regards—I had almost impudently said 'love' to Mrs Huxley.[7] But even if I did it would be a mere Irishism[8] and valued accordingly.

Ever Yours | John Tyndall.

I ought to be at Hooker's to night. To night also to be at Moore's,[9] but I have forsworn society till I get some of this horrible weight off my brain.

RI MS JT/1/TYP/9/2885
IC HP 8:31
Typed Transcript Only

———

1. *[28 November 1856]*: the date is given based on Tyndall's letter to Rudolf Clausius of 27 November 1856 (a Thursday), in which he also requested information about Agassiz's map (letter 1302), and Clausius's reply of 2 December (letter 1305), suggesting that this letter (written on a 'Friday night') falls in between.

2. *Agassiz's 'Etudes sur les Glaciers'*: L. Agassiz, *Nouvelles études et expériences sur les glaciers actuels* (Paris: Victor Masson, 1847).

3. *Aar*: see letter 1305, n. 2.

4. *A map*: see letter 1302, n. 11.

5. *an engineer named Wild*: Johannes Wild, see letter 1302, n. 10.

6. *the 'Pavillion'*: in 1844 Daniel Dollfuss-Ausset (see letter 1312, n. 6) constructed a hut on the Aar glacier known as Pavillon Dollfuss (*HLS*). On 22 August 1856 Tyndall and Huxley spent the night at the hut; Tyndall noted in his journal that he 'listened all night to Huxley snoring. Had no sleep' (Journal, RI MS JT/2/13c/882–84; *Glaciers of the Alps*, pp. 18–20, 112).

7. *Mrs Huxley*: Henrietta Huxley.

8. *Irishism*: the fact or quality of being Irish (*OED*).

9. *Moore's*: James and Harriet Moore and their family Harriet Jane, Julia, John, and Graham.

From Thomas Henry Huxley   [late November 1856][1]   1304

My dear Tyndall,

This is the only drawing[2] I can find likely to be of much use—

Agassiz[3] contains much important matter—I have put marks on a lot of passages which will facilitate your referring to it.

Ever Yours | T. H. H.

RI MS JT/1/TYP/9/2886
IC HP 8:32
Typed Transcript Only

———————

1. *[late November 1856]*: the month and year are given by relation to letter 1303, after 28 November but possibly early December

2. *only drawing*: in letter 1303, Tyndall had asked Huxley: 'Could you lend me those sketches of the crevasses that you took underneath the "Pavillion"?'. The enclosed sketch is missing.

3. *Agassiz*: L. Agassiz, *Nouvelles études et expériences sur les glaciers actuels* (Paris: Victor Masson, 1847).

From Rudolf Clausius   2 December 1856   1305

Zürich, d. 2 ten Dec. 1856.

Lieber Tyndall,

Ich beeile mich, Ihre Anfrage wegen der Carten, welche Desor erwähnt, zu beantworten, weil ich aus eigener Erfahrung weiss, wie unangenehm es ist, wenn man in einer wissenschaftlichen Untersuchung dadurch aufgehalten wird, dass man das nöthige Material nicht zur Hand hat.

Die Carten sind veröffentlicht, zwei Carten vom ganzen Aare-Gletscher und eine von einem schmalen Querstreifen, auf welcher unter anderemn verzeichnet ist, wie eine Linie, welche ursprünglich gerade war, sich in vier darauf folgenden Jahren allmälig durch das Vorrücken des Gletschers gekrümmt hat. Ausserdem gehören dazu noch eine Anzahl anderer Tafeln mit Zeichnungen in Bezug auf die Theorie der Gletscher. Diese /Tafeln und Zeichnungen bilden den/ zu einem Werke, welches von den hiesigen Geologen als eins der wichtigsten der Gletschertheorie betrachtet wird. Der vollständige Titel ist:

Système glaciaire ou Recherches sur les Glaciers, leur Mécanisme, leur ancienne extension et le rôle qu'ils ont joué dans l'histoire de la terre; par M. M. L. Agassiz, A. Guyot et E. Desor. Première Partie.

Dieser erste Theil, der einzige, welcher überhaupt erschienen ist, hat ausserdem noch folgenden besonderen Titel:

Nouvelles études et expériences <u>sur les glaciers actuels,</u> leur structure, leur progression et leur action physique sur le sol; par L. Agassiz. (1 Vol. avec un Atlas de 3 Cartes et 9 Planches) Paris, Victor Masson 1847.

Damit Sie ungefähr sehen können, was in dem Werke enthalten ist, habe ich einen Auszug des Registers gemacht, welcher auf den einliegenden Blättern enthalten ist, wobei ich zugleich die Seitenzahl, welche auf jedes Capitel verwandt ist, hinzugefügt habe.

Sollte dieses Werk auf den naturwissenschaftlichen Bibliotheken London's nicht vorhanden sein, so ist es gewiss zur Anschaffung sehr geeignet. Den Preis kann ich Ihnen nicht genau sagen. Escher von der Linth meinte ungefähr 50 francs.

Ausserdem möchte ich Sie noch auf ein anderes kleines Buch aufmerksam machen:

Die Gletscher der Jetztzeit von Albert Mousson, Zürich, 1854.

Es ist nur eine populär gehaltene Zusammenstellung, und enthält wahrscheinlich keine Thatsachen, die Sie nicht schon kennen, aber es sind sehr viele Citate darin, die Ihnen vielleicht nützlich sein können. Ich bin so frei, Ihnen dasselbe auf Buchhändler-Wege zuzusenden, mit der Bitte es als unbedeutendes Zeichen meiner Freundschaft freundlich anzunehmen. Ich würde es mit der Post schicken, aber soviel ich weiss kann man nach England von hier aus keine Bücher unter Kreuzband senden.

Sollte ich Ihnen in irgend einer anderen Beziehung nützlich sein können, so wird es mich jedesmal sehr freuen, und ich bitte Sie, sich mit allen Aufträgen, welche in der Schweiz besorgt werden können, ohne weiteres an mich zu wenden.

Seien Sie so gut Dr Hirst freundlich von mir zu grüssen, und auch ihm für die Uebersetzung meiner Abhandlungen meinen Dank auszusprechen. Ich habe bei diesen Uebersetzungen eine Bemerkung gemacht, wie schon früher bei anderen, welche Sie angefertigt hatten, dass sie an manchen schwierigen Stellen klarer & verständlicher sind, als das Original. Das liegt vielleicht zum Theil an der eigenthümlichen Kürze & Präcision der englischen Sprache, zum grossen Theile aber ist es jedenfalls das Verdienst des Uebersetzers.

Mit herzlichem Grusse | Ihr | Clausius.

Zurich, 2nd Dec. 1856.

Dear Tyndall,

I shall make haste to answer your inquiry about the maps which Desor mentions,[1] because I know from my own experience how unpleasant it is when one is held up in a scientific investigation by the fact that one does not have the necessary material at hand.

The maps have been published, two maps of the entire Aare Glacier[2] and one of a narrow horizontal strip, on which is drawn, among other things, how a line which was originally straight has gradually become crooked in four successive years through the advance of the glacier. Along with these there are also a number of other plates with drawings regarding the theory of glaciers. These [plates and drawings form part of] the work which is considered by geologists here to be one of the most important on the theory of glaciers.

The complete title is:

Système glaciaire ou Recherches sur les Glaciers, leur Mécanisme, leur ancienne extension et le rôle qu'ils ont joué dans l'histoire de la terre; par M. M. L. Agassiz, A. Guyot[3] et E. Desor. Première Partie.[4]

This first part, the only one which has ever appeared, also has the following special title:

Nouvelles études et expériences <u>sur les glaciers actuels</u>, leur structure, leur progression et leur action physique sur le sol; par L. Agassiz. (1 Vol. avec un Atlas de 3 Cartes et 9 Planches) Paris, Victor Masson 1847.[5]

So that you can see roughly what is in the book, I have made a summary of the index, which is on the enclosed pages,[6] on which I have at the same time added the number of pages which is spent on each chapter.

Should this work not be available in London's natural science libraries, then it will certainly be a very suitable acquisition. I cannot tell you the price exactly. Escher von der Linth thought approximately 50 francs.

In addition to all this, I would like to draw your attention to one other small book:

Die Gletscher der Jetztzeit by Albert Mousson, Zürich, 1854.[7]

It is only a popular compilation and probably does not contain any facts that you do not already know, but there are a great many citations in it that can perhaps be of use to you. I am taking the liberty of sending it to you by way of a bookseller, with the request that you kindly accept it as a small token of my friendship. I would send it by regular mail, but as far as I know, one cannot send books to England from here in a postal wrapper.

Should I be able to be of service to you in any other regard, then it will always be a great pleasure for me, and I ask you to turn to me without further ado with any tasks which can be taken care of in Switzerland.

Be so good as to give Hirst my kind regards, and to give him my thanks as well for the translation of my papers.[8] I have noticed with these translations, as earlier with others which you had prepared, that in some difficult passages they are clearer & more understandable than the original. Perhaps that is due in part to the characteristic brevity & precision of the English language, but in a large part it is, at any rate, the credit of the translator.

With kind regards | Your | Clausius.

RI MS JT/1/TYP/7/2224–24a
LT Typescript Only

---

1. *your inquiry about the maps which Desor mentions*: see letter 1302, n. 11.

2. *Aare Glacier*: or Aar, a system of glaciers located at the sources of the river Aare in the Bernese Alps, consisting of the Unteraargletscher (formed by the confluence of the Lauteraargletscher and Finsteraargletscher) and the Oberaargletscher.

3. *A. Guyot*: Arnold Henri Guyot (1807–84), a geographer and geologist born in Switzerland who studied among other subjects the movement of glaciers. He emigrated to the United States in 1848 and was appointed professor of geography at Princeton University (*CDSB*).

4. *Système glaciaire . . . Première Partie*: L. Agassiz, A. Guyot, and E. Desor, *Système glaciaire ou Recherches sur les Glaciers, leur Mécanisme, leur ancienne extension et le rôle qu'ils ont joué dans l'histoire de la terre,* was a planned multivolume work, the first part of which was the only one to be published (see n. 5).

5. *Nouvelles études et expériences <u>sur les glaciers actuels</u> . . . Paris, Victor Masson 1847*: L. Agassiz, *Nouvelles études et expériences sur les glaciers actuels, leur structure, leur progression et leur action physique sur le sol* (Paris: Victor Masson, 1847). Agassiz's *Nouvelles études* was the first and only published volume of the *Système glaciaire* (see n. 4).

6. *enclosed pages*: missing.

7. *Die Gletscher der Jetztzeit by Albert Mousson, Zürich, 1854*: A. Mousson, *Die Gletscher der Jetztzeit* (Zürich: Fr. Schulthess, 1854).

8. *translation of my papers*: see letter 1302. Hirst translated two of Clausius's papers for publication in the *Phil. Mag.*

To Thomas Archer Hirst   4 December 1856   1306

London, 4th. Dec. 1856

My dear Tom,

I promised to write to you today and now set about fulfilling that promise though I fear my doing so will be merely lost time as far as the subject of the promise is concerned. A gentleman[1] called upon me today to know if I could recommend any one to a school of very high standing, and which contained some of the children of the nobility, merely for four hours a week, to prepare a class of 12 pupils for the public examinations for appointments in the Royal Artillery and Engineers.[2] The salary would be £100 a year, only 9 months of the year being thus occupied, and it was said that more would be given sooner than miss a good man. I said I knew nobody but yourself, stated your peculiar situation but said at the same time that I would write to you, which the gentleman earnestly requested me to do. I have no doubt that were you in London you would soon have three or four things of this kind which would enable you to buy bread and cheese; in writing thus I simply fulfill my promise as you must judge whether the state of Anna's[3] health is such as to permit you to accept any thing of the kind. The school is kept by a M[r] Hopkirk,[4] Eltham Kent,[5]—not far from Blackheath,[6] and I am assured it is an establishment of a superior order. Indeed the fees of the students, 100 guineas each yearly, insures this. Write to me when you receive this as I have promised to give an answer soon. Write so that I may forward your letter as the reply.

I have been working very hard at glaciers lately. I had no idea when I commenced the subject that it would open out so well. Forbes of Edinburgh[7] will assuredly think that I have some design on him, although I have not, and have been led by the purest accident, if such there be, to tread in his footsteps. My object first was to examine the structure of glacial ice of which Forbes has given a most ingenious theory,[8] and which has been the subject of most lively discussion between several men of great eminence.—Well I think I have so smashed up this theory; so utterly annihilated it, that its author will hardly acknowledge it. Of course this may be all self delusion—others must judge of the force of my arguments, but unless I labour under a hallucination altogether new, the theory is killed without the hope of recovery. It is therefore more than an Irish 'killing'[9] from which people recover sometimes. Well I have not been altogether destructive for I think I have made compensation for what I have pulled down by shewing that the structure of the ice of glaciers and the cleavage of slate rocks are to be attributed to one and the same cause. How we are led forward in this life. How little did I think as I expounded my little hypothesis of slaty cleavage[10] to you on the summit of Broughton Hill[11] two years ago that it would carry such fruit. Another subject which has caused much discussion are certain discoloured bands observed first by Forbes on the surface of the Mer de Glace.[12] These I think I have also succeeded in explaining, and can produce them experimentally. The next point regarding which I did not intend to say anything, or to learn anything was the motion of the glaciers downwards. The so called viscous theory has been propounded by Forbes to account for the phenomena, and this theory commands the assent of almost all English geologists and others at the present day. D[r]. Kame[13] in his Arctic Voyage[14] has even named certain places after Forbes in testimony of his being the only one to unravel the mystery of glacial motion: on this point I have also something to say. The glacier does certainly exhibit many phenomena which are exactly such as would be produced by a plastic mass, but I think I have succeeded in shewing[15] that all the phenomena which it exhibits are perfectly reconcilable with those properties of ice which experiment reveals to us, and that it is not viscosity, but a property known to some for years, but wholly overlooked by glacial theorists that these phenomena are due. Give my love to Anna, and respect to M[r] Martin—as ever—John

RI MS JT/1/T/634
RI MS JT/1/HTYP/484–84a

1. *A gentleman*: not identified.

2. *Royal Artillery and Engineers*: from 1856–57, Tyndall wrote and graded examinations in scientific topics for commissioned officers of the Royal Artillery and Royal Engineers at the Royal Military Academy at Woolwich. It appears he was recommending Hirst for a teaching position to prepare students for such examinations. Tyndall stopped being an examiner when the War Office decreased the pay in April 1857 (see letters 1375 and 1376). In January 1858, he again became an examiner for commissioned officers of the Royal Artillery and Royal Engineers at the Royal Military Academies at both Woolwich and Sandhurst for the newly-established Council of Military Education. For more on Tyndall's and others' role as examiners for military education, see J. Elwick, 'Economies of Scales: Evolutionary Naturalists and the Victorian Examination System', in G. Dawson and B. Lightman (eds) *Victorian Scientific Naturalism: Community, Identity, Continuity* (Chicago: University of Chicago Press, 2014), pp. 131–56.

3. *Anna's*: Anna Hirst.

4. *Hopkirk*: Thomas Hopkirk (1819–81), principal of the preparatory Military Academy in Eltham, Kent. See 'Obituary', *Monthly notices of the Royal Astronomical Society*, 42 (1882), p. 145.

5. *Eltham Kent*: then a town southeast of London bordering the county of Kent.

6. *Blackheath*: an area of southeast London to the west of Eltham.

7. *Forbes of Edinburgh*: James Forbes.

8. *a most ingenious theory*: Forbes's theory of viscous glacier motion that Tyndall and others disagreed with. Forbes published his theory in *Travels through the Alps of Savoy and Other Parts of the Pennine Chain with Observations on the Phenomena of Glaciers* (Edinburgh: Adam and Charles Black; London: Longman, Brown, Green, and Longmans, 1843); and in three papers for the RS: J. D. Forbes, 'Illustrations of the Viscous Theory of Glacier Motion Part I. Containing Experiments on the Flow of Plastic Bodies, and Observations on the Phenomena of Lava', *Phil. Trans.*, 136 (1846), pp. 143–55; 'Illustrations of the Viscous Theory of Glacier Motion Part II. An Attempt to Establish by Observation the Plasticity of Glacier Ice', *Phil. Trans.*, 136 (1846), pp. 157–75; and 'Illustrations of the Viscous Theory of Glacier Motion Part III', *Phil. Trans.*, 136 (1846), pp. 177–210.

9. *an Irish 'killing'*: a stage death.

10. *my little hypothesis of slaty cleavage . . . would carry such fruit*: Tyndall explained glacier motion by the *fracture* of ice (similar to his theory of slaty cleavage, which he lectured on at the RI on 6 June 1856, 'Comparative View of the Cleavage of Crystals and Slate Rocks') and *regelation* (the refreezing of water between two pieces of ice). In *Glaciers of the Alps,* Tyndall described regelation as when 'Two pieces of ice at 32° Fahr., with moist surfaces, when placed in contact freeze together to a rigid mass'. He continued, 'When the attachments of pressed ice are broken, the continuity of the mass is restored by the regelation of the new contiguous surfaces. Regelation also enables two tributary glaciers to weld themselves to form a continuous trunk; thus also the crevasses are mended, and the dislocations of the glacier consequent on descending cascades are repaired. This healing of ruptures extends to the smallest particles of the mass, and it enables us to account for the continued compactness of the ice during the descent of the glacier' (p. 423).

11. *summit of Broughton Hill*: a hill near Queenwood College, Hampshire.

12. *Mer de Glace*: a large glacier on the north side of Mont Blanc, France.

13. *Dr. Kame*: Elisha Kent Kane (1820–57), an American explorer and naval medical officer who was present on two Arctic expeditions in search of Sir John Franklin (*ANB*).

14. *Arctic Voyage*: E. Kane, *Arctic Explorations: The Second Grinnell Expedition in search of Sir John Franklin, 1853, '54, '55*, 2 vols (Philadelphia: Childs & Peterson, 1856).

15. *shewing*: alternative spelling of *showing*.

From Samuel Haughton   6 December 1856   1307

Trin: Coll: Dub:[1] | 6 December 1856.

My dear Sir,

I send you with this note an account of my experiments on Irish Granites—Part of the first portion on the Leinster granites[2] was published in the Phil. Mag.[3]—I wish to direct your attention to the curious transformation of ordinary granite into Anorthite and Hornblende syenite in Carlingford,[4] by the junction with limestone. I would send an abstract of the second part of my paper,[5] viz on the Granites of the North East of Ireland—only I am prohibited by the rules of the Geol. Soc. London.[6]

The change in the order of the Artillery Examination, will, in my opinion, have the effect of stopping the study of Geology and Mineralogy by the Candidates altogether.

Our Dublin Candidates are now reading <u>Moral Sciences</u> instead of <u>Natural</u>,[7] as they know well that they will 'pay better' at the Examination.

The Natural Sciences should be joined with the Experimental, like Greek and Latin, Mathematics and Mechanics, & both together only counted against the Candidate as one subject.

I am, | Yours very sincerely | Saml. Haughton

RI MS JT/1/TYP/2/483
LT Typescript Only

1. *Trin: Coll: Dub::* Trinity College, Dublin.
2. *first portion on the Leinster granites*: S. Haughton, 'On the Granites of Leinster', *Proceedings of the Royal Irish Academy*, 6 (1855), pp. 230–40.
3. *published in the Phil. Mag.*: S. Haughton, 'Notes on Minerology I.—No. 1. On the Chemical Composition and Optical Properties of the Mica of the Dublin, Wicklow, and Carlow Granites', *Phil. Mag.*, 9 (1855), pp. 272–75.
4. *Carlingford*: a coastal town in northern Ireland.
5. *second part of my paper*: S. Haughton, 'Experimental researches on the granites of Ireland. Part II. On the granites of the North-east of Ireland', *Quarterly Journal of the Geological Society*, 14 (1858), pp. 300–305.
6. *Geol. Soc. London*: Geological Society of London.
7. <u>*Moral Sciences*</u> *instead of* <u>*Natural*</u>: possibly a reference to the East India Company's civil service examinations (begun in 1855) which included, among other subjects, English composition, English literature and history, mathematics and Sanscrit. Natural science and moral science were weighted equally in the exams. See 'The Civil Service Examinations and the Study of History', *Gentleman's Magazine*, 44 (1855), pp. 595–600. For Houghton's work training students for such examinations see N. D. McMillan, 'Samuel Haughton, 1821–1897', in M. McCartney and A. Whitaker (eds), *Physicists of Ireland: Passion and Precision* (Philadelphia: Institute of Physics Publishing, 2003), pp. 106–15, on p. 107–8.

To Edward Frankland   8 December 1856   1308

Royal Institution. | 8th, Dec, 1856.

My dear Frankland

I am overjoyed to find that I shall have an opportunity, of afflicting you every Saturday for 7 weeks after Easter.[1] I am so glad that you have got over your terror of Dyspepsia![2] I had a talk with Bence Jones over the matter, and though my hopes were gloomy in the extreme I recommended him to write to you. By George I am a cubit[3] taller since I read your letter[4]—The opportunity of pinching you and the luxury of kicking you if I please although still merely prospective is highly edifying and delightful to me The best way to respond to your questions is—and I do it by return—to send you the syllabusses of the last courses. There are <u>no</u> chemical courses before Easter this time so that there is no fear of your being forestalled. Huxley,[5] John Phillips[6] the Geologist and your implacable foe the writer of this note[7] are the lecturers before Easter. Give my kind remembrances to Mrs Frankland,[8] she shall have the luxury of nursing you every week after your return from London.[9]

Ever Yours | John Tyndall

RI MS JT/1/TYP/12/3968
LT Typescript Only

---

1. *Every Saturday for 7 weeks after Easter*: Frankland gave a series of seven Saturday lectures at the RI on the 'Relations of Chemistry to Graphic and Plastic Art', which began on 25 April 1857 and ran through 6 June 1857, 23 May being the specific mention here. See RI Guard Book, vol. 2, p. 99, and Russell, *Edward Frankland,* p. 187.

2. *Dyspepsia*: indigestion (*OED*).

3. *cubit*: a unit of measurement roughly equal to the length of a man's forearm, between 18 and 22 inches, or 45 cm (*OED*).

4. *your letter*: letter missing.

5. *Huxley*: Huxley presented at the RI 'Twelve Lectures on Physiology and Comparative Anatomy'.

6. *John Phillips*: John Phillips (1800–74), an English geologist. He developed a global geologic time scale and coined the term Mesozoic (*ODNB*). Phillips presented at the RI 'Ten Lectures on Leading Questions in Geology'.

7. *the writer of this note*: Tyndall presented 'Eleven Lectures on Sound', delivered on Thursdays at the RI from 22 January to 2 April 1857. Following the Easter holiday, he delivered another course on sound, eight lectures from 23 April to 11 June. See RI Guard Book, vol. 2, p. 98–99.

8.　*Mrs Frankland*: Sophie Frankland (née Fick, 1821–74), of Cassel, near Marburg in the state of Hesse-Cassel. Edward Frankland met her in 1847, on his first visit to Marburg. They became engaged in 1849, and married in 1851. Two of her brothers were professors in the University of Marburg, and her father, Friedrich W. Fick (1783–1861), was chief engineer of the state. See Russell, *Edward Frankland*, pp. 83–91.

9.　*your return from London*: to Manchester, England, where Frankland was professor of chemistry at Owens College (now University of Manchester).

From Heinrich Gustav Magnus　　　20 December 1856　　1309

Berlin 20 Dec. 56.

Theurer Freund

Hr. Scott ist eingetroffen. Er wird indeß jetzt nicht bei mir arbeiten, sondern wünscht sich noch in Chemie zu vervolkomnen. Seine Absicht war bei Prof. H. Rose zu arbeiten, doch ist dessen Laboratorium durch Americaner (deren jetzt viele hier sind) besetzt; deßhalb wird er bei D^r. Schneider arbeiten. Ich glaube er thut gut daran, denn die Chemie ist eine vortreffliche Vorschule für physicalische Untersuchungen, auch scheint mir daß noch einige Studien in der Physik für ihn wünschenswerth sind.

Wahrscheinlich haben Sie bei Empfang dieser Zeilen das Instrument von Sauerwald bereits erhalten. Er hat dasselbe schon vor mehr als 8 Tagen an Asher hier übergeben; es ist daher ohne Zweifel schon in London. Dasselbe ist so eingerichtet daß es zu 2 verschiedenen Zwecken dienen kann. 1, Um die Schwingungsknoten einer Stahlfeder zu zeigen. 2, um eine Saite in fortwährender Schwingung zu erhalten, und zwar je nach dem man die Saite durch die kleine oder durch die große Oeffnung des Hammers gehn läßt, schwingt sie in einer Ebene, oder so daß jeder Punkt derselben einen Kreis beschreibt.

Diese beiden letzteren Versuche sind, nach meiner Ansicht, ohne besonderes Interesse, aber der erste ist recht hübsch. Ich habe das Ganze so vollständig als möglich herrichten lassen. Die Ständer welche dazu dienen die Saite auszuspannen werden mit Zwingschrauben an einem Tisch befestigt. Bei Dove's Apparat, nach dem Sauerwald gearbeitet hat, fehlen diese Zubehör.

Meine Frau und Kinder tragen mir auf Sie freundlichst zu grüßen, sie hoffen mit mir Sie recht bald hier zu sehn.

An Faraday und M^{rs}. Faraday meinen Respect.

D^r. Häusser, der nach Brasilien geht, wollte Sie aufsuchen, konnte sich aber nur einen Tag in London aufhalten, da er am 24ten. auf dem Steamer in Southhampton sein muß.

Erhalten Sie mir auch in dem neuen Jahre Ihre Freundschaft | von Herzen | Ihr | G Magnus.

Berlin 20 Dec. 56.

Dear friend

Herr Scott[1] has arrived. He will, however, not be working under me now, but rather still wishes to perfect himself in chemistry. His intention was to work under Prof. H. Rose, but his laboratory is occupied by Americans (of which many are here now); he will therefore work under Dr Schneider.[2] I think he is doing the right thing about this, for chemistry is an excellent nursery for investigations in physics, and it also seems to me that some more studies in physics will be desirable for him.

You have probably already received the instrument[3] from Sauerwald by the time you receive these lines. He handed it over to Asher here more than 8 days ago; it will therefore doubtless be in London already. It is constructed in such a way that it can serve 2 different purposes. 1, To show the nodal points of vibration of a steel spring. 2, to keep a string in constant vibration, and indeed, depending on whether one lets the string pass through the small or the large opening of the hammer, it will vibrate in one plane, or so that every point of it describes a circle.

In my view, these last two experiments are without particular interest, but the first one is quite nice. I have had the whole thing set up as completely as possible. The stands which are used to stretch out the string are fastened to a table by means of ferrule bolts. With Dove's apparatus,[4] following which Sauerwald has been working, these accessories are missing.

My wife and children[5] tell me to give you their kindest regards, and, along with me, they hope to see you here very soon.

To Faraday and Mrs Faraday my respects.

Dr Häusser,[6] who is going to Brazil, wanted to call on you but could only stop in London for one day, as he has to be on the steamer in Southampton[7] on the 24th.

Keep me in your friendship in the New Year as well | sincerely | Your | G Magnus.

RI MS JT/1/M/20

---

1. *Herr Scott*: Robert Henry Scott, see letter 1301, n. 4.
2. *Dr Schneider*: probably Ernst Robert Schneider (1825–1900), a German chemist who from 1853 lectured in chemistry at the University of Berlin before becoming an extraordinary professor in 1860. See J. C. Poggendorff, *Biographisch-literarisches Handwörterbuch zur Geschichte der exacten Wissenschaften,* 2 vols (Leipzig: J. A. Barth, 1863), vol. 2, p. 827.
3. *the instrument*: possibly an instrument for demonstrating undulating motion of a string for one of Tyndall's 'Eleven Lectures on Sound' (see letter 1308, n. 7).
4. *Dove's apparatus*: not identified.

5. *wife and children*: Bertha, Anna, Christine, and Paul Magnus.
6. *Dr Häusser*: Jakob Christian Heusser (1826–1909), a Swiss naturalist who earned a doctorate in natural sciences at the University of Berlin in 1851. He became a Privatdozent at the University of Zurich in 1853. In 1856 he travelled to Brazil to examine the complaints of Swiss residents of plantations there (*HLS*).
7. *Southampton*: a city on the south coast of England in the county of Hampshire.

From Jules Lissajous 22 Décembre 1856 1310

Fabrique de tous les instruments | nécessaires à l'étude des Sciences. |Ateliers, rue du Faubourg Saint-Jacques, 73. | 5 Médailles d'or. | Médaille de 1er Classe à l'Exposition Universelle de 1855. | Maison | LEREBOURS ET SECRETAN, | SECRETAN, successeur. | Opticien de S. M. L'Empereur, de l'Observatoire et de la Marine. | Magasins, Place du Pont-Neuf, 13.

22 décembre 1856

Monsieur

Mr Auguste Sécrétan est actuellement absent pour quelques jours, ce qui fait que je profite de cette occasion pour vous écrire à sa place. Vos appareils sont en voie d'exécution et pourront être terminés d'ici à une huitaine de jours. J'en excepte toutefois l'appareil optique destiné à répéter mes expériences. Je désirerais profiter de ce que vous voulez bien répéter ces expériences pour aller, moi-même, les exécuter à l'Institution Royale. Nous nous entendrons ensemble pour l'époque, si vous le permettez; j'espère même d'ici là améliorer encore la disposition des appareils.

Monsieur Sécrétan, me charge de vous dire qu'il n'est nullement nécessaire que vous envoyiez l'argent avant l'expédition des appareils.

Veuillez, Monsieur, recevoir l'assurance de la parfaite cordialité avec laquelle j'ai l'honneur d'être votre très humble et très dévoué serviteur.

J. Lissajous

Manufacture of all instruments | needed for the study of Science. | 5 Gold Medals. | 1st Class Medal at the International Exhibition of 1855. | Workshops, Faubourg Saint-Jacques st., 73 | MAISON LEREBOURS ET SECRETAN | Secretan, successor. | Optician of S. M. the Emperor, of the Observatory and the Navy. | Stores, Place du Pont-Neuf st., 13.

22 December 1856

Sir

Mr Auguste Sécrétan[1] is currently absent for a few days, which is why I am taking advantage of this opportunity to write to you in his place. Your instruments are close to completion and may be finalised within the week. With the exception however of the optical device designed to repeat my experiments. I would like to take advantage of your desire to repeat those experiments to come and perform them, myself, at the Royal Institution.[2] We will agree together about the day, if you please; and, before then, I even hope to have further improved the positioning of the instruments.

Mr Sécrétan asked me to tell you that it is by no means necessary for you to send the payment before the devices are dispatched.

Please do accept, Sir, the assurance of the perfect cordiality with which I have the honour of being your very humble and much devoted servant.

J. Lissajous

RI MS JT/1/L/23

---

1. *Mr Auguste Sécrétan*: Auguste Sécrétan (1833–74), an optical instrument maker in Paris, son of Marc Sécrétan (1804–67) of the firm Lerebours and Secretan. See P. Brenni, '19th Century French Scientific Instrument Makers, III: Lerebours et Secretan', *Bulletin of the Scientific Instrument Society*, 40 (1994), pp. 3–6.

2. *perform them, myself, at the Royal Institution*: on 5 June 1857, Tyndall gave a Friday Evening Discourse at the RI titled 'On M. Lissajous' Acoustic Experiments', an abstract of which is in *Roy. Inst. Proc.*, 2 (1854–58), pp. 441–43. Lissajous was present, as Tyndall wrote to Thomas Hirst (see letter 1404), 'Professor Lissajous made his experiments on acoustics while I expounded them'.

From Hector Tyndale[1]   29 December 1856   1311

Philadelphia 29 December 1856

My dear John

I was sorry to see the date of your last letter[2], forasmuch as I knew by it that my friend Rothermel[3] would not see you, which I had hoped he would do; and his hopes of the same were also disappointed as he told me after in a letter from Paris. He is a good fellow, plain, not very demonstrative, indeed not at all so, but of an exceeding good <u>nature</u> and <u>temper</u>—primarily—with a much broader inlet than outlet. He is an American, I don't mean a fourth of July, starred and striped[4] one, but one who has grown out of the needs and circumstances of America. But better than that he is a Man. And therefor I wished him to know you. He wrote me from Rome a short time since, of Art and the Past men and his letter was very refreshing—as Walt Whitman[5] would say 'with no ducking or deprecating about it'.[6] My friend Read[7] has perhaps seen you before this, as, in a letter dated Manchester[8] a couple of weeks ago, I rec'd from him advisement that he would call on you again, for the third time, before he left England. Read is, I suppose, painting in the town of Manchester having been induced to remain thus a part, at least, of this Winter. I think he will exhibit one or more of his works in London before he leaves for Italy. He tells me he had seen Leigh Hunt[9] and Alfred Tennyson and was very much pleased with the former, as, indeed, is every one who meets him. Read is quietly demonstrative and demonstratively quiet, with more apprehension than comprehension yet with more conception or inception than perception—rather contradictory that seems, but it is true nevertheless. In reference to Read himself, of course applying it to his works, I might quote from your heterodox Art-Critic—John Ruskin,[10] who says of Carlo Dolci[11] that 'he polishes his pictures into inanity'.[12] But Read is a pleasant gentlemanly little fellow and will appreciate <u>whatever</u> you do to him. I cannot forbear, tho' at the risk of tiring you, from speaking further of this—Why is it that Read should have so much reputation as he really has here, and to a small extent also in England, when men like Rothermel are scarcely known? He writes pretty verses and paints paintings, to which I shall <u>add</u> no <u>adjective</u> other than pretty; he talks well, is well informed and agreeable. But these are not enough for me, or rather, to satisfy me of the cause of the reputation. I wish to know why the World cannot or does not recognize <u>force</u> and power? Why it cannot point out the loftiest peak of Manliness as well as the loftiest peak of the Alps? Have we really not two lives or rather is there not ourself and our life? And is not this matter of reputation altogether from life conventionally? If so, is it not time we should throw conventionalisms overboard? This it strikes me is Walt Whitman's object in the 'Leaves of Grass'[13]—to make us love our <u>self</u>—<u>myself</u>, <u>your self.</u> I need not say to you that I do not apply this to Read personally, as an isolation, his name and circumstances suggested the thought, and I would not speak of any friend thus but as a given example. As for Read himself, he has many good qualities with certain kinds of power which <u>I</u> could never think of equalling for example continuity and grip and work. But as he will never know of the e.g. I have made, and as you will hardly think me capable of the littleness of personal disparagement why I may let it rest here.

But why is it that the 'World' goes whistling over the mountains and falling down in worship before the little tumuli[14] of heaped up words and dry bones? I know it is written 'by their fruits you shall know them'[15]—but where is it written or taught to know what fruit is? I make no complaint, have none to make, but wish to know the sign when I meet the Master.[16] I have met him and the sign has been given by Jesus in His sayings by Rubens[17] once or twice, by Carlyle sometimes, tho' latterly he gives me no sign nor has given me any since the counterfeit of his 'Quashee' and 'beneficent whip',[18] sometimes too by our queer friend Walt: Whitman, but I wish to be with Him and to know Him always, in the darkness as in the light, not now and then but always. Teach me, if you have discovered it in your laboratory, that I may know manliness when I meet it. Teach me to know—to recognise at least, honesty, courage, love, justice, Catholicism.

I have a hope that America is working nearer to the light than any before. That light is dark enough truly. But I cannot help hoping this—it may be from the acceptiveness, the catholic receptiveness of America. It seems as tho' all the bricks the Divine anger loosed from Babel,[19] which have been by the divine wrath hurling through the world ever since, are falling on America, in time to form a tower high enough to outtop the divine punishment and to smile forever beyond the clouds of his Anger—and this without the slightest reference to the millenium.[20] <u>That</u> is far enough from us here now.

So you condemn our poor Walt: Whitman's 'Leaves' to one of the three mad cells of Carlyle—the cells of damned books Nos: one, two and three.

I don't. If a man were to build a Clipper ship[21] in the middle of Sahara[22] or a Locomotive on an Alpine Glacier he might not uncharitably be thought mad, And yet the Clipper and the Engine might both be models of excellence. Wherever Conventionalism rules, wherever 'Society' is legislator wherever 'I dare not waits upon I would',[23] why there, I take it, Walt: is in his sphere. I confess that <u>I</u> cock my hat[24] with a deeper breath and tie my neck-cloth[25] less scrupulously when Walter[26] talks to me—and I consult the looking glass[27] less frequently. He pats my egotism but knocks my vanity on the sconce.[28] Vanity is a good thing or outcrop of a good thing but we have had more than enough of it. Egotism is not so bad a thing but that we may try a little more of it. I may be as potter's clay but by George[29] I don't wish 'Society' to be the potter in whose hands I shall be moulded. 'Society' owes as much to me as I owe to it, and a little more, and I wish the recognition of the debt. Let us, once in a while, put 'Society' to the expense of a law suit and not be forever acknowledging judgement upon any claim. Stand up then friend Walt: and let the trial proceed—God show the right.

Hurrah for the Alps. What's the use of saying I should so like to have been with you—if it were but to have rested at the foot while you climbed the head of Alps. I regret exceedingly never having gone into Switzerland—when, too, it would have been but two or three weeks time <u>lost</u> to business! But no matter you have seen them and others have seen them and I have heard your voices coming down to me and <u>that</u> is pleasant. Oh 'unproductive' Alps[30] and barren Sahara[31] how poor would the World be without you. The trip was good for you I hope, good for your health I mean? If it was you can afford to bore yourself a little more by writing to me once in a while! 'And so shall it be made productive unto <u>me</u>' Did you 'get drunk' up yonder, that is were you exhilarated by the air? I have been so on our far western prairies,[32] away off towards the Rocky Mountains; where the air is so pure that breathing becomes very joyous and happy and existence itself is delightful.

The Forbes theory[33] you mention, that of the viscous nature of ice as applied to the movement of Glaciers, is a new one to me, but that is not very wonderful. It rather contradicts, however, the received theory of the three states of matter, or at least does so so far as water is concerned, for it denies it the <u>solid</u> form or state. Does it not? For if it has the fluidity even of the mortar,[34] to which you compare it, it cannot be properly be called a solid, can it? Is it not possible to account for the movement of the glaciers on other grounds than this. Can it not be by the percolated streams, or <u>sub-glacieran</u> water courses which harden into ice, <u>solidify</u> as they again approach the air? Or may not the exuding of the water from the surfaces and the consequent freezing by contact with the air, cause the spread of the glacier in a downward direction giving it the appearance of flowing? Or do I understand you in saying that the glaciers themselves, which I have always understood to be <u>ice</u>, are of the consistency of mortar?

But, doubtless, you think me out of my bounds and will refuse to read if I continue whereof I am ignorant. The novelty of the theory however must be my sole excuse, and I shall look to you for a plainer showing.

Did'nt[35] I tell you in my last[36] that I had received your paper on 'Cleavage'[37]? If I did'nt I should have acknowledged it and the pleasure it's simplicity of explanation gave me. That is your style, stick to it I pray you. Be as untechnical as you can and let everybody, besides the F.R.S.[38], know something of that we are sent here to live upon for a while. In doing this it won't be at all necessary that when we look upon a river we shall say, 'Yes, I know you are only HO,'[39] or when we stand unhatted before the Mountains we shall pull out our scales to weigh the Silica and the what not, nor yet that when we gaze, with our souls in the gaze, upon the dear, sweet blue of the Ocean we shall unpack our test tubes and reagents and *[worry]* 1 = 127 out of the poor sea. No, these are not necessary but they <u>are</u> very technical and are the tendencies of the schools and sciences. Pray forgive me, if I touch on ground some might think profound by ignorance. But if the Lord will suffer children and ignorant ones to look upon the Mountains, upon the Seas upon the Storm and thunderbolt, why may not men suffer other ignorant ones to look upon and speak of <u>their</u> works.

Is there a character any where in any of the old plays or books, of the man who felt always that <u>something</u> was about to happen? I have felt so for some years back, I felt so before the Revolutionary emeutes of 1848[40] and I feel so now. I don't know what it is, whether the World is to be shaken, or I am to have a bad nervous disorder! I believe always that 'the King will come to his own again'[41] as firmly as any Bourbonist or Jacobite,[42] and depend upon it, the sneering, mocking, heartless generations will be hardly dealt with, reformatarily, 'when the King comes in glory.'[43] We all go to fine Churches and hear fine words which never have and never will butter any parsnips.[44] I wish Old Cromwell[45] would come back again with his Ironsides[46] and that against him should appear the Scotch Habbakuk Mucklewrath[47] and the rest of 'em,[48] and that they should go at it again that the world might see something else than traders. Lord, how grand they would appear now among the cyphering men who can convince the devil himself that two ones are two, that twice two make four and that No: One is the best number of all.

Here I am at the bottom of six pages and of my inkstand—a patent humanitary inkstand, warranted to give out as little ink as may be required—I thank goodness that you haven't got 'em yet in England, or you'd set it very <u>shallow</u> for me—hey ? But I don't care I've buttonholed[49] you and shall rise a happier man—having 'bestowed all my tedium'[50] on you. Whenever you can, do please write to me, you don't know what a charitable thing you will be doing. 'All your cousins' send their love and regards to you and hope to see you here some time or other.

Will you ever pick Walt: Whitman up again. By the way I forgot to tell you that he has published a second edition.[51]

Affectionately Yours | Hector Tyndale

RI MS JT/1/T/49
RI MS JT/1/TYP/5/1619–24

---

1. *Hector Tyndale*: Hector Tyndale (1821–80), an American businessman, Union general during the Civil War, and cousin of John Tyndall. See J. M Mclaughlin, *A Memoir of Hector Tyndale* (Philadelphia: Collins, 1882).

2. *your last letter*: possibly letter 1274, John Tyndall to Hector Tyndale, 27 August 1856, *Tyndall Correspondence,* vol. 5.

3. *Rothermel*: Peter F. Rothermel (1817–95), an American portrait and historical painter (*ANB*).

4. *starred and striped*: a reference to the stars and stripes of the American flag (at the time there were 31 stars and 13 stripes).

5. *Walt Whitman*: Walt Whitman (1819–92), an American poet and essayist (*ANB*). Several but not all references to 'Walt Whitman' in this letter include a colon after 'Walt', which is an indication that 'Walt' is abbreviated from 'Walter'.

6. *'with no ducking or deprecating about it'*: in a poem included in *Leaves of Grass* (Brooklyn, NY: self-published, 1855), p. 26, titled 'Song of Myself' in a later edition, Whitman wrote: 'I chant a new chant of dilation or pride, We have had ducking and deprecating about enough, I show that size is only development'.

7. *Read*: possibly Thomas Read (1822–72), an American poet and portrait painter (*ANB*).

8. *Manchester*: a large industrial city in northwest England.

9. *Leigh Hunt*: Leigh Hunt (1784–1859), an English poet and essayist (*ODNB*).

10. *John Ruskin*: John Ruskin (1819–1900), an English art critic and social thinker (*ODNB*). Ruskin was a prolific writer and very much interested in scientific topics. Ruskin would later come into dispute with Tyndall over theories of glacier motion and scientific naturalism. See P. L. Sawyer, 'Ruskin and Tyndall: The Poetry of Matter and the Poetry of Spirit', in James Paradis and Thomas Postlewait (eds), *Victorian Science and Victorian Values: Literary Perspectives* (New York: New York Academy of Sciences, 1981), pp. 217–46; and J. Smith, *Charles Darwin and Victorian Visual Culture* (Cambridge: Cambridge University Press, 2006), pp. 281–82.

11. *Carlo Dolci*: Carlo Dolci (1616–86), an Italian painter.

12. *'he polishes his pictures into inanity'*: in *Modern Painters: Their Superiority in the Art of Landscape Painting to All the Ancient Masters* (London: Smith, Elder and Co., 1843), Part I, Section I, Chapter II, p. 13, Ruskin wrote, 'Three penstrokes of Raffaelle are a greater and a better picture than the most finished work that ever Carlo Dolci polished into inanity'.

13. *'Leaves of Grass'*: Walt Whitman, *Leaves of Grass* (Brooklyn, NY: self-published, 1855).

14. *tumuli*: plural of *tumulus,* an ancient sepulchral mound (*OED*).

15. *'by their fruits you shall know them'*: Matthew 7:16.

16. *the Master*: God.

17. *Rubens*: Peter Paul Rubens (1577–1640), a Flemish Baroque painter.

18. *'Quashee' and 'beneficent whip'*: references to Carlyle's essay 'Occasional Discourse on the Negro Question', *Fraser's Magazine,* 40 (1849), pp. 527–38, in which he controversially argued against the abolishment of slavery. On p. 534, 'Quashee, if he will not help in bringing out the spices, will get himself made a slave again (which state will be a little less ugly than his present one), and with beneficient whip, since other methods avail not, will be compelled to work'. *Quashee* is a term for a native of the West Indies.

19. *all the bricks the Divine anger loosed from Babel*: reference to the Tower of Babel of Genesis 11:1–9, in which God stops those living on earth from building a tower and scatters them across the globe confusing their languages so they could not understand each other anymore. The destruction of the tower by God is not mentioned in Genesis, however, but in other ancient sources such as the *Book of Jubilees*.

20. *the millenium*: a reference to the theological expectation of a Second Coming of Christ, which would bring a new Kingdom of God to restore society from its current state.

21. *Clipper ship*: a fast sailing ship of the mid-nineteenth century.

22. *Sahara*: the Sahara desert of northern Africa.

23. *'I dare not waits upon I would'*: W. Shakespeare, *Macbeth,* I.vii, Lady Macbeth, 'Letting "I dare not" wait upon "I would", | Like the poor cat i' the adage?'.

24. *cock my hat*: to tilt one's hat at an angle. A reference to Whitman's *Leaves of Grass* (see n. 13), on p. 25: 'I cock my hat as I please indoors or out'.

25. *neck-cloth*: a cloth worn round the neck (usually by a man); a cravat, neckerchief (*OED*).

26. *Walter*: Walt Whitman, see n. 5.

27. *looking glass*: a mirror. A reference to Whitman's *Leaves of Grass* (see n. 13), on p. 61: 'Can each see the signs of the best by a look in the looking glass? Is there nothing greater or more?'.

28. *sconce*: a jocular term for the head; esp. the crown or top of the head (*OED*).

29. *by George*: used as an exclamation or mild oath; George as in Saint George (*OED*).

30. *unproductive Alps*: perhaps a reference to 'The rocky precipices of and snowy peaks of the Himalaya are converted into the broad plains of Hindustan, the unproductive Alps spread out into the fertile levels of Lombardy', in J. MacCulloch, *Proofs and Illustrations of the Attributes of God, from the Facts and Laws of the Physical Universe: Being the Foundation of Natural and Revealed Religion,* 3 vols (London: James Duncan, 1837), vol. 1, p. 411.

31. *barren Sahara*: perhaps a reference to 'Woe for the age, woe for the man, quack-ridden, bespeeched, bespouted, blown about like barren Sahara, to whom this world-old truth were altogether strange!' in T. Carlyle, 'Memoirs of the Life of Sir Walter Scott, Baronet. Vol. i–vi' (review), *London and Westminster Review,* 6 (1838), pp. 293–345.

32. *our far western prairies*: Hector Tyndale spent two years in Texas during the late 1830s and joined the 1st Dragoons cavalry expedition to Dakota and Montana under the command

of Major Edwin Sumner. See J. M. Mclaughlin, *A Memoir of Hector Tyndale* (Philadelphia: Collins, 1882), p. 6.

33. *Forbes theory*: see letter 1306, n. 8.

34. *mortar*: a pastelike material, originally consisting of sand, lime, and water, now usually a mixture of cement and water, which is applied to form the joints between stones or bricks and which, when set, bonds them together (*OED*).

35. *Did'nt*: misspelled by Tyndale, as well as in the next sentence.

36. *my last*: letter missing.

37. *your paper on 'Cleavage'*: J. Tyndall, 'Comparative View of the Cleavage of Crystals and Slate Rocks', *Phil. Mag.*, 12 (1856), pp. 35–40.

38. *F.R.S.*: Fellow of the RS.

39. *HO*: hydrogen and oxygen, the two elements that make up a water molecule (H_2O).

40. *Revolutionary emeutes of 1848*: in 1848 a series of violent political uprisings swept through Europe including France, Switzerland, Italy, and a number of German states.

41. *'the King will come to his own again'*: possibly a reference to the popular Royalist, or Cavalier, English Civil War ballad 'When the king enjoys his own again'.

42. *Bourbonist or Jacobite*: Bourbonists were the supporters of House of Bourbon, a French royal house until the French Revolution in 1792. The House of Bourbon was restored in 1814–15 until it was again overthrown in 1830 by the House of Orléans. Jacobites were the supporters of the House of Stuart, a Scottish royal house. The movement drew its name from the Latin, Jacobus, for the Catholic King James VI of Scotland, II of England and Ireland who was deposed in 1688 by the English Parliament. During the eighteenth century two major, albeit failed, attempts were made to restore the Stuarts to the English throne in 1715 and 1745.

43. *'when the King comes in glory'*: Matthew 25:31.

44. *butter any parsnips*: nothing is achieved by empty words or flattery (*OED*).

45. *Old Cromwell*: Oliver Cromwell (1599–1658), an English military and political leader (*ODNB*).

46. *Ironsides*: 'Old Ironsides', nickname given to Oliver Cromwell.

47. *Scotch Habbakuk Mucklewrath*: Habakkuk Mucklewrath, a character in Walter Scott's novel *The Tale of Old Mortality* (Edinburgh: William Blackwood and London: John Murray, 1816).

48. *the rest of 'em*: possibly a reference to fictional Presbyterian Covenanter characters who formed the subject of Scott's novel. Covenanters were persecuted by the restored Stuart monarchy in the late seventeenth century.

49. *buttonholed*: button-hold, to take hold of (a person) by a button, and detain him in conversation against his will (*OED*).

50. *'bestowed all my tedium'*: 'Why should I bestow all my tediousness upon you, because I have you in my power, and have ink, paper, and time before me?' in W. Scott, *Rob Roy*, 3 vols (Edinburgh: Archibald Constable and Co. and London: Longman, Hurst, Rees, Orme, and Brown, 1818), vol. 1, p. 6.

51. *a second edition*: W. Whitman, *Leaves of Grass* (Brooklyn, NY: self-published, 1856).

1857

To Thomas Henry Huxley 6 January [1857][1] 1312

6th. Jan.

My dear Huxley,

I was puzzling my head for a name for that property and had asked Hooker to puzzle his, but your 'frangi-flexibility' comes nearer the thing than anything I had thought of. Still it does not quite please me, and for this reason. A flexible solid according to our hitherto notions must be <u>elastic</u>. Flexibility in a solid implies a power of regaining its shape when the strain upon it is removed. Our ice would not do so and therefore it is frangi-plastic rather than frangi-flexible; but you miss the <u>alliteration</u> in the former which to the ear is something. The particles of a plastic body move past each other and form new attachments as our ice does, but the particles of a flexible body move <u>round</u> each other. If you still consider the term applicable I am willing to let it go. Some term is necessary; and we should fix the theory by calling it the 'frangi flexible theory.' I had set down the 'break and make' theory; a better term than either would be one to include the idea of breaking with that of <u>refreezing</u>.[2] I had the account of our travels written much more fully at first but Francis thought it too full and I cut it down greatly. With regard to the arrangement I think it might be improved, but the change in this respect indicated by your heads would be a formidable one. You have only to remember that Stokes is to have the paper <u>this week</u>, and that I cannot lay a hand on it, except to insert half a dozen notes which are beside me ready written. I shall certainly when it comes into my hand cut away something about the tensions and pressures, so as to leave as little hold as possible upon this point. This I have set down among my notes. We must I think avoid diverting the mind too much from the real points at issue. In fact I have been thinking of omitting the whole of that about the Grindelwald glacier[3] as really irrelevant, and if you encourage me in the least I will do so. Your little experimental devices are full of ingenuity but instead of expanding in this direction I have the strongest inclination to retrench. The arrangement is the point I think which the time at our disposal will confine us to. I am sure this can be made more effective and we might probably be permitted to do something at this even after the reading of the paper on the 15th.[4]

Ever Yours | John Tyndall

In regard to the question 'why does the glacier travel faster in wet weather.' I ask in reply: <u>does</u> it travel faster? The observations upon which this conclusion rests appear to me to be next to worthless. But supposing it does, my temporary answer would be 'that it is certain that <u>sliding</u> along the bottom *[1 word illeg]* hence to some extent.' The theory of Saussure coexists to a certain extent with our theory; thus causing the motion to be made up of two true causes. Now the sliding will increase in wet warm weather and this would be an answer sufficient to deprive an objection founded on the faster movement in wet weather of all force. We cannot hope to solve every question at once, and to attempt now to examine the plasticity of ice is quite out of the question. Others have examined it—Parsons[5] & Dolfuss[6] for example—and found it <u>more flexible</u>, but I know what led them to think so, although I have made no mention of it. Indeed numberless things have occurred to me that I have not mentioned, simply because I wanted to present the subject as unencumbered as possible.

RI MS JT/1/TYP/9/2877
IC HP 8:25 (fragment)
Typed Transcript Only

---

1. *[1857]*: the year is given by the reference to Tyndall and Huxley's glacier paper (see n. 4).

2. *Some term is necessary . . . with that of <u>refreezing</u>*: Tyndall settled on *fracture* and *regelation* for his theory of glacier motion. See letter 1306, n. 10.

3. *Grindelwald glacier*: two glaciers, the Lower Grindelwald (Unterer Grindelwaldgletscher) and the Upper Grindelwald (Oberer Grindelwaldgletscher) south of the town of Grindelwald, on the north side of the Bernese Alps.

4. *reading of the paper on the 15th*: J. Tyndall and T. H. Huxley, 'On the Structure and Motion of Glaciers', *Phil. Trans.*, 147 (1857), pp. 327–46. Tyndall read this paper at the RS on 15 January 1857, commenting in his 16 January letter to Hirst, 'I had an hour and a half talking at the Royal Society last night and am very weary after it. In fact I was weary before I commenced, and the papers merely laid an additional pound upon the bending back of the camel' (letter 1319).

5. *Parsons*: Charles Cléophas Person (1801–84), a French professor of physics at Besançon. See C. C. Person, 'Sur la chaleur latente de fusion de la glace', *Paris, Comptes Rendus*, 30 (1850), pp. 526–28.

6. *Dolfuss*: Daniel Dollfuss-Ausset (1797–1870), an Alsatian industrialist who also conducted glacier research; published the multi-volume *Matériaux pour l'étude des glaciers* in the 1860s and 70s (*ADB*).

From Henry Moseley 7 January 1857 1313

Olveston | 7 Jan 1857.

My dear Sir,

I have read the paper of Ch. Desor[1] which you mentioned to me[2] with great interest, and am truly obliged to you for directing my attention to it. I am ashamed to say that I have neither seen the work of De Saussure nor those of Charpentier[3] nor Agassiz. None of them are in the Library of the Royal Society. If you have the books at hand will you oblige me by giving me the names of the publishers and by telling me something about the size that I may judge of the cost. Allow me to thank you lastly for the copy of your paper on Slaty Cleavage[4] which I have now read through with sincere pleasure. I congratulate you on a very successful effort in Philosophical induction

Yours sincerely | H. Moseley.

P.S. I thank you for your kind promise to inform me of any papers which come to your knowledge about glaciers. I cannot feel convinced that the temperature of the depth of the glacier is only zero. Experiments in open holes however deep are I think fallacious—or in any holes into which water drips. I propose to go to Switzerland this year—But unhappily I must be back at Bristol[5] on the 1 July.

RI MS JT/TYP 3/901
LT Typescript Only

---

1. *paper of Ch. Desor*: not identified. Desor published many papers on glaciers in the 1840s and 50s, as well as the book *Excursions et sejours dans les glaciers et les hautes régions des Alpes de M. Agassiz et de ses compagnons de voyage* (Neufchâtel, 1844).
2. *which you mentioned to me*: letter missing.
3. *Charpentier*: Jean de Charpentier (1786–1855), a German-Swiss geologist who studied glaciers in the Swiss Alps (*CDSB*).
4. *your paper on Slaty Cleavage*: see letter 1295, n. 3.
5. *I must be back at Bristol*: in 1854 Moseley became vicar of Olveston, Gloucestershire, a small town ten miles north of Bristol (*ODNB*).

From Lyon Playfair 7 January 1857 1314

Department of Science and Art, | South Kensington, London, S.W.
| 7 day of Jan[y] 1857

My dear Tyndall

Very many thanks for your letter[1] & its suggestions—which shall be acted upon when the List[2] becomes a National one—at present it is merely a Local one for which a Local Committee is responsible. Your doubtful books shall be struck out. You have somewhat misunderstood the general aspect of the Experiment. I quite agree with you that popular lectures produce little result. You know Ireland for very many years has had popular Lectures delivered at the Expense of the State in provincial towns.[3] The Royal Society was pressing thro' the Parliamentary Committee[4] for a similar grant for the same purpose in England—Doubting very much the utility of such an Expenditure, without testing its results I established in Ireland a system of Examination after the popular lectures Expecting that the results would prove next to nothing & that an argument would be provided against a system which I thought unwise. But I am glad to say that the Experience of 15 Courses, attended by these Examinations, makes me doubtful whether I was right in my opinion. The Examinations have forced the pupils to read & study & the answering has been surprisingly good.

But the question in its general national aspect has nothing to do with Lectures. It is this—The State wishes a certain knowledge in Science; it is willing to pay for this. The knowledge may be had in the School, College, or garret,[5] by books, oral demonstration, or Experiment—in any way—if its <u>attainment is effected</u>, we will pay for it.

In its broad Catholicity it is this—

If Queenwood[6] or the Royal Institution or a National School at Battersea,[7] or an Artisan at Poplar[8] produces a certain knowledge of Science in its pupils, we will pay those pupils for proving that they have this knowledge & we will go further & pay the Teacher so much for Every pupil whom he has sent up with the required standard of knowledge.

It is the <u>result</u> which we wish to be attained. How this result is to be attained would be left to private enterprise.

This is the idea which we are trying only Experimentally to find out its weak & strong points & I shall only be too glad to hear them pointed out by good thinkers & Earnest Educationalists like yourself.[9]

Yours very Sin[cere]ly | Lyon Playfair

D[r]. Tyndall | Royal Institution

RI MS JT/1/P/115
RI MS JT/1/TYP/3/982–83

---

1.   *your letter*: letter missing.
2.   *the List*: not identified.

3. *popular Lectures delivered at the Expense of the State in provincial towns*: see E. Leaney, 'Missionaries of Science: Provincial Lectures in Nineteenth-Century Ireland', *Irish Historical Studies,* 34 (2005), pp. 266–88.

4. *Parliamentary Committee*: the Parliamentary Committee of the BAAS. While President of the RS Lord Wrottesley also served as Chairman of the BAAS Parliamentary Committee and presented a report on the government's role in improving science in Britain. The report recommended that the government support affordable scientific lectures that would be followed by examinations. See 'Report of the Parliamentary Committee', *Brit. Assoc. Rep. 1855,* pp. xlvii–lxiii. See also, 'Obituary Notices of Fellows Deceased', *Roy. Soc. Proc.,* 16 (1867–68), pp. lxiii-lxiv.

5. *garret*: a room on the uppermost floor of a house; an apartment formed either partially or wholly within the roof, an attic (*OED*).

6. *Queenwood*: see letter 1298, n. 13.

7. *a National School at Battersea*: National Schools were founded during the nineteenth century in England and Wales by the National Society for Promoting the Education of the Poor with the goal of providing elementary education to the children of the poor. There were a number of National Schools established in the industrial neighborhood of Battersea in South West London including St. John's, Christ Church, and St. Mary's.

8. *an Artisan at Poplar*: Poplar is a neighborhood in East London where the East and West India Docks were located.

9. *Earnest Educationalists like yourself*: for Tyndall's contributions in science education beginning as an examiner in 1853 in the government Department of Science and Arts under the supervision of Playfair, see N. D. McMillan, Ireland and the Reform of the Politics and Government of British Science and Education', in N. D. McMillan, J. Cooke, and D. D. G. McMillan (eds), *Prometheus's Fire: A History of Scientific and Technological Education in Ireland* (Carlow: Tyndall Publications, 2000), pp.481–525.

From Henry Moseley   9 January [1857][1]   1315

Olveston | 9 January

My dear Sir,

I think these questions will do perfectly.[2]

Will you allow me to take the opportunity of consulting you on a scientific question?

In a paper of mine[3] in the Proceedings of the Royal Society, on the descent of Glaciers, I have quoted from the Annales de Chimie[4] the <u>results</u> of some Russian experiments[5] on the expansion of ice. The <u>results</u> only are there given, reference being made for particulars of the experiments to 'Archiv. f. Wissenchaftl Kunde v. Russland, Bd. VII, S. 333'.

Are you acquainted with the particulars of any experiments on this subject? Or can you put me in the way of procuring the volume here referred to, or of seeing it?

Yours truly | Henry Moseley.

RI MS JT/1/TYP/3/902
LT Typescript Only

1. *[1857]*: the year is given based on Tyndall's active writing and lecturing on his glacier research at this time.

2. *these questions will do perfectly*: presumably this letter is a reply to Tyndall's reply (missing) to Moseley's letter of 7 January (1313).

3. *a paper of mine*: H. Moseley, 'On the descent of Glaciers', *Roy. Soc. Proc.*, 7 (1854–55), pp. 333–42.

4. *Annales de Chimie*: *Annal. Chim. et Phys.*, a French journal for chemistry and physics established in 1789.

5. *some Russian experiments*: Moseley quoted in his paper (p. 339) data of [Heinrich Christian] Schumacher, [Uno] Pohrt, and Moritz [von Jacobi] from 'Vide Archiv. f. Wissenchaftl Kunde v. Russland, Bd. vii, S. 333'. See 'Ueber die Arbeiten der Petersburger Academie der Wissenschaften in Jahre 1847', *Archiv für wissenschaftliche Kunde von Russland*, 7 (1849), pp. 322–58, on 333. Moseley misspelled 'Pohrt' as 'Johrt'.

From Thomas Archer Hirst   11 January 1857   1316

Pau | 14 Rue de la Mairie | Jan^ry 11th 1857

My dear John,

(1 P.M.) To day is a dark day for me in every sense of the word. Outside it is stormy, the wind howls and causes the shutters of the windows in every house in the street to flap angrily against the walls. The rain which instead of falling is driven before the wind beats furiously against the windows. The smoke instead of ascending the chimney is driven down by gusts and the fire almost blown out. Worst of all my poor suffering wife[1] lies close to me in bed breathing with difficulty and able to speak only in whispers. We feel that some crisis is approaching. John Martin is also with us but—in this moment of anxiety I naturally take up my pen to write to you rather than talk to him in order to prevent that tendency of mine to ponder too much over misfortunes present and to come.

Up to the close of the past year Anna appeared to be improving wonderfully we were all in good spirits The only drawback to our contentment was the bad health of John Martin, who suffers frequently from bad attacks of Asthma. But at the commencement of this year and in spite of all our precautions we fear that Anna caught fresh cold. At any rate she has been very very poorly ever since. Of course her strength had been diminishing for a long time, nor even before this last attack could we see that she was gaining strength although her cough was certainly improving and her appetite good. Our lodging[2] is on the first floor, when we first came to Pau[3] she could ascend the stairs by herself without very much fatigue. For the last month, however, I have always had to carry her up. Such being her weak state before you may imagine how this new attack has prostrated her. She can with difficulty walk across the floor, every cough exhausts her, she pants for breath afterwards and feels a terrible oppression of her chest. The doctor comes daily now and to day will come twice bringing another physician with him to try if together they can succeed in relieving her. The difficulty is to relieve the oppression on her chest by counter-irritants, and the fear is that by employing such measures they may weaken her too much. What the result will be God only knows. The Doctor does not say there is any imminent immediate danger, but for my part I fear another week like the one she has suffered through would leave her with very little more strength. These, however, are my own fears merely. I console myself by the thought that the Doctor considers her condition less serious that I do.

(9 P.M.) The doctors have had their consultation and both assure me that they do not think there is immediate danger. Their opinion is that she will rally from this attack but that she is in a very precarious condition. It appears to be certain now that both lungs are diseased. Thus although they have some-what relieved me from present anxiety these doctors have darkened my future prospects. For a long time I had clung to the hope that with one lung compar-atively sound, with care, and the beneficial influence of climate she might bear up and improve. Of this old hope but little now remains.

This evening she is a little better I hope and think she is past the worst. Her bedroom opens into our sitting room. I lifted her out of bed a while ago wrapped her in blankets and carried her to the fire in order to let her bed be made afresh. She enjoyed the little change for three-quarters of an hour, during which time I played for her on the piano. I have carried her to bed again she feels no worse but better, and I trust she will sleep quietly. Should any change occur I will write a line immediately. <u>If you hear nothing you may interpret my silence favourably</u>. I have scarcely any inclination to write on other subjects had I such to write about. My spare time I devote as usual to my mathematical work in which I am making some progress but I cannot say much. I am in the midst of difficulties and as fast as I conquer them others rise. One thing is certain my subject increases in breadth as I work at it and will surely be worth something when I have done with it.

But it is long since I heard of or from you. I have looked forward to receiving a Phil. Mag. for the last two months, thinking I should there hear of you, but none has arrived. I wrote some time ago to Francis telling him to send me the last <u>two numbers,</u> but strange to say they have not yet come to hand. <u>Tell him this if you please.</u>

I have not the time to write another letter tonight so perhaps you will have the goodness to send this to Wright. Although it contains bad news I know both M[r] and M[rs] Wright[4] would be sorry to be kept in ignorance of this unfortunate change in Anna's health.

Write soon to

Yours affectionately | T. A. Hirst.

RI MS JT/1/H/230
RI MS JT/1/HTYP/485–86

––––––––––

1.   *my poor suffering wife*: Anna Hirst.
2.   *our lodging*: not identified.
3.   *first came to Pau*: see letter 1298, n. 4.
4.   *M[rs] Wright*: Fanny Wright.

To Richard Dawes[1]   13 January 1857   1317

Royal Institution of Great Britain [stamped]

My dear M[r] Dean,

I have no doubt that I have to thank your kindness for the little pamphlet on 'Effective Primary Instruction,'[2] which I hope soon to be able to give myself the pleasure of reading. It is a strange country this England of ours—and must in the matter of education be dealt with after strange methods. No continental model will suffice us. We must learn the conditions of the question by actual experiment; and if this is earnestly persisted in, the problem will yield as sure as any other problem in experimental science. That it <u>will</u> be persisted in your actions prove, and the encouragement of your example in this respect will continue after you have ceased to write pamphlets on Effective Primary Instruction.

Ever yours | <u>John Tyndall</u> | 13th <u>Jan.</u> 1857.

Private (permission to publish granted through RI)
RI MS JT/1/TYP/1/319

1. *Mr Dean*: Richard Dawes (1793–1867), Dean of Hereford, an educationalist concerned
    with improving secondary schools in England (*ODNB*). Dawes was Rector and Headmas-
    ter of King's Somborne near Queenwood, where he originally met Tyndall, before being
    translated to Hereford in 1850.
2. *little pamphlet on 'Effective Primary Instruction'*: Dawes published this address as *Effec-
    tive Primary Instruction: The Only Sure Road to Success in the Reading-Room, Library, and
    Institutes for Secondary Education* (London: Groombridge & Sons, 1857).

From Joseph Plateau[1]   15 January 1857   1318

Gand 15 Janvier 1857.

Monsieur,

Bien que nos relations n'aient été jusqu'ici qu'indirectes, et consistant simplement dans l'échange mutuel de nos publications, je n'hésite pas à vous demander un léger service.

Je m'occupe en ce moment à mettre en ordre le manuscrit d'un ouvrage que j'ai commencé il y a une vingtaine d'années: c'est une bibliographie analytique de certains phénomènes subjectifs de la vision, savoir de la <u>persistance des impressions sur la rétine, des couleurs accidentelles, et de l'irradiation.</u> Je dois consigner dans cette bibliographie plusieurs passages des travaux de Newton, et il est un de ces passages, qui ne fait point partie de l'optique de Newton, et dont j'ignore la date ; je sais seulement qu'il a paru originellement dans le N° 88, page 5086, <u>du Philosophical Collections.</u> Comme il m'a été impossible de trouver cet ouvrage en Belgique, et qu'il doit exister nécessairement dans les bibliothèques de Londres, je vous serais bien reconnaissant si vous aviez l'ex-trême bonté de l'y chercher, et de me faire connaître la date de la publication du N° dont il s'agit. J'espère, Monsieur, n'avoir pas compté en vain sur votre obligeance, et j'attends un mot de réponse de votre part.

Nous suivons, en physique, des routes entièrement différentes, mais je ne me tiens pas moins au courant de la plupart de vos belles recherches, et j'ai admiré entre autres les expériences au moyen desquelles vous établissez d'une manière décisive l'existence de la polarité dia-magnétique. J'adopte aussi pleinement les raisons par lesquelles vous combattez l'hypothèse d'un milieu magnétique dans l'espace.

Agréez, Monsieur, l'assurance de tous mes sentiments de haute considération.

J. Plateau | Professeur à l'Université, | Place du Casino 22, | <u>Gand.</u>

Ghent 15 January 1857.

Sir,

Although our relations have been thus far indirect, and consisting simply in the mutual exchange of our publications, I do not hesitate to ask of you a small favour.

I am currently working on organising the manuscript of a book I started nearly twenty years ago: it is an analytical bibliography[2] of some subjective phenomena of the vision, that is to say of the <u>persistence of impressions on the retina, of accidental colours, and of irradiation.</u> I have to record in this bibliography several passages of Newton's works, and there is one of these passages, which is not part of Newton's Optics,[3] and of which the date I don't know. I only know that it was originally published in the No 88, page 5086, of the Philosophical[4] <u>Collections</u>. Because it was impossible to find this publication in Belgium, and that it must inevitably be available in the libraries of London, I would be very grateful if you were so greatly kind as to look for it there, and to let me know of the date of the publication of the N° in question. I hope, Sir, not to have relied on your kindness in vain and I am awaiting a short reply from you.

We follow, in physics, entirely different paths, but I am not any less aware of most of your fine research, and I have admired among other things, the experiments out of which you conclusively establish the existence of the dia-magnetic polarity. I also fully embrace the reasons whereby you fight the hypothesis of a magnetic medium in space.

Please accept, Sir, my highest regards.

J. Plateau | Professor at the University, | Place du Casino 22, | <u>Ghent</u>

RI MS JT/1/P/88

---

1.   *Joseph Plateau*: Joseph Antoine Ferdinand Plateau (1801–83), a Belgian physicist and professor at the University of Ghent who studied optics. In 1832 Plateau invented a device to illustrate moving images called the 'phenakistoscope' (*CDSB*).

2.   *analytical bibliography*: J. Plateau, *Bibliographie analytique des principaux phénomènes subjectifs de la vision, depuis les temps anciens jusqu'à la fin du 18e siècle, suivie d'une bibliographie simple pour la partie écoulée du siecle actuel* (Brussells, 1878).

3.   *Newton's Optics*: I. Newton, *Opticks: Or, a Treatise of the Reflexions, Refractions, Inflexions and Colours of Light* (London, 1704).

4.   *one of these passages . . . of the* <u>*Philosophical*</u>: I. Newton, 'Mr. Isaac Newtons Answer to Some Considerations upon His Doctrine of Light and Colors; Which Doctrine Was Printed in Numb. 80. of These Tracts', *Phil. Trans.*, 7 (1672), pp. 5084–103.

To Thomas Archer Hirst   16 January 1857   1319

16th Dec <u>Jan.</u> 1857

My dear Tom.

I am tired, very tired, but perhaps the quietness of mind which weariness induces is the best suited to reply to your last sad letter.[1] I wished as I read it that I was at your side, and then riper reflection came and demanded what I could do if I were there, and my reply was necessarily 'nothing'. Yet this news which tells me of your gloom and your grief has been a phantom which I have seen sometime in the distance, for I know unhappily too well the insidious notion of that enemy. But I will not dwell upon a subject which must be almost always in your thoughts; and from which it would be well if you could in some degree release them. Those consultations of Physicians Tom, and other incidents of your position in Pau[2] must press heavily upon your means. Let me here press my former offer upon your attention—just use the hint if you need it, and 20, 30, 40, 50 up to 100 pounds shall be promptly at your disposal.

I had an hour and a half talking at the Royal Society last night[3] and am very weary after it. In fact I was weary before I commenced, and the papers merely laid an additional pound upon the bending back of the camel. I saw Francis yesterday—he says the Phil. Mag. has not been sent to you, but that it shall be.[4]

I will try and /scrape/ a few matters together that are likely to interest you in a few days—I ought indeed to have sent you something long ago. Next week I have 3 lectures, one at the London Institution,[5] a 3 oC. lecture at the R.I.[6] and a Friday evening upon the glaciers.[7] This latter subject has been of great service to me by recovering a trust in the power of patience & intellect combined. beyond this it is not quite pleasant to me—at a future day I will tell you why. I have strong hope that your mathematical paper[8] when ready will as you say be worth something. To me it is impossible that you could continue looking for a length of time steadfastly at such a subject without making it worth something. Depend upon it all you need is this persistence—Keep your eye upon your work. I wish you could drive all else as regards the necessity of life from your mind—I have the firmest trust that if you could do so you would find in yourself powers, and in your efforts a success of which you do not yet dream. This trust in yourself, a trust to be awakened only by successful effort is what I want you to feel. It will come someday, and when it comes all my anticipations will assuredly be justified. Give my love to Anna[9]—would that I had the power to set her right again.

ever yours affectionately | John Tyndall.

RI MS JT/1/T/636

---

1. *your last sad letter*: letter 1316, in which Hirst related to Tyndall the deteriorating health of his wife Anna.

2. *your position in Pau must press heavily upon your means*: Tyndall often assisted Hirst in times of financial need, as Hirst had done for him previously. For Pau, see letter 1298, n. 4.

3. *at the Royal Society last night*: Tyndall read his and Huxley's glacier paper at the RS on 15 January 1857, published as J. Tyndall and T. H. Huxley, 'On the Structure and Motion of Glaciers', *Phil. Trans.*, 147 (1857), pp. 327–46. See letter 1312.

4. *he says the Phil. Mag. has not been sent to you, but that it shall be*: see letter 1316.

5. *the London Institution*: on Monday 19 January 1857, Tyndall gave an evening lecture at the LI, 'On the Nature and Phenomena of Light'. The LI was an educational institution founded in 1806 that was a precursor to the University of London, where Tyndall was a visiting lecturer from 1856–59. He would again lecture for the LI on light—'Six Lectures on the Phenomena of Light'—beginning 14 December 1857 and each following week through 18 January 1858.

6. *a 3 oC. lecture at the R.I.*: on Thursday 22 January 1857, Tyndall lectured at the RI 'On Sound'.

7. *a Friday evening upon the glaciers*: on 23 January 1857, Tyndall lectured at the RI on 'Observations on Glaciers', published as J. Tyndall, 'Observations on Glaciers', *Roy. Soc. Proc.*, 8 (1856), pp. 331–38.

8. *your mathematical paper*: Hirst published his paper in two parts: T. A. Hirst, 'On Equally Attracting Bodies', *Phil. Mag.*, 13 (May 1857), pp. 305–24; and T. A. Hirst, 'On Equally Attracting Bodies', *Phil. Mag.*, 16 (September 1858), pp. 161–72.

9. *Anna*: Anna Hirst.

From Mary E. Lyell[1]   18 January [1857][2]   1320

Dear Dr Tyndall

If you <u>can</u> will you kindly let me have two tickets for your lecture next Friday[3] for myself & my sister M[rs] Lyell.[4] Sir Charles[5] has his own admission. I am curious to hear about the Mont Blanc[6] Glaciers. We were only in the Zermatt[7] once last autumn

Very truly y[ou]rs | Mary E Lyell | 53. Harley St. W. | 18. Jany

RI MS JT/1/L/61
RI MS JT/1/TYP/3/844

---

1. *Mary E. Lyell*: Mary Elizabeth Lyell (née Horner, 1808–73), a geologist, conchologist, and wife and scientific partner to Charles Lyell. See M. Ogilvie and J. Harvey (eds), *Bio-*

graphical Dictionary of Women in Science, 2 vols (London: Routledge, 2000), vol. 2, pp. 813–14.

2. *[1857]*: the year is given by the reference to Tyndall's Friday evening lecture on glaciers on 23 January 1857.

3. *your lecture next Friday*: on 23 January 1857, Tyndall lectured at the RI on 'Observations on Glaciers'.

4. *my sister Mrs Lyell*: Katharine Murray Lyell (née Horner, 1817–1915), Mary's younger sister, a botanist, married Charles Lyell's younger brother Henry. In 1870, she published *A Geographical Handbook of All the Known Ferns: With Tables to Show Their Distribution* (London: John Murray). See M. Ogilvie and J. Harvey (eds), *Biographical Dictionary of Women in Science,* 2 vols (London: Routledge, 2000), vol. 2, p. 813.

5. *Sir Charles*: Charles Lyell.

6. *Mont Blanc*: the highest mountain in the Alps, on the border between France and Italy. During the period covered by this volume, Mont Blanc was located within the Duchy of Savoy, a part of the Kingdom of Sardinia, precursor to the unified Italian state established in 1861. Savoy was annexed to France in 1860 through the Treaty of Turin.

7. *Zermatt*: see letter 1297, n. 7.

From Edward Frankland 22 January 1857 1321

Owens College | Manchester Jan. 22/57

My dear Tyndall

You request to have my opinion respecting Bunsen's contributions to science;[1] it is with much pleasure that I give it you, although I feel that any brief sketch of mine can convey only a very imperfect idea of the merits of researches so extensive and embracing so many departments of chemical and physical science, as those for which we are indebted to Bunsen.

In addition to numerous other researches, I may mention the following as highly valuable contributions.

1. Researches on Cacodyl (numerous memoirs)[2]
2. A new system of gaseous analysis (contained in numerous memoirs)[3]
3. Researches on the gases evolved from blast furnaces[4] with suggestions for the useful application of those gases.
4. Researches on the Volcanic Phenomena of Iceland.[5]
5. On an improved method for the determination of nitrogen in organic compounds.[6]
6. Researches on chemical affinity.[7]
7. On a new, cheap & powerful galvanic battery.[8]
8. Bunsen & Berthold on hydrated peroxide of iron as an antidote for Arsenic.[9]

Of the researches on Cacodyl it is impossible to speak too highly. From whatever point of view we regard them, they cannot fail to excite our admiration. The difficulty & danger attending the manipulations with those spontaneously inflammable and highly poisonous volatile bodies, which once nearly occasioned fatal results to the operator, are <u>almost</u>, I may even say <u>quite</u>, unparalleled; especially when we consider the singular perseverance with which the subject was followed out into its most minute ramifications. Further, the accuracy of the results of these researches has been fully established by later investigations; for, although at the time Bunsen's researches were made, none of the generalisations, which have sprung from more recent researches on Organo-metallic bodies, could assist the experimenter in determining rational formulae, yet in no single instance has any formula, he then established, since been found to be incorrect. Finally the results of the investigation were of immense importance to science. The beneficial influence which the discovery of Cacodyl has exerted upon the development of sound views in organic chemistry, and the pursuit of rational paths of research, can scarcely be over estimated, since the support which it gave to the theory of compound organic radicals, enabled that theory to exist and to bring forth abundant fruits, throughout a period in the history of the science when there was but feeble evidence of its truth, and until the researches, both of supporters and antagonists, finally established the fundamental accuracy of that theory.

The importance of Bunsen's contributions to gaseous analysis is well known to all chemists. Previous to his labours in this field, eudiometrical determinations[10] were of a very crude and unsatisfactory character; but his improvements in the construction of eudiometers[11] enabled every operator to construct these instruments for himself with facility, and possessing an accuracy of measurement which could not be exceeded in the workshop of the philosophical instrument maker. These improvements, together with his new methods of applying absorbents to gaseous mixtures, have imparted to eudiometrical determinations a degree of accuracy never exceeded and rarely equalled in other departments of analytical chemistry. It is also perhaps worthy of notice that the subsequent improvements in gaseous analysis have been chiefly if not entirely made by the pupils of Bunsen. (Regnault for instance received special instruction in gas analysis from Bunsen).

The researches on gases evolved from blast furnaces consuming charcoal, coke,[12] and coal, the latter made in conjunction with Lyon Playfair, are masterly specimens of experimental skill and analytical accuracy, and have led to important applications of the waste gases from such furnaces to useful purposes; as for instance at Swansea,[13] to the heating of steam boilers, and in Germany, to the puddling of iron.[14]

Bunsens investigations into the volcanic phenomena of Iceland I need not further allude to, as you are much better acquainted with their peculiar merits than I am, and therefore I will only remark on this head, that they exhibit in a high degree, the combination of acute inductive reasoning with great manipulatory resources.

His mode of determining nitrogen in organic compounds, although somewhat difficult of execution, is undoubtedly one of the best methods known, and, with very few exceptions, it is applicable to every organic compound containing that element.

His researches on chemical affinity open up a new path of enquiry which must yield invaluable results, in fact there can be little doubt, that much of the future of chemistry is most intimately connected with such numerical expressions of the forces engaged in the production of chemical phenomena.

You are yourself intimately acquitted with the working of Bunsen's Carbo-zinc Battery and therefore any remark of mine, on this head, would be quite superfluous.

With regard to the employment of hydrated peroxide of iron as an antidote for arsenic,[15] I must refer you, for its value, to your medical friends. I believe this antidote is now almost universally employed in Germany, and with perfect success, provided it be administered before much of the poison has been absorbed into the system.

In conclusion, it is my deliberate conviction, that even the researches upon Cacodyl <u>taken alone</u>, place Bunsen in the very highest position amongst the cultivators of natural science, and that those researches are, for the reasons already stated, unsurpassed in merit by any others with which chemical science has been enriched.

Believe me, my dear Tyndall | ever yours sincerely | E. Frankland.

RI MS JT/1/F/46

---

1. *You request to have my opinion respecting Bunsen's contributions to science*: letter missing. This letter and 1326 from Heinrich Debus are responses to Tyndall requesting opinions about Bunsen's contributions to science. While Tyndall's letters to each are missing, the purpose is likely due to Bunsen being considered for an RS fellowship or an RS medal. Indeed, he was named FRS in 1858 and received its Copley Medal in 1860, 'For his researches on cacodyls, gaseous analysis, the Voltaire phenomena of Iceland; and other researches'. In his journal for 31 October 1857, Tyndall wrote "I did not push Bunsen forward in opposition to Lyell' for the Copley Medal (Journal, RI MS JT/2/13c/1038).

2. *Researches on Cacodyl (numerous memoirs)*: R. Bunsen, 'Untersuchungen über die Kakodylreihe', *Annalen der Chemie und Pharmacie,* 37 (1841), pp. 1–57.

3. *A new system of gaseous analysis (contained in numerous memoirs)*: see, for example, H. Kolbe, 'Eudiometer', in J. Liebig, J. C. Poggendorff, and F. Wohler (eds), *Handwörterbuch der reinen und angewandten Chemie,* 9 vols (1837–64), vol. 2 (1842), pp. 1050–74.

4. *Researches on the gases evolved from blast furnaces*: R. Bunsen and L. Playfair, 'Report on the Gases evolved from iron furnaces, with reference to the Theory of the Smelting of Iron', *Brit. Assoc. Rep. 1845*, pp. 142–86.

5. *Researches on the Volcanic Phenomena of Iceland*: Bunsen studied the famous 'Giyser' in Iceland in 1846, and developed a theory of geyser eruption. R. Bunsen, 'Ueber den innern Zusammenhang der pseudovulkanischen Erscheinungen Islands', *Annalen der Chemie und Pharmacie*, 62 (1847), pp. 1–59. See also R. E. Oesper and K. Freudenberg, 'Bunsen's Trip to Iceland: As Recounted in Letters to His Mother', *Journal of Chemical Education*, 18 (1941), pp. 253–39.

6. *On an improved method for the determination of nitrogen in organic compounds*: see n. 2, a discussion of 'Stickstoffgehalt', or 'nitrogen content', begins on p. 27.

7. *Researches on chemical affinity*: R. Bunsen, 'Untersuchungen über die chemische Verwandtschaft', *Annalen der Chemie und Pharmacie*, 85 (1853), pp. 137–55; and an English summary, 'Researches on Chemical Affinity', *Quarterly Journal of the Chemical Society of London*, 6 (1854), pp. 82–89.

8. *On a new, cheap & powerful galvanic battery*: Bunsen invented what is now known as the Bunsen cell. R. Bunsen, 'Ueber eine neue Construction der galvanischen Säule', *Annalen der Chemie und Pharmacie*, 38 (1841), pp. 311–13.

9. *Bunsen & Berthold on hydrated peroxide of iron as an antidote for Arsenic*: R. Bunsen and A. Berthold, *Das Eisenoxydhydrat das Gegengift der arsenigen Säure annalen* (Göttingen: Verlage der Dieterichschen Buchhandlung, 1834).

10. *eudiometrical determinations*: eudiometry, the art or practice of using the eudiometer either for ascertaining the purity of the air, or in the analysis of gases (*OED*).

11. *eudiometers*: an instrument for testing the purity of the air, or rather the quantity of oxygen it contains (*OED*).

12. *coke*: the solid substance left after mineral coal has been deprived by dry distillation of its volatile constituents (*OED*).

13. *Swansea*: a coastal city and county in Wales.

14. *puddling of iron*: the process of heating and stirring molten pig iron with iron oxide in a reverbatory furnace, so as to oxidize and remove carbon and other impurities and produce wrought iron (*OED*).

15. *an antidote for arsenic*: see n. 9.

From William Frederick Pollock   24 January 1857   1322

59 Montagu Square | 24th. Jan. 1857.

My dear Tyndall

I was not able to get at you last night to thank you for your lecture[1] & to congratulate you upon having solved the vexed question of glacier movement[2]—& in having escaped not only uninjured but triumphant from the difficult regions . . . 'dove la gelata Ruvidamente un' altra gente fascia'[3]—You, instead of being tortured by the ice, like Dante's worst sufferers, have put the ice to the torture of experiment and it answered your question most satisfactorily—The viscous theory,[4] which was a great stumbling block—and must have been truly repugnant to the feelings of every one, may be now considered as relegated to the same lumber-room of obsolete theoretical apparatus, in which the crystalline spheres of the old astronomy,[5] phlogiston,[6] & various other similar things which have done their temporary work may be supposed to repose.

By the bye the same canto of the Inferno,[7] from which I have quoted contains an authority for the word 'regelation'—At Inf°. 33.v. III you have—

> Levatemi dal viso i duri veli,
> Sì ch'io sfoghi 'l dolor, che 'l cuor m'impregna,
> Un poco pria, che 'l pianto si <u>raggeli</u>—[8]

I have no doubt but that Dante coined the word—& Lammenais in his French translation[9] has adopted it into his own language as '<u>regéler</u>'[10]—

I am sorry that we are not to see you on Tuesday at dinner, but I suppose there is a lady in the case—

I am | Ever yours truly | W. F. Pollock

RI MS JT/1/P/225
RI MS JT/1/TYP/6/1893

---

1.   *your lecture*: on 23 January 1857, Tyndall lectured at the RI on 'Observations on Glaciers'.

2.   *the vexed question of glacier movement*: see letter 1306, n. 8 and 10.

3.   *'dove la gelata Ruvidamente un' altra gente fascia'*: 'Further along we went, to where the ice roughly enswathes another class of people' (Italian), from Canto 33 (line 91–92) of *The Divine Comedy of Dante Alighieri: The Italian Text with a Translation in English Blank Verse and a Commentary by Courtney Langdon*, 3 vols (Cambridge: Harvard University Press, 1918), vol. 1 (Inferno), p. 381 (originally published in 1321).

4.   *The viscous theory*: see letter 1306, n. 8.

5.   *crystalline spheres of the old astronomy*: a reference to the geocentric model of the universe in Greek Antiquity, with Earth at the center and the Sun, planets, and other celestial bodies orbiting on crystalline spheres.

6.   *phlogiston*: a reference to the theory in the sixteenth and seventeenth centuries that combustion could be explained by the presence of a fire-like element (phlogiston) that was contained within combustible bodies. This theory was replaced by the oxygen theory of combustion in the eighteenth century.

7.   *the Inferno*: the first part of Dante's *The Divine Comedy* (1321), see n. 3.

8.   *Levatemi dal viso i duri veli, | Sì ch'io sfoghi 'l dolor, che 'l cuor m'impregna, | Un poco pria,*

che 'l pianto si <u>raggeli</u>—: 'remove for me the hard veils on my face, that I may somewhat vent the pain that fills my heart, before the tears freeze up again' (Italian), from Canto 33 (line 112–14) of *The Divine Comedy of Dante Alighieri.*

9.   *Lammenais in his French translation*: Hugues Felicité Robert de Lamennais (1782–1854), a French priest and philosopher, translated Dante's *The Divine Comedy* as *La Divine Comédie,* but it was not published until after his death (Paris: Didier C and E, 1863).

10.   *'<u>regéler</u>'*: to freeze again (French, *regeler*), in A. G. Collot, *A New and Improved Standard French and English and English and French Dictionary* (Philadelphia: C. G. Henderson & Co., 1856), p. 706.

To William Frederick Pollock   [24 January 1857][1]   1323

Saturday night.

My dear Pollock

I sometimes try to cultivate a little Stoicism with regard to the effect of lectures upon the audience, but I usually find that my nature is too strong for my volition. I must own to the great gratification which I experience from the approval of a mind like yours. I did not hope that the views I ventured to bring forward[2] would take sudden root, and among geologists especially I calculate upon encountering a good deal of inertia; but the more I thought of the matter, the more I felt persuaded that time and further testing would justify me. Expecting opposition, I confess your note[3] has had a cheering and encouraging effect upon me—and I take it very kind of you to have written it. Many thanks for the reference to Dante[4]—it will invest the term with great authority and interest.

I wish I could attribute my confusion about Tuesday[5] to a lady; my blundering brain and nothing else is to blame—You can hardly imagine how ludicrous my forgetfulness is sometimes. But now, thank the gods, my heaviest work is past, and a few days quiet will cause the mental molecules to take up their normal positions.

Believe me | Ever yours | John Tyndall

RI MS JT/1/T/1257
RI MS JT/1/TYP/6/1894

---

1.   *[24 January 1857]*: the date is given by the relation to letter 1322.

2.   *the views I ventured to bring forward*: the topic of Tyndall's lecture at the RI on 23 January 1857; see letter 1306, n. 8 and n. 10.

3.   *your note*: see letter 1322.

4.   *reference to Dante*: in letter 1322, Pollock shared with Tyndall that Dante might have coined the term *regelation*.

5.   *my confusion about Tuesday*: this reads as if Tyndall forgot to join Pollock for dinner (per-
      haps on Tuesday 20 January), yet in letter 1322 Pollock had written 'I am sorry that we are
      not to see you on Tuesday at dinner', indicating that dinner would be the following Tues-
      day, 27 January. Tyndall's journal is incomplete for this time and provides no clarification.

To Thomas Archer Hirst   26 January 1857   1324

26<u>th</u> January 1857

My dear Tom,

I have just returned from Hampshire[1] and before I permit the vortex of duty to lay hold upon me again I will assert my freedom for an hour and write to my old friend. I travelled to Bossington Churchyard[2] yesterday as fast as train and horse could carry me and arrived just in time to see the mourners departing.[3] I stood over the little grave and could not squeeze back my tears. I took the trap[4] on to Queenwood[5] and the bitter sleet that smote us as we crossed the hill along the Roman road[6] gave me an excuse to wrap my plaid[7] around my head I actually wept like a child, and though thick with sorrow and sympathy for poor Wright and his wife I blessed God to feel that I was still human. I slept at their house, and should have liked to talk to them in my own way; but I was forced to fall into the talk which had been established before my arrival and which was light and common place instead of earnest. Such talk of course staves off serious thought and may do a temporary good; but it does not impart strength; it does not arm the soul to cope with sorrow, and when it is past and the trifles are gone the poor thwarted heart will have its revenge. Thus though my going may gratify Wright & his wife as a testi-mony of respect on my part I fear the benefit must end here. As far I could see however they bear up against their calamity bravely. Another death occurred at Queenwood on the same week: that sickly boy Willie Smith[8] has also taken wing. I saw Debus, but had time for only a few minutes conversation with him. Of business matters I only heard a little addition to George's[9] catalogue of follies, and a very unpleasant matter regarding your successor and his wife[10]—To my mind your sorrow and your grief are not to be compared with what that man must suffer. There is something holy and elevating in afflic-tion—but for those who hate each other and whose aim seems to be to render each other miserable to be compelled to live in proximity to each other, to sleep together, to eat together, to travel the same road like two chained sheep pulling in opposite directions, if there be death in life surely this is it. But Wright will have given you an account of these things first hand so I may spare myself the trouble of dwelling upon the subject. It rejoices me to hear of the change in Anna's health;[11]—though the expression is so conventional it warms my heart when she sends me her love. Were I not held here by chains beyond my power to sever I would assuredly set out and soon be at your side. I long to see Anna and to talk to you both. Tom, that thought regarding the diminution of your power is a falsehood and must be trampled out. You had never half the reason that I had to dread such a diminution, and I know now, though worn down by fatigue, that the fear was a delusion. Your body must be strengthened, but this you neglect to do. Last Friday I had to lecture:[12] a year or two ago I should have spent the day puzzling over my experiments—but I left the Institution,[13] took a pair of oars in my hand and gathered physical strength upon the Serpentine;[14] and this proved more effective in tendering my lecture successful than if I had all the mechanisms of my lecture perfect and a broken soul speaking through the mechanisms. At the risk of provoking a smile I will tell you what I have found infinitely advantageous to me. I firmly believe that had I continued to take coffee for breakfast it would have killed me. I have long given it up and now take chocolate. I find it an excellent thing to take a basin[15] of chocolate before going to bed—excellent for the bowels and well worth a trial on your part. But habitual exercise and a dismissal of science from the thoughts is an exceeding gain in the long run. I think a man must not expect a desire to be always strong—there are certain problems which require a certain intensity to crush them, and which can never be conquered by moderate work: but even here a sharp eye must be kept upon the relation of mind and body, and it is only by preserving the latter in a certain state of vigour that the maximum effect is produced. If you feel muddled do not persist—This is Sir Henry Holland's[16] warning to me—it is mine to you. But I wish I was near you: I would often draw the sweat from your brow—one ounce of such sweat would be worth a week's study. I am almost an idea at your side, and hear Anna saying bravo! to me while I drive you from your books. Here I pause. God bless you Tom and God bless Anna.

Ever yours affectionately | John Tyndall

RI MS JT/1/T/637
RI MS JT/1/HTYP/488–89

---

1. *just returned from Hampshire*: a county on the southern coast of England.
2. *Bossington Churchyard*: a church in the village of Bossington in Hampshire.
3. *the mourners departing*: the daughter of Richard Wright, Tyndall's friend from Queenwood, Alice Maud Wright (1855–57). Tyndall wrote in his journal on 26 January 1857, 'Wright has lost his lovely little daughter, I went to Bossington yesterday, and could not squeeze back the tears quite as I stood over the little grave' (Journal, RI MS JT/2/13c/917).
4. *trap*: a small carriage on springs; usually, a two-wheeled spring carriage, a gig, a spring-cart (*OED*).

5. *Queenwood*: see letter 1298, n. 13.

6. *Roman road*: an old Roman road passed near Queenwood College, linking the Hampshire towns of Sarum (now Salisbury) and Winchester. See Russell, *Edward Frankland,* p. 38.

7. *plaid*: a twilled woollen cloth, usually with a chequered or tartan pattern (*OED*).

8. *Willie Smith*: William Singleton Smith (1835–57), nephew of George Edmondson, died on 24 January 1857. See 'Deaths', *British Friend: A Monthly Journal, Chiefly Devoted to the Interests of the Society of Friends,* 15 (1857), p. 56.

9. *George's*: George Edmondson.

10. *your successor and his wife*: according to William Brock regarding Hirst's replacement, 'There was much discussion over his successor and Hirst did persuade Edmondson that he needed "a superior man" on the same salary he had been receiving. Unfortunately, his successor's name is not recorded'. See Brock, 'Queenwood', p. 19.

11. *Anna's health*: see letter 1316, where Hirst relayed to Tyndall Anna's current improvement.

12. *Last Friday I had to lecture*: on 23 January 1857, Tyndall lectured at the RI on 'Observations on Glaciers'.

13. *the Institution*: the RI.

14. *upon the Serpentine*: a recreational lake in Hyde Park, west of the RI.

15. *basin*: a bowl.

16. *Sir Henry Holland's*: Henry Holland (1788–1873), a British physician (notably, to the Royal family) and President of the RI from 1865–73 (*ODNB*).

From Jules Lissajous   29 January 1857   1325

Paris 29 Janvier 1857

Mon cher Monsieur,

J'ai bien des reproches à me faire, de vous avoir laissé si longtemps sans réponse, j'espère néanmoins que vous ne me garderez pas rancune. J'ai eu en effet tant d'occupations depuis un mois que je n'ai pas eu le temps de vous écrire.

J'accepte avec le plus grand plaisir les propositions que vous me faites, et je me mettrai à votre disposition dans le courant du mois de juin pour répéter—devant votre auditoire—ces expériences auxquelles vous avez bien voulu accorder une si bienveillante approbation.

J'espère, d'ici là, donner aux appareils plus de perfection, et apporter aux expériences tout le soin nécessaire pour qu'elles réussissent avec l'éclat qui convient à un auditoire aussi distingué.

Je serai heureux de vous revoir à cette époque ainsi que M. Faraday, votre maître, notre maître à tous. J'ai gardé un bien précieux souvenir de sa visite, on est heureux de se trouver en contact avec un grand savant, surtout quand il réunit, comme Monsieur Faraday, au mérite le plus éclatant, la plus aimable simplicité.

Je suis heureux d'avoir obtenu au début de ma carrière scientifique, l'approbation si bienveillante du plus éminent physicien de notre époque. C'est un précieux encouragement dont je garderai toujours le souvenir. Veuillez le lui dire pour moi, et l'assurer de ma profonde et éternelle reconnaissance.

Veuillez, cher Monsieur, recevoir pour vous-même l'assurance de mon profond dévouement, et de mes sentiments bien affectueux.

Tout à vous | J. Lissajous

Paris 29 January 1857

My dear Sir,

I have much to blame myself for, to have left you without a response for so long,[1] however I hope you will not bear me a grudge. Indeed, I was so busy this last month that I did not have time to write to you.

I accept with the utmost pleasure the propositions you offer me, and I will be at your service in the course of June to repeat—before your audience—these experiments[2] that you have been so benevolently willing to endorse.

I hope, by then, to have improved the instruments, and to give the experiments all the necessary care for them to succeed with the brilliance suitable for such a distinguished audience.

I look forward to seeing you then as well as Mr. Faraday, your mentor, our mentor to us all. I kept a precious memory of his visit;[3] it is fortunate to be in contact with a great savant, especially when, like Mr. Faraday, he combines to the most striking merit the most amiable simplicity.

I am fortunate to have obtained at the beginning of my scientific career the so benevolent endorsement of the most eminent physicist of our time. It is an invaluable encouragement that I will always remember. Please tell him that on my behalf, and assure him of my deepest and lasting gratitude.

Please, dear Sir, accept the assurance of my deepest devotion and my most affectionate regards.

Truly yours | J. Lissajous

RI MS JT/1/L/24

---

1. *left you without a response for so long*: Tyndall's response to letter 1310, to which this letter is a reply, is missing.

2. *at your service in the course of June to repeat—before your audience—these experiments*: see letter 1310, n. 2.

3. *his visit*: perhaps Lissajous and Faraday met in summer 1856. In November 1855, Faraday was awarded the Commander of the Legion of Honour at the closing ceremony of the Universal Exhibition in Paris. He was unable to attend the ceremony, and when in April 1856 it was realized he had not yet received the award, it was sent to him in London. Faraday felt obliged to travel to Paris, and did so in late July and August 1856. See *Faraday Correspondence*, 5, p. xxxvi.

From Heinrich Debus   31 January 1857   1326

Queenwood College | Stockbridge | Jan. 31. 1857.

My dear John,

I enclose a list of Bunsen's <u>principal</u> investigations:—There are now about 80 boys at Queenwood,[1] and 15 or 16 more expected, in the course of this week. In other respects every thing as usual.

Ever yours | H. Debus.

1) On the antidotes of arsenious acid.[2] This research led to the application of sequioside of iron as an antidote against AsO_3.[3] It has been proved to act in the most effective manner and is now universally in use for the purpose mentioned.

2) A new method to determine nitrogen. Very good, only a little complicated and therefore not much in use.

3) Researches on Kakodyle.[4]

A long and very beautiful investigation. It has two chief merits. a) In 1840 or (1841) when it was published[5] the theorie of the so called 'Zusammengesetzten Radikale' was indeed a mere theorie in as much no Radikal of the kind was known. Now in this investigation Bunsen produces such a Radikal and shows that its qualities are like those of potassium or rather like Manganese although it is composed of AsC_4H_6.[6] The consequence was that the above theorie got a sound fundament, was more generally accepted and led to a very fine series of researches by other chemists. If the time is considered when this was done, some 17 years ago, of course, its true value becomes apparent.

b) the masterly manner in which the extraordinary experimental difficulties were overcome. The methode's then employed have become patterns for all subsequent investigators to similar subjects. Messrs Frankland and Kolbe[7] could give testimony on this point. To experiment with a substance which takes fire in air and is of the most poisonous character required the invention of new apparatus and manipulations.

4) Improvement of the methodes used in gas analysis and the invention of new methodes. A very meritorious piece of work. He was the first to perfect Gas analysis to such a degree of exactness as was unparalled[8] at the time.

5) Investigations on Iceland.[9]

6) Invention of the galvanic battery[10]

7) New method of determining Urea.[11]

8) On electrolytical decomposition of chloride of magnesium, Calcium, Lythium, Barium, Chromium, Manganese and Aluminium and preparation of these metals.[12]

9) On chemical affinity.[13] A very good investigation wherein the action of
the mass of matter on the chemical results is examined. It led to the
conclusion that when the quantity of one substance is increased the
chemical influence of this substance is not proportionately increased,
but a jump takes place; on further addition of the same substance no
effect seems to be produced till, when a certain quantity has been added
a second time, a jump takes place &c. &c. .

10) On the law of the absorbtion of gasses.[14]
This investigation confirms by experiment Dalton's[15] law of the absorbtion of gases. It also shows how to use the coefficient of absorbtion as
a means in analysis either to recognise by it a certain substance or even
to determine their quantity. The investigation is a very interesting and
good one.

There are also many researches made in his laboratory which are in
fact his work. You know also his last investigation on the chemical
action of light.[16] Every work of Bunsen's is done in a masterly and admirable matter.

Should not the above not answer your purpose[17] than I shall be delighted
to send you another description if you will state your want.

RI MS JT/1/TYP/7/2418–19
LT Typescript Only

---

1. *Queenwood*: see letter 1298, n. 13 about Queenwood College, where Debus was the current science teacher. Debus previously studied under Bunsen and served as his assistant in
Marburg.

2. *arsenious acid*: inorganic compound with the formula H_3AsO_3. Arsenic-containing compounds are highly toxic, and Bunsen found that iron oxide hydrate as a precipitating agent
was the most effective antidote to arsenic poisoning, which remains the method today. See
letter 1321, n. 9.

3. AsO_3: arsenite.

4. *Kakodyle*: cacodyl, a compound radical composed of carbon, hydrogen, and arsenic.

5. *when it was published*: R. Bunsen, 'Untersuchungen über die Kakodylreihe', *Annalen der
Chemie und Pharmacie*, 37 (1841), pp. 1–57.

6. AsC_4H_6: the chemical formula for cacodyl (or kakodyle).

7. *Messrs Frankland and Kolbe*: Edward Frankland and Hermann Kolbe (1818–84), a student
of Bunsen's who succeeded him as Professor of Chemistry at the University of Marburg.

8. *unparalled*: this misspelling is in LT's typescript and might have been misspelled by Debus
in the original letter. Tyndall remarked to Hirst, in letter 1440 of this volume, that Debus's
English language skills are 'a great hinderniss to him'.

9. *Investigations on Iceland*: see letter 1321, n. 5.

10. *Invention of the galvanic battery*: see letter 1321, n. 8.

11. *New method of determining Urea*: R. Bunsen, 'Ueber quantitative Bestimmung des Harnstoffs', *Annalen der Chemie und Pharmacie,* 65 (1848), pp. 375–86.

12. *On electrolytical decomposition of . . . these metals*: R. Bunsen, 'Notiz über die elektrolytische Gewinnung der Erd- und Alkalimetalle', *Poggend. Annal.,* 92 (1854), pp. 648–51.

13. *On chemical affinity*: see letter 1321, n. 7.

14. *On the law of the absorbtion of gasses*: R. Bunsen, 'On the Law of Absoprtion of Gases', *Phil. Mag.,* 9 (1855), pp. 116–30, and 9 (1855), pp. 181–201; and R. Bunsen, *Gasometrische Methoden* (Braunschweig: F. Vieweg, 1857).

15. *Dalton's*: John Dalton (1766–1844), an English chemist and physicist, who studied gas pressure (*ODNB*). Dalton's 'law of partial pressures' states that when non-reacting gases are mixed, the pressure of each individual gas contributes to the sum of the total pressure of the mixture.

16. *chemical action of light*: R. Bunsen and H. E. Roscoe, 'Photo-Chemical Researches. —Part I. Measurement of the Chemical Action of Light', *Phil. Trans.,* 147 (1857), pp. 355–80. The paper was read at the RS on 20 and 27 November 1856.

17. *your purpose*: see letter 1321, n. 1. The first instance of the word 'not' is a mistake by Debus.

From Heinrich Gustav Magnus   31 January 1857   1327

Berlin 31 Januar 1857.

Mein theurer Tyndall

Mit Erstaunen bemerke ich daß Ihr lieber Brief vom 6[tn]. d. M. noch unbeantwortet ist.—

Sauerwald versichert daß er eine Rechnung für Sie im Betrage vom 10 Rthl.[1] in die Kiste gelegt hätte, wenn Sie dieselbe nicht gefunden haben so ist sie wahrscheinlich auf dem Costomhouse fortgekommen.

Sie würden mir einen Gefallen thun wenn Sie mir ein Photometer nach Bunsen, in der Art wie sie in London zur Bestimmung der Intensität des Gaslichts angewandt werden, wollten zusenden lassen. Wir könnten dann gegen einander abrechnen. Ich zahle hier an Sauerwald die 10 Rthl. und sende Ihnen den Rest durch Asher.

Sie scheinen sich, mein lieber Tyndall, viel Mühe mit mir geben zu wollen. Es thut mir von Herzen leid daß ich Ihnen viel Zeit kosten soll. Aber es ist mir interessant gewesen auf Ihre Veranlassung die erste Arbeit Regnaults über die Ausdehnung der Gase noch einmal zu sehn. Mir war die Sache aus dem Gedächtniß entschwunden.—Sehen Sie sich ein Mal an: Annales de Chim III Series Bd. IV p 52. Dort finden Sie den Ausdehnungs Coefficienten für alle Gase bis auf die Thausendtheile! Alle zwischen 0,366–0,368 und dann vergleichen Sie damit Bd. V. p. 75. wie verschieden die Ausdehnungscoefficienten sind. p. 77. (in der Mitte) gesteht Regnault zu daß er sich in Betreff der Cyans und der Schweflichten Säure wesentlich geirrt habe, und p. 76. erkennt er sogar die kleine Verschiedenheit in Betreff des Wasserstoffs an, die ich gefunden hatte. Für die übrigen Gase erwähnt er mich nicht, auch wäre das überflüssig gewesen da meine Abhandl. bereits im Bd. IV. 330 der Annales de Chim französisch veröffentlicht war. Denn sie war gleichzeitig mit Regnault erschienen. Beide eröffnen den Jahrgang 1842. Seine Abhandl in den Annal de Ch. meine in Pogg.

In Bezug auf die Spannkraft der Wasser *[1 word illeg]* verhält sich die Sache ähnlich. Obgleich ich meine Arbeit schon im Decemb 1842 der Academie vortrug; dagegen Regnault die seinige im April 1844. Meine findet sich in Pogg. 1844. Febr. seine in Annales de Chim 1844 July.—Doch die Prioritäts=Sachen scheinen mir von sehr geringem Interesse. Daß die Facta gefunden sind! ist die Hauptsache, wer sie gefunden hat ist gleichgültig.

Wer ist denn D^r. Harley der meine Respirations Theory angreift (beiläufig gesagt das beste was ich gemacht habe.) Ich kann ihm zwar nicht sogleich antworten aber es freut mich daß ich Gelegenheit habe auf diese Sache noch ein Mal zurück zu kommen.

Wollen Sie denn nicht ein Mal wieder zu uns kommen? geben Sie uns doch einige Hoffnung!

À propos! Was sagen Sie zu Buff's Brief an mich, der in Pogg. abgedruckt ist. Ich habe ihm vorher geschrieben daß ich nicht hätte aus dem Brief entnehmen können was er meint. Mehr konnte ich nicht thun! er hat ihn doch abdrucken lassen.

Ich sehe so eben daß Sie nach allen meinen Publicationen fragen. Sie finden sie alle in Pogg. mit Ausnahme einer Arbeit über die Ernährung der Pflanzen, die mir jetzt die Ehre eingetragen hat daß ich Associé étranger de la Société Imperiale de l'Agriculture geworden bin.

Geben Sie sich keine Mühe mit mir, das ist verlorene Zeit!!

Grüßen Sie alle Bekannte und Freunde dort. Vor Allem Faraday, Hoffmann, Wheatstone, Gassiot, Francis und wer sich sonst meiner erinnert.

Ihre Schülerin Anna würden Sie schwerlich wieder *[erkennen]*, drum kommen Sie und sehen Sie sich dieselbe an.

Meine Frau grüßt von Herzen. Poggendorff läßt jetzt den hundertsten Band seiner Annalen *[drucken]*. Seine hiesigen Freunde wollen ihm eine kleine Aufmerksamkeit erweisen, und lassen ihn Lithographiren.

Von Herzen | Ihr | G Magnus. | D^r. Tyndall.

Berlin 31 January 1857.

My dear Tyndall

I notice with astonishment that your kind letter[2] of the 6th of this month is still unanswered. —

Sauerwald[3] assures me that he had put a bill for you in the amount of 10 Reichsthaler[4] in the crate; if you have not found it, then it has probably disappeared at the customhouse.

You would do me a favour if you would have a photometer[5] following Bunsen's design[6] sent to me, of the type as they are used in London to determine the intensity of gaslight. We could then settle our expenses. I shall pay the 10 Reichsthaler here to Sauerwald and send you the difference through Asher.

You seem, my dear Tyndall, to be willing to go to a great deal of trouble with me. I am sincerely sorry that I should cost you a great deal of time. But it has been interesting for me, at your prompting, to have another look at Regnault's first work on the expansion of gases.[7] The thing had slipped my memory. —Have a look some time at: Annales de Chim III Series Vol. IV p 52.[8] There you will find the expansion coefficient for all gases up to the thousandth part! All of them between 0,366–0,368, and then compare with this, in Vol. V. p. 75,[9] how different the expansion coefficients are. On p. 77 (in the middle), Regnault admits that he was fundamentally mistaken with respect to the cyanogen and the sulphuric acid and, on p. 76, he even acknowledges the small difference with respect to hydrogen that I had found. He does not mention me for the remaining gases, but this would have been superfluous, as my paper had already been published in French in Vol. IV 330 of the Annales de Chim.[10] For it had appeared at the same time as Regnault. Both start the year's issues for 1842. His paper in the Annal. de Ch.,[11] my one in Pogg.[12]

The whole affair is similar with regard to the tension force of water drops. Although I had already presented my work to the Academy in Decemb. 1842,[13] Regnault, on the other hand, did his one in April 1844. My one can be found in Pogg. Feb. 1844,[14] his in Annales de Chim. July 1844.[15]—But questions of priority seem to me to be of very little interest. That the facts are found! is the main thing, who found them does not matter.

Who is this Dr Harley[16] who is attacking my theory of respiration (by the way, the best that I have done). I cannot reply to him right away, but I am pleased that I have the opportunity to come back to this issue one more time.

Do you not want to come here again some time? do give us some hope!

À propos! What do you think about Buff's letter to me,[17] which is printed in Pogg. I wrote to him beforehand that I could not gather from his letter what he meant. I could not do anything more! He has had it printed nevertheless.

I see just now that you are asking after all my publications.[18] You will find them all in Pogg. with the exception of one work about the feeding of plants,[19] which has now brought me the honour of becoming an Associé étranger de la Société Imperiale de l'Agriculture.[20]

Do not go to any trouble with me; it is a waste of time!!

Give all my acquaintances and friends there my regards. Above all Faraday,[21]

Hoffmann,[22] Wheatstone, Gassiot, Francis and whoever else remembers me.

You would hardly recognise your pupil Anna[23] again, so come and see her for yourself.

My wife[24] gives you her sincere regards. Poggendorff is now having the hundredth volume of his Annals printed. His friends here want to bestow a little something on him, and are having a lithograph of him done.

Sincerely | Your | G Magnus. | Dr. Tyndall.

RI MS JT/1/M/21

---

1. *Rthl.*: in the handwritten letter, '10' was followed by the standard symbol for 'Reichsthaler' (as it was written in Tyndall's time). Here we have inserted the standard abbreviation 'Rthl' in place of a character for the symbol.

2. *your kind letter*: letter missing, however, it is undoubtedly Tyndall's reply to letter 1309.

3. *Sauerwald*: in letter 1309, Magnus informed Tyndall that the instrument Sauerwald sent was probably already in London.

4. *Reichsthaler*: a silver coin issued by Prussia from 1750 which became a common currency throughout much of Germany in the mid-nineteenth century before being replaced by the Vereinsthaler in 1857.

5. *photometer*: an instrument used to measure the intensity of light (*OED*).

6. *Bunsen's design*: probably a grease-spot photometer as designed by the German chemist Robert Bunsen in 1843. See X. Chen, 'Visual Photometry in the Early 19th Century: A "Good" Science with "Wrong" Measurements', in J. Buchwald and A. Franklin (eds), *Wrong for the Right Reasons* (Dortrecht: Springer, 2005), p. 178.

7. *Regnault's first work on the expansion of gases*: V. Regnault, 'Recherches sur la dilatation des gaz', *Annal. Chim. et Phys.*, 4 (1842), pp. 5–63.

8. *Annales de Chim III Series Vol. IV p 52*: see n. 10.

9. *Vol. V. p. 75*: V. Regnault, 'Recherches sur la dilatation des gaz; deuxième mémoire', *Annal. Chim. et Phys.*, 5 (1842), pp. 52–83.

10. *Vol. IV 330 of the Annales de Chim*: G. Magnus, 'Sur la dilation des gaz', *Annal. Chim. et Phys.*, 4 (1842), pp. 330–36.

11. *his paper in the Annal. de Ch.*: see n. 10.

12. *my one in Pogg*: G. Magnus, 'Ueber di Ausdehnung der Gase durch die Wärme', *Poggend. Annal.*, 55 (1842), pp. 1–27.

13. *the Academy in Decemb. 1842*: Magnus read his paper on water vapour on 21 December 1842 at the Königliche Akademie der Wissenschaften zu Berlin (founded 1700), which is now the Berlin-Brandenburgische Akademie der Wissenschaften. See G. Magnus, 'theilte neue Versuche mit über die Spannkräfte des Wasserdampfs für die Temperaturen zwischen−6,6 un + 104,6 C', *Bericht über die zur Bekanntmachung geeigneten Verhandlungen der Königlich Preussischen Akademie*, 7 (1842), pp. 282–99.

14. *My one can be found in Pogg. Feb. 1844*: G. Magnus, 'Versuche über die Spannkräfte des Wasserdampfs', *Poggend. Annal.*, 61 (1844), pp. 225–47.

15. *his in Annales de Chim. July 1844.*: V. Regnault, 'Mémoire sure les forces élastiques de la vapeur d'eau', *Annal. Chim. et Phys.*, 11 (1844), pp. 273–335.

16. *Dr Harley*: George Harley (1829–96), a Scottish physician and chemist who received an MD from the University of Edinburgh in 1850, and afterwards trained in Paris, Würzburg, Berlin, and Vienna before taking a lectureship at University College, London, in 1855 (*ODNB*). For Harley's critique of Magnus see G. Harley, 'On the Condition of the Oxygen absorbed into the Blood during Respiration', *Phil. Mag.*, 12 (1856), pp. 478–81.

17. *Buff's letter to me*: Johann Heinrich Buff (1805–78), a German physicist and chemist, who from 1838 was professor of physcis at the University of Giessen (*ADB*). Buff's letter to Magnus was published as J. Buff, 'Schreiben an Hrn. Prof. G. Magnus vom Prof. Buff', *Poggend. Annal.*, 100 (1857), pp. 168–71.

18. *I see just now that you are asking after all my publications*: in Tyndall's missing reply to letter 1309, he may have asked for a list of Magnus's publications. Around this time Tyndall also requested information about Robert Bunsen's research from Heinrich Debus and Edward Frankland. Their replies can be found in letters 1321 and 1326.

19. *one work about the feeding of plants*: G. Magnus, 'Ueber die Ernährung der Pflanzen', *Journal für praktische Chemie*, 50 (1850), pp. 65–75.

20. *Associé étranger de la Société Imperiale de l'Agriculture*: the French Academy of Agriculture (l'Académie d'agriculture de France), which was titled the Société impériale et centrale d'agriculture from 1853–59. Magnus was made a foreign member on 4 June 1856. See 'Séance du 4 juin 1856', *Bulletin des séances de la Société impériale et centrale d'agriculture*, 11 (1855–56), p. 446

21. *Faraday*: Michael Faraday.

22. *Hoffmann*: August Wilhelm Hofmann (1818–92), a German chemist who had studied under Liebig. From its establishment in 1845, Hofmann was the director and professor of the Royal College of Chemistry, London, which in 1853 became part of the Government School of Mines. Hofmann was one of the most influential chemists of the mid-nineteenth century (*ODNB*).

23. *your pupil Anna*: Anna Magnus.

24. *My wife*: Bertha Magnus.

From Joseph Dalton Hooker [January 1857][1] 1328

Dear Tyndall

Many thanks for the Enclosed.[2] What an intelligent note it is; as to 'regeler' how true it is that there is nothing new under the sun! You fairly put me to the blush by quoting me for so very small an affair as a <u>name</u>:[3] you are indeed delicately honest. I must confess to feeling a pride in standing in any position however humble, to a series of investigations which I venture to consider brilliant, & can quite afford to blow the bellows for such an organ.

I cannot Envy you any discussion with the geologists on the subject, if they are as obtuse in appreciating Physical as they are Biological Evidence. Darwin is the only one of them I know who has any accurate conception of the right or wrong of a Botanical problem; though Lyell has a fair grasp of the philosophy of a question in Nat. Hist., too.

I met a man[4] the other night who spent weeks with Agassiz & Forbes in the Aar Glacier,[5] he was charmed with your investigations & confirmed all I have said of Forbes' conduct to Agassiz. I think you have convinced Babbage, Grove & Darwin. Oddly enough Ball haggles at the lamination only![6]

I grieve to think of your melancholy errand[7]—do come to Kew[8] next Sunday,—a sight of the children will do you good; & you may find some solace in sweet innocent faces—what better does this world afford to temper our afflictions—Tis not this their best function?

Ever dear Tyndall Y[ou]rs | JD Hooker.

RI MS JT/1/H/504

1. *[January 1857]*: the year and month are given by reference to Tyndall mentioning Hooker in a lecture (see n. 2 and n. 3). The day is possibly 26–27 January, as Tyndall returned to London on Monday 26 January after having attended the funeral of Alice Wright on 25 January (see letter 1324, n. 3). Per Hooker's invitation in this letter, Tyndall spent Sunday 1 February 1857 at Kew (Journal, 6 February 1857, RI MS JT/2/13c/918).

2. *Enclosed*: likely a manuscript of Tyndall's Friday evening lecture at the RI on 23 January 1857 which was published as J. Tyndall, 'Observations on Glaciers', *Roy. Inst. Proc.*, 2 (1854–58) pp. 320–27.

3. *quoting me for so very small an affair as a <u>name</u>*: Tyndall credited Hooker with suggesting the term *regelation* to describe the freezing together of thawing pieces of ice upon contact (see letter 1306, n. 10 and letter 1312). Though not mentioned in the abstract of the lecture (see n. 2), an account in the *Saturday Review* of 31 January 1857 stated: 'This phenomenon the lecturer attributed, not to viscosity, but to the property which was announced by Dr. Faraday in 1850, and which Dr. Hooker has named "regelation"'. See 'Dr. Tyndall's Theory of Glaciers', *Saturday Review*, 3 (1857), pp. 102–3, on p. 102. Tyndall noted in his journal on 6 February 1857: 'In the Saturday Review of last Saturday is an article on my Friday evening—The writer is a well informed man a friend to and well up in Forbes: nevertheless he admits the formidable nature of some of the objections; towards the conclusion he reasons very philosophically: Hooker remarked that the conclusion of the article cannot have been written by the man who wrote the commencement' (Journal, RI MS JT/2/13c/918).

4. *a man*: John Moore Heath (1808–82), a fellow of and mathematical tutor at Trinity College, Cambridge. He was a climbing companion of Forbes. See A. D. D. Craik, *Mr Hopkins' Men: Cambridge Reform and British Mathematics in the 19th Century* (London: Springer-Verlag, 2008), p. 122.

5. *Aar Glacier*: see letter 1305, n. 2.

6. *Ball*: John Ball disagreed with Tyndall on lamination, perhaps in an unidentified letter, but later in 'Observations upon the structure of glaciers', *Phil. Mag.*, 14 (1857), pp. 481–505.

7. *your melancholy errand*: probably a reference to Tyndall's attending the funeral of Alice Maud Wright (1855–57), daughter of Richard and Fanny Wright, on 26 January 1857. See letter 1324, n. 3.

8. *Kew*: a London district where the Royal Botanic Gardens, Kew are located. Hooker served as the garden's Assistant Director from 1855 and followed his father William Jackson Hooker as Director from 1865–85.

From Charles Lyell [January 1857][1] 1329

My dear Sir

Before you kindly reminded me of your lecture, I had asked a party to meet some American friends who were in town for a few days as it was the only day I could have them tho' I wished much to hear your lecture[2] on a geological subject in which I am much interested.

I hope you will send me the abstract when printed.

I have given your cards[3] to persons who will profit by them & tell me of the lecture—Lady Lyell[4] hopes to see you after you have finished your task on Friday if you can come here

Believe me | /truly yrs/ | Cha Lyell

RI MS JT/1/L/60
RI MS JT/1/TYP/3/846

1. *[January 1857]*: the month and year are given by the reference to Tyndall's lecture on glaciers at the RI on 23 January 1857.
2. *Your lecture*: on 23 January 1857, Tyndall lectured at the RI on 'Observations on Glaciers'.
3. *your cards*: probably a reference to Lyell's tickets of admission to Tyndall's lecture (see letter 1320).
4. *Lady Lyell*: see letter 1320, n. 1.

From Charles Darwin 4 February [1857][1] 1330

Down Bromley Kent | Feb. 4th

Dear Tyndall

I am <u>very</u> much obliged to you for your note.[2] My only excuse for having troubled Huxley,[3] was my very great curiosity to hear something more of your views. —

I am as ignorant of mechanics as a pig as you will have perceived; but Glaciers for years & years have interested me greatly.[4]

I am so very glad to hear that you are continuing your experiments on ice; & I hope to hear that you will explain about the freezing together of ice under the freezing point. —

I can fancy a man so ignorant of nat. History as to advise Owen[5] to compare a skull with a vertebra;[6] on exactly same principle, I hope that you will squeeze together pieces of ice quite dry as far as water is concerned, but wetted with something which will not freeze. There is a valuable suggestion for you!!

I wish you all sorts of good fortune in your most interesting investigations; & the Lord have mercy on you, when Forbes answers you[7] is my prayer

Most truly yours | C. Darwin

It is most beautiful your having given cleavage to ice.[8]

The Correspondence of Charles Darwin, vol. 6[9]
RI MS JT/1/TYP/9/2801
Typed Transcript Only

---

1. *[1857]*: the year is given by the reference to Tyndall's research on cleavage in glacial ice. Darwin had attended part of Tyndall's lecture at the RS on 15 January 1857, on 'Observations on Glaciers'.

2. *your note*: letter missing.

3. *having troubled Huxley*: in two letters to Huxley, Darwin had discussed Tyndall's investigations of glacial ice. See Charles Darwin to Thomas Huxley, 17 January [1857], and Charles Darwin to Thomas Huxley, 3 February [1857], in F. Burkhardt, et al. (eds), *The Correspondence of Charles Darwin*, 24 vols (New York: Cambridge University Press, 1991), vol. 6 (1856–57), pp. 322–23, 329–30.

4. *Glaciers for years & years have interested me greatly*: in his book Geological Observations on the Volcanic Islands visited during the voyage of H.M.S. Beagle (London: Smith, Elder and Co., 1844), Darwin compared the lamination he observed in volcanic rocks to the zoned structure in glaciers as described by James David Forbes (pp. 70–71).

5. *Owen*: Richard Owen (1804–92), a British comparative anatomist and paleontologist, superintendent of the natural history department of the British Museum and later founder of the Natural History Museum in Kensington, London. He was especially known for his skill at interpreting fossil remains (*ODNB*).

6. *compare a skull with a vertebra*: the idea that the bones of the skull were modified vertebrae. While originating with the German naturalist Lorenz Oken (1779–1851), Owen developed his own views on the origin of the vertebrate skull and used it as support for his theory of homology. Huxley would attack Owen's view in his Croonian lecture at the RS on 17 June 1858. See N. A. Rupke, *Richard Owen: Biology without Darwin* (Chicago: University of Chicago Press, 2009), pp. 108–11, 134–35.

7. *when Forbes answers you*: at this point Forbes did not really know how much under attack
   he would be. He had not heard the glacier lecture nor seen the actual paper, for which he
   had to wait some time. A month later, Forbes requested from Tyndall the published paper
   (see letter 1349).
8. *given cleavage to ice*: see letter 1295, n. 3, and 1306, n. 10.
9. *The Correspondence of Charles Darwin, vol. 6*: our transcription has been checked against
   that of the Darwin Correspondence Project. See Charles Darwin to John Tyndall, 4 February [1857], in F. Burkhardt, et al. (eds), *The Correspondence of Charles Darwin*, vol. 6, p.
   330–31.

From Oscar Schulze 4 February 1857 1331

Paulinzelle, den 4ten Febr. 1857.

Hochgeehrtester Herr Professor!

Durch dieses beehre ich mich, Ihnen ergebenst anzuzeigen, daß ich vor
einiger Zeit einen Apparat zur Versinnlichung der Schall-Wellen an Ihre
geehrte Addresse aufgegeben habe, in dessen Besitz Sie also in Kurzem gelan-
gen werden. Ich bitte den Apparat als Zeichen meiner Hochachtung und meines Dankes anzunehmen und mir dadurch einen Beweis Ihres fortdauernden
Wohlwollens und Ihrer Freundschaft zu geben.

Den Apparat habe ich mich bemüht, in einigen Stücken noch zu ver-
vollkommnen, namentlich darin, daß man außer der Interferenz-Welle über
dem Deckel des Apparates immer auch die *[jene]* componirenden, primären
Wellen übersieht. Außerdem habe ich mich bestrebt, den Apparat so viel als
möglich zu vereinfachen, um ihn besonders für Gymnasien, Realschulen etc.
billiger herstellen zu können. ich kann den Apparat jetzt von 40 Rthl.[1] an
ablassen.

Ich wünsche nun sehr, daß der Apparat in seiner neuen Gestalt Ihren Bei-
fall findet. Den Gebrauch desselben kennen Sie ohne Zweifel noch, eventuell
werden Sie darüber in Poggendorffs Annalen bald das Nöthige finden. Wenn
Sie zur Empfehlung desselben etwas thun können, so bin ich überzeugt, daß
Sie dies gern thun werden und sage Ihnen im Voraus dafür meinen schönsten
Dank.

Mein älterer Bruder war eben in Orgelbau-Angelegenheiten in England
und hat in Doncaster wegen des Baues einer großen Orgel von 60 Stimmen
[Accord] abgeschlossen. Vielleicht komme ich in gleicher Ursache auch bald
einmal dahin und werde dann nicht verfehlen Ihnen in London meine Auf-
wartung zu machen.

Mit dem Wunsche, daß Sie sich recht wohl befinden mögen, empfehle ich
mich bestens und zeichne | mit voller Hochachtung | ergebenster

Oscar Schulze | Addr. Joh. Fr. Schulze & Söhne

[handwritten annotation on price table 1]
Schallwellen-Apparat mit Wellenschrauben | zum Drehen—   100 Rthl.
  ”   mit Wellenleisten | zum Schieben   40 Rthl. und darüber.

Paulinzelle,[2] 4th Febr. 1857.

Most esteemed Herr Professor!

Through this I have the honour of advising you most humbly that some time ago I posted to your esteemed address an apparatus to represent sound waves,[3] which you should shortly have in your possession. I ask you to accept the apparatus as a sign of my deep respect and my thanks and thereby to give me proof of your continuing goodwill and your friendship.

I have endeavoured to perfect the apparatus further in some respects, particularly in that, besides the interference wave above the lid of the apparatus, one always also sees the two component primary waves. In addition, I have strived to simplify the apparatus as much as possible, in order to be able to produce it more cheaply, especially for grammar schools, modern general secondary schools, etc.[4] I can now sell the apparatus starting at 40 Reichsthaler.[5]

I now wish very much that the apparatus will meet with your approval in its new form. You will no doubt still be familiar with its use, but as may be the case, you will soon find the necessary information about it in Poggendorffs Annalen.[6] If you can do anything to recommend it, then I am convinced that you will gladly do this, and give you my very many thanks for this in advance.

My elder brother[7] has just been in England on business to do with organ building and has concluded /an arrangement/ in Doncaster[8] about the building of a large 60-voice organ.[9] I shall perhaps also go there some time soon on the same account, and shall then not fail to pay my respects to you in London.

With the wish that you may remain very well, I give you my best compliments and sign | with full respect | your very humble

Oscar Schulze | c/o Joh. Fr. Schulze & Sons

[handwritten annotation on price table 1]
Sound wave apparatus with wave screws | for turning—   100 Reichsthaler.
  ”   with wave strips | for sliding   40 Reichsthaler and above.

RI MS JT/1/S/59

---

1. *Rthl.*: see letter 1327, n. 1.

2. *Paulinzelle*: a small town located in Thüringen, Germany.

3. *an apparatus to represent sound waves*: the device is described in O. Schulze, 'Akustischer Wellen-Apparat', *Poggend. Annal.*, 100 (1857), pp. 583–89.

4. *grammar schools, modern general secondary schools, etc.*: in Germany the Gymnasium is the most advanced of the secondary schools and are intended to prepare students who will attend university. The Gymnasium is comparable to the grammar school (UK) and

the preparatory school (US). *Realschule* is another form of the secondary school which emphasizes practical subjects and the study of science and modern languages.

5. *Reichsthaler*: see letter 1327, n. 4.

6. *Poggendorffs Annalen*: see n. 3.

7. *My elder brother*: Edmund Schulze (1824–78), brother of Oskar Schulze, and head of the Schulze Brothers organbuilding firm after the death of their father, Johan Friedrich Schulze (1793–1858). See B. Hughes, *The Schulze Dynasty: Organbuilders 1688–1880* (St. Leonard's on Sea: Musical Opinion Ltd., 2006), pp. 11, 33–34.

8. *Doncaster*: a town in South Yorkshire, England.

9. *the building of a large 60-voice organ*: in 1853 the 12th-century parish church in Doncaster was destroyed when it caught fire. Between 1854–58 the present St. George's Minster was constructed in its place. Edmund Schulze was selected to construct the church's new organ, which was completed and installed in 1862.

From César-Mansuète Despretz   [*c.* 6 February 1857][1]   1332

Mon cher Mons. Tyndall

Je vous ai écrit, il y a un certain temps. J'ai mal mis votre adresse. J'ai écrit. Institution Royale de la Grande Bretagne Pall Mall Street au lieu de Albemarle Street.

Je vous prierais de m'envoyer la note de vos principaux mémoires, avec l'indication des journaux dans lesquels ils se trouvent et quelques lignes (7 à 8 à chaque mémoire ou davantage) pour faire ressortir le fait fondamental démontré dans chaque mémoire. Je voulais essayer de vous admettre parmi les candidats à une place de correspondant. La liste est déjà nombreuse, elle contient douze noms.

Comme *[j'écrivais]* que vous n'avez pas reçu ma lettre, je vous écris de nouveau.

Rappelez-moi, je vous prie, au bon souvenir de Mr Faraday.

Votre dévoué | C. Despretz | Rue Cassette 20

Il faudrait que j'eusse la note dimanche.

My dear Mr Tyndall,

I sent you a letter[2] some time ago. I miswrote your address. I wrote. Royal Institution of Great Britain on Pall Mall Street,[3] instead of Albemarle Street.

I would ask you to please send me the memorandum of your main essays, with the indication of the journals in which they appear and a few lines (7 or 8 for each essay or more) to highlight the fundamental outcome demonstrated in each essay. I wanted to try and enrol you among the candidates to a seat as a corresponding member.[4] The list is already long, it includes twelve names.[5]

As I wrote that you did not receive my letter, I am writing again.

I ask you to please send my regards to Mr Faraday.

Faithfully yours | C. Despretz | 20 Rue Cassette

I must have the memorandum by Sunday.

RI MS JT/1/D/126

1. *[6 February 1857]*: the estimated date is given based on a note left on the letter by LT, and Tyndall's mention of the letter in his journal for 6 February 1857 (Journal, RI MS JT/2/13c/918).

2. *I sent you a letter*: letter missing.

3. *Pall Mall Street*: a street in central London not far from Albemarle Street (location of the RI).

4. *among the candidates to a seat as a corresponding member*: probably the corresponding member seat of the French Académie des Sciences, left vacant by the death of Macedonio Melloni (1798–1854).

5. *The list is already long, it includes twelve names*: Despretz presented a list at the meeting of the Académie des Sciences on 11 May 1857 and on 18 May Matteucci was elected a corresponding member. Tyndall's name did not appear on the list. See *Paris, Comptes Rendus*, 44 (1857), pp. 1007, 1013.

To Rudolf Clausius   9 February 1857   1333

9th Feb. 1857

My dear Clausius

At length I have time to write you a line to thank you for your last friendly letter[1] and for the book on glaciers[2] which you sent to me. I found both very useful. The book is highly interesting and written by a man who has evidently taken pains to make himself acquainted with the phen/omena/ Since my return to England I have been at work upon the question. It opened out to a degree beyond my expectations, and finally embraced the viscous theory of Forbes,[3] the veined structure of the ice, and the 'Dirt Bands' upon the surface of the glacier. On all three points I have been obliged to differ from Forbes. A few evenings ago I read a paper[4] before the Royal Society /on the/ subject, and I find that a short account of the communication has got into the weekly newspapers. I send you an extract from one of these which will give you some idea of the nature of the enquiry. As far as I can see at present the theory which I have ventured to propose is in accordance with all the phenomena <1 word missing> seems to meet all the cases adduced by /M./ Mousson at page 17 of his book[5] in proof of the viscosity of the ice. I am still at work upon the subject and hope before I give it up to *[clear]* away much of the obscurity which has hitherto surrounded it.

I think the structure of the glacier *[ice]* as revealed by the infiltration experiments of Agassiz is more in accordance with *[the]* theory which I have given than with the *[viscous]* theory. In connexion with this point I took a piece of common ice dug a cavity in it, and filled the cavity with a strong infusion of cochineal.[6] The ice was perfectly impervious to the liquid. The same ice was placed beneath a press and squeesed gently on takeing[7] it out and making the same experiment the coloured liquid freely diffused itself through the capillary fissures and presented an appearance exactly resembling that figured by Agassiz in the Atlas which accompanies his important work.[8]

In communicating the paper to the Royal Society I took the liberty of reading before the Fellows that portion of your letter[9] to <*1 word missing*> which referred to your own observations. Indeed Forbes himself expressly and repeatedly states that the veined structure is at right angles to the *[lines]* of maximum pressure. It is surprising how fruitful Forbes was in observation <u>before</u> he formed his theory, but having pledged himself to it he made facts bend to harmonize with it and I believe fell into great error.

With best wishes my dear Clausius | Ever sincerely yours | John Tyndall
I <u>may</u> go to the Mer de Glace[10] <u>this year</u>.
Herrn Professor Clausius | Zurich[11] | <u>Switzerland</u>[12]

RI MS JT/1/T/172

---

1. *your last friendly letter*: see letter 1305.
2. *book on glaciers*: A. Mousson, *Die Gletscher der Jetztzeit* (Zürich: Druck und Verlag von Fr. Schulthess, 1854).
3. *viscous theory of Forbes*: see letter 1306, n. 8.
4. *I read a paper*: on 15 January 1857, Tyndall lectured at the RS on 'Observations on Glaciers', published as J. Tyndall, 'Observations on Glaciers', *Roy. Soc. Proc.*, 8 (1856), pp. 331–38; and J. Tyndall and T. H. Huxley, 'On the Structure and Motion of Glaciers', *Phil. Trans.*, 147 (1857), pp. 327–46. He lectured on the same topic at the RI on 23 January 1857.
5. *his book*: see n. 2.
6. *cochineal*: a dye-stuff consisting of the dried bodies of the insect *Coccus cacti* (*OED*).
7. *takeing*: misspelled by Tyndall.
8. *the Atlas which accompanies his important work*: L. Agassiz, A. Guyot, and E. Desor, Système glaciaire ou recherches sur les glaciers (Paris: Victor Masson, 1847).
9. *your letter*: see n. 1.
10. *Mer de Glace*: see letter 1306, n. 12.
11. *Zurich*: a city in north central Switzerland.
12. *Herrn Professor Clausius | Zurich | Switzerland*: written as the address on the envelope.

From Samuel Haughton   10 February 1857   1334

Trin: Coll: Dub:[1] | 10 February 1857.

My dear Sir,

I am not a Fellow of the Royal Society[2] & have been deterred hitherto from seeking that honour by pecuniary considerations. I have been told, however, that one may diminish the expense by being the author of a paper published in the Transactions[3]—I have been engaged for some years, during my leisure hours, (which are few) in carrying out the experiments on reflected polarized light, which I gave a description of in the Phil Mag.[4] some years ago—I have already collected a valuable mass of observations on the Metallic Reflexion of Polarized light, particularly as compared with the Chemical Composition of a large series of Alloys of Zinc and Copper, Tin and Copper.

I have also succeeded in reducing under the domain of Mathematical Theory the Laws I published in the Phil Mag.[5] as to the Circular Polarization of light, by Reflexion from Transparent bodies. All my results are, I believe, new, and many of them interesting—but I cannot bring myself to publish them in their present imperfect condition, and I cannot see my way to their speedy completion at present, as I have been directed by my medical adviser,[6] to abstain from all 'works of supererogation'[7] for some months. What are the conditions for the bestowal of the Royal Medals[8]—they appear to have been given for <u>very different</u> degrees of merit hitherto.

I feel much obliged for your kindness,[9] which, for the present, I am compelled to decline to avail myself of

I am, sincerely yours | Saml. Haughton.

RI MS JT/1/TYP/2/484
LT Typescript Only

---

1. *Trin: Coll: Dub:*: Trinity College, Dublin.
2. *Royal Society*: Haughton was elected FRS in 1858.
3. *the Transactions*: Phil. Trans.
4. *a description of in the Phil. Mag.*: S. Haughton, 'On the reflexion of polarized light from the surface of transparent bodies', *Phil. Mag.*, 6 (1853), pp. 81–88.
5. *I published in the Phil Mag.*: S. Haughton, 'On some new laws of reflexion of polarized light', *Phil. Mag.*, 8 (1854), pp. 507–20.
6. *my medical adviser*: not identified.
7. *works of supererogation*: any work or good deed done over and above what is required (*OED*).
8. *Royal Medals*: at this time the RS awarded annually since 1731 the Copley Medal for

'outstanding achievements in research in any branch of science' and twice a year since 1826 the Royal Medal for 'the most important contributions to the advancement of natural knowledge' (a third Royal Medal began being awarded each year from 1965).

9. *I feel much obliged for your kindness*: possibly this is in response to a letter from Tyndall, which is missing.

From Oscar Schulze   11 February 1857   1335

Paulinzelle, den 11ᵗⁿ Febr. 1857.

Hochgeehrtester Herr Professor!

Indem ich wünsche und voraussetze, daß Sie die Zusendung des Schall-wellen-Apparates mit Nachsicht und Wohlwollen aufgenommen haben, überweise ich Ihnen *[beistehend]* die Beschreibung des Apparates, wie ich sie für Poggendorffs Annalen ausgearbeitet hatte, und füge derselben in Folgendem noch eine etwas practischere Zusammenstellung der mit dem Apparate anzustellenden Versuche bei.

1, Entstehung und Fortpflanzung der Wellenbewegung überhaupt.

Man läßt auf dem einen verschiebbaren Stücke die kurze Leiste mit gerader Kante u gebraucht blos auf der anderen die Wel*/*len*/*leiste 1.a (deren Wellenlänge in Bezug auf die andere = 1 angenommen, und *[beispielsweise]* der Stimmgabel a entsprechend gesetzt wird). Bei der Erklärung der longitudinalen Schwingungen kehrt man den Apparat um und verdeckt durch einen *[Schirm]* (eine *[1 word illeg]* des Apparates) die transversale Reihe.—

2, Interferenz zweier einfacher Wellen von gleicher Wellenlänge, ohne Gang-Unterschied *[oder]* mit beliebigem Gang-Unterschied.

Man bedient sich der schmalen Wellenleiste 1.a *[und]* der ihr gleichen breiten. Durch das Querstück können *[die]* Wellenleisten in selbstverständlicher Weise in jedem beliebigen Gang-Unterschied combinirt werden.

3, Darstellung der stehenden einfachen Wellen.

Man bedient sich derselben Wellenleisten wie im vorigen Versuche, indem man das gezahnte Rad zwischen die Zahnstangen einsetzt.

4, Interferenz zweier einfacher Wellen von verschiedener Wellenlänge:

Man vertauscht die schmale Wellenleiste 1.a mit einer der übrigen: 3/4 d.i die Quarte d, 2/3 das ist die Quinte e, 3/5 das ist die große Sexte fis.

5, Interferenz zweier zusammengesetzter Wellen.

Man wählt die schmale Wellenleiste 2+1, A+a und die ihr gleiche breite, welche aus 2 einfachen von dem Wellenlängen-Verhältniß 2:1 oder der Octave construirt sind. Man kann sie durch das Querstück so verbinden, daß sie durchaus parallel sind; durch eine Verschiebung findet man eine Stelle, wo die eine der constituirenden einfachen Wellen durch einen Gang-Unterschied von 1/2 Welle aufgehoben wird und nur die andre einfache Welle übrig bleibt. Ein *[andres]* Mal hebt diese sich auf und die erstere einfache Welle bleibt übrig.

Wenn man die schmale Wellenleiste 2+1 A+a, umkehrt, das *[vordere]* *[1 word illeg]* nach *[hinten]* nimmt, so ist sie die Gegenwelle zu der breiten Wellenleiste. In einem bestimmten Falle heben sie sich nun ganz auf, in anderen wird nur die eine oder die andere der constituirenden einfachen Wellen aufgehoben; *[u]* neben diesen Fällen giebt es noch *[unendlich]* viele, wo sich nichts aufhebt.

Wenn man die schmale Wellenleiste 2+1 mit der im Verhältniß 4/3 *[oder]* 4/5 zusammengesetzten vertauscht, u diese mit den breiten 2+1 combinirt, so erhält man die Wellen des Dreiklangs A, e, a, cis.

6, Darstellung der stehenden zusammengesetzten Wellen.

Man bedient sich der schmalen Wellenleiste 2+1, A+a und der ihr gleichen breiten nach Einsetzung des gezahnten Rades.

Sie werden noch einen *[Rechen]* gefunden haben in Natursalzfarbe. Diesen braucht man beim Auseinandernehmen des Apparates zum Fortfalten der longitudinalen Nadeln.

Ich wünsche nur, daß der Apparat gut angekommen ist und Ihren Beifall gefunden hat und empfehle mich, Ihrer *[gefälligen]* Antwort <*1 word illeg*>

Hochachtungsvoll | ergebenst | Oscar Schulze.

N.A.

Vor ganz Kurzem haben wir den Bau einer großen Orgel von 64 Stimmen für *[neue]* Pfarrkirche in Doncaster übernommen. Mein älterer Bruder war deßhalb in England und ist erst diese Woche wieder nach Hause gekommen. Die Orgel wird erst in 1½ bis 2 Jahren zu bauen angefangen und in 3 Jahren fertig sein. Jedenfalls wird mir das auch Veranlassung geben, nach England zu reisen, und ich werde mir dann die Erlaubniß nehmen, Sie zu besuchen.

Unterdeß Gott befohlen! O.S.

Paulinzelle,[1] 11th Febr. 1857.

Most esteemed Herr Professor!

Wishing and assuming that you have received the delivery of the apparatus for sound waves with forbearance and goodwill, I am sending you *[for your assistance]* the description of the apparatus as I had prepared it for Poggendorffs Annalen,[2] and append to it in the following a further somewhat more practical summary of the experiments that can be performed with the apparatus.

1, Origin and propagation of the wave movement generally.

One places on one sliding piece the short strip with the straight edge and uses on the other one only the corrugated strip 1.a (whose wavelength in relation to the other is assumed = 1, and is set *[for example]* corresponding to the tuning fork a). In explaining the longitudinal vibrations, one turns the apparatus around and covers the transverse row with a *[screen]* (a *[1 word illeg]* of the apparatus). —

2, Interference of two simple waves of equal wavelength, without path-difference *[or]* with any path-difference.

One uses the narrow corrugated strip 1.a *[and]* the broad one like it. Using the cross piece, *[the]* corrugated strips can be combined in any path-difference in a self-evident manner.

3, Illustration of standing simple waves.

One uses the same corrugated strips as in the previous experiment by inserting the cogged wheel between the cogged racks.

4, Interference of two simple waves of different wavelength:

One substitutes the narrow corrugated strip 1.a with one of the remaining ones: 3/4, i.e., the fourth d, 2/3 that is the fifth e, 3/5 that is the major sixth f sharp.

5, Interference of two composite waves.

One selects the narrow corrugated strip 2+1, A+a and the broad one like it, which are constructed from 2 simple ones of the wavelength-ratio 2:1 or the octave. One can connect them using the cross piece in such a way that they are quite parallel; by sliding them, one finds a point where one of the constituent simple waves will be neutralised by a path-difference of 1/2 wave and only the other simple wave remains. An*[other]* time, the latter will be neutralised and the first simple wave will remain.

If one turns the narrow corrugated strip 2+1 A+a round, and moves the *[front]* *[1 word illeg]* *[backwards]*, then it will be the opposing wave to the broad corrugated strip. In one specific case, they will now be entirely neutralised, in others, only one or the other constituent simple waves will be neutralised; *[and]* in addition to these cases there are still *[infinitely]* many more where nothing is neutralised.

If one exchanges the narrow corrugated strip 2+1 with the composite one in the ratio of 4/3 *[or]* 4/5 and combines that with the broad ones 2+1, then one obtains the waves of the triad A, e, a, c sharp.

6, Illustration of standing composite waves.

One uses the narrow corrugated strip 2+1, A+a and the broad one like it after inserting the cogged wheel.

———

You will have found also a *[rack]* in a natural salt colour. One needs this to fold away the longitudinal needles when dismantling the apparatus.

I wish only that the apparatus has arrived safely and has met with your approval, and commend myself, *[awaiting]* your favourable reply[3]

Yours faithfully | most humbly | Oscar Schulze.

P.S.

A very short while ago, we took on the building of a large 64-voice organ for *[the new]* parish church in Doncaster.[4] My elder brother[5] had been in England because of this, and has only this week come home again. The organ will only start to be built in 1½ to 2 years and will be finished in 3 years. At any rate, that will also give me occasion to travel to England, and I shall then permit myself to visit you.

In the meantime, God be with you! O.S.

RI MS JT/1/S/60

---

1. *Paulinzelle*: a small town located in Thüringen, Germany.
2. *description of the apparatus . . . for Poggendorffs Annalen*: O. Schulze, 'Akustischer Wellen-Apparat', Poggend. Annal., 100 (1857), pp. 583–89.
3. *[awaiting] your favourable reply*: Tyndall's reply is missing, but based on Schulze's next letter (1370) Tyndall appears to have not accepted the instrument.
4. *a large 64-voice organ . . . in Doncaster*: see letter 1331, n. 9.
5. *My elder brother*: Edmund Schulze, see letter 1331, n. 7.

From Walter Pollock   16 February [1857][1]   1336

59 Montague Square | Feb. 16th.

My dear Mr Tyndall,

I beg the pleasure of your presence next Saturday[2] which is my birthday[3]

I hope you will find time to answer this letter.

Yours affectionately | Walter H. Pollock

RI MS JT/1/TYP/6/1896
LT Typescript Only

---

1. *[1857]*: the year is given by the reference to Tyndall having given a gift to Walter Pollock (see letter 1343), and that Walter's birthday fell on a Saturday in 1857.
2. *next Saturday*: 21 February.
3. *my birthday*: Walter Pollock (born 1850) was about to celebrate his seventh birthday. In letter 1337, his mother Juliet clarified plans for the party.

From Juliet Pollock   [16 February 1857][1]   1337

My dear Mr Tyndall,

Walter's invitation[2] is not a very definite one so I must inform you that he expects his young friends from <u>half past six to ten o'clock next Saturday evening</u>[3] and that we are to have charades, children's games, and then at half past nine, supper.

One or two <u>grown ups</u> good naturedly join us in our attempts to amuse the children and I feel that if it is in your power to add to that number, you will

We were vexed to miss seeing you last . .[4]

You must take care of your health in earnest and I think you should avoid late crowded parties though society in a quieter, earlier, and as I think better form, the society of intimate friendship may be a useful relaxation for you— you have given me many lectures:[5] now I have given you one.

Yours always most truly | Juliet Pollock

We hope still to have some walks with you

You shall have a copy of Keats[6] when I next see you. I must beg you to return the little Julien[7] as it is Fred's[8] property.

RI MS JT/1/TYP/6/1896
LT Typescript Only

---

1. *[16 February 1857]*: the date is given based on relationship to letter 1336, and most likely posted together.
2. *Walter's invitation*: letter 1336.
3. <u>*next Saturday evening*</u>: 21 February.
4. *seeing you last . . .* : the typescript has no day inserted here.
5. *you have given me many lectures*: Juliet Pollock was a regular attendee of Tyndall's public lectures at the RI.
6. *copy of Keats*: referring to a work by John Keats (1795–1821), the English Romantic poet (*ODNB*).
7. *the little Julien*: J. Pollock, *Stories of Julian and his Playfellows* (London: Grant and Griffith, 1852).
8. *Fred's*: Frederick Pollock, Walter's brother.

To Juliet Pollock   [17 February 1857][1]   1338

Tuesday

My dear M[rs]. Pollock

I called last night just to tell you what I had been doing with myself in the Country, and to get some instructions from you in the preparation of cocoa.[2] But my maid has already made a trial & succeeded so I hope to derive whatever modicum of peace cocoa is capable of conferring upon an over active brain. Indeed I must do as you say. I was quite exasperated with the state of my head on Friday. Nor is it well now. The country, it is true, broke the terrible spell of wakefulness and made me wonderfully strong in two days, but I find even now a slight return of my enemy. I will however lay siege to him & dislodge him. In fact I think the speed with which I recover from these things when I give myself rest is sometimes a snare to me; but a recovery swiftly wrought I do not find to be durable & I really will try and recover my normal self who was a very tough & elastic little fellow.

I quite agree with you regarding going out & I have absolutely declined a great many invitations of late—I think I must go to M[rs]. Drummond's[3] this eveng, she is so good & kind to me. If I stay away it is the fear of my head that will keep me.

Ever yours most Sincerely | John Tyndall

I look forward to our little excursion[4]—I will send you Julian[5] very soon.

RI MS JT/1/T/1189
RI MS JT/1/TYP/6/2113

------

1. *[17 February 1857]*: the date is given by its relation to letter 1337, dated 16 February 1857, in which Juliet Pollock requests from Tyndall to have the book *Stories of Julian and his Playfellows* returned. Juliet Pollock's response to this letter is 1343, dated 25 February 1857, and a Tuesday falling between was 17 February.

2. *cocoa*: cocoa or hot chocolate was thought of as having health benefits. Known in Tyndall's time as 'Dutch cocoa', chocolate was either powdered or machine-pressed then shaved to make a hot drink.

3. *M[rs]. Drummond's*: Maria Drummond.

4. *our little excursion*: in letter 1343, Juliet Pollock thanks Tyndall 'for your escort to the glaciers'. Possibly Tyndall had gone with the Pollocks to an exhibit or lecture on glaciers, or possibly English author and mountaineer Albert Smith's (1816–60) very popular 'Ascent of Mont Blanc' diorama and performance at Egyptian Hall, which ran from 1852–58. See Peter H. Hansen, 'Albert Smith, the Alpine Club, and the Invention of Mountaineering in Mid-Victorian Britain', *Journal of British Studies*, 34 (1995), pp. 300–324, on pp. 304–9.

5. *Julian*: see letter 1337, n. 7.

To William Frederick Pollock   [18 February 1857][1]   1339

Wednesday

My dear Pollock,
    I am rejoiced to be able to say
    Y E S![2]
    Ask Walter[3] to match that if he can.
    Ever yours | John Tyndall

RI MS JT/1/TYP/6/1897
LT Typescript Only

———————

1. *[18 February 1857]*: letter dated based on relationship to 1336 and 1337, and that 18 February 1857 fell on a Wednesday.
2. *I am rejoiced to be able to say Y E S!*: Walter Pollock had invited Tyndall to attend his seventh birthday party on 21 February. See letter 1336.
3. *Walter*: Walter Pollock, the birthday boy.

From Carlo Matteucci   19 February 1857   1340

Pise. 19 Fevrier 1857

Mon cher Monsieur.
    Je vous remercie beaucoup de votre article sur les glaciers. Je vois avec plaisir que nous sommes a peu près d'accord sur la nature de la glace de glacier. En rendant compte de l'idée de Mosely et de la reponse de Forbes, j'avais bien cru ou admis que la critique geologique du glacier conduisait a expliquer ses proprietes en faisant l'hypothèse de Forbes—Mais j'avais ajouté que l'etat <u>semi fluide</u> de la glace etait plutot celui d'<u>une masse solide brisée</u> et <u>dont les fragments etaient tenus ensemble</u>, <u>reunis par l'eau</u> congelee . . . C'est, a peu pres, il me semble votre idée. J'ai tout de suite fait l'extrait de votre notice pour le Nuovo Cimento. Faraday m'ecrivait dernièrement—<u>je vois avec plaisir</u> que <u>malgré</u> la <u>différence</u> d'<u>opinions</u> sur des sujets scientifiques vous et M Tyndall, vous etes de bons amis . . . Mais certainement . . . ce n'est pas la polarite diamagnetique en vrai qui peut le moins du monde alterer mes sentiments de respect et d'amitie pour des qualites de caractere et d'intelligence que j'admire.
    Mais—enfin—nous ne sommes pas d'accord sur la deduction des experiences sur cette polarite. Vous verrez dans les <u>Compt Rendus</u> l'extrait de mes longues recherches sur le diamagnétisme. D'abord, j'ai montre l'accord des <u>plans diamagnetiques</u> et <u>d'induction</u> en presence du pole d'un aimant. J'espere

que vous trouverez bien la methode de determiner exactement les plans de ces forces d'induction. [soit 1 et 1 2 et 2 etc]. Les repulsions diamagnetiques dans ces plans sont dans le meme rapport. Il est difficile d'ebaucher [ces] demonstrations [avec] des [poles] de formes differentes et cela se concoit: il faut une pole grande surface—Je ne vais pas, comme je crains que M. Verdet l'ai fait trop à la hate, a [conclure] de l'induction, que la resultante magnetique est proportionnelle à l'induction et au diamagnetisme . . . On peut pas passer d'element à la somme des elements et d'un mouvement elementaire a [celui] d'un conducteur. Il se peut que l'experience ne decouvre pas de petites differences—mais enfin ma methode est exacte et a l'abri d'objections. Les circuits induits ne se deplacent pas et ils se font equilibre . . .

La 2ème partie de mes recherches est la demonstration que le pouvoir diamagnetique d'un <u>corps purs</u> augmente [aussi] son etat de division—argent . . . et cela d'autant plus que [ce] corps est meilleur conducteur. Quand a la polarite, j'ai verifie toutes vos experiences avec les spirales et les melanges de bismuth en poudre et resine. J'ai varie la distribution—J'ai mesure ces forces. Je ne trouve pas [qu'on ne] puisse expliquer tout avec la simple experience de Reich—Je vais en decrire une—ab cylindre [diamagnetique] [suspendu] M une spirale conique—W une autre quelque part ce point du cylindre par lequel agit A . . . toujours lorsque les deux spirales agissent [attraction][apparait] vers W.

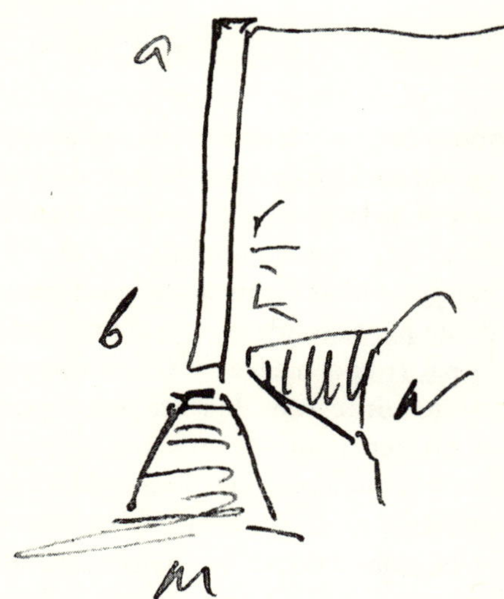

Si les poles sont du meme [force] et repulsion ils sont de [cour/nom] [continu]—cela du moins jusqu'a un certain rythme de M.

Vous verrez mon extrait et je vous prie de m'en ecrire quelque chose.

J'attends l'appareil de [Leipsig . . . Leiser].

Si vous avez l'occasion de voir M. Faraday, je vous prie de lui dire bien des choses de ma part et de lui demander si il a jamais eu l'occasion de faire solidifier son <u>verre /passant/</u> entre les poles d'un grand electroaimant. J'avais eu dans les prismes qu'il m'avait donne un morceau et en profitant d'une fabrique de porcellaine que j'ai a côte de ma maison de campagne, j'ai fait l'experience—et je l'ai fait trois fois—Malheureusement, la masse de verre se fend en se solidifiant (elle trop petite) et je n'ai pu chercher s'il restait le /pouvoir rotation/ sans aimant que sur des petits morceaux—et je ne puis plus repeter l'expérience—Mais j'ai eu deux fois des resultats qui encourageaient a repeter l'expérience.

Tout a vous | C Matteucci
Prof. Tyndall | Royal Institution | Albemarle St. W | London

Pisa. 19 February 1857

My dear Sir,

Thank you very much for your article on the glaciers.[1] I am pleased to see that we are more or less in agreement on the nature of the glacial ice. When reporting on Moseley's idea[2] and its response from Forbes,[3] I had certainly thought or admitted that the geological critique of the glacier led to an explanation of its properties, assuming Forbes' hypothesis—But, I had added that the <u>partially fluid</u> state of the ice was rather that of a <u>solid mass</u>, <u>broken</u>, of <u>which the fragments were held together</u>, <u>joined by frozen water</u> . . . This is approximately your idea, I think. I immediately summarised your explanatory note for the Nuovo Cimento.[4] Faraday recently wrote to me—I am pleased to see that despite the difference in opinion regarding scientific subjects, you and Mr Tyndall are good friends . . .[5] but of course . . . in truth, the diamagnetic polarity cannot in the least spoil my feelings of respect and friendship for some qualities of character and intelligence I admire.

But—anyway—we disagree on the conclusions drawn from the experiments on this polarity. You will see in the <u>Comptes Rendus</u> the summary of my extensive research on diamagnetism.[6] First, I showed the tuning between the diamagnetic planes and the inductive ones in the presence of a magnet's pole. I hope that you will be satisfied with the method to determine accurately the plane of these forces of induction. /that is to say 1 and 1 2 and 2 etc/. The diamagnetic repulsions in these planes are in the same ratio. It is difficult to sketch /these/ demonstrations with batteries of different shapes and this is understandable: a battery of a large surface area is necessary. I will not, as I fear that Mr Verdet did it too hastily, /conclude/ of the induction, that the magnetic outcome is proportional to the induction and to the diamagnetism . . . It is not possible to move from an element to the sum of elements and from a basic movement to /that/ of a conductor. It is possible that the experiment might not resolve small differences—and yet my method is accurate and safe from refutation. The inductive circuits do not move and they balance each other.

The 2nd part of my research[7] is the demonstration that the diamagnetic force of a <u>pure body</u> *[also]* increases its state of division—silver . . . all the more that *[this]* body is a better conductor. As for the polarity, I verified all your experiments with the spirals and the mixtures of powdered bismuth and resin. I varied the distribution—I measured these forces. I cannot find that one cannot explain everything with the simple experiment from Reich[8]—I will describe one of them—ab *[suspended]* *[diamagnetic]* cylinder M a conical spiral W another somewhere this point of the cylinder by which A acts.

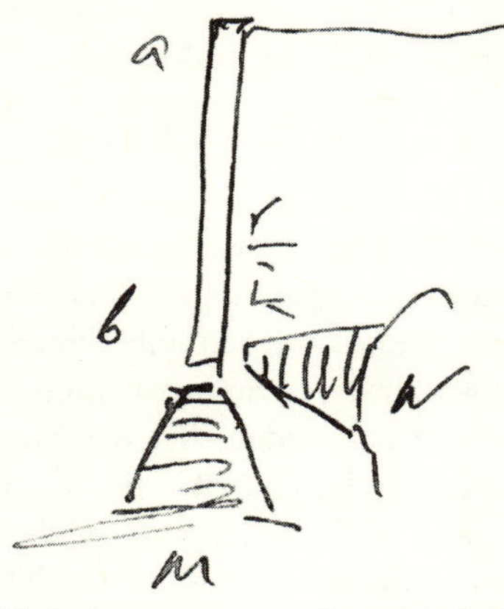

If the batteries are of an equal *[strength]* and repulsion they are of *[1 word illeg]* *[continuous]*—This at least until a certain rhythm of M.

You will see my summary and I am asking you to write something to me about it.

I am waiting for the device of *[Leipsig . . . Leiser].*[9]

If you happen to see Mr. Faraday, I would ask you to send him my regards and ask him if he ever had the opportunity to solidify his <u>glass *[suspended]*</u> in between the poles of a large electromagnet. I had found in the prisms he had given me a piece and taking advantage of a ceramic factory close to my summer house, I did the experiment—and I did it thrice—Unfortunately, the glass mass cracks when solidifying (it is too small) and I could not search if the *[rotational force]* remained without a magnet only on small pieces—and I could not repeat the experiment—But twice I obtained results encouraging the repetition of the experiment.

Truly yours | C Matteucci

Prof. Tyndall | Royal Institution | Albemarle St. W | London

RI MS JT/1/M/62

1. *your article on the glaciers*: J. Tyndall and T. H. Huxley, 'On the Structure and Motion of Glaciers', *Phil. Trans.,* 147 (1857), pp. 327–46.

2. *Moseley's idea*: see H. Moseley, 'On the Descent of Glaciers', *Roy. Soc. Proc.,* 7 (1855), pp. 333–42.

3. *its response from Forbes*: J. D. Forbes, 'Remarks on the Rev. H. Moseley's Theory of the Descent of Glaciers', *Roy. Soc. Proc.,* 7 (1855), pp. 411–17.

4. *summarised your explanatory note for the Nuovo Cimento*: 'Sulla Teoria delle Ghiacciaje, del Prof. Tyndall; Estratto di una Lettera al Prof. Matteucci', *Nuovo Cimento,* 5 (1857), pp. 68–70.

5. *I am pleased to see . . . you and Mr Tyndall are good friends . . .* : here Matteucci is quoting Faraday.

6. *summary of my extensive research on diamagnetism*: C. Matteucci, 'Recherches expérimentales sur le diamagnétisme', *Comptes Rendus,* 44 (1857), pp. 242–44, 331–35, 625–28.

7. *2nd part of my research*: see n. 6.

8. *Reich*: Ferdinand Reich (1799–1882), a German chemist who was professor of Physics at the Freiberg University of Mining and Technology from 1827 to 1860. In 1863 Reich and Theodor Richter isolated indium (*NDB*).

9. *[Leipsig . . . Leiser]*: the instrument maker Leyser of Leipzig, Germany, who constructed an apparatus for the demonstration of diamagnetic polarity.

From Charles Babbage   21 February 1857   1341

Dear Tyndall

Edward Scheutz[1] and Mr Gravatt[2] breakfast with me tomorrow (Sunday) at 9½ to examine my models[3] and also the Calc^g machines of Lord Stanhope.[4]

The Trigonometrical machine of Sir Saml Morland[5] has also turned up.

I shall be glad if you can join them

Very Truly Yours | C Babbage

Dorset S^t | Manch^r Sq. W | 21 Feb 1857

RI MS JT/1/B/1
RI MS JT/1/TYP/1/136

1. *Edward Scheutz*: Edvard Scheutz (1822–81), a Swedish inventor who with his father Per George Scheutz (1785–1873) built a difference engine in the 1830s and 1840s based on Babbage's design. See K. Freeman, 'Scheutz, George', in L. Day and I. McNeil (eds), *Biographical Dictionary of the History of Technology* (London: Routledge, 1996), p. 627.

2. *Mr Gravatt*: William Gravatt (1806–66), an English civil engineer and scientific instrument maker, assisted Per George Scheutz in promoting his difference engine in London. D. Greenfield, 'Gravatt, William', in P. S. M. Cross-Rudkin, et. al. (eds), *Biographical*

Dictionary of Civil Engineers in Great Britain and Ireland, 3 vols (London: Thomas Telford Publishing, 2008), vol. 2 (1830–90), pp. 346–48.

3. *my models*: Babbage designed two calculating machines (a difference engine and an analytical engine), but they were too big to build in his lifetime. His design for the analytical engine led to the design of modern computers.

4. *Calcr machines of Lord Stanhope*: Charles Stanhope (1753–1816), third Earl Stanhope, a British statesman who studied mathematics and electricity, and invented two calculating machines. He was elected FRS in 1772 (*ODNB*).

5. *Trigonometrical machine of Sir Saml Morland*: Samuel Morland (1625–95), an English mathematician and inventor, built several calculating machines, including one that performed trigonometric functions (*ODNB*).

From William Snow Harris   21 February 1857   1342

Athenaeum | 21 February 1857.

Dear Dr. Tyndall,

I have much pleasure in placing in your hands a copy of the Parliamentary Papers relative to Shipwrecks by Lightning[1] and which I venture to hope you will find not altogether without Scientific interest. If you will bear with me, I would beg to be allowed to offer for your information a few observations upon these papers—First the Question of the effectual Security of Ships against Atmospheric Electricity is one of vast importance not only to this Country but to every Commercial and Maritime power.

I turned my attention to this question so long since as the year 1818—The loss with which we were then and had been afflicted in Men, money, and Ships, was something enormous. I thought this rather a reflection upon our national science. The state of our knowledge of the operation of Lightning Rods (as they have been called) both Theoretical and Practical was at that time very imperfect—Amongst Sailors and the most extraordinary superstitions prevailed—and discussions in the Scientific world continually arose relative to the <u>danger</u> of the <u>principle itself</u>. It was said that we tempted the vengeance of Heaven by inviting Lightning to our ships and Buildings:—no wonder that in such a confusion of sentiment the country lost from 6 to 10 thousand a year by damage to our Navy. Having been mixed up with Naval matters all my life, knowing the construction of a ship's mast &c. &c., and having acquired in Edinburgh and Dublin a great love of scientific pursuits,[2] I was led to take a broad general view of this matter. I conceived the idea of fixing metallic conductors on ships in such way as to bring the whole fabric into that comparatively passive or non resisting state as concerned Electric discharge—as it would assume supposing the whole structure was one continuous mass of metal—this I did by a peculiar mechanical arrangement of

Conductors of copper plate applied to the mast[3] and throughout the Hull.[4] It was I can assure you no easy matter: all this;—Of course I had to contrive such an arrangement as would meet all the variable conditions of a ship's spars[5]—stand well under all the flexure incidental to the terrible pressure of the sails—which should likewise be true scientifically under all changes in the position of the spars—so as to be consistent with certain laws of Electrical discharge which I had to investigate and demonstrate by Physical experiment. The old sailors of the Admiralty of that day pronounced the application quite <u>impracticable</u>. On the other hand I was accused of a visionary scheme by which men and ships would be 'sent to perdition'.[6] The Royal Navy was to be blown up by Lightning &c. . You had no idea what a gauntlet I ran for 20 years—and although such men as the late Dr. Wollaston,[7] Davy and many other men of that stamp came at length into my views and after going into the subject with me experimentally and hearing all I had to say reported favorably to the Government still an iron-hearted & inveterate prejudice carried out in every unfair way for the time triumphed. And so we went on year after year suffering tremendous losses in men, money, and ships. At last Sir G. Cockburn[8] the First Naval Lord of the Admiralty and Sir Byam Marten[9]—then the Comptroller of the Navy—in consequence of accounts which were received of the effects of lightning on our Fleets in the Mediterranean and in the Pacific—thought something should be done especially as men such as Davy and Wollaston had given my plan their countenance. So <u>ten</u> ships were ordered to be fitted on trial—I won't tire you by detailing <u>all</u> that subsequently occurred up to the year 1842, when my plan was ordered to be generally carried out in every ship of the Navy—you will find a full history of the whole question in the pages of the Nautical Magazine[10]—but I do assure you I was treated with extreme injustice. When some of my violent opponents were literally obliged to confess that I had succeeded, then they turned round upon me and said; that after all I had done little more than fix <u>a little copper plate to the ships masts</u> and was not entitled to any consideration from the Gov., whatever.

I think if you will do me the honor to look over these papers you will see that much good has resulted to science by my labors. I spent whole days from month to month in rummaging over the old dusty rusty Records of the Navy—You will see by Appendix No.12 that many interesting facts have been brought out.

There are some pretty engravings in the Book[11] worth your having. They were originally drawn by one of our most accomplished Marine Artists.[12]

Pray excuse my having troubled you with all this—but I am sure you will not consider it as any bad compliment to you, when I avow my desire to hold a place in your esteem, so that should the Question ever fall in your way you may fully comprehend its several bearings—Now whatever may have been said by my opponents or by others who judge of a scientific matter as to its merits—<u>after</u> but /not/ <u>before</u> it is worked out—Here is my reply:—In former Periods of our History the Country lost many thousands annually. Its Sailors were killed or hurt in all directions. We lost the full Services of our Fleets and Ships at critical times—no great general system of Metallic Conductors had been ever proposed or imagined, to ward off these evils. Permanently fixed conductors of Electricity were deemed inapplicable to Ships &c &c. &c and much more to this effect. <u>Whereas</u> at this moment Damage of Lightning to H.M. Ships has positively <u>vanished</u> from the records of our Navy.

I remain dear Dr. Tyndall. always very truly yours W. Snow Harris

RI MS JT/1/TYP/2/471–72
LT Typescript Only

---

1. *copy of the Parliamentary Papers relative to Shipwrecks by Lightning*: Shipwrecks by lightning. Copies of papers relative to shipwrecks by lightning, *as prepared by Sir Snow Harris, and presented by him to the Admiralty,* HC 453 (1854), xlii, p. 553. Also see R. B. Forbes, *Protection of Ships from Lightning, According to Principles Established by Sir W.S. Harris, F.R.S.* (Boston: Sleeper & Rogers, 1848), and R. B. Forbes, *Shipwreck by Lightning. Papers relative to Harris's Lightning Conductors* (Boston: Sleeper & Rogers, 1853).

2. *having acquired in Edinburgh and Dublin a great love of scientific pursuits*: after studying medicine at the University of Edinburgh (1811–12), Harris became a member of the Royal College of Physicians of London in 1813. He served as a militia surgeon, so perhaps this service may have taken him to Dublin.

3. *mast*: an upright pole or spar, usually raked, which is fixed or stepped in the keel of a sailing ship in order to support the sails, either directly or by means of horizontal spars (*OED*).

4. *Hull*: the body or frame of a ship, apart from the masts, sails, and rigging (*OED*).

5. *a ship's spars*: the general term for all masts, yards, booms, gaffs, etc. (*OED*).

6. *'sent to perdition'*: a reference to hell in Christian theology.

7. *Dr. Wollaston*: William Hyde Wollaston (1766–1828), an English chemist and physicist, who studied electricity later in his life and served on the British government's Board of Longitude from 1818–28 (*ODNB*).

8. *Sir G. Cockburn*: George Cockburn (1772–1853), First Naval Lord, was a Royal Navy officer who captained ships during the French Revolutionary Wars, Napoleonic Wars, and the War of 1812. Cockburn pushed for raising standards in naval gunneries and was instrumental in the establishment of new technologies for naval ships (*ODNB*).

9. *Sir Byam Marten*: Thomas Byam Martin (1773–1854), Admiral of the Fleet, was a Royal Navy officer who captained ships during the French Revolutionary Wars and Napoleonic Wars. As Deputy Controller of the Navy in the post-Napoleonic period from 1815–31, Marten streamlined the Royal Naval fleet to focus on protection of the British Empire (*ODNB*).

10. *in the Nautical Magazine*: W. Snow Harris, 'Review of the History and Progress of a System of Permanently Fixed Lightning Conductors for H.M. Ships—From the Year 1820,

When First Proposed, to the Present Year 1852, When Fully Adopted into the Public Service', *The Nautical Magazine and Naval Chronicle*, 21 (1852), pp. 250–57 (May); 292–300 (June); 355–65 (July); 437–42 (August); 466–72 (September); and 'Review of the History and Progress of the General System of Lightning Conductors Employed in the Royal Navy', *The Nautical Magazine and Naval Chronicle*, 22 (1853), pp. 131–40 (March).

11. *the Book*: W. Snow Harris, *On the Nature of Thunderstorms; and on the Means of Protecting Buildings and Shipping against the Destructive Effects of Lightning* (London: John H. Parker, 1843).

12. *one of our most accomplished Marine Artists*: not identified.

From Juliet Pollock   25 February 1857   1343

59 Montague Square | Feby. 25th. | 1857

My dear M^r. Tindal[1]

Many thanks for your escort to the glaciers.[2]

I have read the lecture[3] with great interest and in the summer I trust I may remind you of your kind promise to repeat for me some of the experiments.

I omitted when I met you last in Clarges S^t.[4] to say what I particularly intended to say how much pleasure your present[5] had given Walter.[6] When he was unwell, it was what amused him most, and he congratulated himself over and over again upon being the possessor of an <u>apparatus</u>.—

'Marianne' (to the nurse)[7] 'bring me my apparatus: Mama, is it not grand for me to have a real apparatus'?

Yours most truly | Juliet Pollock

RI MS JT/1/P/164
RI MS JT/1/TYP/6/1899

---

1. *Mr. Tindal*: Juliet Pollock also misspelled Tyndall's name as 'Tyndal' in letter 1418, as did William Frederick Pollock in letters 1439 and 1474.

2. *your escort to the glaciers*: see letter 1338, n. 4.

3. *the lecture*: on 23 January 1857, Tyndall lectured at the RI on 'Observations on Glaciers', published as J. Tyndall, 'Observations on Glaciers', *Roy. Soc. Proc.* 8 (1856), pp. 331–38.

4. *Clarges St.*: a street in Westminster, London close to the RI on Albemarle St., location of a home of the Moore family, mutual friends (9 Clarges St.).

5. *your present*: Tyndall attended young Walter Pollock's seventh birthday party on 21 February. See letters 1336, 1337, and 1339. 'Apparatus' not identified.

6. *Walter*: Walter Pollock.

7. *Marianne' (to the nurse)*: Marianne was the common English name for a French children's nurse.

From Henry Clifton Sorby   26 February 1857   1344

Broomfield, Sheffield | Feb. 26/57.

My dear Sir,

I have just received your paper on the ice of glaciers,[1] and I am sure you will believe that I have read it with the greatest pleasure, and take the deepest interest in the subject. What makes me the more pleased is the fact that, what you have proved so completely, corresponds to what I had concluded would turn out to be the case.[2] Since, in the course of a few weeks, I hope to have the pleasure of seeing you in London, and explaining my views more completely, I will not say much now. I shall bring with me a number of my notions, and will show you a structure, that very often occurs in mica schist,[3] that agrees most remarkably with the structure you have described in the ice of glaciers, only that the material is mica and quartz. The theory I had formed was that, when the rock was of such a nature as not to give way as a <u>plastic</u> substance, it had been <u>broken up</u> by the weather, and in this manner there is produced a structure thus[4] when the lines in the direction E F are those of the stratification foliation—the rock having been metamorphosed before pressure occurred. Then, when compressed, not being able to give way as a plastic substance, it becomes fractured in the lines A B, C D. In some cases, as at A B, there is a layer, composed altogether of mica, whilst the rest contains much quartz. There is thus as it were a band, somewhat as in the ice; whereas at C D is only a simple fracture. The distance of such joints from one another varies from 1/1000 of an inch upwards.

You will remember that in my last note[5] I said that there was such a structure in these rocks, and that I fancied that of glacier ice would turn out to be similar, little knowing that you had so clearly proved it to be so.

I am sure you will be much surprised to see some of my thin sections; for they show these things in so clear a manner, and I hope we shall be able to arrange to go over them carefully.

Trusting you are well, and thanking you for your most excellent paper. I remain

Yours very truly | Henry Clifton Sorby.

P.S. I have been working at the structure of cleaved wax, and find that, when so cleaved after pressure, it has exactly the same kind of structure as in fine grained slates which have yielded as <u>plastic</u> substances.

RI MS JT/1/TYP/4/1347–48
LT Typescript Only

1. *your paper on the ice of glaciers*: J. Tyndall, 'Observations on Glaciers', *Roy. Soc. Proc.*, 8 (1856), pp. 331–38.

2. *what I had concluded would turn out to be the case*: that the mechanical theory of slaty cleavage in rocks also applies to glacial ice. See letter 1295, n. 3 and n. 4, and letter 1306, n. 10.

3. *mica schist*: a foliated, crystalline, metamorphic rock composed of alternate layers of mica and quartz and easily split (*OED*).

4. *a structure thus*: at this point in the typescript LT noted there was a diagram in the original letter, but the original letter is missing.

5. *my last note*: letter missing.

To John Frederick William Herschel   2 March 1857   1345

Royal Institution of Great Britain [stamped] | 2nd. <u>March 1857</u>

Dear Sir.

I send you a brief sketch of a lecture on the structure & motion of glaciers,[1] though in all probability you will have already seen the substance of it in the Proceedings of the Royal Society.[2] The time which we spent upon the glaciers was far too short to enable me to do more than express in a general manner the causes to which I believe the phenomena referred to are due. As you very well know such a question always has its 'residual phenomena'. and these I trust will be dealt with in a satisfactory manner before the question is relinquished. There is a great deal to be done towards the development of the physical properties of ice. In the paper I have been compelled to dwell very briefly upon this question, but I am still engaged on it, and hope by the application of polarized light and other tests to lay firmer hold of the subject. As a mechanical question the cleavage of the ice[3] is very interesting; but in the paper I have almost restricted myself to shewing that by bringing the forces acknowledged to be at work in a glacier to bear upon ice, on a small scale, the same results are produced. The minor changes which accompany this result, or rather of which the cleavage is the resultant expression I have not particularly dwelt upon, because they form the subject of my present and future enquiries.

I hope to see the paper of which an abstract is given in the 'proceedings' printed in full in the Phil. Trans.[4] When this is done it will give me pleasure to send you a copy, and that pleasure will be greatly increased if you should find in the memoir nothing unworthy of the calm and Philosophic Spirit in which such a question should be treated, and of which your own writings furnish the best example

Believe me dear Sir | most truly yours | John Tyndall

RS HS 17.385

———

1. *brief sketch of a lecture on the structure & motion of glaciers*: on 23 January 1857, Tyndall lectured at the RI on 'Observations on Glaciers'.
2. *the substance of it in the Proceedings of the Royal Society*: J. Tyndall, 'Observations on Glaciers', *Roy. Soc. Proc.*, 8 (1856), pp. 331–38.
3. *cleavage of the ice*: see letter 1295, n. 3 and n. 4.
4. *printed in full in the Phil. Trans.*: J. Tyndall and T. H. Huxley, 'On the Structure and Motion of Glaciers', *Phil. Trans.*, 147 (1857), pp. 327–46.

From George Biddell Airy   3 March 1857   1346

Royal Observatory, Greenwich | London, SE. [printed]
| 1857 March 3

My dear Sir

There is a branch of Galvanic science, which seems now to be growing into very large proportions, and with which I have no practical acquaintance whatever, namely that of induced galvanism.[1]

Shall you be giving any series of lectures, in which the apparatus for illustrating induction will necessarily be upon the table? If so, I should be glad to have your permission to attend a lecture, with the hope of perhaps picking up a few additional notions from you at the end of it.

I am, my dear Sir, | Faithfully yours | G B Airy

D^r. Tyndall | &c &c &c

RGO MS.RGO 6/471.157
RI MS JT/1/TYP/1/19
RI MS JT/1/A/29

———

1. *induced galvanism*: or *galvanism,* electricity developed by chemical action (*OED*).

To George Biddell Airy   4 March 1857   1347

4th March 1857

My dear Sir

I regret to say that I am at present lecturing upon a subject which lies rather remote from electricity—namely on sound—but I think what you require[1] is capable of easy arrangement. Could we not spend a few hours together over the phenomena of induction. I would propose fixing a day early next month when I might have the suitable apparatus gathered together here and we might gradually climb from the elementary phenomena to the more complicated. If you think the delay till next month too long then I might arrange some day previously—I merely mention next month because I shall then have a few days leisure.

Believe me | dear Sir | very sincerely yours | <u>John Tyndall</u>

RGO MS.RGO 6/471.158–59

---

1. *what you require*: see letter 1346.

From George Biddell Airy 4 March 1857 1348

1857 March 4

My dear Sir

I am very much obliged by your kindness.[1] There is not the least hurry as regards me, and the next month would suit me as well as any other time.

I will therefore hope to hear from you again,[2] at some time (if such ever occurs) when you are perfectly free.

I am, my dear Sir, | Yours very truly | <u>G B Airy</u>
Dʳ. Tyndall

RGO MS.RGO 6/471.160

---

1. *your kindness*: see letter 1347.
2. *hear from you again*: see letter 1373.

From James David Forbes 6 March 1857 1349

Edinburgh | 6 March 1857
Dʳ Tyndall | London

My dear Sir

I received lately the number of the Proceedings of the Royal Society containing an abstract of a paper on glaciers[1] by you & Mʳ Huxley. I am desirous of making some remarks on several parts of this paper; but I should prefer doing so were there a prospect of seeing it published in detail.[2] Will you be so good as to let me know whether it is to be printed; & if so will you oblige me by sending me a copy of it as early as convenient?

I remain dear Sir | Yours truly | James D. Forbes.

RI MS JT/1/F/31
RI MS JT/1/TYP/1/3941

---

1. *an abstract of a paper on glaciers*: J. Tyndall, 'Observations on Glaciers', *Roy. Soc. Proc.*, 8 (1856), pp. 331–38.
2. *prospect of seeing it published in detail*: see letter 1350 for Tyndall's reply. The paper was published in full as J. Tyndall and T. H. Huxley, 'On the Structure and Motion of Glaciers', *Phil. Trans.*, 147 (1857), pp. 327–46.

To James David Forbes   7 March 1857   1350

Royal Institution 7ᵗʰ March 1857.

My dear Sir

The paper to which you allude[1] has already been 'referred' by the Council of the Royal Society, but whether it is to be printed in the Transactions or not must depend upon the decision of the referees. If it should be printed I shall take the earliest opportunity of forwarding you a copy of it.

I would, if permitted, express here the earnest hope that should the paper be published you will find in it no expression at variance with the Philosophic Spirit in which a question of the kind ought to be discussed, nor inconsistent with the fullest recognition upon my part of the great services which your researches upon glaciers have rendered to science

Believe me dear Sir | very faithfully yours | John Tyndall
Prof. J. D. Forbes | Edinburgh

StA JDF Incoming letters 1857, no. 33

---

1. *paper to which you allude*: see letter 1349; J. Tyndall and T. H. Huxley, 'On the Structure and Motion of Glaciers', *Phil. Trans.*, 147 (1857), pp. 327–46.

To Thomas Archer Hirst   7 March 1857   1351

7ᵗʰ. March 1857.

My dear Tom.

When I last wrote to you,[1] indeed each time that I wrote to you I had fifty pounds in my purse and more in the Bank which I could have sent you at a moment's notice. I kept this until I received your last note[2] but one declining it, and then thinking it useless to keep it longer beside me I invested it, together with £150 which I had in the Joint Stock Bank, in an East Indian

Railway.[3] Francis has managed all this for me thus taking all trouble out of my hands. This left me without ready money. But I had completed my series of articles for Hughes[4] and he owed me 45 pounds. I therefore immediately wrote to him for the money and waited 10 days without receiving any reply: I wrote again five or six days ago, but up to the present moment have heard nothing from the man. Today I managed to get twenty five pounds into my hands and have already dispatched the first halves of the notes to Wright.[5] When I receive his acknowledgment I will send him the second halves— so this point is settled. I also sent immediately after receiving your letter 9 pounds to Henry Hill[6] the receipt of which he has acknowledged in a letter to me—which however like an unbusiness like man I have thrown in the fire. There was no need whatever my dear Tom of fencing your request round with those precautions—had you said to me in your last note but one 'I want 100 pounds.' you could have had it without crippling me in the least, and it was your repeatedly declining the thing which induced me at length to invest the money. If you need any more you may have it in May for then my half years salary is paid[7]—and I hope you will not think it necessary for me to pledge my word to any conditions, but that you will just deal with the money exactly as if it were your own. So much for these stupid matters of pounds shillings & pence.

I had a note from Prof. Forbes this morning[8] in which he states that he has seen the proceedings of the Royal Society and that he will have some remarks to make on several parts of the paper upon glaciers, that if the paper were to be published in detail he would defer making those remarks until he should receive it and requested me very courteously to have the kindness to send him a copy of the paper as soon as it should be published. I replied of course with all courtesy. The paper has caused some sensation, and produced a very mixed effect. Some heavy & high authorities have recently given us their adhesion to the new theory. My love to Anna: Huxley is waiting for me

Ever yours | J. Tyndall

RI MS JT/1/T/879
RI MS JT/1/HTYP/490

---

1. *last wrote to you*: letter 1324.

2. *your last note*: letter missing.

3. *Joint Stock Bank, in an East Indian Railway*: the London Joint-Stock Bank, founded in 1836; and the East Indian Railway, founded in 1845, which introduced railways to eastern and northern India.

4. *series of articles for Hughes*: Edward Hughes (1819–59), Headmaster of the Royal Naval Lower School, Greenwich Hospital, edited a series of illustrated Reading Lesson Books

for youth (published in London by Longman, Brown, Green, and Longmans between 1855–58), which featured articles written by eminent men of science and literature in four volumes. See 'Edward Hughes', *Minutes of Proceedings of the Institution of Civil Engineers,* 19 (1859–60), pp. 190–91. Tyndall wrote articles on 'Natural Philosophy' for the series: *First Book* (1855), pp. 296–338; *Second Book* (1855), pp. 289–322; *Third Book* (1856), pp. 238–65; and *Fourth Book* (1858), pp. 229–76. The series did not compete well with a similar series published in Ireland. See D. Layton, 'Britain's first science HMI', *New Scientist,* 75 (1977), pp. 404–6.

5. *Wright*: Richard Wright.

6. *Henry Hill*: a friend of Hirst's.

7. *half years salary is paid*: from the RI.

8. *a note from Prof. Forbes this morning*: letter 1349.

From John Stevens Henslow   12 March 1857   1352

Cambridge | 12 March 1857

My dear Sir,

I have had your name (with sundry others) in my memorandum list, but have really been so thoroughly occupied that I could not write before leaving home—Many thanks for your Farridayan Eulogy[1] which has afforded me very much pleasure, & to every word of which I most heartily subscribe—I do not know the relative merits of OErsted[2] & Faraday[3] in regard to priority of discovery leading to the invention of the Electric Telegraph but OErsted is, I believe, more generally considered to have the credit which quite as probably or more so may belong to Faraday—Pray excuse my seeming neglect of your paper,[4] but I assure you I read it the moment I received it with the greatest pleasure—

Y[ou]^{rs} very truly | J S Henslow

RI MS JT/1/H/79
RI MS JT/1/TYP/2/490

---

1. *Many thanks for your Farridayan Eulogy*: most likely a positive review Tyndall had written of Faraday's research. Tyndall wrote one such article for the *Westminter Review* of January 1856, which praised the third volume of Faraday's *Experimental Researches in Electricity* (1855). See 'Science', *Westminster Review,* 65 (1856), pp. 254–61. Tyndall also noted in his journal, on 26 March 1857, that he 'Corrected a proof for Chapman', who was then editor of the *Westminster Review* (Journal, RI MS JT/2/13c/920). In the following issue of the *Review* an article, which might possibly be attributed to Tyndall, critiqued James Forbes's contribution to the eighth edition of the *Encylopedia Britannica.* In addition to critiquing

Forbes on Faraday's research into induction, the article also included a laudatory discussion of Faraday's 27 February lecture to the RI, 'On the Conservation of Force'. See 'Science', *Westminster Review,* 67 (1857), pp. 584–602, esp. 588–93.

2. *OErsted*: Hans Christian Ørsted (1777–1851), a Danish physicist and chemist credited with the discovery that electric currents create magnetic fields (*CDSB*).

3. *relative merits of OErsted & Faraday*: Michael Faraday's early 1820s research into electricity and magnetism was disputed as lacking originality as compared to Ørsted's research. See G. Cantor, D. Gooding, and F. A. J. L. James, *Faraday* (London: MacMillan, 1991), p. 49.

4. *your paper*: see n. 1.

To Auguste de la Rive   13 March 1857   1353

13th, March, 1857.

My dear Sir

I have long thought of writing to you but have been so busy, and at times having such a bad head, that I have put the thing off from day to day thus far. The reading of the proofs[1] is progressing with due regularity. Nearly three hundred pages have already been read over. The style of writing is not quite English. Mr Walker[2] probably thinks it better to adhere to your own forms of expression; in minor cases, when I can accomplish it, I make alterations in the style, which I trust will have the effect of softening it a little.

Since I last wrote to you[3] I have been at work upon the glaciers, and in conjunction with Mr Huxley, my companion in Switzerland, communicated a paper[4] to the Royal Society upon the subject some weeks ago. I now send you an abstract of a lecture given at the Royal Institution[5]—the paper will I trust appear in due time in the Philosophical Transactions[6] and when you see the whole of the discussion I am not without a hope that you will think I have differed on sufficient grounds from the views of Prof Forbes.

It is my intention to visit, if possible, the Mer de Glace[7] this year. I am desirous of making a complete examination of this glacier, and this will carry me to Geneva.[8] Would you have the kindness, when you are at leisure, to inform me on the following points:—Would it be possible to borrow a small hydraulic press in Geneva to take with me for two or three weeks to Chamouni.[9] I want to conduct my experiments on the compression of ice upon the glacier itself, and to take a press with me from London would be laborous and expensive. A small model of the press is all I should require. 2.—Would it not be possible to obtain Logwood[10] or some other highly coloured infusion at Geneva for the purpose of making infiltration experiments upon the ice? 3.—Would it be possible to borrow a theodolite[11] in Geneva and a surveyor's chain?[12] I am accustomed to the use of both, and would do them no injury.[13] But in all these cases, supposing the instruments to be lent to me by the makers of them, I should be very glad to guarantee their safe return, and to pay for the use of them.

Mr Faraday I am happy to say is very well, and would doubtless send his greetings if he knew that I was writing to you. He gave us a lecture a few days ago on the Conservation of Force. I do not agree with his conclusions but the lecture delighted me greatly. There were united in it such a bold and courageous statement of his own views and so much toleration for the views of others.

Grove I have not seen for some time. Wheatstone has just completed a beautiful little telegraph[14] which will probably be much used for domestic purposes. Sabine is well, and gave us a paper on the Colonial Observatories[15] last week. Austria is about to send an expedition round the world.[16] A savant[17] who accompanies the expedition was here last week to obtain instruments and instruction in the use of them. I accompanied him one day to Woolwich[18] to see the ship's compasses.[19] I had the pleasure of meeting a lady, a friend of yours,[20] at Sir Benjamin Brodie's[21] a few days ago. M. Matteucci I find has not yet made up his mind regarding diamagnetic polarity.[22] He will be convinced at last.

Ever Yours | John Tyndall

RI MS JT/1/TYP/1/347
LT Typescript Only

---

1. *reading of the proofs*: see letter 1293, n. 16.
2. *Mr Walker*: Charles Walker.
3. *Since I last wrote to you*: letter 1299.
4. *communicated a paper*: John Tyndall, 'Observations on Glaciers', *Roy. Soc. Proc.,* 8 (1856), pp. 331–38.
5. *a lecture given at the Royal Institution*: on 23 January 1857, Tyndall lectured at the RI on 'Observations on Glaciers'.
6. *the paper will I trust appear in due time in the Philosophical Transactions*: J. Tyndall and T. H. Huxley, 'On the Structure and Motion of Glaciers', *Phil. Trans.,* 147 (1857), pp. 327–46.
7. *Mer de Glace*: see letter 1306, n. 12.
8. *Geneva*: see letter 1293, n. 3.
9. *Chamouni*: Chamonix, a town in the valley north of Mont Blanc, located in Upper Savoy, France. During the period covered by this volume, Chamonix was located within the Duchy of Savoy, a part of the Kingdom of Sardinia, precursor to the unified Italian state established in 1861. Savoy was annexed to France in 1860 through the Treaty of Turin.
10. *Logwood*: extract of logwood (an American tree) used for colouring or dyeing (*OED*).
11. *theodolite*: portable surveying instrument for measuring horizontal angles (*OED*).

12. *surveyor's chain*: a 66 foot chain, of 100 links, used as a measuring device for land surveys; also known as a Gunter's chain, after Edmund Gunter (1581–1626).

13. *do them no injury*: in his youth Tyndall was a surveyor. See G. Cantor and G. Dawson (eds), *The Correspondence of John Tyndall, Volume 1: The Correspondence, May 1840–August 1843* (Pittsburgh: University of Pittsburgh Press, 2016); and *Tyndall Correspondence*, vol. 2.

14. *a beautiful little telegraph*: in 1858, Wheatstone patented a version of his 1840 ABC telegraph.

15. *a paper on the Colonial Observatories*: on 5 March 1857, Sabine read to the RS the following paper: 'On what the Colonial Magnetic Observatories have accomplished', *Roy. Soc. Proc.,* 8 (1856), pp. 396–413.

16. *an expedition round the world*: the international Austrian expedition of 1857–59.

17. *savant*: Ferdinand von Hochstetter (1829–84), a German-Austrian geologist (*ADB*). Tyndall noted in his journal on 6 February 1857 that he accompanied Hochstetter to Woolwich for the day (Journal, RI MS JT/2/13c/918).

18. *Woolwich*: a military and industrial area in SE London.

19. *ship's compasses*: the expedition's ship was the SMS *Novara*.

20. *a lady, a friend of yours*: not identified.

21. *Sir Benjamin Brodie's*: Benjamin Collins Brodie (1783–1862), 1st baronet, an English physiologist, surgeon, and manager of the RI (*ODNB*).

22. *not yet made up his mind regarding diamagnetic polarity*: see letter 1340.

To unidentified[1]   13 March 1857   1354

My dear Friend.

After my departure today I read that little note and walked afterwards mechanically forward; half unconscious of the world around me. The rain fell and I thought of an omnibus,[2] but one left its station before my eyes and I had not presence of mind to hail it. Thus I marched to <*1 word missing*> whence I had shelter home.

What put that strange request[3] in your head? Was it the idle talk of Sunday evening? Well, be it as thou wilt: You shall assuredly know it in the manner you desire when the time comes. I almost smile at the thought; the contingency in my case is so remote. Indeed I often look forward with tranquil eye to a ripe old bachelorhood, much opposed as this appears to be to my constitution. I shall always love children, and I have no doubt that there will always be two or three upon the face of the earth that will love me. Hence my condition can never be like that of those forlorn singlemen whose want ... years of intercourse,[4] I do not fear one of the changes will be a lowering of your esteem.

RI MS JT/2/7/396–98
RI MS JT/2/13c/918–19

1. *unidentified*: no recipient is given for this letter within the typescript journal. Possibly Maria Drummond or one of her daughters, as around this time Tyndall was enamoured with either Mary or Emily Drummond (see letters 1404, n. 5, and 1406).

2. *omnibus*: a large, horse-drawn public vehicle carrying passengers by road, running on a fixed route and typically requiring the payment of a fare (*OED*).

3. *strange request*: not identified.

4. *gentlemen whose want . . . years of intercourse*: LT noted in the typescript journal that a page had been torn out.

From Thomas Archer Hirst   15 March 1857   1355

Pau. | March 15th 1857

My dear John.

Thanks for your last letter[1] and the loan of £25. To all your previous letters containing offers of money I answered as I felt at the time. It was not therefore from any hesitation at accepting your offer of help but simply because I did not see far enough into the future circumstances that I did not accept the same when first made. By so doing it appears I have caused you some trouble which might have been prevented. It is past however and I will say no more about it but feel thankful that I have a friend who will ever help me as a brother and from whom I can unhesitatingly accept such help. You will not be surprised to hear that since Anna's illness[2] I have had many such offers that I could not accept. This very day in fact I have had to express a few firm words to John Martin on the subject, and they have not pleased him. John Martin has many amiable and admirable traits of character as I have had many opportunities of seeing during the winter. His unvarying tenderness and kindness to Anna were there nothing else would be sufficient to secure my respect. As men, however, there is little sympathy between us and as you may easily conceive, far easier than I can explain, I cannot accept pecuniary assistance from him, especially when I know that I am just as well able to give Anna every little luxury she desires as he is to assist me. It is long since he first offered me money and I know my constant refusal displeases him, but what can I do. There is or ought to be in every man a certain <u>feeling</u> of independence which, even at the risk of displeasing others, it is a sacred duty to keep untarnished. It is of no avail to call it pride and argue that if I accept help from one I may do so from another; the one who accepts assistance or the one to whom it is offered alone can judge of what he ought to do in the matter. If Anna's comfort or health were at stake, I should not hesitate a moment, but I do not know that she has had a desire that I was not able to satisfy not a want that I was not able to supply nor shall she have as long as I have a penny. At the same time I will not be defrauded of my unquestionable right to use my own pence for this object which of all others is closest and dearest to me. Hitherto I have had no burden to bear my income has sufficed even for our daily wants, but of course I am prepared to make any efforts when the time shall arrive and I must demand from many good people the liberty to bear my own burden even though my refusing their assistance may pain them. Yes, in many respects I am truly sorry John Martin has pressed his offer and having pressed it that I cannot accept it. Small as it may seem this has been a source of some anxiety to me. If you had been near me I should have talked to you about it. You will no doubt wonder what Anna's feelings on the subject are for of course the gift was <u>nominally</u> to her, and I took care to leave her free to choose. As you may guess she would not decide without knowing my feelings and consequently I did not dissemble them. I am glad to say she felt exactly as I did in the matter that as long as our means were sufficient to gratify every wish she preferred being free. At the same she felt more keenly perhaps than myself that our refusal would hurt her brother. I have filled my letter with the subject and yet have expressed myself badly but you will understand me.

Anna still continues better and daily improves a little. As for myself owing to the changeable weather I have a nasty cough lingering about me, it will neither make advances nor quit possession and did give me a little uneasiness a week ago by threatening to settle in my chest. By care however I hunted it away from that quarter of my body and I no longer care about it. I did expect to be able to send you a short memoir for the Phil. Mag. During the last week I gathered up my material and put it into shape but I find by doing so that there are still a few things I must examine further before I let it go away from me. A great step is made however towards publishing and I trust before long to dedicate (privately) and to forward to <u>you</u> my dear John this my first production.[3] It will be short though I fear too purely mathematical to interest you much. I will when sending it to you give you a short abstract of what it contains and by what it will be followed.

Did you ever read 'Eulers letters to a German Princess'[4] I bought them a short time ago and like them very much There are some admirably written little scientific and metaphysical essays in them.

Anna sends her love with mine. | <u>T. A. Hirst</u>

RI MS JT/1/H/231
RI MS JT/1/HTYP/491–92

---

1. *your last letter*: letter 1351.
2. *since Anna's illness*: see letter 1316.
3. *this my first production*: see letter 1319, n. 8. Hirst had originally written 'act of creation' but crossed it out and replaced it with 'production'.
4. *'Eulers letters to a German Princess'*: possibly this first English edition: L. Euler, *Letters of Euler on Different Subjects in Physics and Philosophy Addressed to a German Princess*, 2 vols, trans. H. Hunter (London: Henry Hunter & H. Murray, 1795). Originally published in French in three volumes between 1768–74.

From John Frederick William Herschel   17 March 1857   1356

(Copy) Collingwood | March 17/57

Dear Sir

Allow me to thank you for your lecture on Glacial cleavage.[1] There is one thing which a little puzzles me. <u>How is it</u> that regelation takes place when two masses of ice <u>at 32°</u> are placed in contact? 1st. are they <u>really</u> at 32°? is not the interior of every lump of ice some infinitesimal of a degree colder then the surface? Paradoxical as it may appear it is not impossible. On the occasion of one of the Arctic expeditions[2] I suggested (and it was done) that boring rods should be taken to bore into the great Ice-blocks to try the internal temperature on this notion—viz: that any degree of <u>cold</u> might be conducted into ice, but no <u>heat</u> above 32°; since a higher heat applied at the surface would run off in water—Thus ice collected in winter below 32° <u>might</u> retain some little cold below 32° in its interior. In fact the Wenham Lake blocks packed for shipment to Calcutta[3] are purposely shipped at a temperature very much below 32°—

2^{ndly}. If this be <u>not</u> the cause, what is? Can it be possible that the crystalline forces should determine a movement of heat so as to <u>thaw</u> one part of the ice in order to <u>freeze</u> another? and if this were possible would it not at least require that the crystals should be juxta-posed with their axes parallel?

Have you examined the polarizing powers of your thawed and re-cemented ice—that for instance which exhibited a cleavage like gypsum?[4]

Excuse my troubling you with these crudities & believe me Y[ou]^{rs} very truly | (signed) J. F. W. Herschel

P.S. The failure of the 'Ice Calorimeter' as reported by Wedgewood[5] in Phil. Trans.[6]—is in point of the regelation of Ice;—a'propos of Calorimeters I should very much like my modification of this instrument (Cape Obs[ns])[7] to have a fair trial. Babinet[8] wrote me word that he found it give <u>perfectly satisfactory results</u> but I have not seen them in print.

Prof^r. Tyndall

RS HS 17.386
Transcript Only

———

1. *your lecture on Glacial cleavage*: on 23 January 1857, Tyndall lectured at the RI on 'Observations on Glaciers'; later published as J. Tyndall and T. H. Huxley, 'On the Structure and
Motion of Glaciers', *Phil. Trans.*, 147 (1857), pp. 327–46.
2. *Arctic expeditions*: not identified.
3. *Wenham Lake blocks packed for shipment to Calcutta*: a lake in Massachusetts (United
States), where ice blocks were harvested for shipment to countries across the Atlantic from
the 1840s and into the twentieth century.
4. *gypsum*: hydrous calcium sulphate, the mineral from which plaster of Paris is made (*OED*).
5. *Wedgewood*: Josiah Wedgwood (1730–95), founder of the highly successful Wedgwood
pottery company, abolitionist, and maternal grandfather to naturalist Charles Darwin.
Wedgwood was a noted experimental chemist (*ODNB*).
6. *in Phil. Trans.*: J. Wedgwood, 'An Attempt to Compare and Connect the Thermometer
for Strong Fire, Described in Vol. LXXII. Of the Philosophical Transactions, with the
Common Mercurial Ones. By Mr. Josiah Wedgwood, F.R.S. Potter to Her Majesty', *Phil.
Trans.*, 74 (1784), pp. 358–84.
7. *Cape Obs[ns]*: refers to 'Suggestion of an improvement on the Ice Calorimeter' in Appendix C of J. Herschel, *Results of astronomical observations made during the years 1834, 5, 6,
7, 8, at the Cape of Good Hope* (London: Smith, Elder and Co., 1847), pp. 446–47.
8. *Babinet*: Jacques Babinet (1794–1872), a French physicist and mathematician who focused
on the study of optics (*CDSB*).

To John Frederick William Herschel   20 March 1857   1357

20th. <u>March</u> 1857

My dear Sir

    I intend to subject some of the questions which you have proposed[1] to me
to a very searching examination. Hitherto my attention has been directed to
the change of <u>form</u> which a glacier undergoes, and I find that change, taking
glacial ice as it actually is, to be accounted for by the incessant fracture—
sometimes between very minute masses—and regelation of its parts. I am
now working on the physical properties of ice and shall deal fully with those
points to which you refer. With regard to the supposition of internal coldness
the following fact may have some value. Suppose two flat pieces of ice at 32° to
be placed together, with a moist film of the thinnest gold leaf between them,
the film will, after a time, attain the temperature of the interior of the mass,
which will equalize itself by what you call the conduction of cold, if cold exist,
and the film will finally be frozen up between the two pieces. But this is never

observed: the thinnest film of a good conductor which interrupts the contact of the ice masses themselves completely prevents their regelation. If you suspend a crystal of alum[2] in a dilute solution of the salt, as the concentration increases by evaporation the alum will deposit itself on the crystal which will continue to <u>grow</u> while not a particle of the salt is deposited on the sides of the containing vessel. Does not this indicate a special attraction on the part of the crystal for particles of its own kind?, and may not this attraction, in the case of ice, be augmented by the juxtaposition of the two surfaces? I merely throw out these as conjectures at present, but before a year rolls over I hope to be able to give you a more satisfactory account both of the properties of ice and of the further bearing of those properties upon glacial phenomena. I wish I dare ask you to invent a term for me which would include the idea of fracture & renewal to which I believe the bendings of the glaciers are due. I have not seen Wedgewood's[3] paper,[4] but will look for it.

I remain dear Sir | very faithfully yours | John Tyndall

RS HS 17.387
RI MS JT/1/TYP/2/504

---

1. *questions which you have proposed*: see letter 1356.
2. *alum*: an astringent mineral salt, typically occurring as colourless or whitish crystals, that is used as a mordant for dyeing, in tanning, for sizing paper and fireproofing materials, in water purification, and in medicine (*OED*).
3. *Wedgewood's*: Josiah Wedgwood; see letter 1356, n. 5.
4. *paper*: see letter 1356, n. 6.

From William Frederick Pollock   21 March 1857   1358

59 Montagu Square | 21ˢᵗ. March 1857

My dear Tyndall

I did not like the word 'regelation'[1] although I gave you something like an authority for it, & I am glad that you think of discarding it, and are seeking for a better—

Greek is the great source of all scientific nomenclature, and to it accordingly one turns in the first instance, when in want of a name—

The verb πήγνυμι[2] is the one to which I first had recourse—you wish to express the action of a solid body passing through a phasis[3] of disintegration or disruption, and re-assuming a solid form under pressure—in a different shape—and, altho' the immediate phenomenon to be described is exhibited by ice, you do not wish to engage yourself to an assumption of thawing and freezing again—You want a word which will define mechanical changes independently of its association in the mind connected with the transformation of ice into water, and of water into ice—The meaning of πήγνυμι satisfies this condition, because altho' it has the special sense of—<u>to freeze</u>—its primary & general meaning is—<u>to make solid</u>, <u>hard</u>, or <u>stiff</u>—to <u>crystallize</u>—to <u>curdle</u>—It is in fact the same word as the Latin—pingo—there is also the verb συμπήγνυμι, which is still more to the purpose; and from it there is the *[*substantive*]* σύμπηξις—<u>a forcible coming or bringing together or solidification</u>—but I don't like this in English—<u>Sympexis</u> (the existing Greek word)—or <u>Anasympexis</u>, which we may fairly make for ourselves to express the full idea of <u>breaking up</u> and <u>reconstruction</u>, are neither of them nice words, nor would they, I think, immediately suggest their intended meaning—There is another form of πήγνυμι, namely παγόω, which is used more specially for <u>freeze</u>—from which we may very legitimately coin a substantive πάγωσις (as necrosis, νέκρωσις from νεκρόω) which also looks badly—pagosis—and still worse—<u>anapogosis</u>—and <u>synpagosis</u>—or <u>anasympagosis</u>—Notwithstanding its very appropriate signification therefore, I am forced to abandon my πήγνυμι, & turn to something else—From συντείνω—<u>to strain or stretch together</u>—<u>draw tight</u>—we have the existing substantive σύντασις to express that action—which gives us <u>syntasis</u>, and as before we may make for ourselves—<u>anasyntasis</u>—but I don't like this either—The most likely Greek name I can suggest is <u>symplasis</u> (an authentic word) from συμπλάσσω—<u>to mould together</u>—which furnishes the substantive σύμπλασις—<u>a moulding together</u>[4]—to which again we may prefix the other preposition and form <u>anasymplasis</u>—which most accurately describes the process exhibited by you at the lecture,[5] both with the help of the hydraulic press—I cannot say however that I love this substantive—but the corresponding adjectives may perhaps be worth adoption—<u>symplastic</u> and <u>anaplastic</u> and <u>anasymplastic</u> are good words, which could not fail to be understood, and commend themselves to use by the familiar sound of their terminal syllables—Regretfully I leave the Greek fountains, and go lower down the stream of language to look for a word in the Latin—<u>Pingo</u> as I have noticed, is the same as πήγνυμι—& we have its compounds, <u>compingo</u>, and <u>recompingo</u> (not quite a classical word but vouched for from Tertullian)[6]—We have no such english word as <u>compaction</u>—but there is <u>compacture</u> for which Johnson quotes Spenser[7]—He also gives a verb <u>to recompact</u>—in a physical sense—with an authority from Donne[8]—You may choose then between <u>recompacture</u> or <u>recompaction</u>—the latter having the advantage of the more usual termination—<u>Anacrymosis</u>, which just occurs to me, w^d. be a mere Greek equivalent for <u>regelation</u>—The root has already done scientific duty for Wollaston[9] when he named the <u>Cryophorus</u>[10]—on the whole I venture to recommend 'recompaction' as an exact and intelligible word[11]—& which may be formed by a very slight extension of the language—

Y[ou]<u>rs</u> truly | W. F. Pollock

RI MS JT/1/P/226
RI MS JT/1/TYP/6/1900–1901

---

1. *the word 'regelation'*: for more on discussion of the term *regelation,* see letters 1322, 1323, and 1328.
2. πήγνυμι: the translation for this Greek word is 'pegnumi'.
3. *phasis*: a phase (*OED*).
4. *a moulding together*: Pollock provided a footnote at this point: 'there are also αναπλάσσω and ανάπλασις—to <u>remould</u> and a <u>remoulding</u>'.
5. *the lecture*: on 23 January 1857, Tyndall lectured at the RI on 'Observations on Glaciers'.
6. *Tertullian*: Quintus Septimius Florens Tertullianus (*c.* 155–*c.*240 AD), an early Christian author.
7. *Johnson quotes Spenser*: later editions of Samuel Johnson's (1709–84) *A Dictionary of the English Language* (originally published in 1755) cite English poet Edmund Spenser (1552–99) as the source for the word *compacture,* in his epic poem 'The Faerie Queene' (1590–96).
8. *authority from Donne*: the English poet and clergyman John Donne (1572–1631) was the source for the word *recompact* in Johnson's *Dictionary,* from the poem 'A Valediction of My Name, in the Window' (*c.* 1590s).
9. *Wollaston*: William Hyde Wollaston, see letter 1342, n. 7.
10. *Cryophorus*: a device made of glass for demonstrating freezing of water through evaporation. See W. H. Wollaston, 'On a Method of Freezing at a Distance', *Phil. Trans.,* 103 (1813), pp. 71–74.
11. *to recommend 'recompaction' as an exact and intelligible word*: despite Pollock's recommendation of the word *recompacture,* Tyndall stuck with *regelation.*

From William Hopkins   23 March 1857   1359

Cambridge | March 23rd 1857.

My dear Sir,

I believe that the course which you propose I should take[1] respecting my own memoir on Glacial Motion, may be the best, both as regard my own claims on the subject, and as likely to afford the most effective support to your views. I have felt unwilling to write again on the subject in any way which might involve me in controversy with such men as Forbes and Whewell,[2] neither of whom I am convinced can carry on a controversy without a large spice of ill-humoured feeling whether it be openly displayed or not. Forbes will as a matter of course either attack your theory[3] or claim a large part of it as involved in his own,[4] and Whewell is not unlikely to back him up, especially if he find <u>me</u> supporting <u>you</u>. However whatever I think right to be said for the advancement of scientific truth I will say. At the same time I should be sorry that it should in any degree <u>appear</u> that I was claiming any thing which you and Huxley were not previously and fully prepared to allow me. I think, therefore, it would be better just in a single passage to express your belief that your results and theory would be found in strict accordance with the results which I had obtained by strict mechanical reasoning on the internal pressures and tensions of glaciers.

I am glad to hear you are going to visit the Mer de Glace.[5] It will probably enable you to meet Forbes's arguments much more effectively in detail than you could otherwise do. I omitted to return you his note[6] which I have now enclosed. I hope he may now have discovered that one might reasonably expect to assert one's own opinions and dissent from those of others on scientific subjects without being supposed to give just cause of offence. <u>Nous verrons.</u>[7] I can hardly guess what arguments he will bring against your theory, but so long as you keep within your own proper defences, I am satisfied of your impregnability.

Believe me | Yours very truly | W. Hopkins.

RI MS JT/1/H/505
LT Typescript Only

---

1. *the course which you propose I should take*: letter missing, but likely referring to Hopkins' lengthy paper published a few years later: W. Hopkins, 'On the Theory of the Motion of Glaciers', *Phil. Trans.*, 152 (1862), pp. 677–745.
2. *Whewell*: William Whewell (1794–1866), an English philosopher and historian of science, Master of Trinity College, Cambridge, and author of, among many other books, *History of the Inductive Sciences*, 3 vols (London: John W. Parker, 1837) and *The Philosophy of the Inductive Sciences, Founded Upon Their History*, 2 vols (London: John W. Parker, 1840). In 1833, Whewell coined the term *scientist* (*ODNB*).
3. *your theory*: see letter 1306, n. 10.
4. *in his own*: see letter 1306, n. 8.
5. *Mer de Glace*: see letter 1306, n. 12.
6. *his note*: letter 1349.
7. <u>*Nous verrons*</u>: we shall see (French).

To Michael Faraday   24 March 1857   1360

Royal Institution | 24th. March 1857.

My dear Mr. Faraday

I think I ought to let you know that my feelings with regard to M. Plücker are, that he is not dealing with me in an open and upright manner. I can hardly imagine myself writing of him as he does not scruple to write of me.[1] My relation to him has been altogether of a public kind, and he ought to deal with it in a public manner. If the facts do not justify what I have done let him shew this, and I am willing to make reparation. But M. Plücker must feel that a reference to facts would only prove that I have been very tender of his reputation as an experimenter, and hence he resorts to a private canvass of my 'motives'.[2] What guarantee have I that he does not write to all his friends as he has written to Wheatstone, where he broadly insinuates that I have unfairly influenced you against him[3]—a charge which you know to be as unjust as it is unwarranted. Instead of dealing with facts in a philosophic spirit, M. Plücker deals in suspicions regarding me which arise purely out of his own constitution. I think I have reason to complain of this. If he wishes to influence you in his favour let the case be laid fully before you and I pledge myself to abide by your decision. If you can call to mind your own impressions regarding his meaning when he experimented with you here,[4] I venture to say they will be substantially the same as mine.

Ever yours, | John Tyndall

RI MS JT/1/TYP/12/4067
Typed Transcript Only

---

1. *I can hardly imagine myself writing of him as he does not scruple to write of me*: the dispute between Tyndall and Plücker revolved around Tyndall's 1855 'On the Nature of the Force by Which Bodies Are Repelled from the Poles of a Magnet', published in *Phil. Trans.*, 145 (1855), pp. 1–51. In the paper, Tyndall quoted an English translation of Plücker's 'Ueber die Abstossung der optischen Axen der Krystalle durch die Pole der Magnete', *Poggen. Annal.*, 72 (1847), pp. 315–43, published in English as J. Plücker, 'On the Repulsion of the Optic Axes of Crystals by the Poles of a Magnet', *Scientific Memoirs*, 5 (1852), pp. 353–75. In March 1856 Plücker wrote to Faraday that the English translation was incorrect and that Tyndall was 'fighting against a theory, which never was mine' (*Faraday Correspondence*, 5:3109). Plücker eventually replied to Tyndall's critique in a paper communicated through Faraday to the RS, writing: 'the theoretical views imputed to me by Professor Tyndall … are not mine, and have never been mine'. See *Faraday Correspondence* 5:3251; J. Plücker, 'On the Magnetic Induction of Crystals', *Phil. Trans.*, 148 (1858), pp. 543–87, on p. 545. In

letter 1299 of this volume, Tyndall wrote to de la Rive of Plücker's unfriendliness toward him. Their unfriendly relationship likely extended before this. When Tyndall was chosen for the RS's Royal Medal in 1853 'For his paper on diamagnetism and magne-crystallic action, published in the Philosophical Magazine in 1851', some council members doubted his priority in the original research, insisting that Plücker deserved the award instead. On account of this dispute, Tyndall declined the medal (see letter 0828, John Tyndall to Samuel Hunter Christie, 15 November 1853, *Tyndall Correspondence,* vol. 4). See also, R. Jackson, 'John Tyndall and the Royal Medal that was never struck', *Notes and Records: The Royal Society Journal of the History of Science,* 68 (2014), pp. 151–64; and R. Jackson, 'John Tyndall and the Early History of Magnetism', *Annals of Science,* 72 (2015), pp. 435–89, on pp. 461–62 and 477–79.

2. *he resorts to a private canvass of my 'motives'*: Faraday may have shared Plücker's letter of 2 January 1857 with Tyndall. Plücker wrote to Faraday: 'I have no animosity against M T. as I think he has none against myself [.] I will not examine the motives he had, when he suggested to me ideas, which never were mines, and which I think absurd' (*Faraday Correspondence,* 5:3220).

3. *he has written to Wheatstone, where he broadly insinuates that I have unfairly influenced you against him*: in his journal entry for 1 April 1856, Tyndall wrote that Wheatstone had read to him a portion of a letter from Plücker. Tyndall wrote 'there are few men in his position that could be converted into a greater scientific scarecrow that this same M. Plücker—but this I have always abstained from doing, and will continue to abstain unless he pushes me altogether too far' (Journal, RI MS JT/2/13c/832–33). Tyndall also wrote to Hirst about the letter in April 1856 (see letter 1208, John Tyndall to Thomas Hirst, 6 April 1856, *Tyndall Correspondence,* vol. 5).

4. *when he experimented with you here*: Faraday wrote in his diary for 25 August 1848: 'Today Plücker shewed me for the first time some of his experiments'. See *Faraday's Diary: being the various philosophical notes of experimental investigation made by Michael Faraday,* ed. T. Martin 7 vols (London: G. Bell and Sons, Ltd., 1934), vol. 5 (1847–51), p. 59.

To Thomas Archer Hirst [25 March 1857][1] 1361

My dear Tom.

It is a warm genial evening, and I having prepared my lecture for tomorrow[2] may throw anxiety on that score to the winds and follow my natural inclination in replying to your last note[3] which reached me this morning. Of course I understand it all and cheer you in each of your determinations. If that be 'pride' then I have it to my spinal marrow, and thus give another proof that I am bone of your bone and flesh of your flesh.[4] Why should you think for an instant of foreign aid when your own old father who taught you to measure a base line,[5] who was at your side in the forests of Westphalia,[6] and who rejoiced with joy unspeakable when he found that the bowl of punch administered by that godlike little waiter at Dillenburg,[7] had the effect of loosening your bowels; who eat for years the same spiritual food and drank the same spiritual drink—who moreover consumed his son's substance when he was a wanderer on the banks of the Lahn,[8] and a sojourner amid the Lindens of Berlin.[9] Who in fact made your treasury his treasury—By Heaven it would be against nature if you turned your face towards any one else during this temporary cessation of your money getting labours. So I will not beg nor entreat, but I command you, with a father's authority raised to the fifteenth power, I command you sir to write to me instantly if you need any of the '<u>Devils Dust</u>'.[10] This word brings the great old Carlyle to my mind: I called upon his wife[11] who has been long ill a few Sundays ago: she is a wonderfully clever woman. She asked me to come down next evening and see Carlyle. I did so. The man fed me milk, tea, toast and marmalade; and we had three hours of human cordial conversation. He suffers much from ill health, and I think intended to warn me in an indirect manner from going too far. He talked a good deal of science unto me among other things. I differed from him in his estimate of some scientific men and held unswervingly to my opinion. He laughed two or three times and I was delighted with the depth of his laugh—I was in good health at the time and could laugh deeply myself. Miss Jewsbury[12] came in at last, Carlyle came down to the door with us looked up at the stars and said something to them or of them and so I bade him goodnight.

My lectures are going on well. The audience is very large, and is perfectly sustained. I have many very fair maidens among my hearers and Anderson tells me they are in love with me! but I have not yet found it out myself. I have found myself ever since I came to London slowly and surely striking root among the people and this simply because I never sought their friendship but did my work. And this I shall continue to do by Gods help. As a man I rejoice in the good will of my fellow men, and perhaps I sometimes experience a feeling akin to pride to find the great and the eminent holding out the hand of fellowship & equality to me; but if it be pride it is but of a moment's duration—I fall back upon my natural instincts and am at peace. I should not wonder however if in the midst of it all I should make up my mind some fine morning to get married!

Well Tom I will not insult you by saying that your intellectual firstborn[13] (though it is not such) will be more precious to me than the offer of a Knighthood. And I prize it the more as I know it is the first of a progeny, which will nail themselves into the memory of men. This is my unswerving faith—I defy you to look constantly at any one thing and come to any other result. Forbes has written to me,[14] but not bitterly: Hopkins of Cambridge a great authority on glaciers has written to me[15] and given in his complete adhesion to my view.

I intend to go the Chamouni[16] & the Mer de Glace[17] this year & so make a clean & masterly finish—Love to Anna.[18] Love to Tom.

Ever. John.

RI MS JT/1/T/638
RI MS JT/1/HTYP/493

---

1.   *[25]*: the date is given by the reference to Tyndall mentioning having received Hopkins's letter of 23 March (letter 1359), and his lecture the next day. In his journal for 26 March 1857, he mentioned receiving Hopkins's letter and that 'I lecture to day', making this letter 25 March (Journal, RI MS JT/2/13c/919). LT had annotated the letter with 18 March 1857 as the date.

2.   *my lecture for tomorrow*: the tenth of Tyndall's 'Eleven Lectures on Sound' (see letter 1308, n. 7).

3.   *your last note*: letter 1355.

4.   *bone of your bone and flesh of your flesh*: a reference to Genesis 2:23, where Adam recognizes the creation of Woman, Eve.

5.   *to measure a base line*: a reference to Tyndall and Hirst's time working together surveying for the construction of railway lines in the mid-1840s.

6.   *the forests of Westphalia*: Westphalia was a province of the Kingdom of Prussia for most of the nineteenth century until the mid-twentieth. Tyndall and Hirst explored forests nearby the University of Marburg, where they both studied in the late 1840s and early 1850s.

7.   *bowl of punch administered by that godlike little waiter at Dillenburg*: on 8 April 1851, Tyndall and Hirst stayed at the Zum Hirsch gasthaus in Dillenburg (a German town nearby and to the east of Marburg), where they were served in their room a fine meal and a 'glass of grog', served from a punchbowl. Tyndall wrote in his journal, 'it was choice' (Journal, RI MS JT/2/13b/530–31). During this period, Hirst suffered from extreme constipation, which often required medical attention, and he and Tyndall would occasionally discuss the issue in their letters.

8.   *consumed his son's substance when he was a wanderer on the banks of the Lahn*: referring to conversations between Hirst and Tyndall while walking along the Lahn River, a tributary of the Rhine River passing through the University of Marburg, during a week's holiday in 1851.

9.   *the Lindens of Berlin*: Unter der Linden, a street in the German capital that is lined with linden trees.

10. *'Devils Dust'*: traditional, the flock to which old cloth is reduced by the machine called a devil (*OED*), but it seems that Tyndall is using the phrase to mean 'money'.

11. *his wife*: Jane Carlyle (née Welsh, 1801–66). Thomas and Jane were married in 1826 (*ODNB*).

12. *Miss Jewsbury*: Geraldine Jewsbury (1812–80), a novelist and book reviewer, friend of the Carlyles (*ODNB*).

13. *your intellectual firstborn*: see letter 1319, n. 8.

14. *written to me*: letter 1349.
15. *written to me*: letter 1359.
16. *Chamouni*: see letter 1353, n. 9.
17. *Mer de Glace*: see letter 1306, n. 12.
18. *Love to Anna*: Anna Hirst.

From James Harrison[1]   26 March 1857   1362

44 Albert Street | Mornington Crescent | March 26, 1857

Sir,

I herewith enclose copy of results of my last experiment with freezing machine,[2] in the hope that it will be found suggestive of more accurate
experiments.

I have removed my machine to No. 4, Red Lion Square;[3] where I shall
have it again in operation, in about a fortnight.

If you will intimate to me the time when you will be prepared to commence experiments, I shall endeavour to have everything in readiness.

I am | Sir | Your obed[ien]^t serv[an]t | J. Harrison
Professor Tyndall

RI Uncat box 1929

---

1. *James Harrison*: James Harrison (1816–93), a Scottish-Australian printer and inventor, is
   credited with developing the first refrigeration apparatus used for commercial purposes.
   See D. Pike (ed.), *Australian Dictionary of Biography, 1788–1850*, 2 vols (Victoria: Mel-
   bourne University Press, 1966), vol. 1, pp. 520–21.
2. *results of my last experiment with freezing machine*: Harrison travelled to London in 1856
   where he sought patents for his refrigeration and ice-making apparatuses.
3. *No. 4, Red Lion Square*: a square in Holborn, central London. No. 4 was from 1846–50
   the lithographic print shop of Isaac Basire (*ODNB*); later it was the place where James
   Harrison demonstrated refrigeration in 1858 (UCL Bloomsbury Project).

From Henry Clifton Sorby   27 March 1857   1363

Broomfield, Sheffield | March 27/57.

My dear Tyndall,

I am extremely obliged to you for your kind note,[1] but am sorry that circumstances will prevent me seeing you as you name. It was all arranged for us
to leave here this week, and I should then have been with you before now; but

we were compelled to put off till next Monday, when, after staying a day or two with David Forbes[2] in Birmingham,[3] we shall proceed to Cornwall[4] and shall return by London towards the end of May. I am sorry I could not see you, because I much wished to explain to you some new principles of research which I have been working at the last half year, and are such as I think will much interest you, and will lead to some very striking results. However to partly make up for not being able to explain them personally I will just give you an outline of them; for I should much like you to know of them, and they have never been published, and but few know of them.

When crystals are formed from any solution, it very often happens that portions of it are <u>caught up in the crystal</u>, and may be seen with a high magnifying power. Now suppose the liquid was water or an aqueous solution, and the temperature elevated, there would be a cavity containing <u>water</u> and which by contracting or becoming cool would leave a vacuity or <u>bubble</u>. The appearance then is thus

When such are very small it often happens that the bubble <u>moves about</u>, apparently spontaneously, in a very curious manner. If strongly heated, crystals containing such fluid cavities, fly to pieces or loose the water more gently, and then none is seen and <u>no bubble</u> exists in the cavities. But if the crystal had formed from a fused glassy solvent, instead of water being caught up, small drops of this glass become enclosed in the crystal, and, on cooling more, the glass often contracts and forms a spherical vacuity, which of course <u>never moves about</u> and still remains <u>after</u> strong heating. In other cases, on the cooling of the glass, small crystals are deposited from it in and on the sides of the cavity; thus in glassy feldspar[5] These two kinds of cavities, viz. aqueous and glassy, are characteristic of crystals formed from solution or from fusion— artificial crystals deposited from <u>water</u> having cavities with <u>water</u> and those formed from <u>fusion</u> in slag &c. containing <u>glass</u> &c. We therefore have here a most excellent guide in studying mineral crystals in rocks, whether they were formed from <u>solution in water</u> or <u>solution in glass</u>—whether they be of <u>aqueous</u> or <u>igneous</u> formation.

Now you will at once perceive that I have thus got hold of a most excellent method of research, applicable to the study of rocks, and I will just give you a slight outline of what I have made out so far.

A power of 400–800 liniar[6] is required. You perhaps are aware that the igneous origin of some volcanic minerals, even of [leucite][7] has been questioned.[8] When such are however examined, from Vesuvius,[9] they are seen to contain no aqueous cavities but glassy, passing into crystalline, just as in the crystals formed in slags;[10] whilst the calcareous spar[11] of Vesuvius contains aqueous cavaties, indicating an aqueous origin. In the same manner when their sections of greenstones[12] &c are examined they show most admirable examples of glass cavities, but no aqueous, except in some of the zeiolites[13] found in them by aqueous action after the partial cooling of the previously melted rock. We can thus clearly determine which minerals were formed by the cooling of the <u>fused</u> rock and what afterwards produced by <u>aqueous meta-</u><u>morphism</u>. So far as I have explored, basalts[14] and greenstones, and such rocks have a structure that closely agrees with modern volcanic products and artificial slag and which proves most conclusively that they were once in a <u>fused</u> condition but have certainly been very greatly altered by <u>subsequent aqueous</u> <u>action</u>.

Now applying the same principles to granitic rocks,[15] which differ so very much from any modern volcanic product, what is the result? Probably the feldspar has been formed from fusion, but the quartz (the mineral whose presence distinguishes the rock so much from modern lavas,) is full of cavities containing <u>water</u>—it has every appearance of the quartz of veins, which nearly every one believes to have been formed by aqueous action and, so far as I at present can see, there is no good reason to come to any other conclusion than that it was formed from <u>solution in water</u>. Now, unless water has some properties not now understood, (we must discuss this when I see you) the quartz must have been deposited at about 300°F, for it expands so as to fill the cavities at about that, whereas if it had been present when at a red heat, the cavities should have been only about half full, instead of about 15/16.

Now, from what I have said, you will see I have got hold of some very curious principles, which, when thoroughy worked out and applied, will unquestionably clear up some of the most difficult points in the history of rock masses. However the subject is quite in its infancy. When you examine thin sections of some aqueous rocks it appears at first as if it was a new world, and everything requires to be learned, so that, though I have even now made out many very curious facts, I am sure far more remains to be done. It is to work out this problem that I am going into Cornwall.

There are several other general principles that are of very great value in proving mineral changes, that I will explain when I see you, involving optical properties.

Yours very truly | Henry Clifton Sorby
My address will be as usual.

RI JT/1/TYP/4/1349–51
LT Typescript Only

1. *your kind note*: letter missing; possibly a reply to Sorby's letter to Tyndall on 26 February (letter 1344).

2. *David Forbes*: David Forbes (1828–76), a chemist and mining engineer, was a partner in the nickel-smelting firm of Evans & Askin in Birmingham. He was the younger brother of naturalist Edward Forbes (1815–54) (*ODNB*).

3. *Birmingham*: a city in England central to industrial-era manufacturing.

4. *Cornwall*: a county in southwest England, in which the underlying rock is mainly granitic.

5. *feldspar*: a name given to a group of minerals, usually white or flesh-red in colour, occurring in crystals or in crystalline masses (*OED*).

6. *liniar*: probably linear, as in linear magnification.

7. *[leucite]*: mineral composed of potassium and aluminium tectosilicate.

8. *the igneous origin of some volcanic minerals . . . has been questioned*: probably a reference to the 'basalt controversy' of the late eighteenth century, when geologists debated the origin of basalt and related rocks and minerals as volcanic (Huttonian Plutonism) or aqueous (Wernerian Neptunism). See Sally Newcomb, *The World in a Crucible: Laboratory Practice and Geological Theory at the Beginning of Geology. Special Paper 449* (Boulder, CO: Geological Society of America, 2009), chapter 10; and Emile Den Tex, 'Clinchers of the Basalt Controversy: Empirical and Experimental Evidence', *Earth Sciences History*, 15 (1996), pp. 37–48.

9. *Vesuvius*: Mount Vesuvius, a volcano in western Italy south of Naples that decimated the Roman towns of Pompeii and Herculaneum in AD 79.

10. *slags*: a rough clinker-like lump of lava (*OED*).

11. *calcareous spar*: lime, chalk (*OED*).

12. *greenstones*: any of various rocks (typically altered igneous rocks) which are coloured green by feldspar, hornblende (or augite), and other minerals (*OED*).

13. *zeiolites*: zeolite, generic name for a large and varied group of minerals, consisting of hydrous silicates in which the bases are alumina and the alkalies and alkaline earths; commonly found in the cavities of igneous rocks (*OED*).

14. *basalts*: a greenish- or brownish-black rock, igneous in origin, of compact texture and considerable hardness, composed of augite or hornblende containing titaniferous magnetic iron and crystals of feldspar (labradorite), often lying in columnar strata (*OED*).

15. *granitic rocks*: granite, a granular crystalline rock consisting essentially of quartz, orthoclase-feldspar, and mica (*OED*).

From Henry Moseley   29 March [1857][1]   1364

Olveston | 29 Mar.

My dear Sir,

You will see from the enclosed[2] that we cannot have Prof. Christison.[3]

My only doubt about Prof. Henfrey[4] arises from our having already one King's Coll. Professor in our list of Examiners[5] and must perhaps have a second. Prof Henfrey would make a third.[6] As we must be above the <u>suspicion</u> of partiality for any particular Institution, I fear this would not do? Perhaps he does not know Zoology. I think however that would easily be learned in London.

Is the Mr Babington[7] of whom you speak of St. John's Coll Cambridge?
Yours truly | Henry Moseley.

RI MS JT/1/TYP/3/887
LT Typescript Only

1. *[1857]*: the year is given based on Tyndall being an examiner in scientific topics for commissioned officers of the Royal Artillery and Royal Engineers at the Royal Military Academy at Woolwich, from 1856–57. See letter 1306, n. 2.
2. *the enclosed*: letter missing.
3. *Prof. Christison*: possibly Robert Christison (1797–1882), a Scottish physician and toxicologist who worked in a variety of positions at the Royal Infirmary of Edinburgh for over half a century (*ODNB*).
4. *Prof. Henfrey*: Arthur Henfrey (1819–59), a botanist who succeeded naturalist Edward Forbes in the chair of botany at King's College, London in 1854. In 1857, he published *An Elementary Course of Botany* (London: John Van Voorst) (*ODNB*).
5. *one King's Coll. Professor in our list of Examiners*: Moseley and Tyndall were both examiners for commissioned officers of the Royal Artillery and Royal Engineers at the Royal Military Academy at Woolwich, and were tasked with recruiting other examiners. For more on Tyndall as an examiner, see letter 1306, n. 2.
6. *Prof Henfrey would make a third*: in addition to Henfrey, Robert William Browne (1809–90), professor of Classical Literature at King's College, London, also served as an examiner for the Royal Military Academy at Woolwich in 1857. See *Report on Examination for Admission to Royal Military Academy at Woolwich, June 1857*, HC 196 (1857–58), xxxvii, p. 627. In 1856 the King's College professors Richard Chenevix Trench (1807–86) and Adolphus Bernays (1794–1864) also served as examiners. See War Department, *Report on the Examination for Appointments to the Royal Artillery & Practical Class of the Royal Military Academy at Woolwich* (London: Harrison, 1856).
7. *Mr Babington*: Charles Cardale Babington (1808–95), a botanist and archaeologist from St. John's College, University of Cambridge (*ODNB*).

From Joseph Dalton Hooker [2 April 1857][1] 1365

Dear Tyndall

Do you remember at the big hall at Vienna[2] an announcement of fellows elected into the Imp. Acad. Cæs. Nat. Cur. Leopoldina[3]—well that sapient body, (of which all I knew then and now I told you, and which was that it was the oldest scientific body in Europe) want to make a member of you, and have asked me if I can supply your age and place of birth.

I could not get to R.S. Council to day having another engagement:

Willy[4] is improving at Hastings,[5] wearing out his shoes and learning to throw stones—Mrs H.[6] will I think go on to Hastings next week.

I shall go to Brighton[7] to see them on Sunday

Ever yours | J. D. Hooker.

Also can you tell me Faraday's age and where he was born?

RI MS JT/1/TYP/8/2542
LT Typescript Only

---

1. *[2 April 1857]*: the date is given based on a note on the letter from LT. Tyndall was indeed asked to become a member of the Academia Caesarea Leopoldino Carolina Naturae Curiosorum in 1857, proposed by Berthold Seemann and Hooker. See letter 1427.

2. *the big hall at Vienna*: Tyndall visited Vienna in 1856 for a meeting of the Deutsche Akademie der Naturforscher (German Naturalist Academy), held from 16–19 September at the Polytechnic Institution (Journal, 15–19 September 1856 RI MS JT/2/13c/912–14).

3. *Imp. Acad. Cæs. Nat. Cur. Leopoldina*: the Academia Caesarea Leopoldino Carolina Naturae Curiosorum, founded in 1652 and later renamed Deutsche Akademie der Naturforscher Leopoldina. Sometimes referred to simply as the Leopoldina, it is now the Nationale Akademie der Wissenschaften, Leopoldina.

4. *Willy*: William Hooker, aged four.

5. *Hastings*: perhaps Willy (b. 1853) was sent to the seaside town of Hastings for health reasons. Frances Hooker mentioned to Tyndall later this same year that her unwell mother would be going to St Leonards-on-Sea, a part of the town of Hastings (see letter 1432).

6. *Mrs H.*: Frances Hooker, Joseph Hooker's wife.

7. *Brighton*: a resort town on the southern England coast, close to Hastings.

To Joseph Dalton Hooker   [3 April 1857][1]   1366

Royal Institution. Friday

My dear Hooker

Are you a member of this antique body?[2] if so I shall be glad to be at your side. I take it for granted however that you are. Well I was born at Leighlin Bridge Ireland and am 36 years old.[3] It may be a year more or less for aught I know. I was so broken down yesterday that I was unable to go to the Council.[4] Mrs Hooker[5] has written to me for Hyperion[6]—I will send it to you, and another book[7] which I think will interest her—and you I know will be good enough to bear them with you on Sunday. I am rejoiced to hear of Willy's[8] wellfare, no doubt his present vocation[9] is much to his taste, Kiss the little man for me when you see him.

Thank the gods my last lecture before Easter is over.[10] I am quite tired out, and purpose next week filing my chest with pure oxygen and my muscles with vigour along the southern rim of the Isle of Wight.[11] I wish you were at my side.

Is there such a thing as a Bath chair[12] to be had in the neighbourhood of Kew?[13]

I have copied the following from a brief autobiographical sketch[14]
'Michael Faraday was born on Sep 22nd 1791 at Newington Surrey'[15]
Newington is, as you know, a portion of London. Ever yours | John Tyndall

RI MS JT/1/TYP/8/2543
LT Typescript Only

---

1.   *[3 April 1857]*: the year is given by the reference to Tyndall's going to the Isle of Wight with Debus from 5–15 April, and the last of his 'Eleven Lectures on Sound', which commenced on 22 January and continued every week thereafter on Thursdays until 2 April (see letter 1308, n. 7).

2.   *this antique body*: Imperialis Academia Cæsariana Naturæ Curiosorum. See letter 1365, n. 3.

3.   *36 years old*: the general introduction in the first volume of *The Correspondence of John Tyndall* notes that while older scholarship, and Louisa Tyndall, give Tyndall's birth year as 1820, this is doubtful and 1822 is more likely. Either way, this would make him 34 or 36 at the beginning of our volume and 36 or 38 when our volume ends.

4.   *Council*: the council of the RS.

5.   *Mrs Hooker*: Frances Hooker.

6.   *Hyperion*: H. W. Longfellow's *Hyperion: A Romance*, 2 vols (New York: Samuel Colman, 1839).

7. *another book*: an unidentified book by Emerson, which Frances Hooker found 'rather tough reading' (see letter 1395).

8. *Willy's*: William Hooker.

9. *his present vocation*: see letter 1365, n. 5.

10. *my last lecture before Easter is over*: Easter fell on 12 April in 1857. LT wrote on the typescript that the last lecture was on 2 April.

11. *Isle of Wight*: located in the English Channel, the largest island in England was a popular destination for holidays. Tyndall wrote in his journal entry of what is likely 5 April 1857: 'Started with Debus in due time and reached Bishopstoke' (Journal, RI MS JT/2/13c/920). The village of Bishopstoke is on the way to the Isle of Wight from London. That this line from Tyndall falls under the entry for 4 April, but also under a note from LT that a page had been torn out, as well as Tyndall's mention to Hirst on 4 April that 'tomorrow I steer towards Debus' (see letter 1367), suggests that these lines were written on 5 April.

12. *Bath chair*: a large chair on wheels for invalids (*OED*).

13. *Kew*: see letter 1328, n. 8.

14. *a brief autobiographical sketch*: not identified.

15. *Newington Surrey*: now a district of central London.

To Thomas Archer Hirst   4 April 1857   1367

4th April 1857

My dear Tom,

I have this week finished my course before Easter[1] and as usual feel very tired after the conclusion of it all. For the last few days my head has been bad—I have been giddy as I walked the street; still thank the Gods I had sufficient utterance, sufficient mastery of my own thoughts granted to me on Thursday, to enable me to finish my course with comfort, if not with 'joy'. Tonight—Saturday night; I am on the rim of a flight into the country— tomorrow I steer towards Debus, and feel the human blessedness of having that spark of friendship to turn to at a distance from London.[2] Sitting in the reading room of the Institution[3] and the books and papers of the day I turn over a volume of Bacon's Essays[4] and find one of them on Friendship. The wise man explains 'There were princes that had wives, sons, nephews, yet all these could not supply the comfort of friendship'.[5] One of the uses of friendship which he renumerates is that it 'redoubleth joys and cutteth griefs in half; for there is no man that imparteth his joys to his friend but he joyeth the more, and no man that imparteth his griefs to his friend, but he grieveth the less'.[6] One fragment sentence struck me much 'those that want friends to open themselves to are cannibals of their own hearts'.[7] And I softened as I read these words, and thought why should I not write to Tom now, and exalt my little gladnesses by communicating them to him. I thought of the conclusion of my course, and of the real kindly feelings that seemed to have established themselves between me and my audience. I thought of the friendly grasp of men's hands, of sweet smiling eyes and fair faces. I remembered that an old member said to me that since he had come to the Institution he had never heard lectures to equal those. That other members with cool heads and not enthusiastic temperaments had called them 'splendid'. Well I really do not set an overweening value upon them myself. To me they did not appear splendid, but I was satisfied, considering my environments and the impediments which often beset me. But all this appreciation and recognition—what am I to do with it? Shall I cast it utterly away and let it have no weight in the formation of my happiness; or shall I not rather accept it as I do the sunshine of nature, and permit it, though external, to gladden my heart. We are men and must, upon occasion, be able to live through a stormy sunless day; we must not faintly shiver when the sun is absent, but this is not incompatible with the enjoyment of his beams when they appear. Bacon has rather tended to the development of this softer side of human nature within me tonight, and seems to justify the satisfaction derived from the external approbation granted to my work, and not only so but tempts me in accordance with his theory to augment the little pleasure by communicating it to my friend. I am a being of slow growth, and ever since I came to London I have found my roots striking slowly deeper; slowly gathering, without haste or strain, kindly feelings and friendly relations round me. With one old gentleman of fine benevolent disposition whose name has been recently given to the highest mountain in the world[8] I had a wholesome cordial dinner a few days ago. One mild cultivated countenance which has met me often of late, I find to be that of Admiral Codrington.[9] Cardwell[10] the ex president of the Board of Trade[11] was in my last lecture,[12] and came to me with his friendly hand at the conclusion. Faraday was not there being engaged out of doors, but his wife[13] saw that his message was 'remember me to him'. There was M[rs]. Drummond[14] too, the wife of Drummond[15] of the Drummond light[16]—Under Secretary of State for Ireland, and who uttered the immortal sentence which roused the brutal roar of the landlords of Ireland 'Property has its duties as well as its rights'.[17] I name her relations thus particularly because she is so kind towards me; and her fair daughters[18] are very pleasant blossoms, both physically & intellectually. Last night I had a chop with my friend Pollock[19] the son of the Chief Baron,[20] a man of fine cultivated mind, and whose wife[21] overflows, by her excellence, the measure of those beautiful ideals that we used to admire together long ago in the earlier poems of Tennyson.[22] The Moores[23]— nephews & nieces of Sir John Moore[24] who fell at Corunna[25] all treat me with the most brotherly kindness. I told you too I think that they had talked of putting me on the list to be presented to the Academy of Sciences.[26] I was written to by a member of the Academy upon the point. Of course I am no <u>candidate</u>, as the sole value of such things consists in their being spontaneous. Today I had a note from my friend Hooker[27] saying that he had been asked for my age and birthplace with a view to my election into the oldest scientific body in Europe.[28] But I feel my heart getting cold as I talk of these honours; let me turn once more to the thoughts suggested by Bacon. Do I afflict you? Why do I ask the question? If I had a suspicion of the kind it would tear Bacons theory to rags, or more truly would cut away its application in my case. But I do not think that even Bacon unaided would have caused me to write as I have written; my heart was overflowing and he simply transacted the mechanism of turning the top. Added to all the kindnesses, and friendly gladnesses which it has been my lot to experience within the last ten days your memoir with its dedication came to me this morning.[29] I have noticed that I am like a reed shaken with the wind when my eyes wander over anything that pleases me extremely. It seems as if my spirit could not calmly drink in its happiness, but fluttered tremulously like a bee which had discovered a well of honey at the bottom of a flower, and which can hardly taste the treasure for joy. Well Tom I have handed over your paper to Francis, and to your private copies he will have your dedication attached—that is the usual way. There it shall stand, a remembrancer and refresher to my old age, if this heart which now feels so warm and fresh can ever grow old. With what treble comfort can I read the Essays of Bacon feeling as I do that I possess in you all that a man ever possessed in a friend. There is a certain wildness in my heart when I think of our relationship to each other; I mean that wildness which would make self sacrifice cheap if offered in the cause of that holy friendship. If you ever felt that 'fine excess' you will understand me: if not it must be an Irish feeling which words would not explain. Ever & always John.

RI MS JT/1/T/639
RI MS JT/1/HTYP/494–95

---

1. *finished my course before Easter*: Tyndall's 'Eleven Lectures on Sound' (see letter 1308, n. 7).

2. *a distance from London*: on 5 April, Tyndall headed toward the Isle of Wight with Debus, through 15 April (see letter 1366, n. 11).

3. *the Institution*: the RI.

4. *Bacon's Essays*: Philosopher Francis Bacon's *Essays* was first published in 1597, with additions to new editions in 1612 and 1625. Tyndall's edition not identified. For the commitment of Irish scientists to Bacon's ideas and philosophies, see N. D. McMillan, 'Bacon and Ireland', *Baconiana*, 67 (1984), pp. 37–46; and 68 (1985), pp. 88–103.

5. *'There were princes that had wives, sons, nephews, yet all these could not supply the comfort of friendship'*: from Bacon's *Essays,* edition not identified. The quote occurs in 'Book XXVII, Of friendship'.

6.	*'redoubleth joys . . . grieveth the less'*: from Bacon's *Essays,* edition not identified. See n. 5.

7.	*'those that want friends to open themselves to are cannibals of their own hearts'*: from Bacon's *Essays,* edition not identified. See n. 5.

8.	*highest mountain in the world*: George Everest (1790–1866), a Welsh surveyor and geographer. Everest openly opposed the naming of the peak after himself, first proposed by Andrew Waugh, his predecessor as Surveyor General of India. The Royal Geographical Society officially adopted the name Mount Everest in 1865 (*ODNB*).

9.	*Admiral Codrington*: Henry Codrington (1808–77), a Royal Navy officer, promoted to rear admiral in March 1857, then to full admiral in October 1867 and to Admiral of the Fleet in 1877 (*ODNB*).

10.	*Cardwell*: Edward Cardwell (1813–86), a British politician who served as the Secretary of State for War from 1868–74, and as President of the Board of Trade from 1852–55 (*ODNB*).

11.	*Board of Trade*: a British government department concerned with making policy and instituting regulations regarding economic activity in Britain and its empire.

12.	*my last lecture*: see n. 1.

13.	*his wife*: Sarah Faraday.

14.	*M^{rs}. Drummond*: Maria Drummond.

15.	*Drummond*: Thomas Drummond (1797–1840), a Scottish civil engineer who worked on the Irish Ordnance Survey and served as the Irish under-secretary from 1835–40 (*ODNB*).

16.	*Drummond light*: limelight; a lamp which produces an intense light by the incandescence of a piece of lime when it is heated, formerly much used in theatres to light up important actors and scenes, and so direct attention to them (*OED*).

17.	*'Property has its duties as well as its rights'*: in a 22 May 1838 letter to the Landlords of Tipperary, Drummond wrote, 'Property has its duties as well as its rights; to the neglect of those duties in times past is mainly to be ascribed that diseased state of society in which such crimes take their rise'. J. F. M'Lennan, *Memoir of Thomas Drummond, R.E., F.R.A.S., Under Secretary to the Lord Lieutenant of Ireland, 1835 to 1840* (Edinburgh: Edmonston and Douglas, 1867), pp. 315–24, on p. 322.

18.	*her fair daughters*: Mary, Emily, and Fanny Drummond.

19.	*my friend Pollock*: William Frederick Pollock.

20.	*Chief Baron*: Jonathan Frederick Pollock (1783–1870), a British lawyer and politician, appointed Lord Chief Baron of the Exchequer in 1841. He also studied mathematics and was a FRS (*ODNB*)

21.	*whose wife*: Juliet Pollock.

22.	*Tennyson*: Alfred Tennyson.

23.	*The Moores*: Harriet Jane Moore who painted scenes of Faraday at work at the RI, and her siblings Julia, John (a field geologist), and Graham.

24.	*Sir John Moore*: John Moore (1761–1809), British Army general and uncle to Harriet Moore and her siblings.

25. *who fell at Corunna*: John Moore died while leading the British to victory against the French in the Battle of Corunna during the Peninsular War on 16 January 1809 (*ODNB*).

26. *Academy of Sciences*: César Despretz wrote Tyndall about submitting his name to become a corresponding member of the French Académie des Sciences, see letter 1332.

27. *a note from my friend Hooker*: letter 1365.

28. *oldest scientific body in Europe*: the Academia Caesarea Leopoldino Carolina Naturae Curiosorum, see letter 1365, n. 3.

29. *your memoir with its dedication came to me this morning*: in his journal entry of 4 April 1857, Tyndall wrote 'Tom has sent me his first intellectual product' (Journal, RI MS JT/ 2/13c/920). See letter 1319, n. 8.

From Frances Hooker [5 April 1857][1] 1368

My dear Mr Tyndall

You will probably come to the conclusion before long that I am an intolerable bore, but really if you undertake to teach you must forgive your pupils if they come for explanations if they do not understand. I enjoyed Thursday's lecture[2] extremely, as I think all did. Did you not <u>feel</u> that your audience were interested? I think I understood all but the last part, and I am perfectly willing to believe that non-comprehension to be entirely my own fault. Still I feel so anxious to understand, and have been trying so hard to make it out, that if you <u>could</u> spare a little time to explain it a little more fully I should be very grateful. You were so quick, that I could not follow you. The part which puzzles me is your last diagram. Your calculations were so rapid that I could not see where the figures came from, and I could not make out how you deduced the length of a wave from them. In short, I am very stupid, but I do think I could understand it if you could go over it again for me. Am I <u>very</u> bothering? I wonder whether you thought of the drive to Schönbrunn[3] when you were drawing that diagram of the interference of waves!

When shall we see you again here?

With many apologies for worrying you

I remain | Yours most sincerely | F. H. Hooker.

RI MS JT/1/TYP/8/2602

LT Typescript Only

---

1. *[5 April 1857]*: the date is given by the reference to Tyndall's RI lectures on Sound (see n. 2). His last lecture on 2 April was on the interference of sound waves (RI MS JT/4/6a/4, 27), and upon his return from his vacation on the Isle of Wight from 5–15 April, he noted in his journal on 16 April that he had received a letter from Frances Hooker (Journal,

RI MS JT/2/13c/926). If Frances attended the lecture on 2 April, she likely wrote this letter to Tyndall soon thereafter (she wrote, 'I enjoyed Thursday's lecture,' implying it was very recent) and before he left for the Isle of Wight on 5 April. We date it as 5 April with the understanding that it could have been written anytime from 3 April to 14 April.

2. *Thursday's lecture*: Tyndall's 'Eleven Lectures on Sound' (see letter 1308, n. 7). In his journal for 26 January, he wrote: 'Now however my weightiest work is over, and my Thursday lectures will I think be easy and amusing to me. I found today that I could project the figures of the kaleidophone beautifully, and this indicates a wide field of illustration' (Journal, JT/2/13c/918).

3. *drive to Schönbrunn*: Tyndall, the Hookers, and Thomas Huxley, after exploring in the Alps in August, attended the Vienna meeting of the Deutsche Akademie der Naturforscher (German Naturalist Academy), held from 16–19 September 1856 at the Polytechnic Institution (Journal, RI MS JT/2/13c/877, 912–14). They must have also visited Schönbrunn, a former Habsburg palace in Vienna.

To Lyon Playfair   7 April 1857   1369

7th April 1857.

Sir

I have to acknowledge the receipt of your circular of the 15th of January,[1] and in reference to the subject of it—the Educational Museum[2]—would beg to make the following remarks.

Two things I imagine to be essential to establish a healthy relationship between such a museum and the general education of the country. The collection should enhance the machinery of public instruction; and teachers should possess the knowledge requisite to enable them to make proper use of these tools. With reference to the first point, as far as it bears upon physics, manufacturers of instruments and diagrams are very imperfectly acquainted with what may be done in the way of efficient experimental demonstration; and hence instructions and suggestions, relative to the nature of the objects with which it would be desirable to supply the museum will be necessary.

Let me illustrate my meaning by reference to the subject of Acoustics, which stands first upon your list under the general head of physics. It is possible to go to considerable *[expense]* in the experimental illustration of this subject by resorting to the instruments usually constructed for such illustration; but it is equally possible, by making use of those simple means which reveal themselves in the practice of those who make experimental science their especial study, to effect the same object at a small fraction of the usual outlay. With half a crown's worth of India rubber tubing,[3] for example, the phenomena of progressive and stationary undulations; the nodes and neutral segments of vibrating cords; the division of a string into its harmonic parts;

the coexistence of vibrations; the so called sympathetic communication of vibrations, might all be rendered visible to a class in a more effectual and instructive manner than by means of the apparatus usually sold to illustrate these things at, perhaps, ten times the cost. In short the longitudinal and transverse vibrations of strings, rods, and plates, with their associated musical sounds; the tones of disks, bells &c. might be illustrated at an expense which, I am inclined to think, would be deemed incredible by those who are unaccustomed to create their own means of experimental illustration.

I have taken the case of Acoustics as an illustration simply, because I found it first upon the list. The principal phenomena of Light, Heat, Electricity, & Magnetism, might be illustrated by similar simple means.

But it is manifest that the power of rendering such means vital for instruction must reside in the teachers themselves. It is useless to place four feet of India rubber tubing in the hands of a schoolmaster, without teaching him how to extract its virtues. Hence the necessity of some means of instruction by which teachers should be shown the use of the instruments in the collection; not merely by seeing the experiments made, but by being taught to make the experiments themselves.

The museum then may be regarded as standing midway between the manufacturer and the teacher. The former will require definite instructions [as] to the nature of the objects required, leaving of course an ample margin for the application of individual skill. [These] instructions, if they embrace all that is possible at the present day, will have the effect of introducing many new means of illustration. But to render these means available teachers must be instructed in the proper use of them. These [two] conditions, as far as physics are concerned, appear to me essential to the efficiency of the Museum, and by attending to them great good may, I think, be accomplished.

I can hardly conclude, however, without throwing out the suggestion that too much may be expected from the <u>machinery</u> of education. No machinery, however perfect, can supply the spirit which ought ever to animate the practice of the true teacher. Give two men the selfsame machinery—one will cause his school to prosper, while the other will just as certainly produce the opposite effect. <u>Means</u> are essential, but <u>men</u> are far more so; and except to those practically acquainted with the fact it is scarcely possible to conceive the amount of incompetence, and worse than incompetence, which is afloat in this Country under the general designation of School Master. But further reference to this subject would not be pertinent here and I therefore subscribe myself

Your obedient servant | John Tyndall
Lyon Playfair &c. &c. &c.

RI MS JT/1/T/1125

1. *receipt of your circular of the 15th of January*: letter and circular missing. Playfair wrote to Tyndall on 7 January 1857 (letter 1314), and discussed the role of the state in providing science instruction to its citizens. The circular is possibly an extract of the appendices of the Department of Science and Art, *Fourth Report,* Command Paper 2240 (London: Her Majesty's Stationery Office, 1857), in particular pp. 27–30. Playfair was the secretary and author of the report.

2. *Educational Museum*: the Educational Museum was one of several museums housed in the South Kensington Museum (now the Victoria and Albert Museum), which opened on 22 June 1857. Beyond the art museums, these included 'the Patent Museum, the Museum of Animal Products, the Food Collection, the Museum of Construction and the Educational Museum'. See B. Robertson, 'The South Kensington Museum in context: an alternative history', *Museum and Society*, 2 (2004), pp. 1–14, on p. 6.

3. *half a crown's worth of India rubber tubing*: tubing made from natural rubber or caoutchouc, the coagulated latex of certain trees and other plants of South America, Africa, Indonesia, etc., which forms a highly flexible substance (*OED*).

From Oscar Schulze   7 April 1857   1370

Paulinzelle den 7ten April 1857.

Hochgeehrtester Herr Professor!

Ihr Brief vom 21. Mrz ist mir sehr werth, hat mich aber wehmüthig gestimmt. So liebevoll und schonend Sie auch Nein sagen, so fühle ich doch nur das 'Nein' und ich hatte auf ein 'Ja' ganz sicher gerechnet. Es war freilich eine große Kühnheit von mir, daß ich Ihnen, Herr Professor, anbot, ein Andenken von mir anzunehmen, ich that es im Vertrauen auf die Nachsicht und das Wohlwollen, mit welchem Sie mir in den verschiedenen Verhältnissen, in welchen wir in Berührung kamen, *[immer]* begegneten; ich wollte Ihnen damit auch ein Zeichen meiner Dankbarkeit geben, da ich wohl wußte, daß ich die Auszeichnung, welche der auf der Pariser Ausstellung befindliche Apparat davon trug, nur Ihrer Empfehlung schuldig war. Sie waren so freundlich, sich specieller, als der Apparat es vielleicht verdiente, für denselben zu interessiren, und mir selbst noch mehre gute Winke zur Verbesserung und Vervollkommnung desselben zu geben, die ich bei den später gearbeiteten Exemplaren nicht vergessen habe, nach Möglichkeit zu benutzen und zu befolgen. Ich war deßhalb aber gerade auch sehr erwartungsvoll, Ihr Urtheil über die vorgenommenen Verbesserungen zu hören. Dies Alles *[1 word illeg]* mich Ihnen einen Apparat zu senden; ich wollte Ihnen aber damit keine Mühe machen, da ich wohl weiß, wie kostbar Ihre Zeit ist. Sie sollten nur bei Gelegenheit, in einer Erholungsstunde das Ding vornehmen und sich mit dem Spiele desselben ein wenig unterhalten. Wenn ich Sie bat, denselben in England zu empfehlen, so meinte ich das nicht anders, als daß Sie bei der vielen und täglichen Gelegenheit, die Sie haben, mit den Herrn Professoren der Schulen zu verkehren, für welche der Apparat ein Erleichterungsmittel des Unterrichtes sein soll, ein gutes Wort über den Apparat sagen sollten. Ich selbst würde, wenn ich auch über Lang oder Kurz nach England komme, wenig thun können, da ich zwar englisch lesen, aber nicht sprechen und Gesprochenes nicht gut verstehen kann.—Die Kosten, welche ich mir mit dem Apparate gemacht habe, sind so gering, daß Sie deßhalb kein Bedenken zu tragen brauchen, denselben anzunehmen. Sie wissen, daß ich aus der Anfertigung der Apparate kein Gewerbe mache, von dem ich leben wollte; es ist mir eine Nebenbeschäftigung für den Feierabend, womit ich mir die langen Winterabende verkürze, da ich in einem Dörfchen von nur 15 Häusern wohne, und mit welcher ich mich in die schönen Zeiten zurückträume, wo ich in der reinen Wissenschaft Studien machte, und mich für Ihre Arbeiten so sehr interessirte, als ich wünschte und mit großem Danke anerkennen würde, daß Sie es nur zum kleinen Theil für meine Spielereien thäten; auch die Rücksicht brauchte Sie daran nicht zu hindern, daß wir uns bald einmal wieder als Preisrichter und *[Aussteller]* begegnen könnten. Das nächste Mal schicken wir eine schöne große Orgel. So sehr ich wünsche und darauf rechne, daß Sie sich auch für diese interessiren werden, so glaube ich doch nicht, daß Sie jemals der Physic untreu werden und das Orgelfach ergreifen werden; und so wäre nach meinem Dafürhalten *[gar]* kein *[objectives]* Bedenken vorhanden und es hinge blos davon ab, ob Sie einem Orgelbauer die Vertraulichkeit *[gestatten]* wollen, Ihnen ein Erinnerungszeichen zu offeriren und ob Sie demselben die Liebe und Freundschaft erweisen wollen, es zu *[acceptiren]*, worin ich Ihnen aber keinen Zwang anthun will.

Neulich habe ich einen Apparat an H. Professor Magnus geschickt, der mir sein besonderes *[Gefallen]* über die Leistungen desselben ausgesprochen und gleich einen Auftrag auf einen 2ten Apparat für H. Professor Dove und auch *[1 word illeg]* ein 2tes Exemplar für sich *[1 word illeg]* mit Schrauben-Construction gegeben hat. Ich hatte keinen Apparat mit Wellenschrauben (statt der Wellenleisten) vorräthig und arbeite sie auch wegen ihrer großen Schwierigkeit nicht gerne, sonst würde ich Ihnen einen solchen geschickt haben. Die Leistungen derselben sind in manchen Beziehungen vollkommner, weil die Wellenbewegung eine kontinuirliche ist, es ist aber schwer, dieselbe Genauigkeit der Sinuslinie der Welle zu erreichen, wie bei den Wellenleisten.

Ich habe mit großem Bedauern in Ihren Worthen gelesen, daß Sie nicht ganz wohl sind. Ich wünsche, daß das Frühjahr, welches sich bei uns sehr schön anläßt, Ihnen volle Gesundheit bringt, und empfehle mich Ihnen mit der Versicherung | vollkommenster Hochachtung

Ihr ergebenster | Oscar Schulze.

KOENIGSEE 8/4 1857 | EISENACH | 9 4 • [II] | HALLE
Sr. Wohlgeboren | Herrn Professor Dr Tyndall | am Royal Institution |
<u>London</u>

Paulinzelle[1] 7th April 1857.

Most esteemed Herr Professor!

Your letter from 21 Mar.[2] is very dear to me, but it has left me in a melancholy mood. As very kindly and gently as you say No,[3] I nevertheless feel only the '<u>No</u>', and I had quite confidently been counting on a '<u>Yes</u>'. It was, admittedly, a great temerity on my part to suggest to you, Herr Professor, that you accept a memento from me. I did so trusting in the forbearance and goodwill with which you have [always] met me in the various circumstances in which we came into contact.[4] In doing so, I also wanted to give you a token of my gratitude, as I knew very well that the prize won by the apparatus displayed at the Paris Exposition was due solely to your recommendation.[5] You were so kind as to take more particular interest in the apparatus than it perhaps deserved, and to give me several further good suggestions on how to improve and perfect it, which I did not forget to use and to follow where possible with later examples I built. I was, therefore, very eager also to hear your opinion about the improvements that I had undertaken. All of this [1 word illeg] me to send you an apparatus. I did not want to cause you any bother with it, though, as I know well how valuable your time is. You should only have a look at the thing when you have the opportunity, in a leisure hour, and amuse yourself a little by having a play with it. When I asked you to recommend it in England, I meant only that you should say a good word about the apparatus on the many and daily occasions which you have for associating with professors in schools, for whom the apparatus should be a means of making teaching easier. Even if I do come to England sooner or later, I would be able to do little myself, as, while I can read English, I cannot speak it and cannot understand it well when spoken.—The costs which I have incurred with the apparatus are so little that you therefore need not have any concerns about accepting it. You know that I do not make the construction of these apparatus a trade from which I would want to live. For me, it is a hobby for my leisure hours with which I while away the long winter evenings (I reside in a little village of only 15 houses), and with which I dream back to the good times when I did studies in pure science and was so very interested in your works, that I wished and would acknowledge with great thanks, that you would do so only a little for my gadgets. Nor need any consideration of this impede you from our being able to meet again some time soon as jurors and [exhibitors]. Next time, we shall send a beautiful large organ. As much as I wish and count on the fact that you will be interested in this too, I nevertheless do not believe that you will ever become unfaithful to physics and take up the organ; and thus there would not in my opinion be any objective concern, and it would depend solely on whether you are willing to *[grant]* an organ-builder the familiarity of offering a you a reminder, and whether you were willing to show him the affection and friendship of *[accepting]* it, in which regard, however, I do not want to put you under any duress.

I have recently sent an apparatus to Professor Magnus, who expressed his special *[liking]* for its performance and immediately placed an order for a 2nd apparatus for Professor Dove and also *[1 word illeg]* a 2nd instrument for himself, one with a bolt construction. I did not have an apparatus with wave bolts (instead of wave strips) in supply, and do not like making them either due to their great difficulty, otherwise I would have sent you one. Their performance is in some respects more accomplished, because the wave movement is a continuous one, but it is difficult to achieve the same accuracy of the sine line of the wave, as it is with the wave strips.

I have read with great regret in your words that you are not entirely well. I wish that the spring, which is off to a very nice start here, will bring you full health, and I commend myself to you | with the assurance of | my most complete respect

Your very humble servant | Oscar Schulze.

KOENIGSEE 8/4 1857 | EISENACH | 9 4 • *[II]* | HALLE

Professor Dr Tyndall | Esquire | at the Royal Institution | <u>London</u>

RI MS JT/1/S/61

---

1. *Paulinzelle*: a small town located in Thüringen, Germany.

2. *Your letter from 21 Mar.*: letter missing.

3. *As very kindly and gently as you say No*: Schulze sent Tyndall one of his instruments and in letter 1331 asked that Tyndall recommend it. Tyndall apparently declined to do so.

4. *the various circumstances in which we came into contact*: not identified. Neither the previous letters from Schulze to Tyndall in this volume (1331 and 1335) nor Tyndall's journal give any information about how they first came in contact.

5. *the prize won by the apparatus . . . your recommendation*: Tyndall served as a juror for the 1855 Paris Exposition Universelle in class VIII 'Arts relating to the exact Sciences, and to Instruction'. Schulze and Sons were awarded a first place medal for an instrument for visualizing sound waves. The committee's report on the instrument can be found in *Rapports de jury mixte international: Exposition Universelle de 1855*, 2 vols (Paris: Imprimerie Impériale, 1856), vol. 1, p. 438.

## From Mary Anne Coxe			12 April [1857][1]			1371

Edinbro. | April 12th

One single line dear Dr Tyndall, to thank you for yours of yesterday[2]—and to say how cordially I hope you will indeed lay in a stock of health that will stand you in good stead for many a long day.

James[3] desired me to tell you, you must look sharp and convert the writer[4] of that article in the Daily—Scotsman.[5] There is no time to lose, for he is already 75!

I am sure you will be sorry to hear that my brother[6]—who was with us for a week lately, on his way to the far North[7] in pursuit of his favourite enjoyment—fishing—met with an accident in getting out of the steamer at Wick[8]—the sea being extremely rough, he fell and broke one of his ribs—thus not only losing all chance of what he went so far, and under most disagreeable circumstances, to obtain—but—I fear, laying himself up for a considerable time, away from all the amenities of life—in a most outlandish region, with neither books—society nor any one thing to mitigate the tedium of confine-ment. Close too, to a river swarming with salmon—and this he feels the sorest trial of all! His enthusiasm in the 'gentle art'[9] exceeds any thing I ever saw—An absolute passion.

Are <u>you</u> among those who rail at Sir David Brewster's marriage?[10] I—for one applaud it highly—and, advocating marriage as I do at <u>all</u> ages, at none does it seem to me so important as at that, when one can no longer find enjoy-ment in, or welcome from the world—When one's own fire side is the place where one wants to be cheery and happy. He may yet well live 10 years—and should he be doomed to pass these like 'a crazy owl—moping in solitary dark-ness'?[11] When ... [12]

RI MS JT/1/TYP/1/1/273
LT Typescript Only

1.	*[1857]*: the year is given based on Tyndall's journal entry of 15 April 1857: 'A pleasant note from Mrs Coxe. She tells me the gentleman who wrote that stricture in the Daily Scotsman must, if he be converted at all, be converted soon as he is already 75' (Journal, RI MS JT/2/13c/925).

2.	*yours of yesterday*: letter missing.

3.	*James*: James Coxe.

4.	*the writer*: probably David Brewster, mentioned later in this letter regarding his age at his second marriage (see n. 10). He had turned 75 years old on 11 December 1856, four months before this letter. In a 13 December 1861 letter to Tyndall, William Hopkins wrote, 'Sir David Brewster told me at Manchester that he (Sir D.) was a convert to my, (or rather to our), views [on glacier motion]' (RI MS JT/1/TYP/2/641).

5. *that article in the Daily—Scotsman*: not identified.

6. *my brother*: William Fullerton Cumming (b. 1804–92), a physician and author of *Notes of a Wanderer in search of Health, through Italy, Egypt, Greece, Turkey, up the Danube, and down the Rhine* (London: Saunders and Otley, 1839) and *Notes on Lunatic Asylums in Germany: and other parts of Europe* (London: J. Churchill, 1852). See 'Obituary', *Lancet*, 140 (1892), pp. 349–50. In his journal entry of 15 April 1857, Tyndall wrote, 'Dr Cumming had broken a rib through a fall from a steamer in Scotland' (Journal, RI MS JT/2/13c/925).

7. *far North*: the far north of Scotland, most likely to the River Wick.

8. *steamer at Wick*: steamer service, probably from Iverness, to Wick, a town in northern Scotland where the River Wick feeds into Wick Bay and the North Sea beyond.

9. *'gentle art'*: fishing.

10. *Sir David Brewster's marriage*: Brewster's first wife Juliet Macpherson died in 1850, and he remarried on 27 March 1857 at the age of 75 to Jane Kirk Purnell (b. 1827), daughter of Thomas Purnell of Scarborough who was 46 years his younger. They had met on 17 November 1856 while Brewster was travelling to Cannes on account of his health. In January 1857 Brewster followed her to Nice, where they were married. Brewster and Lady Jane Kirk Brewster had one daughter, Constance Marion who was born 27 January 1861. See M. M. Gordon, *The Home Life of Sir David Brewster*, 2nd edn (Edinburgh: Edmonston and Douglas, 1870), p. 270.

11. *'a crazy owl—moping in solitary darkness'*: possibly a reference to the poem 'Elegy Written in a Country Churchyard' (1751) by Thomas Gray (1716–71): 'Save that from yonder ivy-mantled tow'r | The moping owl does to the moon complain | Of such, as wand'ring near her secret bow'r, | Molest her ancient solitary reign'.

12. *When . . .* : fragment only, letter is incomplete.

From Anna Hirst 12 April 1857 1372

14, Rue de la Mairie | Pau, Sunday 12ᵗʰ April | /57

My dear M͟r Tyndall

You will no doubt be rather surprised to have a letter from me, but I am about to trouble you with a commission.

Wednesday the 22ⁿᵈ will be Tom's[1] birth day, and I wish to give him a little present. Will you therefore be so kind as to send me by post M͟rs Barrett Browning's[2] last poem,[3] I forget its name. We shall most probably meet in Paris during the summer when I can pay you for both book and postage.

Tom is pretty well and looks quite fat. I wish he would take more exercise. We had a very nice little trip lately to L'estelle[4] a village at the foot of the mountains, the change did us all good.

John[5] is still here, the very showery bad weather prevents his setting out on a foot tour.

I am pretty well at present, but far from as strong as I was before that terrible long illness which commenced on New Years Day.[6] I shall be glad to leave Pau[7] though it is a lovely place. I am tired of every thing about it. We hope to leave for Paris next month and there I hope to meet my sister, M<u>rs</u> Simpson,[8] her husband[9] and family, which you may be sure will be a great joy to me.

We have no French acquaintances whatever and I have no opportunity of speaking except to the servant whose accent is not quite the thing as she is a Bearnaise.[10]

I hope you and D<u>r</u> Debus had good weather for your walk.[11]

Pray excuse the trouble I give you, and also this long note, but we Martins[12] know not where to stop when we begin writing.

Believe me | Dear M<u>r</u> Tyndall | Very sincerely yours | <u>Anna M. Hirst</u>

PS | Please address the book to <u>me</u> as I do not want Tom to know until he receives it.

RI MS JT/1/H/124
RI MS JT/1/HTYP/496

---

1. *Tom's*: Thomas Hirst.
2. *M<u>rs</u> Barrett Browning's*: Elizabeth Barrett Browning (1806–61), an English poet popular in Britain and the United States and wife of the English poet and playwright Robert Browning (*ODNB*).
3. *last poem*: E. Barrett Browning, *Aurora Leigh* (London: Chapman and Hall, 1857). See letter 1387.
4. *L'estelle*: Lestelle, a village at the foot of the Pyrenees in southern France close to Pau, now known as Lestelle-Bétharram. In his journal for 5 April 1857, Hirst wrote 'Yesterday, the weather being tempting, we left Pau for a day or two, and directed our way to Lestelle' (Brock & MacLeod, *Hirst Journals*, p. 1255).
5. *John*: John Martin.
6. *that terrible long illness which commenced on New Years Day*: see letter 1316.
7. *Pau*: see letter 1298, n. 4.
8. *M<u>rs</u> Simpson*: Mary Simpson (née Martin, 1816–*c.* 1900), Anna's sister who married Maxwell Simpson in 1845.
9. *her husband*: Maxwell Simpson.
10. *Bearnaise*: of or relating to the French province of Béarn in the Pyrenees.
11. *good weather for your walk*: Debus and Tyndall went to the Isle of Wight in early April. See letter 1366, n. 11.
12. *we Martins*: Mary Simpson (née Martin), John Martin, Anna Hirst (née Martin), and six other siblings were the children of Samuel Martin, Sr. (1749–1831) and Jane Harshaw (*c.* 1787–1847).

To George Biddell Airy   17 April 1857   1373

My dear Sir

If it would suit your convenience I should be happy to make arrangements for our experiments[1] with the Induction coil on some day of the week after next. I should like if possible to shew you the most powerful arrangement yet devised and to effect this should have to communicate with the deviser. Would you kindly tell me on what days among the following you would be at liberty to come to the Royal Institution—Monday 27th. Tuesday, Friday or Saturday? I can then fix upon upon the particular day which will be most convenient to Mr Bentley[2] who has introduced the most recent improvements on the coil.

very sincerely yours | <u>John Tyndall</u>
17th April 1857

RGO MS.RGO 6/471.161

1. *make arrangements for our experiments*: see letters 1346, 1347, and 1348. For Airy's reply, see letter 1374.
2. *Mr Bentley*: Charles A. Bentley, who later in 1857 disputed with Jonathan Nash Hearder (1809–76) through letters to the *Phil. Mag.* over priority of improvements to induction coils.

From George Biddell Airy   17 April 1857   1374

London, S.E. | 1857 April 17

My dear Sir

It is very kind in you to bear in mind my request.[1]—Of the days which you mention, Monday is not good for me (it is a very busy day at the Observatory[2]), but Tuesday, Friday, Saturday, are perfectly free. I would therefore <u>provisionally</u> propose Tuesday 28th. And, as you have not specified any hour, I would <u>provisionally</u> say that an early hour is best for me—the earlier the better. But I beg you to consider these as truly provisional, and as subject to alteration for the slightest reason.

Will you bear in your friendly recollection, that the <u>whole</u> subject is practically new to me, and therefore that the instruction to be derived from an experiment of a low order will be, to me, a very valuable introduction to the higher.

I am, my dear Sir, | Very truly yours | <u>G. B. Airy</u>
Professor Tyndall | &c. &c. &c.

RGO MS.RGO 6/471.162–63
RI MS JT/1/TYP/1/20

---

1. *my request*: this letter responds to letter 1373.
2. *the Observatory*: the Royal Observatory, Greenwich, where Airy served as Astronomer Royal from 1835–81.

From Henry Runciman Drewry[1]   17 April 1857   1375

War Office Pall Mall S.W. 17th April 1857

Sir,

It having been found necessary to reduce the number of gentlemen employed to examine candidates for admission into the Royal Artillery Academy[2] and also the rates of remuneration which they receive I am directed to request that you will acquaint me for the information of Lord Panmure[3] whether you are prepared to undertake the examination on Experimental and Natural Sciences, in June next, with a remuneration of £25[4] in conjunction with M[r] Huxley

I am | Sir | Your obedient servant | H. R. Drewry

RI MS JT/2/13c/927
RI MS JT/2/7/423–24

---

1. *Henry Drewry*: Henry Runciman Drewry (1801–86), a clerk in the War Department.
2. *Royal Artillery Academy*: from 1856–57, Tyndall wrote and graded examinations in scientific topics for commissioned officers of the Royal Artillery and Royal Engineers at the Royal Military Academy at Woolwich. See letter 1306, n. 2.
3. *Lord Panmure*: Fox Maule-Ramsay.
4. *a remuneration of £25*: Tyndall did not agree with this decrease in pay, and thus resigned as examiner. See letter 1376 for his response to the War Department.

To Henry Runciman Drewry[1]   18 April 1857   1376

Sir.

The duties of an Examiner of Candidates for the Artillery and Engineers[2] are, in my estimation, of so responsible a character, and the labour necessary to the efficient discharge of those duties (especially with such a number of candidates as may be expected to present themselves next June) is so great, that I feel compelled in reply to your letter of yesterday[3] to state that I am not prepared to undertake the examination in Experimental Science for a remuneration of £25.[4]

I write thus with regret, and I would avail myself of this last opportunity to thank Lord Panmure[5] for the great consideration which I have experienced at his Lordship's hands; and to express at the same time my earnest conviction that were the resources of Experimental Science in relation to military matters better understood, they would be more highly appreciated than they now appear to be.

I am Sir | Your obedient servant | <u>John Tyndall</u>

RI MS JT/2/13c/927
RI MS JT/2/7/424–25

1. *Henry Drewry*: see 1375, n. 1.
2. *Examiner of Candidates for the Artillery and Engineers*: see letter 1375, n. 2.
3. *your letter of yesterday*: letter 1375.
4. *remuneration of £25*: evidently a decrease in pay.
5. *Lord Panmure*: Fox Maule-Ramsay.

To Michael Faraday 18 April 1857 1377

Saturday, 18th. Apr 1857.

My dear Mr Faraday

For some time past the thought of writing to you on a subject of some importance to myself has occurred to me at intervals, and while in the country lately I was only prevented from doing so by the fact of having other matters to deal with which left me no time for writing.

When I concluded my last course of lectures[1] I felt so weary that I looked with some dismay on the course to come after Easter, and did I not know that a change of arrangement at the time would be practically impossible, I should certainly have endeavoured to transfer the burden to stronger shoulders. I am now comparatively rested, but still I think I ought to endeavour to make some other arrangement for years to come.

Far even as they fall below the standard which I should like them to reach, I find 19 lectures to one audience to be a work of such labour that I would willingly shorten it if possible. They consume half the year in delivery, and an additional portion of the year in thinking of them, thus leaving me a comparatively small amount of time, and no great stock of strength for original research. If I thought that the labour of lecturing would sufficiently diminish with practice, I should hardly write thus; but four years experience[2] warns me that the respect which I feel to be due to our audience would prevent me from ever dealing lightly with the lectures, and would always urge me to seek after matter suited to their taste and comprehension.

Under these circumstances I am inclined to think that it would be better for me, and I hope also better for science, if I occupied a position as regards the number of lectures to be given, and the remuneration to be derived from them, similar to that occupied by Mr. Brand[3] during his connexion with the Institution.[4] And if you see nothing objectionable in a proposal to this effect, I should be very thankful to you if, at the proper season, you would have the kindness to lay it before the Managers.[5]

Believe me | dear Mr. Faraday | Yours most faithfully | John Tyndall
[in Faraday's hand]
read to the Managers on (I think) the 27th May following MF | Managers' attention drawn to it again 15th Feb 7 1858. MF

RI MS JT/1/TYP/12/4068–69
Faraday Correspondence, 5:3272
Typed Transcript Only

---

1. *my last course of lectures*: Tyndall's 'Eleven Lectures on Sound' (see letter 1308, n. 7).
2. *four years experience*: Tyndall began working as Professor of Natural Philosophy at the RI in 1853.
3. *Brand*: William Thomas Brande (1788–1866), an English chemist, who in 1812 succeeded Humphry Davy as the chair of chemistry at the RI (*ODNB*).
4. *the Institution*: the RI.
5. *the Managers*: When Faraday read this letter to them on 27 May 1857 the Managers were: John George Appold (1800–65), Benjamin Collins Brodie (1783–1862), Benjamin Bond Cabbell (*c.* 1782–1874), Thomas Davidson (1817–85), Warren de la Rue (1815–89), George Dodd (1808–81), Charles Fellows (1799–1860), Henry Holland (1788–1873), John Carrick Moore, William Frederick Pollock, James Rennie (1781–1867), Joseph William Thrupp (1799–1873), John Webster (1795–1876), James Parke (1782–1868, 1st Baron Wensleydale), and Charles Wheatstone. See 'Annual Meeting, Friday May 1 [1857]', *Roy. Inst. Proc.*, 2 (1854–58), pp. 421–22.

From Auguste de la Rive   18 April 1857   1378

Naples, le 18 avril 1857.

Mon cher Monsieur,

Il m'a été impossible de répondre plus tôt à votre lettre du 13 mars; je profite maintenant d'un instant de liberté pour venir le faire sans plus tarder:

Avant tout que je vous remercie de la peine que vous avez prise; j'en suis excessivement reconnaissant; je reconnais toute la vérité de ce que vous me dites au sujet des <u>gallicismes</u> de Mr Walkers, c'était déjà frappant dans les premiers volumes; mais qu'y faire ? On ne peut changer cela, car ce serait toute la traduction à remanier. Heureusement que c'est clair, ce qui est un point essentiel. Bientôt je serai de retour à Genève & je pourrai pour la fin du volume vous décharger de la tâche que vous avez bien voulu prendre pour me rendre service, ce que je n'oublierai jamais.

Maintenant répondant à vos questions, je prendrai les divers points les uns après les autres.

1⁰) Nous n'avons point de <u>presse hydraulique</u> adaptée au but que vous vous proposez, il faut donc que vous en portiez une avec vous; mais, comme depuis longtemps je désirais en avoir une, je serais extrêmement content de me charger au prix coûtant de celle que vous apporterez, après que vous en avez fait tout l'usage qui vous sera nécessaire. Loin d'être un service que je vous rendrais, c'est un service au contraire que vous me rendrez, car il y a longtemps que j'ai le désir de posséder une presse hydraulique du genre de celle dont il s'agit. Choisie par vous, employée par vous, elle me sera doublement précieuse. Ainsi voilà une affaire entendue.

2⁰) Quant à l'infusion fortement colorée, je ne doute pas que vous ne puissiez vous procurer à Genève celle dont vous aurez besoin, mais je ne puis cependant en répondre. Vous feriez bien peut-être d'en apporter un peu; toutefois, comme je serai surement de retour à Genève avant que vous quittiez l'Angleterre, je pourrai vous répondre d'une manière plus positive sur ce point.

3⁰) Vous trouverez à Genève un théodolite, une chaine d'arpenteur, en un mot tous les instruments nécessaires pour l'arpentage & la triangulation; il sera facile de vous les confier; vous n'avez donc aucun besoin d'apporter avec vous ce genre d'instruments. Il en existe soit au Cabinet de Physique, soit à l'Observatoire, soit au Dépôt de la Carte Suisse dirigé par le Général Dufour, & *[1 word illeg]* il nous sera facile aux uns & aux autres, de vous en procurer.

J'ai passé à Rome un hiver fort agréable, voyant souvent le Père Secchi dont l'Observatoire est très bien monté, & qui est lui-même un observateur très actif & très intelligent. J'ai vu aussi fréquemment le Prof. Volpicelli qui vient d'achever quelques recherches intéressantes sur l'induction électrostatique. C'est un bon physicien, peut-être un peu prompt à conclure, mais ingénieux, bon calculateur & observateur perspicace & intelligent.

Mille remerciements pour les détails que vous me donnez sur nos amis de Londres & en particulier sur Faraday dont j'apprends avec joie que la santé continue à être bonne. Je n'ai lu qu'un extrait très abrégé de son dernier mémoire sur l'or réduit en lames minces & sur les métaux en général. J'espérais y trouver quelques phénomènes électriques ou magnétiques; mais il m'a paru qu'il ne s'agissait que de phénomènes chimiques & moléculaires; mais n'ayant pu me procurer qu'un extrait du mémoire, il est possible que je n'aie pas pu voir tout ce qui y était contenu; ce sera pour mon retour. En attendant ayez la bonté de faire mes compliments les plus affectueux à notre digne et excellent ami.

Vous avez une bien bonne idée de venir faire cet été une course dans nos glaciers. Je vous prie de vous rappeler que vous avez chez moi une chambre qui vous attend pendant votre séjour à Genève, & j'espère bien que vous me ferez le grand plaisir d'élire votre domicile chez moi. Je vous demanderai seulement d'avoir la bonté de me dire un peu d'avance l'époque approximative de votre arrivée à Genève, afin que je combine dans cette idée le moment d'une course que je serai probablement appelé à faire à Vichy pour ma santé pendant l'été.

Merci de votre imprimé sur les glaciers, vous aurez vu qu'on a imprimé dans la <u>Bib. Univ.</u> votre travail sur ce sujet intéressant, je suis impatient de m'en entretenir avec vous & de connaître le résultat de vos observations faites sur les lieux mêmes. Vous piquez ma curiosité en me parlant d'une dame de nos amies que vous avez rencontrée chez Sir Benjamin Brodie; vous auriez dû me dire son nom. Qui est-elle? Je suis tout-à-fait de votre avis au sujet de Mr Matteucci; il faut qu'il vienne à la polarité diamagnétique. Pourriez-vous, à ce sujet, quand vous viendrez à Genève, m'apporter encore <u>un</u> ou <u>deux</u> échantillons de bismuth comprimé? Vous me ferez un très grand plaisir.

Adieu, très cher Monsieur & bon ami, votre dévoué et affectionné | A. de la Rive.

Naples, April 18th 1857.

My dear Sir,

I could not respond sooner to your letter of March 13;[1] I am taking now advantage of a moment of freedom to do so without further ado:

—First and foremost, let me thank you for the trouble you took; I am extremely grateful for it; I recognize the whole truth of what you say on the subject of Mr Walker's <u>Gallicisms</u>,[2] it was already striking in the first volumes; but what can be done about it? We cannot change this, because that would entail reworking the entire translation. Fortunately, it is clear, which is an essential aspect. I will return soon to Geneva[3] & I will be able to relieve you for the end of the volume from the task that you accepted to take up to help me, which I will never forget.

Now to answer your questions, I will address the different issues one after the other.

10) We do not have a <u>hydraulic press</u> adapted to your intended purpose, so you will have to bring one with you; but since I have desired to own one for a long time, I would be extremely happy to cover the cost price of the one you will bring, when you are done using it for your intended requirements. Far from being a service I will provide you, you will on the contrary do me a favor, since I have

wanted to own a hydraulic press of this particular type for a long time. Chosen by you, used by you, it will be twice as precious to me. This matter is settled.

20) With regards to your highly tinted solution, I have no doubt you will get hold of the one you need in Geneva, but I cannot guarantee it. You would perhaps be better off bringing a sample of it. However, since I will probably be back to Geneva before you leave England, I will be able to provide you with a more concrete response on this matter.

30) In Geneva, you will find a theodolite,[4] a surveyor's chain,[5] in a word all the instruments necessary for land survey & triangulation; it will be easy to entrust them with you; so, you do not need to bring these types of instruments with you. They might be either in the office of Physics,[6] or in the Observatory,[7] or in the Map Repository of Switzerland headed by the General Dufour,[8] & *[1 word illeg]* it will be easy for everyone, to provide them for you.

I spent a very pleasant winter in Rome, having often seen Father Secchi,[9] whose Observatory[10] is well equipped, & who is, himself, a very active & very intelligent observer. I have also often met Prof. Volpicelli[11] who just completed some interesting research on electrostatic induction. He is a good physicist, perhaps hasty in his conclusions, but ingenious, a good arithmetician & a perceptive and intelligent observer.

Many thanks for the details that you give me regarding our friends in London & especially about Faraday, who is still in good health, as I found out with much joy. I read only a much-abbreviated excerpt[12] from his latest essay on the reduction of gold in thin strips & on metals in general.[13] I hoped to find there some electric or magnetic phenomena, but it seemed to me that it was just chemical & molecular phenomena. However, since I was only able to find one excerpt of the essay, it is possible that I could not see everything that it contained; it will have to wait for my return. Meanwhile, be kind enough to send my most affectionate compliments to our worthy and excellent friend.

It is a great idea for you to come for an excursion in our glaciers this summer. Please remember that there is a room ready for you at my place for your stay in Geneva & I hope that you will give me the great pleasure of taking up residence at my home. I will only ask of you to be kind enough to tell me a little in advance about the approximate date of your arrival in Geneva, so that I can combine in this plan the time for another excursion for my health in Vichy[14] that I will probably be obliged to take during the summer.

Thank you for your print on the glaciers;[15] you would have seen that your work on this interesting subject was printed in the <u>Bibl. Univ.</u>[16] I am eager to discuss it with you & to know the results of your observations that were done at the sites themselves. You pique my curiosity by telling me of this lady[17] among our friends that you have met at Sir Benjamin Brodie's[18] home; you should have told me her name. Who is she? I completely agree with you in regards to Mr Matteucci; he must accept diamagnetic polarity.[19] On this matter, when you come to Geneva, could you bring me again <u>one</u> or <u>two</u> samples of compressed bismuth? You would do me a great pleasure.

Farewell, dear Sir & good friend, your devoted and affectionate | A. de la Rive.

RI MS JT/1/D/91
RI MS JT/1/TYP/348–49

1. *your letter of March 13*: letter 1353.

2. *Mr Walker's <u>Gallicisms</u>*: Charles Walker translated de la Rive's third volume of *Traité de l'Électricité Théorique et Appliquée* (see letter 1293, n. 16). Tyndall reviewed the proofs, and in letter 1353 commented on the 'style of writing' as 'not quite English. Mr Walker probably thinks it better to adhere to your own forms of expression'. A 'Gallicism' is an idiom or mode of expression belonging to the French language, esp. one used by a speaker or writer in some other language (*OED*).

3. *Geneva*: see letter 1293, n. 3.

4. *theodolite*: see letter 1353, n. 11.

5. *surveyor's chain*: see letter 1353, n. 12.

6. *office of Physics*: at the Academy of Geneva, where de la Rive was professor of physics and chemistry.

7. *the Observatory*: an observatory in Geneva, founded in 1772.

8. *Map Repository of Switzerland headed by General Dufour*: Guillaume Henri Dufour (1787–1875), a Swiss army officer, bridge engineer, and topographer, was in charge of efforts to produce maps of Switzerland and its states (*HLS*).

9. *Father Secchi*: Pietro Angelo Secchi (1818–78), an Italian astronomer and Director of the Observatory at the Pontifical Gregorian University, Rome, then known as the Roman College (*CDSB*).

10. *Observatory*: see n. 9.

11. *Prof. Volpicelli*: Paolo Volpicelli (1804–79), an Italian physicist and Professor of Experimental Physics at University of Rome. See 'Paolo Volpicelli', *Nature,* 20 (1879), pp. 126–27.

12. *much-abbreviated excerpt*: not identified.

13. *his latest essay on the reduction of gold in thin strips & on metals in general*: M. Faraday, 'Experimental Relations of Gold (and other Metals) to Light', *Phil. Trans.,* 147 (1857), pp. 145–81.

14. *Vichy*: a resort town in central France.

15. *your print on the glaciers*: an offprint of J. Tyndall and T. H. Huxley, 'On the Structure and Motion of Glaciers', *Phil. Trans.,* 147 (1857), pp. 327–46.

16. *printed in the <u>Bibl. Univ.</u>*: J. Tyndall, 'Sur la théorie des glaciers', *Archives des sciences physiques et naturelles* 34 (1857), pp. 177–85. The Swiss journal *Archives des sciences physiques et naturelles* was previously known as *Bibliothèque universelle de Genève*.

17. *this lady*: not identified.
18. *Sir Benjamin Brodie's*: see letter 1353, n. 21.
19. *he must accept diamagnetic polarity*: see letters 1340, 1353, and 1483.

To George Biddell Airy   20 April 1857   1379

R.I. <u>20 April</u>. | 1857

My dear Sir

I am happy to inform you that I hope so to arrange matters[1] as to have everything ready for you on Tuesday the 28th—From half past 9 A.M. downwards I am at your service.

very sincerely yours | <u>John Tyndall</u>

RGO MS.RGO 6/471.165

————

1. *to arrange matters*: see letters 1373 and 1374.

From George Biddell Airy   21 April 1857   1380

1857 April 21

My dear Sir

If I do not hear from you to the contrary, I will wait on you[1] on Tuesday the 28th, about 10 to 10 ½ A.M.

I am, my dear Sir, | Yours very truly | <u>G B Airy</u>
Professor Tyndall

RGO MS.RGO 6/471.164

————

1. *I will wait on you*: see letter 1379.

To Thomas Henry Huxley   22 April [1857][1]   1381

22nd. April.

My dear Huxley

I take it for granted that you have received the enclosed,[2] it reached me on Saturday and I wrote to say that I was not prepared to undertake the examination.

On looking into my journal this morning the following case of homology[3] struck me: it is quoted from Henry Moore,[4] who himself is quoting from 'Chymists'[5] whom he considers as 'more or less palpably mad'.[6]

'The pure blood in man answers to the element of fire in the great world, his heart to the earth, his mouth to the Arctic pole, and the opposite orifice to the Antarctic pole!'[7]

Ever Yours John Tyndall | Kind regards to Mrs Huxley.[8]

RI MS JT/1/TYP/9/3157
IC HP 8:282a
Typed Transcript Only

1. *[1857]*: the year is given by the reference to letter 1375.
2. *the enclosed*: letter 1375, in which Tyndall is asked to continue being an examiner for commissioned officers of the Royal Artillery and Royal Engineers at the Royal Military Academy at Woolwich for less pay.
3. *homology*: sameness of relation; correspondence (*OED*).
4. *Henry Moore*: Henry More (1614–87), an English philosopher who wrote many works, especially on religious topics.
5. *'Chymists'*: More was referring to Paracelsus and alchemists or iatrochemists generally.
6. *'more or less palpably mad'*: H. More, *Enthusiasmus Triumphatus: Or, a Discourse of the Nature, Causes, Kinds and Cure of Enthusiasme* (Flesher, 1656), on p. 41, 'But I have observed generally of Chymists and Theosophists, as of severall other men more palpably mad, that their thoughts are carried much to Astrology'.
7. *'The pure blood in man . . . to the Antarctic pole!'*: More, Enthusiasmus Triumphatus, p. 43. In the original passage, the words Arctic and Antarctic are spelled as Arctick and Antarctick.
8. *Mrs Huxley*: Henrietta Huxley.

From Joseph Dalton Hooker   [28 April 1857][1]   1382

Kew Tuesday

Dear Tyndall

I was concerned to miss you from the Club[2] yesterday and more so to hear that you were poorly—Will you not come here?[3] I am altogether alone. Father,[4] and Mother[5] both away and you can have perfect quiet and such grub as I have got. What folly it is of you to work yourself into these conditions. You are not strong—and know it—and yet cannot exercise the simplest of all the virtues of self governance—taking care of yourself. It is a weak short-sighted policy. Your spirit drives you to it, true enough, and so it does other men to worse things—who yet are no worse than you in yielding to their evil spirits. All your friends remark how haggard and worried you look and you won't have pity on them either. Take heart of grace and be reasonable. Come here and rusticate with books and papers and a room to yourself if you please.

Ever yours | J. D. Hooker

RI MS JT/1/TYP/8/2545
LT Typescript Only

1. *[28 April 1857]*: the date is given based on relation to letter 1383, and the date of the referenced meeting of the Philosophical Club of the RS (see n. 2).
2. *the Club*: Philosophical Club of the RS (see letter 1299, n. 7), Hooker was in attendance at a meeting on 27 April 1857. See T. G. Bonney, *Annals of the Philosophical Club of the Royal Society, Written from Its Minute Books* (London: Macmillan and Co., 1919), p. 48.
3. *come here?*: while Hooker lived at 55 Kew Green, here he is likely inviting Tyndall to stay at his parents' house, the director's residence at 49 Kew Green. See R. Desmond, *Kew: The History of the Royal Botanic Gardens* (London: Harvill Press/Kew, 1995), p. 416.
4. *Father*: William Jackson Hooker (1785–1865), botanist, director of the Royal Botanical Gardens, Kew, and father of Joseph Hooker (*ODNB*).
5. *Mother*: Maria Hooker (née Turner, 1797–1872), wife of William Jackson Hooker (married in 1815) and mother of Joseph Hooker.

To Joseph Dalton Hooker 28 April 1857 1383

28th, April, 1857.

My dear Hooker

A certain class of philosophers to whom for some years I lent a willing ear, and from whom I really derived a great deal of benefit, are in the habit of urging their disciples to cultivate the virtues of the Stoic,[1] to draw upon their own inner resources and to render themselves independent of external influences. I never, I confess, even when I admired it most, could find myself completely at home in this creed. The hand of a friend, the laugh of a child, not to say anything of the 'rare and radiant maidens'[2] were always sources of happiness to me, and so far therefore my happiness was of external origin. Afterwards I began to see that this was natural, and that it was a truer philosophy to conform to nature than to stifle and pervert her by stoicism however noble and selfdenying it might appear. I confess therefore to the pleasure which your kind, and hospitable, and admonitory and inculpatory little note[3] gave me. But really this time I am not quite so much to blame. I was well on Saturday— very well—but I happened to be lightly clad and the bitter east wind sought out my soft points and smote me. It was this and not hard work that laid me low. But I am up again like the giant Antaeus,[4] filled with new strength from the contact with my mother earth. If you knew how tough I used to be; what an amount of physical labour I was capable of going through, what I endured in the mental line during my stay in Germany, you would not wonder at my discontent to find myself staggering under burdens which seven short years ago I should have shaken from me as the apostles[5] did the dust from their sandalled feet whenever a village refused them grub and lodging. But our lectures are wearying—my head is never out of school, and the work, though really light, has something anxious and straining about it which renders it far more trying than the hardest original thought. I have been trying to shorten it, and hope this may be accomplished. Think of two years of dyspepsia,[6] nightmare, flatulency and blue devils,[7] being inflicted on Frankland by the last course of lectures he gave here![8] If I feel it needful I will really shelter myself under your friendly roof: at all events I will swoop down upon you sometimes for a dinner. I know I am haggard and that my spirit often writes its weariness in my face, but come to the Mer de Glace[9] with me this year and you will cause my face to shine. Send Willy[10] a kiss from me: kindest remembrances to Mrs Hooker when you write and believe me always

J Tyndall.

RI MS JT/1/TYP/8/2546
LT Typescript Only

---

1. *the Stoic*: one of a school of Greek philosophers (founded by Zeno, fl. *c.* 300 BC), characterized by the austerity of its ethical doctrines for some of which the name has become proverbial (*OED*).

2. '*rare and radiant maidens*': from Edgar Allen Poe's poem *The Raven* (1845), ii.11 and xvi.95: 'rare and radiant maiden whom the angels name Lenore'.

3. *your . . . little note*: letter 1382.

4. *the giant Antaeus*: the son of Poseidon and Gaia ('mother earth') in Greek mythology, a giant who would challenge others to a wrestling match.

5. *the apostles*: the twelve disciples of Christ.

6. *dyspepsia*: see letter 1308, n. 2.

7. *blue devils*: feelings of depression or melancholy (*OED*).

8. *last course of lectures he gave here!*: in 1853, Frankland gave a series of ten Thursday lectures at the RI on 'Technological Chemistry', which began on 7 April. See RI Guard Book, vol. 2, p. 77, and Russell, *Edward Frankland,* p. 217.

9. *Mer de Glace*: see letter 1306, n. 12.

10. *Willy*: William Hooker.

From George Biddell Airy   28 April 1857   1384

Royal Observatory Greenwich | 1857 April 28

Dear Sir

I am sorry to hear of your illness.[1] This weather is most unfavorable for any thing like bronchitis, and I should be sorry to hold the shadow of an engagement[2] over you during its continuance.

If in better weather and in your better health an opportunity of meeting should occur, I should be grateful for it.

I am, dear Sir, | Yours very faithfully | G B Airy

D[r]. Tyndall | &c &c &c

RGO MS.RGO 6/471.167

---

1. *to hear of your illness*: Tyndall wrote in his journal for 26 April 1857: "'The cold smote my unprotected throat.' This morning I felt the effects from it. The boundary between throat and chest was sore and irritated.' He noted that it worsened throughout the day and his entry for the following day stated that he was in bed all day (Journal, RI MS JT/2/13c/929).

2. *the shadow of an engagement*: see letter 1380.

To George Biddell Airy   2 May 1857   1385

Saturday | 1857 May 2

My dear Sir

I regret very much the disappointment which my illness[1] must have caused you. It is now passed away and on next Tuesday I shall have every thing ready for you. Do not give yourself the trouble of replying to this unless you are unable to come on Tuesday.

very sincerely yours | John Tyndall

RGO MS.RGO 6/471.168

---

1. *my illness*: see letter 1384, n. 1.

## From James Prescott Joule[1]			2 May 1857			1386

Oak Field, Moss Side | Manchester May 2/57

My dear Tyndall

Many thanks for your kind letter[2] announcing the selection of Smith[3] by the Council.[4] I am sorry you took the trouble to correct the slip of pen you made as we all know that Selection by the Council is in fact Election by the Society, and it is a gratifying proof of the confidence of the Society in its managers that this can be so confidently said. I communicated the intelligence to Dr Smith this morning who expressed much gratification and is very thankful for the kind interest you have taken in his behalf. I much regretted hearing of your illness,[5] but trust you are now fully restored. I hope you will not be tempted by the ardour of philosophical research to overstep the limits of exertion which your health & strength can accomplish. I have seen so much of this that I always advise my friends to take things easily, which is also an advice I have been obliged to take myself.

Believe me | Yours always truly | James P Joule

RI MS JT/1/J/137
RI MS JT/1/TYP/2/661

1.	*James Prescott Joule*: see letter 1297, n. 3.
2.	*your kind letter*: letter missing.
3.	*Smith*: Andrew Smith (1797–1872), a Scottish surgeon and zoologist who studied natural history in South Africa and led an expedition into Central Africa. He authored *Illustrations of the Zoology of South Africa*, 5 vols (London: Smith, Elder and Co., 1838–50) (*ODNB*).
4.	*Council*: Council of the RS. Smith was elected FRS on 5 May 1857. See *Roy. Soc. Proc.*, 8 (1857), p. 395.
5.	*your illness*: see letter 1384, n. 1.

## From Thomas Archer Hirst			3 May 1857			1387

Pau, | May 3rd 1857

My dear John.

I have often read your letter of the 4th of April,[1] and in my heart thanked both Bacon[2] and yourself. I never read his Essays[3] but I have already a great respect for him as a venerable old thinker, and someday out of gratitude I will know more of his thoughts. You were quite right in judging that in thus innocently enjoying your successes by communicating them to me you would also give me pleasure. As I have often told you I feel a brotherly pride in all your victories I recognize in them the fulfilment of an old prophecy of my own, I experience a peculiar satisfaction, indeed, when I notice that qualities of character to which I have pinned my faith actually possess the force for which I give them credit, and are able to make their way in and wring acknowledgement from the world. Indeed you may permit yourself to enjoy the sunshine of sunny days, there is no fear of <u>your</u> being sunburnt. I know that all days will find you working and intent upon your work for its own sake. It is thus you will keep up the necessary warmth even were the sun to absent himself, and it is thus you will protect yourself against every moral "coup de soleil".[4] The world little thinks how in paying you homage it gladdens another about whom it knows nothing but so it is.

By this time you will be once more deep in your London work and again storing up new honours. I hope you have used your vacation well and have returned with health to your toil. I have heard of your pale looks, and some of our friends have as usual prophecied that some day you will be 'carrying things too far.' But I have strong faith in that reserve of prudence which I know you to possess, and I hope by its exercise you will prove these prophets to be false. You have a strong though a peculiar constitution. I have often noticed how a few days rest of your own sort, namely a strong and unhesitating determination to get strength, has worked wonders; though I have also noticed that the strength thus rapidly gained seldom remains very long. Sometimes I have been inclined to think that with better management you might preserve the necessary equilibrium between health and work more perfectly. But after all, this equilibrium is perhaps not so essential or desirable for you as it is for me, for instance, at any rate you know best.

Francis tells me the R.S. have granted you a sum of money to finish your glacier researches,[5] and I am glad of it for two reasons. First because we shall probably see you as you pass through Paris, whither we shall go if all be well in a few weeks, and secondly because I cannot but think that of all subjects of research with which you have heretofore occupied yourself this must be healthiest.

Anna is tolerably well just now. As usual, her state of health is very unstable, I live in continual fear and hope, for although well to-day I know she may be suffering to-morrow. I have found change of scene and life to be very beneficial for her, and we have already made one or two little excursions towards the mountains. As yet we have been unfortunate as regards the weather, and have been compelled to return to Pau[6] instead of completing our excursion. At present we are holding ourselves in readiness to start on our last little excursion as soon as the weather promises to become settled. Our way of travelling is very pleasant. I hire a horse and phaeton[7] for five francs a day, and drive myself. The phaeton is conveniently built, it has a cover which can be drawn over Anna's head to protect her against the sun or rain. John Martin prefers walking and contrives to meet us at different points of our journey. To take care of Anna at the hostels, as well as to give her pleasure too, we take our French servant with us. She is a good, honest and faithful girl as I mentioned once before I believe. Her affectionate nursing of Anna through the winter has pleased me so much that if I were rich enough I would free her from the necessity of toiling for the rest of her days—for she is growing old herself and not very strong. For my part I enjoy these trips exceedingly on account of their good effect on Anna; under other circumstances my enjoyment would be weakened or destroyed by the consciousness of lost time whereas now it is heightened by the conviction that I am lengthening and brightening the few days that still remain for my poor wife. Yes—struggle as I may—I can no longer shut my eyes to the terrible fact that she is slowly wasting away before my eyes. There are moments when the thought that no force of will no prayers no power on earth can arrest the progress of her terrible disease almost drives me to despair and yet God be thanked I have so disciplined myself that in her presence I can often forget this, and rejoice in her comparative free- dom from suffering and share with her her few & small pleasures. It is strange what power of accommodation we possess, how after our sphere of hopes and desires have been gradually diminished we can still find enough to live upon. Two years ago I would have thought it impossible to exist under the circum- stances by which I am now surrounded, and yet I not only live, but have also moments of great joy and gladness.

Anna sends her love, and thanks for your assistance in procuring for me the small surprise on the morning of my birthday.[8] "Aurora Leigh"[9] arrived safely, and in due time.

Yours affectionately, | <u>T. A. Hirst</u>

RI MS JT/1/H/232
RI MS JT/1/HTYP/491–92

---

1. *your letter of the 4ᵗʰ of April*: letter 1367.

2. *Bacon*: Francis Bacon (1561–1626), an English natural philosopher who espoused that knowledge should be gained through empirical, inductive means. Often regarded as the father of the scientific method (*ODNB*). Tyndall quoted from Bacon's essay 'On friend- ship' in letter 1367.

3. *his Essays*: see letter 1367, n. 4.

4. *"coup de soleil"*: sunburn (French).

5. *to finish your glacier researches*: Tyndall began in the summer of 1856 to study the structure

and movement of glaciers in the Alps. See letter 1295, n. 3 and n. 4, and 1306, n. 8 and n. 10.

6. *Pau*: see letter 1298, n. 4.
7. *phaeton*: a type of light four-wheeled open carriage, usually drawn by a pair of horses, and having one or two seats facing forward (*OED*).
8. *the small surprise on the morning of my birthday*: see letter 1372.
9. *"Aurora Leigh"*: see letter 1372, n. 3.

From George Biddell Airy   4 May 1857   1388

Royal Observatory Greenwich | London S.E. | 1857 May 4

My dear Sir

In thanking you for your note of Saturday,[1] I regret to say that I cannot take advantage of your freedom on Tuesday, for I am engaged to be in Suffolk[2] on that morning.

But if Thursday or Friday should be available, or any day of next week after Monday, I would gladly profit by your kindness. Yours my dear Sir

Very truly | G B Airy
Professor Tyndall

RGO MS.RGO 6/471.169

1. *your note of Saturday*: letter 1385.
2. *Suffolk*: a county in eastern England. In 1845, Airy purchased a cottage in Playford, Suffolk, where he and his family spent their holidays (*ODNB*).

From Henry Matthew Witt[1]   4 May 1857   1389

ON THE TEMPERATURE OF FOAM. | *To Dr. Tyndall, F.R.S. &c.*
Museum of Practical Geology, | May 4, 1857.

DEAR SIR,

YOUR remarks in the last Number of the Philosophical Magazine,[2] on the Temperature of Foam, recall to my mind an observation of a similar kind made by myself and my friend, Mr. James Simpson,[3] jun., when on a pedestrian excursion in Surrey in December 1855.

Our attention was arrested, on arriving at the last lock on the river Wey,[4] at the distance of about half a mile from the point where it empties itself into the Thames,[5] by the production of an extraordinary amount of foam, in consequence of the falling of the river over a weir.[6]

This foam was not only unusual in amount, but had a peculiar soapy appearance, and on dipping the hand into it was found to be quite warm: fortunately we had a thermometer with us, and were therefore enabled to measure the extent to which the temperature of the foam exceeded that of the water on which it was formed.

The temperature of the water was 42°·5 F., whilst that of the foam was 45°, so that the foam was 2°·5 warmer than the water. The thought immediately struck me that its warmth must be due to the air enclosed within it; and if such were the case, it should have a temperature intermediate between that of the water and the air. And such really was the case; for the air, though feeling very cold in consequence of a strong easterly wind blowing, did not fall below 50° F.

I may state that the phaenomenon appearing to me very curious at the time, considerable care was devoted to the observation of these temperatures, the thermometer being allowed to remain long enough in each medium to become quite stationary.

I am, Sir, | Yours respectfully, | HENRY M. WITT, F.C.S., | *Assistant Chemist to the Government | School of Mines.*

H. M. Witt, 'On the Temperature of Foam', *Phil. Mag.*, 13 (1857), pp. 467–68

---

1. *Henry Witt*: Henry Matthew Witt (*c.* 1833–58), a pharmaceutical chemist and assistant chemist at the Government School of Mines, who had studied with Hofmann at the Royal College of Chemistry. See 'Melancholy Death of Mr. Henry Matthew Witt', *Pharmaceutical Journal*, 18 (1858), p. 46.

2. *YOUR remarks in the last Number of the Philosophical Magazine*: J. Tyndall, 'Remarks on Foam and Hail', Phil. Mag., 13 (1857), pp. 352–53.

3. *Mr. James Simpson*: possibly James Liddell Simpson (1829–89), eldest son of waterworks engineer James Simpson (1799–1869). In 1857 James Simpson junior joined the family firm Simpson & Co. to work on improvements of pumping machinery. D. G. M. Roberts, 'Simpson, James Liddell', in P. S. M. Cross-Rudkin, et. al. (eds), *Biographical Dictionary of Civil Engineers in Great Britain and Ireland,* 3 vols (London: Thomas Telford Publishing, 2008), vol. 2 (1830–90), p. 716.

4. *the river Wey*: an 87-mile tributary of the River Thames.

5. *the Thames*: the River Thames, a 215-mile river in southern England, passing through London.

6. *weir*: a dam, of which there are various forms, constructed on the reaches of a canal or navigable river, to retain the water and regulate its flow (*OED*).

From Josiah Latimer Clark[1]   6 May 1857   1390

The Electric & International Telegraph Co. | (Incorporated 1846) | Engineer's Office | Lothbury, London. | 6th. May 1857.

Dear Sir,

I have received the enclosed invitation[2] from Mrs Crosland.[3] Unless I hear from you to the contrary I will meet you <u>at the train</u> which leaves London Bridge Station for <u>Greenwich</u>[4] at 6.5 p.m.[5] on Friday evening—Their house being nearer to Greenwich than to Blackheath.[6]

Very truly yours | Latimer Clark

Professor Tyndall.

P.S. | Should I by any possibility not be there, you will of course proceed on without me—you will easily find the house by a little enquiry. | LC

RI MS JT/1/TYP/1/222
LT Typescript Only

---

1. *Latimer Clark*: Josiah Latimer Clark (1822–98), an English civil engineer who worked on, among other subjects, telegraphy and submarine cables. He was assistant, and later chief, engineer at the Electric and International Telegraph Company (*ODNB*).
2. *enclosed invitation*: enclosure missing but the content can be inferred from a letter in Faraday's correspondence in which Clark and Faraday were invited to visit the spiritualist, Camilla Crosland. See *Faraday Correspondence*, 5:3280.
3. *Mrs Crosland*: Camilla Dufour Crosland (née Toulmin, 1812–95), a writer of fiction and poetry who published a book about spiritualism in 1857 (*ODNB*). Tyndall noted in his journal that he had an 'interview with the spirits' on Friday 8 May 1857, and that Sarah Faraday was afraid that he 'might fall into the same trap as Latimer Clarke' (Journal, RI MS JT/2/13c/932–33).
4. *Greenwich*: a district in southeast London, location of the Royal Observatory.
5. *6.5 p.m.*: 6:05 p.m.
6. *Blackheath*: an area of southeast London south of Greenwich.

To George Biddell Airy   7 May 1857   1391

My dear Sir

on Tuesday next at 10 ¼ oC. I shall have every thing ready for you[1]—I hope this arrangement will suit you.

very sincerely yours | <u>John Tyndall</u>

Thursday | 1857 May 7

RGO MS.RGO 6/471.170

---

1. *every thing ready for you*: see letters 1346, 1347, 1348, 1373, 1374, 1379, 1380, 1384, 1385, and 1388.

From Henry Clifton Sorby   7 May 1857   1392

Falmouth | May 7/57.

My dear Tyndall,

I have been thinking of writing to you for some time, but one thing or another has prevented it. Your kind notes[1] were forwarded to me here.[2] and I only got them yesterday on my arrival. Since leaving home I have been able to make out several new facts respecting slaty cleavage, which I will describe to you when I am in London, which will I hope be about the 20th inst. One that is interesting as explaining the connexion of two well known facts, is this. You know that in some cases, when a bed of sandy rock has been enclosed between more yielding material, it has been much contorted, in the way shown in my fig. which you copied in your paper.[3] The other is that in other cases we have no contortion, but the cleavage in the firm grained beds lies more in the plane of stratification than in the sandy, thus

Now I find that these are related thus,[4] so that when the general line of the beds is nearly perpendicular to the cleavage i:e: in the line of the pressure, since one yielded more than the other it became contorted, whereas where

the line of pressure was inclined at an angle to the beds, those that could yield most, yielded this extra in the plane of stratification. Hence their cleavage lies more nearly in it than in the more unyielding, coarser beds, as I have described in my papers,[5] and then no such contortions, as in the other case, are produced. I find that many of the so called clay slate rocks of Cornwall[6] are more correctly very fine grained mica schist,[7] and that their cleavage is not like that of the Welsh slates, but simply what I have described to you as 'crumple joints.' There has been very little or no compression of the rocks—at least nothing like what occurs in Wales—they are merely broken up by more or less close jointing, as if disturbed after having been converted into fine grained mica schist, when they could not yield as plastic substances.

Thanking you most sincerely for all the trouble you have taken on my behalf, and looking forward to the pleasure of seeing you soon

I remain | yours very truly | Henry Clifton Sorby.

RI MS JT/1/TYP/4/1352
LT Typescript Only

1. *Your kind notes*: letters missing.
2. *here*: Sorby was pursuing geological fieldwork in Cornwall. See letter 1363.
3. *my fig. which you copied in your paper*: Sorby's figure appears in H. C. Sorby, 'On Slaty Cleavage, as exhibited in the Devonian Limestones of Devonshire', *Phil. Mag.*, 11 (1856), pp. 20–37, on p. 34. Tyndall's similar figure is in J. Tyndall, 'Comparative View of the Cleavage of Crystals and Slate Rocks', *Phil. Mag.*, 12 (1856), pp. 35–48, on p. 41.
4. *are related thus*: at this point in the typescript LT noted there was a diagram in the original letter, but the original letter is missing.
5. *as I have described in my papers*: see H. C. Sorby, 'On Slaty Cleavage, as exhibited in the Devonian Limestones of Devonshire'; and H. C. Sorby, 'On the Theory of the Origin of Slaty Cleavage', *Phil. Trans.*, 12 (1856), pp. 127–29.
6. *Cornwall*: see letter 1363, n. 4.
7. *mica schist*: see letter 1344, n. 3.

From George Biddell Airy 8 May 1857 1393

London S.E. | 1857 May 8

My dear Sir

I will most gladly avail myself of your kind arrangement[1] for Tuesday morning May 12. And if you will permit me, I will bring with me my friend Professor Gautier[2] of Geneva.[3] He has arrived in England within two days, and I have but just now seen him.—He is not a practical electrician, but is generally informed on the whole subject, and his acquaintance with De La Rive will make the experiments interesting to him.

I am, my dear Sir, | Yours very truly | <u>G B Airy</u>
Professor Tyndall

RGO MS.RGO 6/471.171
RI MS JT/1/TYP/1/21

———

1. *your kind arrangement*: see letter 1391, n. 1.
2. *Professor Gautier*: Jean-Alfred Gautier (1793–1881), a Swiss astronomer and professor of astronomy and mathematics at University of Geneva (*HLS*).
3. *Geneva*: see letter 1294, n. 3.

To George Biddell Airy   9 May 1857   1394

My dear Sir

It will give me great pleasure to see M. Gautier[1]—and I know it will interest my friend Prof. De la Rive to have the verbal testimony of an eyewitness to some of the effects of our English Induction coils.

very sincerely yours | <u>J. Tyndall</u>
Saturday. | 1857 May 9

RGO MS.RGO 6/471.172

———

1. *M. Gautier*: see letter 1393, n. 2.

From Frances Hooker   [12 May 1857][1]   1395

Tuesday

My dear Dr Tyndall

Pray adhere to your yesterday's promise,[2] and come tomorrow: do not be tempted to put it off till Friday, on the chance of papa's[3] being here. For I have just had a letter to say he has made a blunder, and finds he is not to come up till next week! Do not please fail us tomorrow, however, for Dr Lindley[4] and Mr Ball[5] are both coming, and they are both worth meeting—And next week you shall come again, and meet Papa. Or rather, I ought to say I 'hope you will give us the pleasure &c.'

But I know you will, if you can, will you not?

Let the lions[6] growl as much as they choose; they must do without you this time.

I must tell you a conversation I had with Willy[7] yesterday; it will amuse you. I was talking to him about Papa's coming, and who were coming to dine here; and I ended by saying 'which do you like best, Dr Lindley, Mr Darwin, or Dr Tyndall'—'Dr Tyndall'. So I said 'Why do you like him?' 'Because he is very good'. 'Why is he good? what does he do to you?' I expected him to say 'brings me comfits'[8]—instead of which, he replied 'He does like this'—and he folded his arms round the screen, and gave it a hug! He is a warm hearted little fellow, is he not?

If you come early enough tomorrow, I shall ask you to translate for me a speech of Mephistopheles,[9] which has utterly baffled me. I am sorry you do not like Hyperion;[10] but I am not surprised—I think it is a book more suited to a woman than a man, except that man be a poet—Still I fancied you had enough of the poetic element in you to like it.

I get on slowly with *[it]* Emerson[11]—that is, I read but little each day— in truth, I find it rather tough reading—That is, I often have to read it two or three times before I fully understand it. Still I like some parts of it very much—

Yours very sincerely | Frances H. Hooker.

RI MS JT/1/TYP/8/2544
LT Typescript Only

---

1. *[12 May 1857]*: the date is given based on the relation to letter 1366, in which Tyndall mentioned plans to send Longfellow's *Hyperion: A Romance* to Frances Hooker, and the mention of Tyndall attending a Red Lions Club meeting. Since this club met on the second Wednesday of each month, and Tyndall was on the Isle of Wight during the April 1857 meeting, this letter is most likely dated the Tuesday before the meeting that was held on Wednesday 13 May. He wrote in his journal for 13 May 1857 that he spent 'half an hour with the Lions in the evening' (Journal, RI MS JT/2/13c/934).
2. *your yesterday's promise*: not identified.
3. *papa's*: John Henslow.
4. *Dr Lindley*: John Lindley (1799–1865), a botanist and professor at University College, London. Lindley received the RS's Royal Medal later in 1857 'For his numerous researches and works on all branches of scientific botany, and especially for his vegetable kingdom, and his genera & species of Orchideae' (*ODNB*).
5. *Mr Ball*: John Ball.
6. *the lions*: the Red Lions Club of the BAAS, which met on the second Wednesday of each month and every year at the annual BAAS meeting. Tyndall began attending meetings in January 1854.
7. *Willy*: William Hooker.
8. *comfits*: a sweetmeat made of some fruit, root, etc., preserved with sugar (*OED*).
9. *Mephistopheles*: a demon of German folklore, featured as the devil in Johann Wolfgang von Goethe's *Faust*.

10. *Hyperion*: see letter 1365, n. 6.
11. *Emerson*: Ralph Waldo Emerson (1803–82), an American essayist and poet (*ANB*). It is not known to which of Emerson's works Frances Hooker refers.

From Frances Hooker [10 or 17 May 1857][1] 1396

Hitcham Rectory | Bildestone | Suffolk | Sunday

My dear Friend

Thank you many times for the gratification you have procured for me, in sending me Dr Grailich's[2] letter,[3] and the little dedication to yourself, both of which have given me very much pleasure, the latter especially more than I can say. <u>That</u> made me much prouder than all Dr G's hyperbole, which you suggested might have that result. Don't be displeased, I could not help writing it. You once spoke to me of Mr Hirst, and I remembered what you had said as soon as I saw the signature; guessing it to be the same person.

Woman-like, of course I read the last part of Dr Grailich's letter first! and was certainly extremely amused by it. Nevertheless, it is very pleasant to find oneself borne in mind by those one likes. When you write to him, pray tell him that I look back to my Vienna visit with the greatest pleasure: it will be very long before I forget the enjoyment of the ten days spent there, or the kindness and hospitality shown us by kind friends there.

I am very glad the Spirits[4] were baffled by you: it is extremely satisfactory—nevertheless I think they must have bungled, for I do not think you could have hindered the proceedings of a first rate conjurer! M. Robin,[5] for instance. I give them great credit for the happy hit they have made in naming you, though—it is the best thing I ever heard of them! Surely some one must have been hovering invisibly near the flame of gas on the previous Thursday?[6] Possibly even, it might have been his voice you heard—that of the Fire-King,[7] or one of his ministers! I should have liked much to have been present, But I look forward, trusting you will some day show me the experiment—will you not? Anne[8] was as much amused as I was—it sounds so very strange. I always think of you on the Thursday afterwards, and hope you are well enough not to find the lecture a burden—I am very sorry to hear of your having been ill, and shall rejoice heartily to know that you are freed from London smoke and work, and can escape to the pure mountain air of Chamounix[9]—I do not fear your rencontre with Prof. Forbes;[10] and should not wonder at your ending by working amicably together, and giving to the world the result of your investigations, with the substitution of the name of Forbes for that of Huxley.

If you would like to know somewhat more of M. de la Rive's letter,[11] than you seem to have made out for yourself, pray send it me—I shall greatly enjoy the deciphering of it—that sort of enigma is precisely the sort of thing in which I take delight—it somehow seems to suit my turn of mind.

I saw by the Lit. Gaz.[12] that Dr Frankland is lecturing at the Royal Institution;[13] pray remember me very kindly to him—I am sorry I shall not have the pleasure of seeing him.

I find this hot weather very trying—what must it be in London! To the fact of my not being very well to day you must attribute my bad writing; it is difficult to guide one's pen properly when lying on the sofa.

Ever yours sincerely | Frances H. Hooker.

Were you at the Royal Society the evening Mr J. Thomson[14] read a paper 'On the Plasticity of Ice'?[15] if so tell me whether it was worth anything, and whether it was for or against your theory.

RI MS JT/1/TYP/8/2548–48a
LT Typescript Only

---

1. *[10 or 17 May 1857]*: the date is given based on the reference to James Thomson's paper being read to the RS by William Thomson on 7 May 1857 (see n. 15), a visit with a spiritualist on 8 May 1857 (see n. 4), Edward Frankland's lectures on 'Graphic and Plastic Arts' from 25 April to 6 June 1857, 23 May being the specific mention here (see n. 13), and Tyndall's after Easter lectures on 'Sound' on Thursdays from 23 April to 11 June 1857. As this letter was written on a Thursday, it could be 10 or 17 May.

2. *Dr. Grailich's*: perhaps Wilhelm Josef Grailich (1829–59), an Austrian physicist and crystallographer who taught at the University of Vienna (*ADB*).

3. *letter*: see letter 1290, Wilhelm Grailich to John Tyndall, 26 October 1856, *Tyndall Correspondence*, vol. 5.

4. *the Spirits*: a reference to Tyndall's visit, with Latimer Clark, to the spiritualist Camilla Dufour Crosland on 8 May 1857 (see letter 1390).

5. *M. Robin*: Henrik Joseph Donckel (1811–74), a Dutch illusionist who went by the name Henri Robin. During the 1850s he performed in England, including at Windsor Castle before Queen Victoria. See L. Senelick, 'Robin, Henri', in M. Banham (ed.), *The Cambridge Guide to Theatre* (Cambridge: Cambridge University Press, 2000), p. 928.

6. *the previous Thursday*: Tyndall's after-Easter lectures on sound, on Thursdays. See letter 1308, n. 7.

7. *Fire-King*: possibly a reference to Ivan Chabert (1792–1859), who called himself the Fire King, an illusionist who swallowed poisons, boiling oil, and molten lead.

8. *Anne*: Anne Henslow Barnard (1833–99), botanical illustrator, youngest daughter of John Henslow, and sister of Frances Hooker (née Henslow).

9. *Chamounix*: see letter 1353, n. 9.

10. *Prof. Forbes*: James Forbes.

11. *M. de la Rive's letter*: probably letter 1378, dated 18 April 1857.
12. *Lit. Gaz.*: The Literary Gazette, and Journal of Belles Lettres, Arts, Sciences, a British literary magazine from 1817–63.
13. *Dr. Frankland is lecturing at the Royal Institution*: in 1857, Frankland gave a series of seven Saturday lectures at the RI on the 'Relations of Chemistry to Graphic and Plastic Art', which began on 25 April and ran through 6 June 1857, with 23 May being the specific mention here. See RI Guard Book, vol. 2, p. 99, and Russell, *Edward Frankland,* p. 187.
14. *Mr J. Thomson*: James Thomson.
15. *'On the Plasticity of Ice'*: J. Thomson, 'On the Plasticity of Ice, as Manifested in Glaciers', *Roy. Soc. Proc.,* 8 (1857), pp. 455–58. Read to the RS by his brother William Thomson on 7 May 1857. Thomson further discussed his theory of glacial motion in a second paper, 'On Recent Theories and Experiments Regarding Ice at or near Its Melting-Point', *Roy. Soc. Proc.,* 10 (1859), pp. 151–60. Tyndall discussed Thomson's papers in his *Glaciers of the Alps,* pp. 340–45.

From Jacob Gijsbertus Samuël van Breda[1]   7 June 1857   1397

Harlem le 7 juin 1857

Mon cher Monsieur,

Je vois dans le Cosmos, que vous avez écrit à Mr Moigno, que Mr Faraday vient de faire une découverte très importante; qu'il a trouvé le moyen d'éclairer les rues, les places etc. au moyen du Magnéto-Galvanisme.

Comme j'ai proposé, il n'y a pas longtemps, au conseil municipal de Harlem, dont j'ai l'honneur d'être un des membres, de ne pas renouveler le contrat avec la société, qui éclaire la ville par le gaz, au moins pas pour un long intervalle de temps, parce que je prévoiais,[2] que bientôt l'éclairage par le gaz serait remplacé par celui de la lumière galvanique; je prends beaucoup d'intérêt à cette affaire.

D'après moi la seule, mais très grande difficulté, qui s'oppose à l'emploi de cette lumière, consiste dans les dépenses qu'il entraine, et qui dépassent de beaucoup celles de l'emploi du gaz.

Monsieur Faraday aurait-il vaincu cette difficulté?!

Si vous pouviez m'en informer par un mot et si vous vouliez en même temps me donner des nouvelles de la santé de notre illustre ami et de la vôtre vous obligeriez beaucoup celui qui a l'honneur d'écrire avec la plus haute considération.

Votre tout dévoué ami | Jan Breda

Vos recherches sur les glaciers sont bien importantes, j'en ai vu et étudié plusieurs et je vais en examiner d'autres le mois prochain. Mes observations que je n'ai communiquées que dans mon cours au musée Teylers sont bien d'accord avec les vôtres—la masse des glaciers est bien solide.

Harlem June 7 1857

My dear Sir,

I read in the Cosmos[3] that you have written to M. Moigno,[4] that M. Faraday has just made a very important discovery; that he found the way to light up the streets, the squares, etc. by means of Magnetic Galvanism.

Because I suggested, not long ago, to the municipal council of Harlem,[5] of which I have the honour of being one of the members, to not renew the contract with the company lighting the city by way of gas, at least not over a long period of time, because I predicted, that gas lighting would soon be replaced by the lighting from the galvanic light; I am taking much interest in this matter.

In my opinion the sole yet very important difficulty, obstructing the use of this light, consists in the expenses it incurs, and which far exceed those of using gas.

Could Mr Faraday have conquered this difficulty?!

If you could inform me on this subject in a missive, and if you could simultaneously give news of your health and that of our illustrious friend, you would much oblige the one who has the honour to write with the utmost respect,

Your most devoted friend | Jan Breda

Your research on glaciers is very important, I saw and studied several, and I will examine others next month. My observations that I only communicated in my lessons at the Teylers museum[6] are much in agreement with yours—the mass of the glaciers is indeed solid.

RI MS JT/1/B/125

1. *Jacob Gijsbertus Samuël van Breda*: this letter is signed 'Jan Breda'; Jan must have been a nickname or shortened name of the Dutch geologist (1788–1867). See C. J. Matthes, 'Levensberigt van J. G. S. van Breda', *Jaarboek van de Koninklijke Akademie van wetenschappen* (1867), pp. 22–32.

2. *prévoiais*: Breda used an older spelling of the French word *prévoyais,* meaning 'predicted'.

3. *Cosmos*: a French scientific journal founded in 1852 by François-Napoléon-Marie Moigno.

4. *M. Moigno*: François-Napoléon-Marie Moigno (1804–84), a French Catholic priest and physicist. See 'L'Abbé Moigno', *Nature,* 30 (1884), pp. 291–92.

5. *Harlem*: Haarlem, a city in the Netherlands to the west of Amsterdam.

6. *Teylers museum*: the Teyler's Museum in Haarlem, Netherlands, which focuses on art and science, was founded in 1778 and is still in operation.

To Thomas Archer Hirst   [9 June 1857][1]   1398

Tuesday | June 1857

My dear Tom.

I send you a note[2] from Playfair[3] referring to the subject concerning which you wrote to me.[4] Be good enough to burn the note when you have read it—It is quite private.[5]

affectionately yours | John Tyndall

I am no longer an examiner for the Artillery & Engineers.[6] They thought proper to cut down the rates of remuneration, and on asking me whether I was prepared to undertake the examinations at the reduced rate I answered 'No!'[7]—

RI MS JT/1/T/963

---

1. *[9 June 1857]*: the date, written on the letter but presumably not by Tyndall, is based on an exchange of letters between Tyndall and Hirst from 31 May and 14 June (see **n.** 4). It seems unlikely to be written Tuesday 2 June since Tyndall would have had to wait for the reply from Playfair that was enclosed with this letter.

2. *a note*: letter missing.

3. *Playfair*: Lyon Playfair. Playfair oversaw the Museum of Irish Industry, a Dublin museum founded in 1854 for displaying material from the Irish mining and manufacturing industries, in his role as secretary of the Department of Science and Art during the 1850s. See n. 4 and n. 5.

4. *you wrote to me*: Hirst wrote in his journal on 31 May 1857: 'On arriving at Paris I found a letter from Mrs Simpson informing me that a mathematical master was about to be appointed at the Museum of Irish Industry, if I cared to apply. I enclosed her letter to Tyndall, and asked his advice'. Hirst noted on 14 June that he received a letter from Tyndall and replied the same day (Brock & MacLeod, *Hirst Journals,* pp. 1265, 1267). For Hirst's reply, see letter 1402.

5. *quite private*: Hirst wrote in his journal on 14 June 1857: 'Tyndall tells me there is to be no Professor of Mathematics at the Museum of Irish Industry this year, otherwise he would have probably used his influence for me. He has also declined continuing his examinership for military appointments in consequence of a proposition to diminish the remuneration. He is quite right' (Brock & MacLeod, *Hirst Journals,* pp. 1267–68).

6. *Artillery & Engineers*: from 1856–57, Tyndall wrote and graded examinations in scientific topics for commissioned officers of the Royal Artillery and Royal Engineers at the Royal Military Academy at Woolwich. See letter 1306, n. 2, and letters 1375 and 1376.

7. *I answered 'No!'*: see letters 1375 and 1376.

To Christian Gerling   12 June 1857[1]   1399

25<u>th</u> June 1857

My dear Friend.

Your card which was handed to me a few days ago by D[r]. Huter[2] lies before me. It gave me sincere pleasure to recognize on it your well known hand so clear and without a tremor as to assure me that the writer was in good health. Unfortunately I lost D[r]. Huter in the confusion of one of our meetings and have not had the pleasure of seeing him since. This night closes our season, when Faraday will give us the concluding lecture.[3] You would love Faraday if you knew him. He is as good as a man as he is great as a philosopher. He is more like a father to me than anything else, and the longer I know him the more I love him. His mind and feelings are as fresh as a boy's; I sometimes tell him that he is the young man and I the old one.[4] Indeed I regard my connexion with him, and the solid warm friendship which has grown up between us, as the greatest blessing of my life. Next month I purpose making a journey to Switzerland, and intend to stay with M[r] de la Rive in Geneva[5] for a day or two. I want to complete that investigation on the glaciers which I referred to at Vienna. I was greatly pleased to see you there, and the kindness of the Vienna people has left a very agreeable impression of the meeting on the minds both of my friends and of myself. Can you tell me what the average thickness of the ice is upon the Lahn[6] during winter, and what is the temperature of the air when the thickness reaches its maximum? I ask you this question because I remember seeing the masses of ice heaped up on the banks of the river from the windows of your auditorium[7] in the winter of 1848–49. The question interests me at present as I am at work upon the glaciers.

And now I have to ask you to convey my warmest remembrances to M[rs]. Gerling[8] and to both the Misses Gerling.[9] I have not yet forgotten the kindness which I experienced from them in Marburg.[10] Indeed the old town and the goodnatured people it contains are to me very pleasant objects of recollection.

Goodbye my dear Friend | and believe me always | Yours most sincerely | John Tyndall.

UB Mbg 319:739

---

1. *25 June 1857*: despite Tyndall having given 25 June as the date, he must have been mistaken as mention of Faraday's 'concluding lecture' the same day this was written dates it as 12 June 1857 (see n. 3).

2. *D[r]. Huter*: possibly Karl Christoph Hüter (1803–57), a German physician, professor of medicine, and Dean of the Medical Faculty at the University of Marburg before his sudden death in August 1857 (*ADB*).

3. *concluding lecture*: see letter 1404, n. 11.

4. *he is the young man and I the old one*: Tyndall repeated this same phrase in another letter to Gerling (see letter 1576).

5. *Geneva*: see letter 1293, n. 3.

6. *Lahn*: see letter 1361, n. 8.

7. *your auditorium*: at the University of Marburg.

8. *M^rs. Gerling*: Christiane Gerling (née Suabedissen), wife of Christian Gerling (married in 1814).

9. *Misses Gerling*: Christian and Christiane had five children: Emma Wilhelmine Gerling (b. 1815), Marie Friederike Platner (née Gerling) (b. 1817), Caroline Mathilde Gerling (b. 1819), Christian Ludwig Gerling (b. 1821), and Caroline Maria Gerling (b. 1823).

10. *Marburg*: Tyndall attended the University of Marburg from 1848–50, studying under Robert Bunsen, Hermann Knoblauch, Friedrich Stegmann, and Christian Gerling.

From William Snow Harris   13 June 1857   1400

Athenaeum | 13 June 1857

My dear Tyndall,

Will you favor me by accepting the accompanying pages[1] being a sort of Supplement to the Parliamentary Papers[2] and which I think you may find interesting—A few copies only were got up for my private friends—I want very much to see you relative to some Scientific Papers I should like to publish in the Phil. Mag. I did not think it worth while to make any observations on Riess'[3] last paper.[4] I am always ready to make a bridge for a flying enemy. If you will say when you can favor me with 20 minutes I will endeavor to suit your convenience. I will come to the Royal Institution to save you all undue inconvenience.

Most truly yours | W. Snow Harris.

Professor Tyndall, F.R.S.

RI MS JT/1/TYP/2/473
LT Typescript Only

---

1. *the accompanying pages being a sort of Supplement*: not identified.

2. *Parliamentary Papers*: see letter 1342, n. 1.

3. *Riess'*: Peter Theophil Riess (1804–83), a German physicist who studied the electricity of friction and magnetism (*ADB*).

4. *last paper*: probably a reference to Riess' 'On Electric Pauses', *Phil. Mag.*, 13 (1857), pp. 261–67. The previous year Harris and Riess were engaged in a dispute in the pages of the magazine.

To Juliet Pollock   [14 June 1857][1]   1401

Sunday.

My dear Mrs Pollock,

I do not know whether Mr Barlow has written to you but lest he has not I write to say that on Tuesday morning at half past 10 o'clock I shall be making those experiments on ice, and a few on other subjects for a little party. If you could join us it would give me great pleasure to see you—but if not my promise to yourself personally still remains. A stray remark from somebody leads me to imagine that you are at Richmond[2]—if so this will hardly reach you in time. I hope this glorious weather has an invigorating effect on you. It will on me as soon as I escape from London, but I believe not before. I hope the little man[3] is doing well, and I look forward to the time when we shall be playfellows. I have just been looking into Mrs Crosser's Memorial of her Husband[4]—One of the most pleasant parts of the book is that where the love of his mother for her 'little Andrew'[5] breaks out so refreshingly. I thought of you, and of your theory of heaven, and its local proximity to the baby.[6] By the way it is quite unorthodox in you to fix heaven there. We have <u>one ordinate</u> towards the determination of its position in space, and that is a line perpendicular to the earth's surface at Jerusalem*.[7] If we had a second ordinate we should determine the place exactly. At present we know its <u>direction</u> merely, but the intersection of the second line would give its position. Mr Pollock[8] I am sure will feel much pleasure in explaining these terms to you; and he will also assure you of the mathematical truth of what I say. Do not show this to any pious people, for they, unfortunately, are for the most part totally impervious to mathematical reasoning.

always yours sincerely | John Tyndall.

RI MS JT/1/TYP/6/1895
LT Typescript Only

---

1. *[14 June 1857]*: the date is given by the reference to William and Juliet Pollock's new baby (see n. 6) and Tyndall's journal entry for Saturday 13 June 1857 (see n. 4). Tyndall's 'escape from London' is likely a reference to his leaving London for Europe on 1 July to study the glaciers of the Alps.

2. *Richmond*: a small town close to London.

3. *little man*: Walter Pollock, for whom Tyndall seemed to have a special affinity.

4. *Mrs Crosser's Memorial of her Husband*: not identified. Possibly a Mrs. Cooper, see Tyndall's entry in his journal for 13 June 1857, 'worked some, and read Mrs Cooper's memorial of her husband' (Journal, RI MS JT/2/13c/941).

5.	*'little Andrew'*: not identified.

6.	*the baby*: Maurice Pollock, third son of Juliet and William Frederick Pollock, born 28 March 1857.

7.	*at Jerusalem**: in the original letter, this asterisk does not include a corresponding footnote.

8.	*Mr Pollock*: William Frederick Pollock.

# From Thomas Archer Hirst	14 June 1857	1402

26 Rue de Lacépède. Paris. | June 14th. 1857

My dear John.

Your letter[1] closing definitively all further thought about the Museum of Irish Industry[2] was at any rate satisfactory For another year at any rate I am free and if the Gods are good to me I hope before the expiration of that time to be in a better position to offer myself as a candidate for any situation that may present itself. You will be surprised to hear that we have found lodgings in the Quartier Latin.[3] It has not certainly the reputation of being a very healthy quarter but we met with a couple of rooms so pleasantly situated in the middle of a large and beautiful garden half way in a direct line between the Pantheon[4] and the Jardin des Plantes[5] that we were tempted to try, at least for a month, whether or not it would be suitable for Anna. We no longer keep house as at Pau,[6] but are boarded and lodged for 100 francs each per month. It is in fact a 'pension bourgeoise'[7] occupied for the most part by Old maids, Old widows, Old Bachelors and old widowers, all french, with a sprinkling of English, Scotch, and Irish, and one or two steady students. We can either live apart from the rest and take our meals in our own rooms, or we can find company in the Salle à Manger[8] and breakfast and dine at the Table d'hote.[9] So far it appears to suit Anna very well, she cannot of course take her meals with the 'locataires',[10] but in the Garden she has as much company as she pleases when my work takes me away from her. The ladies, and indeed every body, are very kind to her; she begins to chat in French with much greater ease, and altogether I think it suits her better than would the quiet solitude of an apartment of our own. She has at present a homeopathic doctor. She has strange faith in doctors, and to look to them for support and help has always been such a solace to her that, although I was myself without faith, I could not but agree to her consulting them frequently. Nevertheless, it was always a source of great anxiety to me to see them continually administering opiates and other strong drugs for merely temporary relief. I was not sorry, therefore, when after our arrival in Paris she expressed a wish to consult Dr Simon,[11] the homeopath,[12] of whose reputation she had heard much of from her french ladies in the garden. At any rate, I thought, his remedies will be less pernicious, and if he is a sensible man his consolation and advice will be equally beneficial to her. My anticipations proved correct; since we came to Paris she has left off all her opiates, for the first time for a year, and she is better rather than worse for so doing. Her cough and the distressing cholic,[13] which it appears have one and the same origin, are decidedly better; no matter whether the improvement be due to the absence of allopathy[14] or the effect of homeopathy I am heartily thankful for the change. Nevertheless, as I well knew, Dr Simon candidly tells <u>me</u> that he can do but little, he says her left lung is as good as destroyed, and her right lung in a very poor condition. And yet, worn as she is almost to a skeleton, it would surprise you, and make you sad too, to see with what tenacity she clings to life and the hope, almost amounting to certainty when she does not suffer, that by care she will grow a little stronger and not leave me yet. Before I have time to check the tears that <u>will</u> rise sometimes she detects them and the reaction on her is painful indeed. She no longer hopes to recover & I, of course, cannot now encourage such a hope, her sole hope is, as she says, to be a <u>little</u> better and stronger, and to be less a source of sorrow and anxiety to me; to be more to me like she once was. May God grant her wish; and may he also gently loosen the ties that bind her to life and me as the time approaches when she must leave both. But as usual I have allowed myself to think and write too much on this sad subject, and it will require some time before I can meet Anna again face to face she is so quick at detecting me. I see her in the garden chatting pleasantly to her neighbours.

I have called on Verdet and renewed my acquaintanceship with him. On the 15th of next month he intends to start on a tour through London to Manchester[15] and thence to Scotland. He wishes to know whether at that time he will find you in London. And in case of his not doing so I have promised him to ask you to send to me, for him, letters of introduction to Frankland, Forbes of Edinburgh (?) and any other people whom you think he would wish to know or who might be of service to him. He knows Thompson.[16] My Trustees the other day sent me £100 it is more than I like to keep by me when I do not want it. I find by my notebook that I have more than £70 of yours so I was induced to get a draft for £50 which I enclose; very possibly I may some day ask for it again, but that time appears at present to be so far distant that I must ask you to consent to make use of what I have not the least necessity for at present. How are you? The first wet Sunday that stops you from your walk write me a letter. There's a good fellow. When do you come to Paris?

Yours affectionately | <u>T. A. Hirst</u>

You did not say a word about the bill for £10 but of course it was presented.

1. *your letter*: letter missing.
2. *Museum of Irish Industry*: see letter 1398, n. 4.
3. *Quartier Latin*: the lively Latin Quarter of Paris, situated on the left bank of the Seine and location of the University of Paris.
4. *Pantheon*: originally a church in the Latin Quarter, this building was changed to a mausoleum to inter distinguished French citizens in the late eighteenth century.
5. *Jardin des Plantes*: the national botanical garden of France, a department of the Muséum national d'histoire naturelle.
6. *Pau*: see letter 1298, n. 4.
7. *'pension bourgeoise'*: boarding house (French).
8. *Salle à Manger*: dining room (French).
9. *Table d'hote*: a fixed-price low-choice menu (French).
10. *'locataires'*: lodgers (French).
11. *Dr Simon*: Léon Simon (1798–1867), a Parisian homeopathic physician.
12. *homeopath*: a practitioner of homeopathy, a system of alternative medicine using diluted substances to treat illnesses, that historically and now receives criticism for its claims to heal.
13. *cholic*: of or pertaining to bile (*OED*).
14. *allopathy*: orthodox medical treatment or practice (*OED*).
15. *Manchester*: see letter 1311, n. 8.
16. *Thompson*: William Thomson.

From Jules Lissajous   14 June 1857   1403

Paris 14 Juin 1857

Mon cher Monsieur

J'ai eu beaucoup d'affaires à mettre au courant depuis mon retour, c'est ce qui m'a empêché de vous écrire, comme je le voulais, pour vous donner de mes nouvelles, et vous remercier encore une fois de votre cordiale hospitalité.

Je suis arrivé à Paris mardi soir à 10^h ½ sans avoir éprouvé de retard, mais très fatigué. La traversée de Douvres à Calais a été assez pénible pour moi, le temps était beau, mais la mer était forte, j'ai même été trempé des pieds à la tête par une vague qui a inondé le pont au milieu du voyage. J'ai lutté avec succès, mais avec une grande fatigue contre le mal de mer.

Je regrette maintenant de n'avoir pas emmené mes caisses avec moi, mes appareils me manquent, et leur absence me gêne beaucoup dans mes travaux. Je pense que vous avez eu l'obligeance de me les renvoyer, néanmoins je n'en ai pas encore eu de nouvelles. Si par hasard vous n'aviez pu encore les adresser au chemin de fer, je vous prie instamment de le faire de suite, je vous en serais très reconnaissant. Peut-être vous demandera-t-on une déclaration écrite de ce que contiennent les caisses avec les prix correspondants; en voici la liste :

Quatre diapasons—10f
Un interrupteur électromagnétique—35f
Un diapason
Un électromoteur à diapason—25f
Un comparateur des mouvements vibratoires—40f
Un stéréoscope—10f
Total—120f

Si vous voyez Monsieur Barlow, je vous prie de lui dire combien je regrette d'être parti sans avoir pu le rencontrer. Veuillez être assez bon pour lui demander la liste des hommes *éminents* qui ont assisté à la séance du vendredi 5 juin. Je serais heureux de l'avoir, afin de faire savoir en France comment l'Angleterre accueille les découvertes scientifiques.

J'espère aussi que vous voudrez bien être auprès de Monsieur Faraday l'interprète de ma profonde sympathie. J'ai été on ne peut plus touché de l'accueil qu'il a bien voulu me faire. Il est si rare de trouver un aussi grand talent réuni à tant de simplicité et d'affabilité qu'en voyant votre illustre et excellent maître on se demande ce qu'on doit le plus admirer de son génie ou de son cœur. Dites lui bien combien je suis heureux et fier d'avoir vécu quelques jours dans son intimité.

Quant à vous, mon cher ami, je conserve précieusement le souvenir de votre cordiale hospitalité. Ne manquez pas de m'avertir à l'avance du jour où vous passerez par Paris afin que je puisse vous recevoir dignement et vous réunir à quelques uns de nos jeunes savants.

Je compte sur votre obligeance pour me rassurer au sujet de mes caisses, et me donner quelques détails sur la manière dont les journaux Anglais ont parlé de notre séance.

Adieu, mon cher Monsieur Tyndall, croyez au dévouement bien sincère et à l'amitié inaltérable | de votre affectionné | J. Lissajous

Paris June 14 1857

My dear Sir,

I had many things to attend to since my return,[1] this is what kept me from writing to you, as I wished, to give you some news and thank you once again for your warm hospitality.

I reached Paris on Tuesday night at half past ten, without experiencing any delay, but very tired. The crossing from Dover[2] to Calais[3] was rather tiresome for me, the weather was fair, but the sea was rough, I was even soaked from head to toe by a wave that flooded the deck mid-journey. I successfully fought, against seasickness but it took a lot of efforts.

I regret now not taking my trunks with me, I miss my instruments and their absence much disturbs my work. I trust you had the courtesy of sending them back to me, however, I have not heard of them yet. If by chance you have not been able to send them to the railway, I urge you to do so right away, I will be very grateful. You might be asked for a written declaration of the content of the trunks and their corresponding prices; here is the list:

Four diapasons[4]—10f

One electromagnetic switch—35f

One diapason

One electromotor with diapason—25f

One comparator of vibrating movements—40f

One stereoscope—10f

Total—120f

If you see Mr Barlow, I beg you to tell him how much I am sorry for leaving without having been able to meet him. Please be kind enough to ask him for the list of the eminent men who attended the session on Friday, June 5.[5] I would be happy to have it, so as to make it known in France how England welcomes scientific discoveries.

I also hope that you will kindly send on my behalf my deepest sympathy to Mr Faraday. I could not have been more touched by the welcome that he so kindly gave me. It is so rare to find such great talent united to such simplicity and affability that as one sees your illustrious and excellent teacher he wonders what to admire most of, his genius or his heart. Tell him how pleased and proud I am to have lived a few days in his proximity.

As for you, my dear friend, I treasure the memory of your warm hospitality. Be sure to let me know in advance about the day when you will come by Paris, so that I may fittingly receive you and present to you some of our young savants.

I count on your kindness to reassure me about my trunks, and to give me some details about the manner in which English newspapers spoke about our meeting.

Farewell, my dear Mr Tyndall, believe in the truly sincere devotion and the unalterable friendship | of your affectionate | J. Lissajous

RI MS JT/1/L/25

---

1.   *since my return*: see letter 1310, n. 2.

2.   *Dover*: a coastal town in England on the English Channel, where travelers would cross by ferry to France.

3.   *Calais*: a coastal town in France on the English Channel, where travelers would cross by ferry to England.

4.   *diapasons*: a rule or scale employed by makers of musical instruments in tuning (*OED*).

5.   *the session on Friday, June 5*: see letter 1310, n. 2.

To Thomas Archer Hirst   [15 June 1857][1]   1404

My dear Tom.

Your letter[2] has reached me and been read by me within the last quarter of an hour. It has stirred up my feelings and while they are loose I will write to you instead of waiting for a wet Sunday. Your reference to poor Anna[3] brought tears to my own eyes, and I divined before you mentioned it that you could never have written thus without tears. It is sad, very sad, and still not without its own high and holy uses. 'Whom god loveth he chasteneth'[4] says the eastern seer and I believe it is true that the best qualities of the human heart would lie latent if sorrow did not call them forth. But all I have to say upon this subject you have already woven into your experience and I will therefore say no more. With regard to the money it is a matter of supreme indifference to me whether it comes now or at some future day, but I should be most gratified if it never came at all. When balancing accounts with the world it is an item which never entered my leger: it was to me just so much money that I had spent upon myself, for which I had my equivalent and which was therefore dismissed altogether from my thoughts. Well I am willing to act as your banker, and when you need money you have only to write to me.

I thought of you this morning as I put aside my journal, and laughed at the prospect of the impression which its perusal will produce on

<Hereafter LT Typescript Only>

you at some future day. It will doubtless interest you to learn how I had fallen in love, and doubly interest to observe how I had got out of it again![5]

My lectures are over:[6] and I believe successfully over. Last Friday week was my evening, when Professor Lissajous made his experiments on acoustics while I expounded them.[7] There was a fine audience and they all appeared delighted. The Duke and Duchess of Northumberland,[8] the Marquis of Lansdowne,[9] Lord Wrottesley the President of the Royal Society, and numbers of 'that ilk' were present. Duboscq was there and took charge of his own lamp,[10] which however he did not manage with any thing like the promptness which I have attained myself. I was thus at times forced to extemporize, and fill up chasms which had not been anticipated. This was observed by many, and manifestly took their fancy. The lecture was received enthusiastically and a vote of thanks passed by acclamation to Lissajous and Duboscq. Last Friday Faraday lectured;[11] and I took charge of the electric lamp and made the experiments while he spoke. His allusions to me were extremely kind. 'His friend Tyndall, for we were friends and helped each other (applause) would take charge of the lamp.' At the conclusion when thanking them for their patience and attention he added 'and I know you all thank Tyndall in your hearts.' It falls to the lot of few to have such a noble affectionate scientific father as your friend John possesses in Faraday. Indeed I know of no material advantage which I could set for an instant in competition with the friendship of that unspoiled and beautiful soul. I may say that the experiments in his lecture passed off admirably. The working of the lamp was excellent; my head was clear and my hand steady and Faraday was never more fresh and vigorous in his life.

This is the day of the Handel Festival[12] at the Crystal Palace.[13] The weather is glorious beyond description. I believe I might have been there in company with some young and beautiful people but I have a few experiments to make for a little party tomorrow which require preparation to day. I sometimes think of you when matters appear to shine upon me, God knows how gladly I would spare half of my sunshine to abolish your shade. Neither am I without my own drawbacks: I have been sometimes greatly broken down, and I have noticed an invariable connexion between this breaking down and the nature of my sleep. If I sleep well I am always strong next day—if ill—which is often the case the avenger is sure to come.

I hope to see you in Paris long before the 15th. of July: I hope to be able to get away by the first of next month. It would give me the greatest pleasure to be the guide of Verdet in London, for he is a man for whom I entertain a solid esteem. I will take care to prepare whatever letters I think may be of use to him.

Did I tell you that my connexion with the examinations of the Artillery and Engineer candidates has terminated. They offered me £25 for the examinations and I declined it.[14]

And now my dear Tom I bid you for the present good bye. God bless you my dear fellow—give my kindest love to Anna

Ever affectionately yours | John Tyndall

RI MS JT/1/T/640
RI MS JT/1/HTYP/500–501

1.	*[15 June 1857]*: the date is not on the original letter but is given as such on LT's typescript, and Tyndall referred to the Handel Festival, the first day of which occurred on 15 June 1857 (see n. 12).

2.	*Your letter*: letter 1402.

3.	*Your reference to poor Anna*: Hirst had discussed Anna's illness and how she felt about it, and the prospect of her not recovering.

4.	*'Whom god loveth he chasteneth'*: Hebrews 12:6.

5.	*I had fallen in love, . . . I had got out of it again!*: Tyndall had fallen in love with one of the daughters of Maria Drummond, either Mary or Emily (see letter 1406). In June 1857 Tyndall's journal entries show that he was undecided and conflicted about his feelings for Miss Drummond (Journal, RI MS JT/2/13c/939–42).

6. *My lectures are over*: Tyndall had finished lecturing at the RI for the season, and would travel in the Alps during the summer.

7. *my evening, when Professor Lissajous made his experiments on acoustics while I expounded them*: on 5 June 1857, Tyndall presented a Friday Evening Discourse at the RI, 'On M. Lissajous' Acoustic Experiments', an abstract of which is in *Roy. Inst. Proc.*, 2 (1854–58), pp. 441–43. In December 1856, Lissajous suggested to Tyndall that he take part in the lecture (see letter 1310).

8. *Duke and Duchess of Northumberland*: Algernon Percy (1792–1865), 4th Duke of Northumberland, a British naval commander and explorer, FRS, and President of the Royal Institution from 1842 to 1865; and his wife Eleanor Percy (née Grosvenor, 1820–1911), Duchess of Northumberland (*ODNB*).

9. *Marquis of Lansdowne*: Henry Petty-Fitzmaurice (1780–1863), 3rd Marquess of Lansdowne, a British statesman (*ODNB*).

10. *his own lamp*: with Léon Foucault in 1851, Duboscq developed the first commercially manufactured electric lamp, which he controlled during the current lecture's experiments.

11. *Last Friday Faraday lectured*: on 12 June 1857, Michael Faraday gave a Friday Evening Discourse at the RI titled, 'On the Relations of Gold to Light', closing lectures for the season. An abstract of this lecture was published as 'On the Relations of Gold to Light', *Roy. Inst. Proc.*, 2 (1854–58), pp. 444–46.

12. *Handel Festival*: the first of a series of celebrations to mark the centennial of the death of the German but London-based composer George Handel (1685–1759), held at the Crystal Palace on 15, 17, and 19 June 1857.

13. *Crystal Palace*: the iron and glass structure erected in Hyde Park for the Great Exhibition of 1851 and later rebuilt in south London. See J. A. Auerbach, *The Great Exhibition of 1851: A Nation on Display* (New Haven: Yale University Press, 1999).

14. *I declined it*: see letters 1375 and 1376.

From John Pringle Nichol[1]   15 June 1857   1405

Observatory | Glasgow | 15th June | 1857

Dear Dr Tyndall.

Some time ago I desired a copy of a 'Cyclopaedia of the Physical Sciences'[2] to be sent you. And presuming that you have got it & may have turned over some of its pages, I venture to ask of you a very great favor. The book necessarily abounds with defects, nor can the list of absolute errors be a slight one. I am going over it now very carefully, with a view to a Second Edition[3] next winter. Would you oblige me by calling my attention however briefly to the defects & errors you may have noticed: and what is of the greatest importance, will you consent to send me any statement that you may deem requisite in reference to cases where I may have misunderstood, or not adequately represented your own most valuable researches? Notes of this sort from you, however rough will vastly assist me in the attempts to produce a work that may for a number of years be a useful & safe guide to the student.—I feel most strongly that I have no right to intrude on you thus, further than the right which you will recognize, given by the fact, that assuredly without a hope of personal pecuniary recompense, I am desirous to hasten the student of Physical Science.

Ever very faithfully yours | <u>J. P. Nichol</u>
D. Tyndall

RI MS JT/1/N/11

1. *John Nichol*: John Pringle Nichol (1804–59), a Scottish astronomer, political economist, and professor of astronomy at the University of Glasgow. Nichol was widely known for his popular accounts astronomy (*ODNB*).
2. '*Cyclopaedia of the Physical Sciences*': J. P. Nichol, *A Cyclopædia of the Physical Sciences* (London: Richard Griffin and Company, 1857).
3. *Second Edition*: J. P. Nichol, *A Cyclopædia of the Physical Sciences,* 2nd edn (London: Richard Griffin and Company, 1860).

To Mary or Emily Drummond[1]   28 June 1857   1406

I declare <u>that</u> strikes myself as quite equal to the sorrows of Werter:[2] I am manifestly in the surges once more, but I have already told you that I am a good swimmer, and I shall doubtless reach the shore at last. I know not whether to laugh or cry over the matter; but laughter is wholesome so I will laugh. One thing is manifest, that science would not get much out of me if my life continued in this fashion. But the daring deed in which I am now engaged being once accomplished I shall return dutifully to my allegiance to science. Goethe[3]—remember I am not like Goethe; not, at least, like your idea of Goethe—says that he always got rid of his misery by writing about it. By some mystic process the statement of misery puts it <u>outside</u> of a man. This is perhaps the clearest account that I can render to you of my present act. I could not go to the <u>glaciers</u>[4] without letting you know something of this matter: there one false step might deprive you for ever of the record (a consummation, you will probably say, devoutly to be wished.) I purposed calling to bid you good bye, but this I will not do, for I see too clearly the result of all this. <u>What will your mother</u>[5] <u>think of me</u>? I know too well that I am cutting her friendship adrift. I lose a limb, but I save my intellectual life. Having said so much I now take my leave of you with a lightened heart. I shall soon right myself & find solace in my old delights. I turn with a love which knows no weariness to that great mother[6] who has been always so kind to me. She will heal her son's wounds and give him the clear joy of the intellect as a rich compensation for that of the heart . . . [7]

RI MS JT/2/7/464–65
RI MS JT/2/13c/943–44

---

1. *To Mary or Emily Drummond*: this letter fragment, found in Tyndall's journal, was his best recollection of the last paragraph of a letter to either Mary or Emily Drummond, whom he referred to in 1857 as Δ'. Tyndall wrote in his journal on 28 June 1857: 'Posted at half past 9 o'clock P.M. this day the extracts of my journal to Δ' I have not quite followed my own natural bent in this matter: that bent would undoubtedly be to make make [sic] sure of the maiden herself first' (Journal, RI MS JT/2/13c/943–44). Tyndall's feelings were not shared by Miss Drummond (Journal, 1 July 1857, RI MS JT/2/13c/944–45). In his journal he would describe this declaration of his affections as his 'folly' (Journal, RI MS JT/2/13c/952, 960, 1057, 1058).
2. *sorrows of Werter*: *The Sorrows of Young Werther*, a 1774 novel by Johann Wolfgang von Goethe that catapulted him to literary celebrity status.
3. *Goethe*: Johann Wolfgang von Goethe (1749–1832), the outstanding German literary figure of the late eighteenth and early nineteenth centuries. His extensive writings included his theory of colours and a study of the structure and function of plants. During his time in Marburg (see letter 1399, n. 10), Tyndall became familiar with some of Goethe's works.
4. *go to the glaciers*: leaving England for continental Europe in the summer, to spend time mountaineering and studying glaciers in the Alps.
5. *your mother*: Maria Drummond.
6. *that great mother*: nature, 'mother earth'. See n. 4.
7. *for that of the heart*: fragment only, letter is incomplete.

To Auguste de la Rive 30 June 1857 1407

30th June 1857

My dear Sir

It is not without some self reproach that I find your last friendly letter[1] to me still unanswered, but I have been so occupied, and my health at intervals so indifferent, that I have been withdrawn in a great measure from correspondence. I now write to say that I purpose starting tomorrow for Paris, where I expect to remain a day or two, and afterward I shall proceed with all possible speed to Geneva.[2]

I return your many thanks for your full and truly kind replies to my questions. With regard to the small hydraulic press we have one here which I have been accustomed to use and which I intend to take with me. Circumstances have also placed a very convenient theodolite[3] at my disposal, and this I also intend to take with me. So that all I shall need at Geneva will be some material of which to make coloured solutions for experiments on infiltration. Logwood[4] and ink will I think be the most suitable substances, and they are easily obtained.

Very many thanks for your hospitable offer of shelter.[5] My stay in Geneva will be very short and it will hardly be worth while to derange[6] your house for such a brief sojourn. Nevertheless in this matter I am quite willing to place myself at your kind disposal.

I have already given directions to ascertain the price of a hydraulic press of convenient dimensions; I will bring you an account of this, so that if you should desire to have the instrument at anytime it may be sent to you.

Both Mr Walker and myself have consumed all our proof.[7] He informed me that he has written to you upon the subject. He has also confided to my care a time table which I am to give to you.

And now my dear Sir I will bid you, for a brief period, good bye, hoping to have the pleasure of seeing you face to face before many days have elapsed.

Believe me | Most sincerely yours | J. Tyndall

RI MS JT/1/T/370
RI MS JT/1/TYP/1/350

---

1. *your last friendly letter*: letter 1378.

2. *Geneva*: see letter 1293, n. 3.

3. *theodolite*: see letter 1353, n. 11.

4. *logwood*: see letter 1353, n. 10.

5. *your hospitable offer of shelter*: Tyndall visited de la Rive at his country home in Geneva on 11 July 1857 (Journal, RI MS JT/2/13c/954). See also letter 1413.

6. *derange*: to disturb or destroy the arrangement or order of (*OED*).

7. *all our proof*: see letter 1293, n. 16.

From Heinrich Gustav Magnus   [June 1857][1]   1408

1857

Mein theurer Tyndall.

Mir ist es so ergangen wie Ihnen. Ich bin auch in dieser Zeit sehr beschäftigt gewesen und habe dadurch die Beantwortung Ihres Briefes von einem Tage zum andern verschoben; was um so mehr unrecht ist als ich in Ihrer Schuld bin. Ich sende Ihnen anbei den Betrag für das Photometer £ 3./11. nämlich eine Anweisung von Asher auf £ 2/1. und eine Quittung von Sauerwald über

10 *[Rthl].*[2] Empfangen Sie zugleich meinen aufrichtigen Dank für Ihre gefäl-
lige Besorgung. Leider ist das Photometer für meine Zwecke zu groß. Man hat
dergleichen portatif in einem kleinen Kasten. Die Entfernung wird bei diesen
Instrumenten an einem getheilten Bande *[abgelesen].* Ein solches hatte ich
im Sinne. *[Ich]* habe das Instrument das Sie so gut waren mir zu senden an
die hiesige Gas Anstalt abgegeben; doch will ich Sie jetzt nicht wegen eines
neuen incommodiren. Ich hoffe Sie werden uns die Freude machen in diesem
Jahre zu uns zu kommen, und dann können wir das Nähere verabreden. Sie
dürfen diesmal nicht Deutschland verlassen ohne ihre Berliner Freunde gesehen zu haben.

Bis zum 7 August bleibe ich bestimmt in Berlin, aber schwerlich viel
länger, wohin ich gehen werde weiß ich noch nicht, vielleicht auch in die
Schweiz. Lieb wäre mir daher könnte ich erfahren wo Sie um diese Zeit sind.
Sollten Sie freilich zur Versammlung nach Dublin gehen, so würden Sie wohl
so früh zurück müssen daß nicht zu hoffen ist daß Sie zu uns *[kommen].*

Was haben Sie zu der Wahl von Matteucci in der Pariser Acad. gesagt? Ist
so etwas erhört? Die Pariser Acad hat sehr an Bedeutung verloren und verliert
immer mehr!

Von hier wüßte ich nicht viel Neues zu berichten. Ich habe meine Electrolytischen Untersuchungen jetzt beendet, oder wenigstens für einige Zeit
ausgesetzt, und hoffe daß das wenige, das bis jetzt fertig geworden, noch vor
meiner Abreise soll gedruckt werden.

Ihre Nachricht über Dr Brewster hat uns alle sehr überrascht, wir wuß-
ten gar nichts von diesem Schritt! Der arme Mann! oder die arme Lady!—
Eine andere erfreulichere Nachricht von hier kann ich Ihnen mittheilen. Die
Tochter von H. Rose (Hannchen) hat sich verlobt (zum zweiten Mal) mit D[r].
Karsten, Botaniker, der viele Jahre in Süd America gelebt hat. Ebenso ist die
älteste Tochter von Ehrenberg (Helene) verlobt mit D[r]. Hanstein, ebenfalls
Botaniker und Lehrer an einer hiesigen Gewerbe-Schule.

Grüßen Sie Faraday vielmals von mir, und ebenso alle Freunde die sich
meiner erinnern.

Viele herzliche Grüße von meiner Frau und den Kindern; sie würden sich
sehr freun Sie hier zu sehn, drum machen Sie uns diese Freude.

Ihr | G Magnus.

D[r]. Tyndall.

1857

My dear Tyndall.

Things have been going the same for me as they have for you. I have also
been very busy lately and because of this have postponed answering your letter[3]
from one day to the next; which is all the more wrong, as I am in your debt. I am

sending you enclosed the amount for the photometer[4] £ 3./11., namely a money order from Asher for £ 2/1. and a receipt from Sauerwald worth 10 *[Rthl]*. Please accept at the same time my sincere thanks for helpfully doing this errand. Unfortunately, the photometer is too big for my purposes. There are portable versions of them in a little case. With these instruments, the distance is *[read]* from a tape divided into parts. I had one of those in mind. *[I]* have given the instrument that you were so good as to send me to the gasworks here; but I do not want to bother you now about a new one. I hope that you will give us the pleasure of coming here this year, and we can then arrange things in greater detail. You will not be allowed to leave Germany this time without having seen your friends in Berlin.

I will definitely be staying in Berlin until 7 August, but scarcely for much longer; where I shall go, I still do not know, perhaps even to Switzerland. I should therefore be glad if I could find out where you will be around this time. Of course, should you go to the meeting in Dublin,[5] then you would probably have to come back so early that there is no hope of you *[coming]* here.

What did you say about Matteucci's election in the Paris Acad.?[6] Is that sort of thing heard of? The Paris Acad. has lost much of its importance and is losing more and more!

I do not have much in the way of news to report from here. I have finished my electrolytic investigations[7] now, or at least deferred them for a while, and hope that the little that has been finished up until now should still be printed before my departure.

Your news about Dr Brewster[8] has taken us all very much by surprise, we knew nothing at all about this step! The poor man! Or, the poor lady![9]—I can give you another more pleasant piece of news from here. H. Rose's daughter (Hannchen)[10] has got engaged (for the second time) to Dr. Karsten,[11] a botanist, who has lived in South America for many years. Likewise, Ehrenberg's eldest daughter (Helene)[12] is engaged to Dr. Hanstein,[13] also a botanist and teacher at a vocational school here.

Give Faraday my best regards, and likewise all friends who remember me.

Many warm greetings from my wife and the children;[14] they would be very pleased to see you here, so do us this pleasure.

Your | G Magnus.

Dr. Tyndall.

RI MS JT/1/M/40

---

1. *[June 1857]*: the month and year are given based on Carlo Matteucci's election to the French Académie des sciences in May 1857 (see n. 6) and that Tyndall would be traveling to the Alps in July 1857.

2. *[Rthl]*: see letter 1325, n. 1.

3. *your letter*: letter missing.

4. *photometer*: see letter 1327, n. 5.

5. *the meeting in Dublin*: the BAAS held its twenty-seventh meeting in Dublin, Ireland in August and September 1857. Tyndall did not attend this meeting.

6. *Matteucci's election in the Paris Acad.*: Carlo Matteucci was elected a corresponding member of the French Académie des sciences on 18 May 1857. 'Nominations', *Paris, Comptes Rendus,* 44 (1857), p. 1013.

7. *electrolytic investigations*: G. Magnus, 'Elektrolytische Untersuchengen', *Poggend. Annal.,* 102 (1857), pp. 1–54.

8. *news about Dr Brewster*: probably news of David Brewster's marriage to a much younger woman. See letter 1371, n. 10.

9. *the poor lady*: see letter 1371, n 10.

10. *H. Rose's daughter (Hannchen)*: Hannchen Karsten (née Rose), daughter of Heinrich Rose.

11. *Dr. Karsten*: Hermann Karsten (1817–1908), a German botanist and geologist who taught in Berlin from 1856 before becoming a professor of plant physiology at the University of Vienna in 1868 (*NDB*).

12. *Ehrenberg's eldest daughter (Helene)*: Helene Hanstein (née Ehrenberg, 1834–90), daughter of Christian Gottfried Ehrenberg (1795–1876), the German botanist and paleontologist. Helene married Johannes Hanstein in Berlin in 1857 (*CDSB, NDB*).

13. *Dr. Hanstein*: Johannes Ludwig Emil Robert von Hanstein (1822–80), a German botanist who taught botany at the University of Berlin from 1855, and became professor of botany at the University of Bonn (*NDB*).

14. *my wife and the children*: Bertha, Anna, Christine, and Paul Magnus.

To Juliet Pollock   [1 July 1857][1]   1409

Wednesday morning

My dear Mrs Pollock,

I am just on the point of starting and the sight of your present[2] as I locked it up brought you to my mind and brought with it at the same time the memory of the awkward manner in which I accepted it from you. Now should I fall into a crevasse, or be thrown into one by Forbes (the mechanical and physiological result being all the same) I do not, trifling as the matter is want you to associate the thought of clownishness with my reception of your gift. I see my own defects as clearly as any body. Do not therefore attribute these defects to any intrinsic boorishness of my character, but merely to the want of that tact which habit alone gives, and which prevents me often from realising in the external world what the spirit sees to be fit and proper. My journal entry for the 21st of June contains the following—'My awkwardness last night in accepting a gift from Mrs Pollock acts like a jarred string upon my memory. But even this jar is an admonition to me. If when men make mistakes no jar, no inner rebuke were felt, one great incentive to improvement would be cast away. Let me therefore resolve this little unhappiness into the determination to do more fitly in future.'

Thus you see I have got a little philosophical honey out of the circumstance. And now I will again write 'good bye' to you, and sincerely hope that when I return I shall see vigorous in body and spirit and all around you as happy as you can wish them to be

Most sincerely yours | John Tyndall.

RI MS JT/1/TYP/6/1905
LT Typescript Only

---

1. *[1 July 1857]*: the date is given based on relation to letter 1418. Tyndall left London for the Alps on 2 July 1857 (Journal, RI MS JT/2/13c/947).

2. *your present*: in his journal entry for 20 June 1857 Tyndall wrote, having dined with the Pollocks the night before, that his 'awkwardness last night on accepting a gift from Mrs Pollock, presented with reference to my Swiss journey, acts like a jarred string upon my memory' (Journal, RI MS JT/2/13c/942–43).

From John Pringle Nichol[1]   2 July 1857   1410

Observatory | Glasgow | 2. July 1857

My dear D[r] Tyndall.

I beg to offer you many and sincerest thanks for your kind note,[2] and also for a paragraph in which I think I recognize your hand in the Westminster Review.[3] I am aware to the fullness of the extremes of manifold defects in the volume. The only excuse is, who, under our exist[in]g cheap publication plan, would or could have warranted a first edition free from multitudes of errors? You will discern many that you might not have expected, on looking more closely at some of the parts of the Book. I plead in extenuation that <u>it has cost me</u> about £300, and that I wish to correct errors in the next edition which I am glad to say, on that account especially is impending.—Do help me therefore by searching criticism and by your invaluable suggestions. Whatever you may choose to write either as to subjects or persons, shall be held by me in honorable privacy.

As to page 10, I found that a collaborateur[4] had not known some of your own very important researches. Some amendment we made in the Appendix—by no means complete. That shall all be set right. Thanks for the paper just received from you.[5] I have recently seen it—

Ever faithfully yours | J. P. Nichol
Dr Tyndall.

RI MS JT/1/N/12
RI MS JT/1/TYP/3/922

---

1. *John Nichol*: see letter 1405, n. 1.
2. *your kind note*: letter missing.
3. *your hand in the Westminster Review*: referring to a short review of Nichol's *A Cyclopædia of the Physical Sciences* (London: Richard Griffin and Company, 1857) in *Westminster Review*, 68:133 (1857), pp. 276–77. Tyndall and Thomas Huxley oversaw the science section of this journal since 1853.
4. *a collaborateur*: not identified.
5. *paper just received from you*: possibly J. Tyndall and T. H. Huxley, 'On the Structure and Motion of Glaciers', *Phil. Trans.*, 147 (1857), pp. 327–46.

From Carlo Matteucci   7 July [1857][1]   1411

Pise 7 juillet

Mon cher monsieur,

Je vous remercie bien de votre lettre. Je désire que votre séjour en Suisse rétablisse bientôt votre santé—Je voudrais bien être avec vous aux glaciers, mais je ne puis pas et ma santé aussi n'est pas bonne.

Vous verrez donc M. <u>de la Rive</u> et toute sa famille et je vous prie de me rappeler à tous. Vous me parlez d'expériences sur des tons obtenus dans la condensation des gaz. Quand vous en aurez le temps, je vous prie de m'en écrire plus longuement ainsi que sur les glaciers—Cela me promet d'abord une lettre de vous et puis une nouvelle scientifique pour le <u>Nuovo Cimento</u>.

Savez-vous si l'eau des glaciers est pure comme l'eau de la neige ou usée ou bien si elle contient des sels?

Adieu. Croyez-moi [tout dévoué] | C. Matteucci

Pisa 7 July

My dear sir,

I thank you for your letter.[2] I wish your stay in Switzerland will rapidly restore your health—I would love to be with you on the glaciers, but I cannot and my health is not good either.

You will meet then Mr. de la Rive and all of his family,[3] and I ask you to send them my regards. You talk about experiments on the tints obtained in the condensation of gas; when you find some time, please write more extensively about it as well as about the glaciers—This will firstly assure me a letter from you and then a scientific paper for the <u>Nuovo Cimento</u>.[4]

Do you know if glacial water is as pure as water from the snow or spoiled or if it contains salts?

Goodbye. Believe me [yours truly] | C. Matteucci

RI MS JT/1/M/66

———

1. *[1857]*: the year is given based on the mention of Tyndall's plan to visit Auguste de la Rive (see n. 3).
2. *your letter*: letter missing.
3. *You will meet then Mr. de la Rive and all of his family*: Tyndall visited Auguste de la Rive at his country home in Geneva on 11 July 1857 (Journal, RI MS JT/2/13c/954). See letters 1407 and 1413 for Tyndall and de la Rive's discussion of the upcoming visit.
4. *Nuovo Cimento*: see letter 1300, n. 3.

From Anne Wynne   9 July [1857][1]   1412

Pariser Hof | Langen Schwalbach[2] | Duchy of Nassau | 9th July

My dear M[r]. Tyndall,

If this letter arrives before you have left London for your holiday will you do me a favour—I think that from having last year had to do with the examinations for Woolwich, you may be able to find out for us, sooner than the public accounts would tell us, whether Eddy has failed or not.[3] it would be something to us to have the time of suspense shortened—tho' it is hardly suspense, we feel so nearly sure that he has failed. He thinks it himself, says that on almost every point he answered very poorly and greatly below his knowledge—

One thing came against him—we had been pleasing ourselves with the thought that as the examination was to begin with drawing (mechanical drawing he knows nothing of landscape) his success in that would give him a feeling of self confidence, which would be in his favour for the rest of the examination. but on that very point he had a severe disappointment which I will tell you in his words—'it has not been for me a fair examination in drawing in this way. You probably recollect that last time I was given only geometrical drawing, and I thought of course that it would be the same this time as they had merely advised in the last report that if any one <u>wished</u> to be examined in landscape, additional credit would be given for it: but on the drawing day they only gave us geometrical questions which were far too long and difficult to do more than half of them in the time, and in the evening they gave us landscape drawing to do; I asked for a mechanical drawing which was down on the list along with the landscape but they would not give it to me and said I must do the landscape, I tried it but could make nothing of it. Now I think this unfair that after spending a great time not only with a master but also at home in the evenings I should get less marks than I did last time when there were fewer marks given to the subject, whereas I expected and I think reasonably, to double them at least'—

I don't know why I tell you this story, excepting that it seems to have been the only unfair thing in the examination & yet one to tell against the boy— Of course half the fathers and mothers believe that in sending their sons to this examination they are sending in valuable members to the Corps of Artillery and Engineers[4]—I know we feel sure of it in Eddy's instance and yet I am nearly sure he will not be tried. They cannot know what a nice boy he is or how much thought and sense he has. They have only the examination to go by—

I will not write to you any more about the matter, but will you if you can, tell us the decision.

This is our first visit to Germany. We do like the country and much in the character of the people—England is not, I feel sorry to think, in all points the first of the nations—

I wish you a very happy holiday my dear Mr. Tyndall & much refreshment from it.

With Edward when he comes we think of seeing various places & perhaps he may go into Switzerland. I think we shall meet him at Heidelberg[5] on the 18th.—

Do not overwork yourself please—I wish we might meet you somewhere & see you well—I am ever yours | most truly | A. Wynne

RI MS JT/1/W/90
RI MS JT/1/TYP/5/1855–56

---

1. *[1857]*: the year is given on the basis that Edward Wynne took for a second time the entrance examinations to the Royal Military Academy at Woolwich on 20 June 1857, the first time in 1856.

2. *Langen Schwalbach*: a town near Wiesbaden in Germany, renamed to Bad Schwalbach in 1927.

3. *Eddy has failed or not*: Edward Toler Wynne (1837–89), son of George and Anne Wynne, took for a second time the entrance examinations to the Royal Military Academy at Woolwich on 20 June 1857. The first time in 1856, which he failed, was against the advice of his father and Hirst. According to the Royal Military Academy records at Sandhurst College, Wynne was admitted to Woolwich in 1857 and was commissioned as a Cadet in 1858.

4. *Corps of Artillery and Engineers*: students at Woolwich were trained for the British Royal Artillery and Royal Engineers.

5. *Heidelberg*: a city in southwestern Germany.

From Auguste de la Rive   10 July [1857][1]   1413

Genève. Vendredi 10 juillet.

Mon cher Monsieur,

J'ai vivement regretté que vous n'ayez pas laissé votre adresse sur votre carte de visite. Je vais envoyer mon domestique vous chercher dans tous les hôtels. J'espère encore que vous voudrez bien accepter notre hospitalité. Je passe à Genève ce soir avec ma voiture & je pourrais vous prendre à votre hôtel à 9h. du soir. Sinon je vous demande en grâce de venir diner avec nous demain, dimanche ou lundi. Nous dinons à 5h, mais je vous attendrai plus tôt. Je désire vivement connaitre vos projets, plusieurs de nos amis ayant un vif désir de vous voir.

Tâchez, je vous en prie, de venir ce soir avec moi coucher à la campagne. Mon domestique attendra votre réponse.

Votre tout dévoué et affectionné | Aug. de la Rive | Genève Vendredi 10 juillet

Geneva. Friday, July 10.

My dear Sir,

I deeply regretted that you did not leave your address on your calling card. I will send my servant to look for you in all the hotels. I still hope that you will accept our hospitality.[2] I will come by Geneva[3] tonight with my carriage & I could pick you up at your hotel at 9 o'clock in the evening. Otherwise, I beg you to come have dinner with us tomorrow, Sunday or Monday. We will dine at 5 o'clock, but I will be expecting you earlier. I sincerely want to know about your projects, many of our friends having a strong desire to see you.

Please, do try and come with me this evening to sleep in the countryside. My servant will await your reply.

Your fully devoted and affectionate | Aug. de la Rive | Geneva, Friday, July 10

RI MS JT/1/D/106
RI MS JT/1/TYP/352

---

1. *[1857]*: the year is given based on the mention of Tyndall's plan to visit Auguste de la Rive (see n. 2).

2. *accept our hospitality*: Tyndall visited de la Rive at his country home in Geneva on 11 July 1857 (Journal, RI MS JT/2/13c/954). See also letter 1407.

3. *Geneva*: see letter 1293, n. 3.

From Herman Knoblauch[1]   16 July 1857   1414

Halle July 16th 57

My dear friend,

First of all I send you my best thanks for the papers containing your most recent inquiries[2] which you have been kind enough to send to me. I have read them with the greatest interest. The acoustic experiments explain the condition of the phenomenon observed by Schaffgotsch[3] which at first sight seemed so very curious and surprising. It loses all its mysterious qualities being reduced to a natural consequence of the battlements. One only I wonder, that of the many persons who tried to repeat the experiments none succeeded in making it with hydrogen coal. Coal-gas seemed to be undispensible. I don't understand at all this difference (but have not yet repeated the experiments myself), you undoubtedly will know the real cause of it.

The wish to send to you the memoir here joined[4] gives me the immediate impulse to write to you. You find in it a series of experiments which I had already begun in Marburg[5] and which then had been interrupted by the inquiry on the thermic pleochroism of crystals.[6] The present paper, which as you see, is not large of size contains the most difficult and troublesome experiments I ever made. The difference in the quality of radiant heat being reflected from various metals is a very small one and can be discovered and be expressed by numbers to be relied upon, only by the most scrupulous attention and carefulness. But by the practice I believe to have got in experiments on radiant heat and by the 17000 observations I have made with the thermomultiplier[7] only for the results contained in this paper 'on the influence of metals on radiant heat,'[8] I have not a shadow of a doubt in the theses pronounced in it, which differ from all which—by Melloni's[9] authority—until now was settled in the compendium books.

I take the liberty to add one copy for Prof. Faraday as a little mark of my high esteem and veneration, and another for our common friend, Hirst, of whom I have got (besides the memoirs he has been kind enough to send to me and for which I address my best thanks to him) no further news since your stay in Halle.[10] I beg him to take the paper as a testimonial of my unaltered friendship. I have much thought of him. How may be M[rs] Hirst?[11] Has her health allowed her to return to his first abode and his former occupation in England?

I don't know whether you have heard that Prof. Kohlrausch[12] had got a professorship in Erlangen.[13] They have given him no successor in Marburg, but to do at least something for Physicks they have made Prof. Gerling 'Geheimer Hofrath'[14] and have given him an assistant! You see 'good old Timer'[15] has come back to Marburg! O my dear Marburg!

My lecture in Halle is now distributed for the whole of Physicks so that I treat optics and heat in Summer; the rest (mechanics, acoustics, magnetism, electricity) in Winter. I had first 4, now 30 hearers. You must go to <u>Bonn</u>[16] for the Naturforscher Versammlung[17] to counterbalance Plücker[18] (16 Sept, as I know). I can not leave Halle. And now good bye! My little John (Johannes)[19] remembers himself kindly to D[r]. John Tyndall, the friend of his father.

Yours | most faithfully | Herm. Knoblauch.

RI MS JT/1/K/20
RI MS JT/1/TYP/7/2499–99A

1. *Herman Knoblauch*: Hermann Knoblauch (1820–95), a German physicist and chemist known for his contributions to the studies of radiant heat. He co-authored two papers with Tyndall in 1850, Tyndall's first scientific publications (*ADB*).

2. *the papers containing your most recent inquiries*: possibly J. Tyndall and T. H. Huxley, 'On the Structure and Motion of Glaciers', *Phil. Trans.*, 147 (1857), pp. 327–46; and the abstract of Tyndall's most recent Friday evening lecture, 'On M. Lissajous' Acoustic Experiments', *Roy. Inst. Proc.*, 2 (1854–58), pp. 441–43.

3. *Schaffgotsch*: Franz Schaffgotsch (1816–64), a German scientist. See F. B. Schaffgotsch, 'Akustische Beobachtungen', *Poggend. Annal.*, 101 (1857), pp. 471–87; and Count Schaffgotsch, 'Acoustic Experiments', *Phil. Mag.*, 14 (1857), pp. 541–44. Tyndall recounts his and Schaffgotsch's similar experiments in acoustics in *Sound: A Course of Eight Lectures* (London: Longman's, Green, and Co., 1867), pp. 226–30.

4. *the memoir here joined*: missing.

5. *Marburg*: Knoblauch was a teacher of Tyndall's at the University of Marburg (see letter 1399, n. 10), and they published two papers together on diamagnetism. See R. Jackson, 'John Tyndall and the Early History of Diamagnetism', *Annals of Science*, 72 (2015), pp. 435–89, on pp. 443–50.

6. *thermic pleochroism of crystals*: pleochroism is the phenomena in which crystals exhibit different colours depending on the angle at which light reaches the crystal (*OED*). Knoblauch investigated the way heat was transmitted through crystals depending on the angle of incidence. See H. Knoblauch, 'Ueber die Abhängigkeit des Durchgangs der strahlenden Wärme durch Krystalle von ihrer Richtung in denselben', *Poggend. Annal.*, 161 (1852), pp. 169–88; and 'Ueber die Abhängigkeit des Durchgangs der strahlenden Wärme durch Krystalle von ihrer Richtung in denselben', *Poggend. Annal.*, 169 (1854), pp. 161–212.

7. *thermomultiplier*: early name for a thermopile (*OED*).

8. *'on the influence of metals on radiant heat'*: H. Knoblauch, 'On the Influence of Metals upon Radiant Heat', *Phil. Mag.*, 14 (1857), pp. 356–74.

9. *Melloni's*: see letter 1293, n. 12.

10. *Halle*: Knoblauch moved from the University of Marburg to the University of Halle in 1853, where Tyndall must have visited him.

11. *M[rs] Hirst*: Anna Hirst. She had died on 1 July.

12. *Prof. Kohlrausch*: Rudolph Kohlrausch (1809–58), a German physicist who taught at the University of Marburg and the University of Erlangen (*CDSB*).
13. *Erlangen*: University of Erlangen.
14. *'Geheimer Hofrath'*: privy councillor (German).
15. *'good old Timer'*: not identified.
16. <u>*Bonn*</u>: a city in western Germany.
17. *Naturforscher Versammlung*: the 33rd meeting of the Versammlung der Gesellschaft Deutscher Naturforscher und Ärzte was held in Bonn in September 1857.
18. *counterbalance Plücker*: Julius Plücker. Regarding Plücker and Tyndall's relationship, see letters 1299 and 1360.
19. *My little John (Johannes)*: Johannes Knoblauch (1855–1915), mathematician and son of Hermann Knoblauch (*NDB*).

To Michael Faraday   18 July 1857   1415

Chamouni, Montanvert | 18 Jul 1857

My dear Mr. Faraday,

I think I may fairly allow myself the rest and luxury of writing to you this morning for I have had a week's hard work as far as limbs and arms are concerned. I reached Paris just in time to learn that the youthful wife[1] of a young friend of mine,[2] whom you once saw (before he was married) at my lodgings in Islington,[3] was in her coffin. I accompanied my smitten friend to the cemetery of Montmartre,[4] and there accident caused it to fall my lot to shake ashes to ashes, earth to earth, dust to dust.[5] I thought it would never do to leave my friend amid scenes which would incessantly revive the memory of his loss and so persuaded him to come with me. He is now here, and we are working hard together.

At Paris I saw Mr. Biot and he desired me to say to you that he was as happy as man can be. He received me with great kindness. Pouillet[6] and Chevreul[7] I also saw, and they were very cordial. *<1 word missing>*[8] introduced me afterwards to M. Becquerel,[9] but I found him <u>hard</u> and not cordial. I must crave your forgiveness for asking you to write to Dumas,[10] for when it came to the point I could not render to myself a sufficient reason why I should call upon him; I therefore did not do so. I spent a couple of days in Geneva:[11] Those I knew were all in the country, but De la Rive found me out at my hotel and I spent one delightful afternoon at his country house. He is a fine genial loveable man. From Geneva to Chamouni[12] was a day's journey. The weather was glorious: as we passed Mont Blanc[13] a patch of reddish light was thrown upon the mountain snow from some clouds which floated in the west. Around this patch a subjective green glory spread itself to some distance: it was very curious and very beautiful. I stayed 3 days at Chamouni, and rose one morning to see the sun rise on Mont Blanc. My bedroom opened into a corridor from one end of which the east was visible, and the other commanded a view of Mont Blanc and of the west. The east sky was of an amber hue, fading insensibly into a rosy violet, which again blended with the deep blue of the zenith.[14] The morning star was glistening between east and west, and not far from it the moon turned her pale face towards the rising day. The mountain rose chaste, and cold, and white, as the unsoiled snow could make him. I walked to the other end of the corridor and looked eastward, hoping to see the sun lift his disk above the mountains, wholly forgetful for the moment that if I waited until he appeared, it would be sunrise not only for Mont Blanc, but for the whole valley. People do very absurd things thus unconsciously. I walked to the other end of the corridor, and saw the highest summits smitten by the sunbeams. Peak after peak then lost its severity and melted into a golden smile. It was a glorious scene. I watched it till the morning star was quenched and the moon became invisible, and then being very tired went to bed again!

On last Wednesday I had all my things carried up to the Montanvert.[15] There is a kind of hotel erected here which possesses three bed rooms. They are divided from each other by partitions of wood, and the noise of the tramping visitors would render sleep in my case an impossibility. I have therefore chosen a little temple as my habitation which was erected many years ago by an Englishman named Blaire,[16] and dedicated by him 'a la Nature'. Its floor is of stones which are rather wet, its walls of the same material and in the same condition; They have put a bed into it, and given me a goatskin to keep my feet from the flags; and I contrive to drive away a little of the moisture by a pine fire. I never felt more like a philosopher than when I sit there alone at night, listening to the wind moaning over the glacier, and to the distant rumble of the stones upon the moraines[17] as they tumble into the crevasses. Hitherto I have chiefly occupied myself with observing the motion, and obtaining a general notion of the glacier. I have seen Balmat[18] and he informs me that Forbes is on his way over, so that I expect to have the pleasure of seeing him soon. From what I have thus far seen the divergence between us is likely to become <u>wider</u>. But the question is merely shaping itself in my brain, and I can only hope that after a little time it will take definite form. The weather thus far has been so good that I have been every day upon the ice. I am very strong and sometimes admonished by my guide[19] for making use of my strength in jumping the crevasses. I regret that I did not bring a pair of Nichols prisms[20] with me: but I shall have plenty to occupy me without them. With kindest remembrances to Mrs. Faraday and Miss Barnard

Believe me always | Most sincerely Yours | John Tyndall

Would you have the goodness to ask Anderson to put the enclosed[21] into envelopes, with the addresses and post them for me? If he would send me my letters once a week until I write to him he would oblige me

RI MS JT/1/TYP/12/4074–76
Faraday Correspondence, 5:3320
Typed Transcript Only

1. *the youthful wife*: Anna Hirst, died on 1 July 1857.
2. *of a young friend of mine*: Thomas Hirst.
3. *my lodgings in Islington*: Hirst had written in his journal on 21 January 1854 after attending a Faraday lecture with Tyndall: 'As John and I were sitting writing, after tea this evening, Faraday himself paid us a brief visit. I had not before sought to be formally introduced to him, and I am glad of it. Such introductions are always most satisfactory when they are unforced, and occur in the natural course of things' (Brock & MacLeod, *Hirst Journals*, p. 1108).
4. *cemetery of Montmartre*: officially known as Cimitière du Nord, a Paris cemetery opened in 1825.
5. *ashes to ashes, earth to earth, dust to dust*: from the Anglican *Common Book of Prayer*, under 'The Order for the Burial of the Dead': 'FORASMUCH as it hath pleased Almighty God of his great mercy to take unto himself the soul of our dear brother here departed: we therefore commit his body to the ground; earth to earth, ashes to ashes, dust to dust; in sure and certain hope of the Resurrection to eternal life, through our Lord Jesus Christ'.
6. *Pouillet*: Claude Pouillet (1790–1868), a French physicist and author of *Élémens de physique expérimentale et de météorologie*, 4 vols (París: Béchet Jeune, 1827–30) (*CDSB*).
7. *Chevreul*: Michel Eugène Chevreul (1786–1889), a French chemist who received the Copley Medal from the RS in 1857 'For his researches in organic chemistry, particularly on the composition of the fats, and for his researches on the contrast of coulours' (*CDSB*).
8. *<1 word missing>*: Tyndall was probably introduced by Despretz. He wrote in his journal on 6 July 1857: 'Introduced by M. Despretz to M. Pouillet made also the acquaintance of M. Chevreul; both of them were kind and cordial. Afterwards introduced to M. Becquerel: found him hard and not cordial' (Journal, RI MS JT/2/13c/949).
9. *M. Becquerel*: either Antoine-César Becquerel (1788–1878), a French physicist who studied electricity; or his son Alexandre-Edmond Becquerel (1820–91), a French physicist who studied electricity and magnetism, and discovered the photovoltaic effect (*CDSB*).
10. *Dumas*: Jean-Baptiste Dumas (1800–84), a French chemist who received the RS Copley Medal in 1843 'For his late valuable researches in organic chemistry, particularly those contained in a series of memoirs on chemical types and the doctrine of substitution, and also for his elaborate investigations of the atomic weights of carbon, oxygen, hydrogen, nitrogen and other elements' (*CDSB*).
11. *Geneva*: see letter 1293, n. 3.

12. *Chamouni*: see letter 1353, n. 9.

13. *Mont Blanc*: see letter 1320, n. 6.

14. *zenith*: the point of the sky directly overhead (*OED*).

15. *Montanvert*: a lookout point on the mountainside, overlooking the Mer de Glace, a large glacier on the north side of Mont Blanc.

16. *Blaire*: an Englishman who erected a house (or 'temple') in the Alps. According to Tyndall, Blaire's structure was inscribed with 'a la Nature'. However, he may be confusing two different huts, for in J. D. Forbes' *Travels through the Alps of Savoy* (see letter 1306, n. 8), Forbes refers to a cabin 'built by an Englishman named Blair' and a 'small solid stone house … built at the expense of M. Deportes' with the inscription 'A la Nature', as two different structures (p. 73).

17. *moraines*: a mound, ridge, or other feature consisting of debris that has been carried and deposited by a glacier or ice sheet, usually at its sides or extremity (*OED*).

18. *Balmat*: Auguste Balmat.

19. *my guide*: possibly Edouard Simond.

20. *Nichols prisms*: probably referring to the Nicol prism, used to produce a polarized beam of light, invented by William Nicol (1770–1851) in 1828 (*ODNB*).

21. *the enclosed*: not identified, but most likely outgoing letters from Tyndall.

To Thomas Henry Huxley   19 July 1857   1416

Montanvert, Chamoni, Savoy | 19th. July 1857.

My dear Huxley,

Here I sit on the Montanvert[1] with blistered hands and weary legs from trampling over this weary ice sea. It must have been called so from the resemblance of its surface to a billowy ocean. Otherwise a '<u>fleuve</u>'[2] and not a '<u>mer</u>'[3] would be its proper designation. I have a friend[4] with me here whose youthful wife[5] was buried during the short time I spent in Paris, and whom I carried away with me to separate him from scenes which would remind him incessantly of his loss. He is up to all Trigonometrical observations, and therefore very useful to me.

Thus far I have confined myself to the observation of the motion, and to obtaining a general notion of the glacier. I have been up at the Talèfre[6] and down to the end of the glacier des Bois.[7] The whole thing is a puzzle. I have found out the dirt bands described by Forbes.[8] He has represented them very fairly on his map, and they are different from any which I have hitherto seen. There is a kind of hotel here with three bed rooms, one of which is occupied by my friend.[9] To escape the tramping of the travellers I have slung my hammock in a little temple erected many years ago by a Mr Blaire:[10] the walls are wet, the stone floor is wet and I have a goat-skin to keep my feet from the flags.[11] When I sit there at night watching the stars, listening to the moaning of the wind over the glacier, or to the incessant rumble of the stones as they fall into the crevasses, with my pine fire gleaming upon me and my octagonal castle, I feel quite like a philosopher. I am strong, and so will you be, when you come out here. Let the thought of this urge you forward. Tell Mrs Huxley[12] that I will watch over you as she does over young Noel.[13] Give my love to her, nothing less remember and believe me

Always yours | J. Tyndall.

IC HP 8:27
RI MS JT/1/TYP/9/2878
Typed Transcript Only

1. *Montanvert*: see letter 1415, n. 15.
2. '*fleuve*': a river that flows into an ocean or sea (French).
3. '*mer*': sea (French).
4. *a friend*: Thomas Hirst
5. *youthful wife*: Anna Hirst, died on 1 July 1857.
6. *Talèfre*: a glacier on the north side of Mont Blanc.
7. *glacier des Bois*: the lower, northern section of the Mer de Glace which during the nineteenth century reached near the hamlet les Bois.
8. *dirt bands described by Forbes*: Forbes discusses these glacial dirt bands throughout his *Travels through the Alps of Savoy* (see letter 1306, n. 8).
9. *my friend*: Thomas Hirst.
10. *Mr Blaire*: see letter 1415, n. 16.
11. *flags*: a piece cut out of or pared off the sward; a turf, sod (*OED*).
12. *Mrs Huxley*: Henrietta Huxley.
13. *young Noel*: Noel Huxley (1856–60), Thomas Huxley's first son, who died at almost four years old on 14 September 1860 from scarlet fever, was born on 31 December 1856. See A. Desmond, *Huxley: From Devil's Disciple to Evolution's High Priest* (Reading, MA: Perseus Books, 1994), pp. 286–88.

From Michael Faraday   28 July 1857   1417

Royal Institution | 28 Jul 1857

My dear Tyndall,

I received your very welcome letter[1] just on our return from Derbyshire,[2] and thank you heartily for its news. So far you seem to have gone on well, i.e. increasing in health and spirits. I trust it will be so, and that your pursuit of nature and her truth will be such a labour of love as to bring health to both body and mind. So you are at Chamouni,[3] and on the ice, and likely to meet Forbes there. I am not sorry for it. I have trust enough in you as to believe that two such men as yourself, though you may have differed on some conclusions, cannot meet in the face of nature, with like love of truth, without advantage. I shall hope the best. You both have a very important witness and mistress to refer to in nature, and I know she will favour the most right minded.

Since you were at Paris, I have had a very pleasant letter from Biot. It would give you a just pleasure to know the terms in which he speaks of you. What a fine old man he is! I am very glad that you saw him, and that he has seen you.

We have no news just now, except about drains and painting and such like matters. The house is in such a state that I cannot work, and if it were in order, I am too weary. Perhaps I ought to say, too lazy. All I know is that I feel tired in creeping up stairs, and find the sofa and a book the best things for me.

We have many foreigners, visitors here. De Vry[4] has shewn himself, and is now off for Java. Soret[5] of Geneva[6] is here. He brought me a letter from De la Rive,[7] speaking of your visit to him. The brothers Schlagintweit[8] from India are here also, as wiry and active as ever; there is no wearing out about them, and they bring me excellent accounts of Humboldt.[9]

You see what poor things I have to tell you about; but great is nature and will prevail. I wish I had a little <u>Schönbein</u> power,[10] and then you should have some letters fit to class as such. But indeed you do not want to be teased with many. All things are well as they are, if we would only be content.

My dear wife[11] and niece[12] desire their kindest remembrances and wishes. You know we are of one mind in these thoughts towards you. I am glad to hear you are able to have your friend[13] with you. I hope it will be some comfort to him, and some pleasure to you. Pray make my sincere respects to him. I have no right to intrude with sympathy, but I do feel deeply for his loss.[14]

Ever, my dear Tyndall | Most truly yours | M. Faraday

RI MS JT/1/TYP/12/4142
Faraday Correspondence, 5:3324
Typed Transcript Only

---

1. *your very welcome letter*: letter 1415.
2. *Derbyshire*: a county in the East Midlands of England.
3. *Chamouni*: see letter 1353, n. 9.
4. *DeVry*: Willem Hendrik de Vriese (1806–62), a Dutch botanist who beginning in October 1857 researched plants in the Dutch East Indies, Borneo, Sumatra, and the Moluccas.
5. *Soret*: Jacques-Louis Soret (1827–90), a Swiss chemist later known for determining the chemical composition of ozone (*HLS*).
6. *Geneva*: see letter 1293, n. 3.

7.   *letter from De la Rive*: on 11 July Tyndall had visited de la Rive in Geneva during his journey to the Alps, having dinner at his country house with his family (Journal, RI MS JT/2/13c/954). See letter 1378 of this volume for de la Rive's invitation.

8.   *brothers Schlagintweit*: Hermann Schlagintweit (1826–82) and Robert Schlagintweit (1833–85), members of a German family of five brothers, all explorers. Adolf Schlagintweit (1829–57) had gone to India with Hermann and Robert in 1854 to study the territory of the East India Company. Adolf remained exploring in Asia when his brothers returned to Europe and he was beheaded in Kashgar when suspected of being a Chinese spy. The two other brothers were Eduard Schlagintweit (1831–66) and Emil Schlagintweit (1835–1904) (*NDB*).

9.   *Humboldt*: Alexander von Humboldt (1769–1859), a world famous Prussian explorer and naturalist, who went on expeditions in Latin America and Russia. His seven-volume *Personal Narrative of Travels to the Equinoctial Regions of the New Continent during the years 1799–1804,* written with Aimé Bonpland (1773–1858) and published in French between 1814 and 1829, was influential to many other explorers and naturalists including Charles Darwin, who read the English translation by Helen Maria Williams (London: Longman, Hurst, Rees, Orme and Brown, 1818–29) (*CDSB*).

10.   <u>*Schönbein*</u> *power*: Faraday's humorous way of saying he wished he had more energy. Christian Schönbein (1799–1868), a German-Swiss chemist, discovered the principle of the fuel cell in 1838 (*ADB*).

11.   *My dear wife*: Sarah Faraday.

12.   *niece*: Jane Barnard.

13.   *your friend*: Thomas Hirst.

14.   *his loss*: Thomas Hirst's wife Anna had passed away on 1 July.

From Juliet Pollock   5 August 1857   1418

St. Julians. Seven Oaks | August 5th. | 1857
Address 59 Montagu Square | <u>To be forwarded</u>

My dear Mr. Tyndal[1]

You will like to know that your letter[2] was of use in turning our minds for awhile from the contemplation of a fresh and bitter grief. In the few weeks that have elapsed since your departure, I have gone through <u>many</u> tears of sorrow. I have lost in Maria Herries,[3] by a sudden death from sore throat at St. Gervais,[4] one of those dearest to me upon this earth. We were brought up together; and to the tenderness that naturally arises from early associations and the habit of companionship, was added a peculiar and intense sympathy of taste, pursuit, and feeling, such as must have given rise to a friendship of an enthusiastic character, even if we had only met in later days by chance. Her mind was of so uncommon an order that to love her in a merely common way was impossible.

But it is not speak of our affliction that I write, but to tell you that your letter opened a source of interest to me when it was much needed; and it is strange how the opening passages harmonized with the tune of my own feelings, preparing me the better to follow you through the rest. It seemed to me, that I saw with an extraordinary distinctness your wandering in the regions of ice, your singular, solitary place of shelter, and then that the sun appeared to me in its rising among the Alps in its full, peculiar beauty. It seemed that these influences of nature flowed into my mind with a particular force, which may be due either to the vividness of your descriptions, or to the state of excitement of my own feelings.

It was my unhappy office to bring the fatal tidings here to the Aunt,[5] who as a mother brought us all up, and it was necessary for me to exert all my strength both for her sake, and for that of my baby. A deviation from the track of sad thought is of great assistance to such an effort, and so I am indebted to your letter almost as to a fresh breeze among the woods or a stroll upon a mountain top.

Coleridge's Sunrise[6] is, I believe, translated from the German, but I don't know it well, and I will look at it when I return again to my home. I think, myself, that many of his aspiring lines are inflated, and that it is in his slighter, more unambitious pieces, that tenderness and melody are to be found. —

I feel an inward satisfaction that you have atchieved[7] the granite left half told by Forbes,[8] and the satisfaction will be complete, when your ice theory[9] is universally acknowledged as the only thoroughly comfortable one. —

I received a note from you[10] before your start, to which I was unable then to reply: but let me now assure you, that 'clumsiness' and 'ungraciousness' were only in your own imagination that night, and that both my husband[11] and myself were aware only of your kindness. You must scratch out that little sentence from your diary. It is not worthy of being there. —Goodbye. My spirits, whatever effort I make over them, are not such at the present time, as to make me a welcome correspondent, and I have only written because I thought it would please you to know that your conference with me across the Alps had helped to soothe a very solemn grief. —My three boys[12] are with me here: and the two eldest desire their love to you—

My husband is obliged to remain in London till the end of the week.

Yours most truly—Juliet Pollock

RI MS JT/1/P/165

---

1. *Tyndal*: Juliet Pollock also misspelled Tyndall's name this way in letter 1439, as did William Frederick Pollock in letter 1474. She also misspelled it as 'Tindal' in letter 1343.

2. *your letter*: letter missing.

3. *Maria Herries*: Maria Julia Herries (*c.* 1820–57), cousin of Juliet Pollock.

4. *S^t. Gervais*: a commune in the Rhône-Alpes region of southeastern France.
5. *the Aunt*: Isabella Maria Herries. See letters 1420 and 1550.
6. *Coleridge's Sunrise*: the English poet Samuel Taylor Coleridge's (1772–1834) 1802 poem 'Hymn Before Sunrise', in his *The Friend: A Series of Essays* (London: Gale and Curtis, 1812), pp. 174–76. Tyndall had written this poem into his journal on 10 October 1848 (Journal, RI MS JT/2/13b/389–91).
7. *atchieved*: an archaic spelling of 'achieved', still in use in the nineteenth century (*OED*).
8. *granite half told by Forbes*: see letter 1306, n. 8.
9. *your ice theory*: see letter 1306, n. 10.
10. *a note from you*: letter 1409.
11. *my husband*: William Frederick Pollock.
12. *My three boys*: Frederick, Walter, and Maurice Pollock.

To Michael Faraday　　　　　15 [August]¹ 1857　　　　1419

Chamouni | 15 Jul 1857

My dear Mr. Faraday,

I felt very thankful to you for your letter,² which quite connected me with the life of London by making me acquainted with what was going on there. I am at present seated in the house of my guide³ at Chamouni,⁴ with Huxley seated at the opposite end of the table at which I write. Last Sunday and Monday the rain descended heavily at the Montanvert,⁵ and snow fell upon the higher elevations. Yesterday (Friday) clouds collected around the summit of Mont Blanc,⁶ and thunder was the harbinger of heavy rain, which has continued incessantly all this day. I have thus noticed two rainy termini, between which however a sunny interval stretched, a fine clear day and a fine moonlight night, and in this interval it was my fortune, or doom, to make an attempt upon the 'Monarch of Mountains',⁷ wishing to learn something from it if possible. Accompanied by a single guide,⁸ the best part of whose valour was discretion, my friend Hirst and myself set out from the Grand Mulets⁹ at 2 o/c. on Thursday morning, and after 14 hours toil reached the summit of Mont Blanc, made some experiments upon the snow, fired a little extempore cannon, and pledged a toast to some of our absent friends.¹⁰ I thought you would not object to your name being linked with that of Saussure, so I pledged you both together, for to say the truth, my quantity of beverage was too scanty to afford a separate bumper to each. I had also the pleasure of carving Mrs. Faraday's name and those of two or three other ladies upon the atmosphere at that high elevation. Our return to the Grand Mulets occupied upwards of three hours, making the excursion, from this point to the summit and back to the Mulets again, 17 hours in all. It is usually accomplished in 10. But the snow presented a fearful obstacle to our progress; we had to wade through it for hours knee deep, and sometimes actually up to the hips. The toil was excessive, and we reached the Grand Mulets just in time to prevent the peril of darkness from being added to the peril of the crevasses which beset us. Standing upon a rock in the midst of the ice and snow on Wednesday night, I noticed a singular effect of star twinkling which you have probably observed, but which was quite new to me. Supposing this to represent the Great Bear,[11] a line drawn through the two stars a,b, and produced, cut a

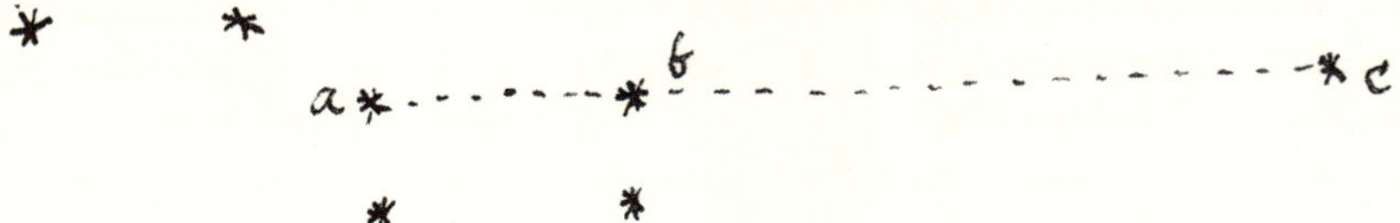

bright star at c. This star I noticed changed its colour incessantly, sometimes it was a bright red and the next instant it changed to a vivid green. In some cases the succession from one colour to the other was very rapid, but in others the colour remained constant for several seconds. I called Huxley's attention to it, and he saw the green and red at exactly the same times as I did. I had also the corroborative testimony of my friend Hirst, who was with me. When Hirst and I were wandering up the ice slopes by moonlight a few hours afterwards, he noticed that the self same star, which had approached the zenith,[12] had ceased to twinkle. Probably all this contains nothing new to you, but the vividness of the colours was to me extremely remarkable. I suppose everybody must have observed the influence of the position of a star upon its twinkling. In the squares and parks of London I have often noticed that stars near the horizon twinkled powerfully, while those in the zenith shone with an almost steady light; I think you once mentioned to me that you had observed this yourself.

Since I wrote to you last,[13] I have been hard at work on the glacier; but the problem is a large one, and it is only now that I am commencing to feel some mastery over the particular form under which it presents itself here. It is always so. It requires long pondering and experiment before I can seize anything with clearness, and I am often disposed to consider myself a very slow coach. But I suppose I must be content with the conditions under which Nature grants me insight, and not repine[14] because I do not possess the quick flashing intellect of other men. Since I last wrote to you I also made the ascent of the Col du Géant,[15] with no guide, but with a little boy,[16] who can climb well, for my companion. Neither of us knew anything of the route; and we found ourselves during part of the time surrounded by considerable perils. We succeeded however in reaching the summit, and in returning to the Montanvert without a single broken bone. The pass is, I believe, considered to be one of the most perilous in the Alps. In this way, by refusing to conform to the rules of the guides, and claiming from the Guide Chef[17] the liberty which ought to be granted to a scientific observer, I have been able to accomplish very heavy excursions for about the tenth part of the ordinary expense.

But my work here is drawing to a close simply through the exhaustion of my funds, so I hope to be in London in about 10 or 12 days from the present time. You would oblige me by asking Anderson not to forward me any letters after the date on which this reaches you. I hope to find Mrs. Faraday and Miss Barnard and yourself well and happy on my return. I am in capital health: burnt brown as an Indian. My guide and friend are both almost blind from ophthalmia[18] produced by the glare of the snow, but this I have escaped, and should have been hard at work today if the weather had permitted it.

With best wishes | Believe me always | Most sincerely yours | J. Tyndall

RI MS JT/1/TYP/12/4070–73
Faraday Correspondence, 5:3327
Typed Transcript Only

---

1. *[August]*: the month is given by relation to letter 1417.

2. *your letter*: letter 1417.

3. *my guide*: Edouard Simond.

4. *Chamouni*: see letter 1353, n. 9.

5. *Montanvert*: see letter 1415, n. 15.

6. *Mont Blanc*: see letter 1320, n. 6.

7. *'Monarch of Mountains'*: Mont Blanc, see letter 1320, n. 6.

8. *a single guide*: Edouard Simond. Simond had requested bringing another guide for this ascent (Auguste Balmat), but they failed to agree on a fee (Journal, 11 August 1857, RI MS JT/2/13c/1004).

9. *Grand Mulets*: a rock island on the north side of Mont Blanc.

10. *our absent friends*: from letter 1420, one of these friends was Juliet Pollock, and from letter 1445, another was Mary Coxe. The others they toasted, as Tyndall listed in his journal entry of 12 August 1857, were de Saussure, Faraday, Henrietta Huxley, Maria Drummond, and Sarah Faraday (Journal, RI MS JT/2/13c/1017).

11. *Great Bear*: Ursa Major, a constellation of the northern hemisphere, also known as the Big Dipper.

12. *zenith*: see letter 1415, n. 14.

13. *wrote to you last*: letter 1415.

14. *repine*: to feel or express discontent or dissatisfaction (*OED*).

15. *Col du Géant*: the high glacier pass between Chamonix and Courmayeur. This climb was famously accomplished by de Saussure and Forbes, which is perhaps why Tyndall desired to do it.

16. *a little boy*: Edouard Balmat.

17. *Guide Chef*: Michel Bossoney.

18. *ophthalmia*: inflammation of the eye (*OED*).

To Juliet Pollock   15 August 1857   1420

Chamounix Savoy 15th Aug. 1857.

My dear M[rs]. Pollock

Your letter[1] reached me half an hour before commencing an expedition to the summit of Mont Blanc,[2] and the feelings which it excited placed me, for a time, quite outside the little world of bustle and commotion which is the usual preliminary to such an undertaking. As my two friends[3] and myself climbed the heights of Charmoz[4] and clambered along an unfrequented and trying route towards the Grand Mulets,[5] we were for a long time silent, each occupied with his own thoughts—mine were occupied with your letter. I think I must have seen Miss Hirries.[6] I had heard of her extraordinary endowments from M[r]. Barlow,[7] and afterwards I think I found myself seated at her side at his dinner table. We talked of many things, and her conversation so much reminded me of you that I think it must be to her that your letter refers. We made some little experiments upstairs after dinner; and to my mind now the expression of her intellectual countenance is vividly present. I think I can appreciate your grief and the grief of your aunt,[8] for even the reflection of it which reached me lent a solemn sadness to the whole journey which I have just concluded. It is instructive to observe the influence of the emotions upon the physical condition of man. During part of my journey the contemplation of your loss made me forget all physical toil, and I climbed ice & granite almost unconscious of the effort required to do so. This is perhaps one of the high uses which it is possible to draw from sorrow—It may overwhelm, but it can also strengthen and purify our souls, and give us a fortitude which happiness cannot bestow. But I speak of my own experience—of the experience of a man whose business it is to develop strength of character and to make even affliction subservient to this end. Still I doubt not, that your practical wisdom, and the contemplation & discharge of the high duties which it falls to your lot to perform in the world, will save you from yielding too much to the pressure of this great calamity. Would that I could lessen it, even by bearing a portion of it myself.

I will tell you at some future day all about my attempt upon the 'Monarch of Mountains'.[9] At present I will only say that my friend[10] and I stood beside each other on the summit, and among others pledged the health of you and yours in the scant bumper which remained to us after our ascent. It cost us seventeen hours of incessant toil to ascend from the Grand Mulets, and to come back there again. This journey is usually accomplished in ten hours; but a heavy fall of snow had taken place three days previous to our ascent, and we had to wade through this usually up to the knees, and sometimes to the very hips. It was fearful labour—we reached the Grand Mulets at night-fall, just in time to save us from the superadded[11] peril of darkness amid the crevasses. We had a single guide[12] along with us.

Since I wrote to you last[13] I have been hard at work upon the Mer de Glace,[14] and its confluent tributaries. The problem is a very large one and unfolds itself by slow degrees. I sometimes attribute this slowness to myself; it requires, in my case, long pondering before I can get a clear glance into the heart of any question at which I am engaged. This period of preliminary doubt is one of dissatisfaction; and perhaps it is the necessity of crossing this chasm which deters many from the prosecution of original work. But it is equally true in science as in poetry that 'Nature never did betray the heart that loved her',[15] and though the precise object at which the investigator aims may not be attained—may indeed be unattainable, still if he be only faithful to his task he is sure to be rewarded according to the method of Nature herself. I was very anxious to examine the state of the insipient ice, and the character of the crevasses upon the Col du Géant[16] and at other considerable elevations, but found that if I conformed to the rules of the Guides, paying five pounds for one excursion, and twenty five for another, that science would come in for but a poor fraction of my funds. I therefore resolved to attempt the ascent of the Col with the little boy alluded to in my last letter.[17] Some preliminary expeditions had given each a confidence in the other, but neither of us knew any thing of the route; indeed we had learned with some satisfaction that there was no route, as the character of the glacier, in its most dangerous and precipitous part, was perpetually changing. In order to be able to commence the expedition early I and my little demon, as I called him, lay together under a rock the night previous. It would hardly be possible for me to give you an idea of the nature of our journey. I have described things in my journal just as they occurred, but my journal is at the Montanvert,[18] and I am at Chamouni;[19] and even if it were here the description would be too long to inflict upon you. The pass I believe is considered one of the most perilous in the Alps, and the orthodox allowance is two practised guides to each <u>voyageur</u>. The two things to be dreaded are the crevasses and ice avalanches—We were within 20 yards of the track of one of the latter as it thundered past us—I fell into one crevasse but escaped without an injured bone. The whole journey indeed proved a strong discipline in caution and fortitude, and both the 'demon' and myself, after we had escaped the seracs[20] and found ourselves amid the ordinary chasms of the glacier, expressed with considerable warmth our 'content" with what we had accomplished. We reached the summit of the Col in two hours less than the time usually accorded to travellers.

I hope you will excuse this misty letter. I hope you will also overlook its dirt and blotches. I am a little muddled today, a kind of half drunkenness being a residual phenomenon to the ascent of Mont Blanc. I am resting myself in the house of my guide, and both he and my friend are almost blind from ophthalmia.[21] I have escaped this, though the glare was at times overpowering. It is pouring rain: had we been a day later clouds and thunder would have caught us on the summit. But kindly nature gave us a glorious, moonlight night and a sunny day to accomplish our journey. My work here is drawing to a close, or at least the time that I can devote to it is nearly finished. So that even if you were disposed to grant me the favour of another letter it would not find me here. I hope to be in London in eight or ten days from the present time—give my love to the boys[22] and say to the eldest[23] that he has forgotten his promise to write to me. With kindest remembrances to M[r] Pollock[24]—

Believe me always | most sincerely yours | <u>John Tyndall</u>

BL 63902–874E–10–51

---

1. *Your letter*: letter 1418.

2. *Mont Blanc*: see letter 1320, n. 6.

3. *my two friends*: Thomas Hirst and Thomas Huxley, though Huxley did not complete the ascent of Mont Blanc with Tyndall, Hirst, and their guide, Edouard Simond. Huxley went up to the Hut, or Cabane, but stopped there, where after having to jump over a crevasse, decided to give up. He emptied his flask of brandy, and said: 'Tyndall I am quite exhausted … I have determined not to attempt the ascent, I thought as I sat beside that crevasse that it was hardly fair to those at home to incur such peril, and I then and there made up my mind not to persist' (Journal, 12 August 1857, RI MS JT/2/13c/1008).

4. *Charmoz*: a mountain to the east of Mont Blanc, south of Chamonix.

5. *Grand Mulets*: see letter 1419, n. 9.

6. *Hirries*: Maria Julia Herries (*c.* 1820–57), a cousin of Juliet Pollock who had recently died. See letter 1418.

7. *M[r]. Barlow*: John Barlow.

8. *aunt*: Isabella Maria Herries. See letters 1418 and 1550.

9. *'Monarch of Mountains'*: Mont Blanc, see letter 1320, n. 6.

10. *my friend*: Thomas Hirst.

11. *superadded*: to add to what has already been added (*OED*).

12. *guide*: Edouard Simond.

13. *wrote to you last*: letter 1409.

14. *Mer de Glace*: see letter 1306, n. 12.

15. *'Nature never did betray the heart that loved her'*: from William Wordsworth's 1798 poem, 'Lines Written a Few Miles Above Tintern Abbey, on revisiting the Banks of the Wye During a Tour', in Wordsworth's *Lyrical Ballads* (London: J. & A. Arch, 1798): 'And this prayer I make, | Knowing that Nature never did betray | The heart that loved her'. Samuel Taylor Coleridge also contributed poems to this book.

16. *Col du Géant*: see letter 1419, n. 15.

17. *the little boy alluded to in my last letter*: Edouard Balmat. See letter 1419.

18. *Montanvert*: see letter 1415, n. 15.

19. *Chamouni*: see letter 1353, n. 9.

20. *seracs*: a tower of ice on a glacier formed at a crevasse (*OED*).

21. *ophthalmia*: inflammation of the eye (*OED*).

22. *the boys*: Frederick, Walter, and Maurice Pollock.

23. *the eldest*: Frederick Pollock.

24. *M[r] Pollock*: William Frederick Pollock.

From George Biddell Airy   15 August 1857   1421

> Royal Observatory, Greenwich. | London, S.E.
> | 1857 August 15

My dear Sir,

On returning from an absence of some weeks, I find a copy of your charming paper[1] on the musical sounds of tubes produced by gas flames. The exhibition which you have given to me had enabled me to seize at once on all that I had not actually seen, and I was much delighted with it.

You are so completely master in every thing that relates to interference of undulations that I very much wish I could enlist you to thoroughly study the geometrical and algebraical theory of the phenomena of Depolarization. The most important papers on this subject are some by myself, namely in the Cambridge Transactions 'On the two rays in quartz'[2] and 'On a new Analyzer'[3] and the tract on the Undulatory Theory[4] in my 'Mathematical Tracts.' Our physicists in general and our optical experimenters in particular (always excepting Stokes, the prince of mathematicians) have been such wretched mathematicians that these subjects are sealed to them: I wish greatly that <u>you would enter into</u> them.[5]

The instance of the Atlantic cable[6] and other things make me think that I must begin to look <u>practically</u> to the Galvanic Induction.[7] Is there any philosophical-instrument-maker in London whom you could recommend as a competent and trustworthy maker of induction-apparatus?

I am, my dear Sir, | Yours very truly | G. B. Airy.

Dr. Tyndall | &c. &c. &c.

RGO MS.RGO 6/378.515
RGO MS.RGO 6/471.173
RI MS JT/1/TYP/1/22

1. *your charming paper*: J. Tyndall, 'On the Sounds produced by the Combustion of Gases in Tubes', *Phil. Trans.*, 13 (1857), pp. 476–79.

2. *'On the two rays in quartz'*: G. B. Airy, 'On the Nature of the Light in the two Rays produced by the Double Refraction of Quartz', *Transactions of the Cambridge Philosophical Society*, 4 (1833), pp. 79–123, 199–208.

3. *'On a new Analyzer'*: G. B. Airy, 'On a new Analyzer, and its Use in Experiments of Polarization', *Transactions of the Cambridge Philosophical Society*, 4 (1833), pp. 313–322.

4. *tract on the Undulatory Theory*: G. B. Airy, *Mathematical Tracts on the Lunar and Planetary Theories, the Figure of the Earth, Precession and Nutation, the Calculus of Variations, and the Undulatory Theory of Optics* (Cambridge: J. Smith, 1831), pp. 249–409.

5. *prince of mathematicians . . . <u>enter into</u> them*: this portion of the sentence is missing from the manuscript letter (the last page of which is a handwritten copy) and is transcribed from the typescript.

6. *instance of the Atlantic cable*: the laying of the first transatlantic telegraph cable on 5 August 1857 was unsuccessful, and a lasting cable connection did not occur until the mid-1860s.

7. *Galvanic Induction*: see letter 1346, n. 1.

To George Biddell Airy   26 August 1857   1422

Royal Institution of Great Britain [stamped] | 26th Aug. 1857

My dear Sir

Your letter of the 15[th][1] reached me at the Montanvert[2] near Chamouni[3] two days before I quitted the place to venture to England. It gratified me much to find that you were pleased with the little paper on the combustion of gases in tubes.[4] With regard to the other part to which you refer—the study of your optical memoirs[5]—it is a matter which I have long had in prospect. Hitherto I have been compelled to content myself with a general and I may say superficial acquaintance with them, as magnetism has taken up so much of my attention. But I trust the day is not far distant when I shall be able to commence the calm study of the subject:—

Of all events I should never think of seriously experimenting upon optics without first making acquainted with all that you had written upon the point.

With regard to the induction apparatus I think Ladd of Chancery Lane[6] is a thoroughly competent maker. & I have no doubt that he would be able to execute to your satisfaction any orders you might give him.

For the last 7 weeks I have been residing at an altitude of nearly 7000 feet above the sea. I climbed to the summit of Mont Blanc[7] and discharged an extempore cannon there—I have carried two similar ones with me to London & purpose discharging them in Hyde Park,[8] so as to test the observations of Saussure as to the enfeeblement of sound on the summit of the mountain.

Believe me dear sir | very truly yours | John Tyndall

RGO MS.RGO 6/378.519–20

1.   *Your letter of the 15ᵗʰ*: letter 1421.
2.   *Montanvert*: see letter 1415, n. 15.
3.   *Chamouni*: see letter 1353, n. 9.
4.   *the little paper on the combustion of gases in tubes*: see letter 1421, n. 1.
5.   *the study of your optical memoirs*: see letter 1421, n. 4.
6.   *Ladd of Chancery Lane*: William Ladd (1815–85), an English optician and maker of scientific instruments with a London shop at 31 Chancery Lane from 1857–60. See G. Clifton, *Directory of British Scientific Instrument Makers 1550–1851* (London: Zwemmer, 1995), p. 161.
7.   *Mont Blanc*: see letter 1320, n. 6.
8.   *Hyde Park*: a park in Central London, the largest of four Royal parks.

From Juliet Pollock   26 August 1857   1423

2 Dorset Place. Tunbridge Wells. | Wednesday. 26ᵗʰ. August
| 1857

My dear Mʳ. Tyndall

The object of my present note is to let you know that we are at Tunbridge Wells[1] and that we have a room very much at your disposal. If you liked to come from the <u>Saturday morning to the Monday night</u>, a return ticket would serve your purpose but we should of course be <u>very glad</u> if you could stay longer with us. You will get this on Thursday, tomorrow, and you can let us know by Friday whether you can come next Saturday.

I was very sensible of the kindness of your last letter,[2] and I feel it to be true that it is as much the duty of a woman as of a man to combat with despondency. I have done so, not without success, and I am able to show by my daily cheerfulness that I value those high blessings that are left to me; yet there is a shadow from the past cast over my onward path that must be an abiding one.

This is a very pretty, rural country, and may be pleasant to you as a contrast with the magnificence you have left—it must be as a contrast, for your imagination can hardly be equal to that of Walter,[3] who when he had climbed one of the little rocks with which our common here abounds, shouted out that he believed himself at the top of Mont Blanc![4] After all, when I say that your imagination cannot equal, my true meaning is that your judgment is more informed. It will be a great pleasure both to my husband[5] and me to walk and talk with you here and to listen to all your adventures <u>demonical</u> and otherwise.

I am about now to set off on a drive with my boys.[6] Walter has begun latin.

Yours always, most truly | Juliet Pollock

RI MS JT/1/P/166
RI MS JT/1/TYP/6/1925

1.　*Tunbridge Wells*: a resort town in Kent, England to the southeast of London, where the Pollocks had another residence at 2 Dorset Place.
2.　*your last letter*: letter 1420.
3.　*Walter*: Walter Pollock.
4.　*Mont Blanc*: see letter 1320, n. 6.
5.　*my husband*: William Frederick Pollock.
6.　*my boys*: Frederick, Walter, and Maurice Pollock.

To Juliet Pollock　　　28 August 1857　　　1424

Queenwood near Stockbridge | Hants
| Friday 28th Aug. 1857

My dear Mrs. Pollock

You will have expected to hear from me this morning, and it is not my fault that this has not been the case. In fact your letter[1] did not reach me until today. I arrived in London on Tuesday night and found the next day to be at a boiling temperature. I had a great quantity of writing to get through and thought that I could accomplish this more pleasantly in the country than in London, so I started on Wednesday afternoon and reached the fringe of the Hampshire Downs[2] that same evening. Were I in town it would give me true pleasure to join you and Mr Pollock,[3] and if you remain at Tunbridge Wells[4] any time, and feel no recession of that wave of kindness which prompted you to ask me down, I shall be very glad indeed to spend a Sunday with you. I hope to get up to London towards the end of next week.

It is the characteristic of a great mind to magnify the small and to micrify the vast, and I learn with much interest that Walter[5] possesses the former faculty in such a remarkable degree. It would give me great delight to join the boy in some of his excursions—to share his perils, and if I might be permitted to hope, his victories also. I trust he has got a suitable brandy flask, for this to me would be an indispensable accompaniment. An Alpen stock,[6] with an iron spike at the end, would also be serviceable: but he will best know what is needful for the arduous undertakings in which he is so habitually engaged, and I shall be very content to be advised by him, should

I ever have the pleasure of joining him in one of his excursions. I wish very much he would write to me and say whether it would be necessary to have my shoes well nailed, as I unfortunately left those which bore me through the Alps at Chamouni.[7] With love to him and his brother,[8] and kindest regards to M^rs. Pollock[9]

Believe me | most sincerely yours | <u>John Tyndall</u>

RI MS JT/1/T/1129
RI MS JT/1/TYP/6/1914

---

1. *your letter*: letter 1423.
2. *Hampshire Downs*: region west of London in the county of Hampshire including the town of Stockbridge, near where Tyndall had taught at Queenwood College (see letter 1298, n. 13).
3. *Mr Pollock*: William Frederick Pollock.
4. *Tunbridge Wells*: see letter 1423, n. 1.
5. *Walter*: Walter Pollock.
6. *Alpen stock*: a long iron-tipped staff used in hill and mountain climbing (*OED*).
7. *Chamouni*: see letter 1353, n. 9.
8. *his brother*: Frederick Pollock.
9. *M^rs. Pollock*: Tyndall likely meant to write 'M^r', as this letter is to Juliet herself.

From Thomas Archer Hirst   30 August 1857   1425

26 Rue de Lacépède | Paris. Aug^st 30th / 57

Just a word my dear John before I go out for my walk to tell you that on leaving you[1] and proceeding to the Rue de Lacépède[2] I was fortunate enough to meet with a pleasant little room on the second story which I have every reason to think will suit me for the winter. I have been at work pretty closely since my return completing my journal and so forth, yesterday only I began breaking the way towards my mathematics. To day I must begin to put my good resolutions for maintaining health into practice. I must go somewhere but where I know not, there is the mischief. How are you, and where are you? Has Huxley returned?[3] John Martin starts for Ireland in a day or two and will be the bearer of this and other letters. I found I had forgotten to give you the key of the Theodolite[4] box so I enclose it. Good bye dear John may your good health long remain with you and may I always remain

Yours affectionately | <u>T A Hirst</u>

RI MS JT/1/H/234

1. *on leaving you*: Hirst had joined Tyndall in the Alps since 13 July and they departed each other's company on 25 August in Paris (Journal, 25 August 1857, RI MS JT/2/13c/1031).
2. *Rue de Lacépède*: a street in the Jardin des Plantes in Paris.
3. *Has Huxley returned?*: Thomas Huxley had joined Hirst and Tyndall in the Alps from 10–22 August to summit Mont Blanc, though he did make the summit (Journal, RI MS JT/2/13c/1002, 1030). Hirst and Tyndall left Chamonix on 22 August and Huxley remained to explore more in the Alps with their guide Simond (see letter 1426). He returned to London by 3 September (see *Life and Letters of Thomas Henry Huxley,* ed. L. Huxley, 2 vols (London: Macmillan and Co., 1900), vol. 1, p. 146).
4. *Theodolite*: see letter 1353, n. 11.

From Henrietta Huxley   30 August 1857   1426

115 Esplanade, Deal. | Sunday August 30th 1857.

Dear Dr Tyndall,

In truth you are a good considerate creature to write and give me news,[1] and good news of Hal.[2] I am very sorry that you and Mr Hirst were compelled to leave, especially too, owing to a financial crisis.[3] The guide who offered to lend you twenty pounds was a marvel, but then like every-one else he saw through you at once, and as you remark knew you would pay him. Don't vaunt of carrying your heart in your face, but seek to keep it in its proper concealment. Besides honesty after all is the commonest of virtues and so interested that even at the earliest age we are taught that 'Honesty is the best policy'

After that I will bestow a little praise on you for your kind care and thought of my dear husband, who told me all about it. He must have sadly regretted staying behind at the Grands Mulets,[4] and I appreciate his thoughtfulness for 'those at home' the more because I know how his wishes urged him to proceed. How anxious he must have been at your prolonged absence. To have gained the summit of Mont Blanc,[5] in the face of such hindrances as you had, was triumphant and I think you must have enjoyed it the more for the difficulty. See what you gain by being a bachelor. No vision of a wife to restrain you from glory—or a snowy burial.

This morning I had a letter from Hal from Courmayeur[6] dated last Sunday. He and old Simond[7] were enjoying their tour, the latter I fear must be a usurer,[8] for he wants sadly to lend the husband 500 francs. They were only 6 hrs 40 min going from Chamounix[9] to Mont Bourraud[10] and from thence to Courmayeur nine, rather a different account from Murray's.[11] I calculated that by Thursday, or Friday at the latest, Hal would be again at Chamounix and consequently might be home by the 1st September.[12] It is a promise that he is to come here and pick us up on returning. I and the boy have been here a fortnight with my father and mother.[13] The sea breezes are delicious and have quite restored Noel[14] to health who was very ill and thin when we left London. If I can I will get Hal to stay a week and rest after his journey. You are wise to stay in the country. London as you say would be too hot for your health just now. You must excuse this bit of crossing and allow me to sign myself

Most sincerely yours | Henrietta Huxley.

RI MS JT/1/TYP/9/2879
LT Typescript Only

1. *to write and give me news*: letter missing.
2. *Hal*: nickname for Thomas Huxley.
3. *I am very sorry that you and Mr Hirst . . . owing to a financial crisis*: Thomas Huxley had joined Hirst and Tyndall in the Alps from 10–22 August to summit Mont Blanc, though he did not make the summit (Journal, RI MS JT/2/13c/1002, 1030). Before departing, Tyndall realized that he and Hirst did not have sufficient funds to pay their guide Simond, who offered to loan Tyndall five hundred francs (Journal, 22 September 1857, RI MS JT/2/13c/1030). Huxley returned to London by 3 September (see *Life and Letters of Thomas Henry Huxley*, ed. L. Huxley, 2 vols (London: Macmillan and Co., 1900), vol. 1, p. 146).
4. *Grands Mulets*: see letter 1419, n. 9.
5. *Mont Blanc*: see letter 1320, n. 6.
6. *Courmayeur*: an Italian town at the foot of the south side of Mont Blanc.
7. *old Simond*: Edouard Simond.
8. *userer*: a money-lender, esp. in later use one who charges an excessive rate of interest (*OED*).
9. *Chamounix*: see letter 1353, n. 9.
10. *Mont Bourraud*: LT must have mistyped Nant Bourant, now Nant Borrant, an alpine hut.
11. *Murray's*: J. Murray, *A Handbook for Travellers in Switzerland, and the Alps of Savoy and Piedmont*, 7th edn (London: John Murray, 1856), pp. 367–71. Murray gives the time to travel from Chamonix to Courmayeur as a three day journey of seven, nine, and eight hours each day, stating: 'A stout walker may accomplish this route in two long days, sleeping at Nant Bourant or the châlets of Mont Joie'.
12. *might be home by the 1st September*: Huxley returned to London by 3 September (see *Life and Letters of Thomas Henry Huxley*, ed. L. Huxley, vol. 1, p. 146).
13. *my father and mother*: Henry and Sarah Heathorn.
14. *Noel*: Noel Huxley (1856–60), Thomas Huxley's first son, who died at almost four years old on 14 September 1860 from scarlet fever, was born on 31 December 1856. See A. Desmond, *Huxley: From Devil's Disciple to Evolution's High Priest* (Reading, MA: Perseus Books, 1994), pp. 286–88.

From Academie der Naturforscher   1 September 1857   1427

Breslau den 1 September 1857.
An | Herrn Professor Dr. John Tyndall | Wohlgeboren | zu | London.

Von Dr. <u>Berthold Seemann</u> und <u>Dr. Joseph Dalton Hooker</u> zur Aufnahme in die Kaiserl. Academie der Naturforscher vorgeschlagen, ersuchen wir Ew. Wohlgeboren, zur Ausfertigung des Diploms, uns baldmöglichst <u>alle</u> <u>Ihre</u> <u>Taufnamen</u>, <u>Doctorgrad</u>, <u>Titel</u>, <u>Aemter</u> und <u>Würden</u>, <u>Ort</u> u. <u>Tag</u> der <u>Geburt</u>, sowie <u>alle Mitgliedschaften</u> von <u>gelehrten Gesellschaften und wissenschaftli-chen Vereinen</u> <u>vollständig</u> und <u>womöglich deutsch</u>, einzusenden und behar-ren hochachtungs- und verehrungsvoll
Ew. Wohlgeboren | Die Academie der Naturforscher.

Breslau,[1] 1 September 1857.
To | Professor Dr John Tyndall | Esquire | at | London.

Proposed by Dr <u>Berthold Seemann</u>[2] and <u>Dr Joseph Dalton Hooker</u> for admission to the Imperial Academy of Natural Scientists,[3] we request You Sir to submit to us as soon as possible, <u>in full</u> and <u>if possible in German</u>, for the drawing up of the certificate, <u>all</u> <u>your given names</u>, <u>doctoral degree</u>, <u>titles</u>, <u>offices</u> <u>and honours</u>, <u>place</u> and <u>day</u> of <u>birth</u>, <u>as well as</u> <u>all memberships</u> of <u>learned societies and scientific associations</u>, and remain faithfully and respectfully,
Sir | The Academy of Natural Scientists.

RI MS JT/1/A/1

---

1. *Breslau*: present-day Wrocław, Poland. During the nineteenth century the city was known as Breslau and was within Silesia, Prussia.

2. *Dr <u>Berthold Seemann</u>*: Berthold Seemann (1825–71), a German botanist. The previous year, Hooker and Seemann had recommended Tyndall for membership in the Academia Caesarea Leopoldino Carolina Naturae Curiosorum, or Deutsche Akademie der Natur-forscher Leopoldina, abbreviated as Leopoldina (see letter 1365). Seemann, himself elected in 1852, received an honorary doctor from the Philosophical Faculty of Gottingen in 1853 and soon after was appointed as adjunct for the Leopoldina (*ADB*).

3. *Imperial Academy of Natural Scientists*: see n. 2.

To Joseph Dalton Hooker   3 September 1857   1428

Queenwood near Stockbridge Hants. | 3rd, Sep, 1857.

My dear Hooker

For the last fortnight almost daily the thought of writing to you has occurred to me. Huxley wrote to you from the Montanvert,[1] and I at that time entertained the righteous determination of sending you a note in hot pursuit of his; for indeed I felt, though I had no official indication of the fact, that I ought to congratulate you on the entrance into this world of a little being[2] whom at some future day I hope to number among my playfellows. I hope Mrs Hooker[3] is quite strong again, and that she will continue strong to tend and support these little sprouts of humanity which are now shooting round her. I spent a glorious time upon the Mer de Glace,[4] glorious chiefly on account of the revolution it wrought in this body of mine, it has pushed me years backward *[1 word illeg]* the grave, and given to me to taste in the autumn of my life the joy and vigour of its spring. When I left Chamouni[5] I was tanned to the hue of a North American Indian. The traces of this still remain, and my muscles have undergone such a thorough renovation that although I have been 8 days in England I feel as yet no diminution of their strength. From 10 to 12 hours each day were spent upon the ice, and I loved the work so well that twilight has sometimes come to puzzle me amid the crevasses. I usually wandered alone, furnished with a hatchet, to which I ought to feel grateful, for it has hewn me out of many difficulties. I have been up the Col de Géant[6] with a little boy[7] as my companion; neither of us knew any thing of the route, and as a natural consequence we had a taste of peril. A difference of two minutes in point of time would have placed us in the midst of an avalanche of ice blocks. In fact we had to recede when we heard them coming, and some of them smote the ice within 20 yards of us. On returning from the summit I fell into a crevasse, but helped myself like a cat and escaped without a broken bone. Huxley[8] told you of our attempt on Mont Blanc.[9] That was the hardest day's work I ever went through: heavy snow had fallen three days before; near the Grands Mulets,[10] this was frozen hard and we congratulated ourselves in the moonlight after we had started that we were likely to perform the journey without much labour or peril. This delusion soon vanished, for we had to walk usually up to the knees and in some cases up to the hips. It was terrible toil, and the last slope of the mountain added an effect which is very rare with me; the beating of the heart. We used to walk forward until the beating reached a certain degree of intensity, then pause, leaning on our batons, the subsidence of the heart action being the signal for further progress. It required but 15 or 20 paces to bring it on again, and the ascent was accomplished by adding lengths of this magnitude together. On returning we had one little accident, which had it occurred elsewhere might have been attended with disagreeable consequences. The footing of my friend[11] gave way as we descended a steep slope. He went down like an avalanche: I was tied to him and tried to check him, but my footing yielded, I fell and went down after him. The shock against him at the bottom was by no means agreeable, and we were both almost quite covered up by the quantity of snow which we brought down. We were upwards of 17 hours in accomplishing the journey from the Grands Mulets and back: this is usually accomplished in nine or ten. Huxley during this time was a prisoner at the Mulets: he expected us at three o'clock; and as he saw the hours of the afternoon pass and the twilight fall without any news of us, he began to entertain gloomy surmises regarding our fate.

With regard to the glaciers I have collected a good many observations: but my power over the thing was only really commencing when I was obliged to leave. In a case of the kind one is a long time learning what it is that is really required, and up to this point the observations and measurements are more or less loose and at random. However I have learned a great deal, and am now endeavouring to throw the observations into an organic shape. I shall soon return to town, and hope shortly afterwards to enjoy, the light of your countenance. Just enclose this[12] to Mrs Hooker and remember me to her most kindly—It may amuse her for five minutes.

Ever Yours | John Tyndall

RI MS JT/1/TYP/8/2549–50
LT Typescript Only

---

1. *Montanvert*: see letter 1415, n. 15.
2. *little being*: Maria Elizabeth Hooker (1857–63), Hooker's fourth child and second daughter, who died at age six.
3. *Mrs Hooker*: Frances Hooker.
4. *Mer de Glace*: see letter 1306, n. 12.
5. *Chamouni*: see letter 1353, n. 9.
6. *Col de Géant*: Col du Géant, see letter 1419, n. 15.
7. *little boy*: Edouard Balmat.
8. *Huxley*: Thomas Huxley.
9. *Mont Blanc*: see letter 1320, n. 6.
10. *Grand Mulets*: see letter 1419, n. 9.
11. *my friend*: Thomas Hirst.
12. *enclose this*: share the letter with Frances Hooker.

From Thomas Henry Huxley   3 September 1857   1429

115 Esplanade, Deal | Sepr. 3rd. 1857.

My dear Tyndall

I don't consider myself returned until next Wednesday, when the establishment at N⁰ 14[1] will re-open on its accustomed scale of magnificence—but I don't mind letting you know I am in the flesh and safe back.

The tour round Mont Blanc[2] was a decided success—in fact I had only to regret you were not with me—The grand glacier of the Allée Blanche,[3] and the view of Mont Blanc from the valley of Aosta[4] were alone worth all the trouble—I had only one wet day & that I spent on the Brenva Glacier[5]—for in spite of all good resolutions to the contrary—I cannot resist poking into the glacier whenever I have a chance—You will be interested in my results which we shall soon I hope talk on together at length.[6]

As I suspected Forbes has made a most egregious blunder[7]—What he speaks of and figures as the 'structure' of the Brenva is nothing but a peculiar arrangement of <u>entirely superficial dirt bands, dependent on the structure but not it</u>.—The true structure is singularly beautiful and well marked in the Brenva, the blue veins being very close set and of course wholly invisible from a distance of a hundred yards—which is less than that of the spot whence Forbes's view of the (supposed) structure is taken

I saw another wonderful thing in La Brenva.[8] About the middle of its length there is a step like this

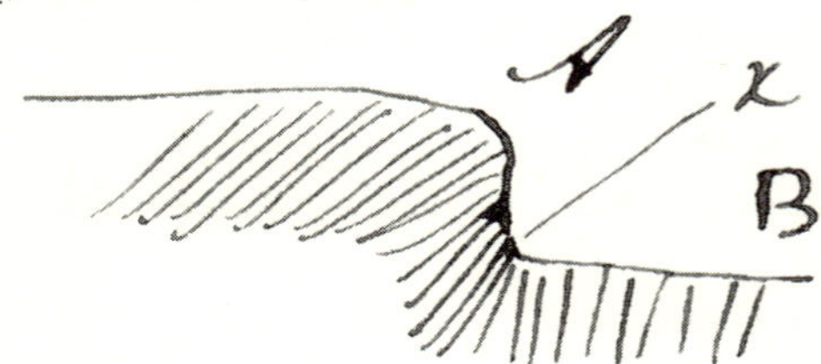

of about 20 or 30 feet in height. In the lower part (B) the structural planes are vertical in the upper A they dip at a considerable angle.—I thought I had found a case of unconformability indicating a slip of one portion of the glacier over another, but when I came to examine the intermediate region (x) carefully, I found the structural planes at every intermediate angle and consequently a perfect transition from the one to the other.

<Hereafter Typescript Only>

I returned by Aosta,[9] the Great St Bernard[10] and the Col de Balme.[11] Old Simond[12] was quite affectionate in his discourse about you, and seemed quite unhappy because you would not borrow his money. He had received your remittance and asked me to tell you so. He was distressed at having forgotten to get a certificate from you, so I said in mine that I was quite sure you were well satisfied with him.

On our journey he displayed his characteristic qualities. 'Je ne sais pas'[13] being the usual answer to any topographical inquiries with a total absence of verve, and a general conviction that distances were very great and that the weather would be bad. However we got on very well, and I was sorry to part with him.

I came home by way of Neuchatel,[14] paying a visit to the Pièrre a Bôt[15] which I have long wished to see. My financial calculations were perfect in theory, but nearly broke down in practice, inasmuch as I was twice obliged to travel first class when I calculated on second. The result was that my personal expenses between Paris and London amounted to 1.50!!, and I arrived at my own house hungry and with a remainder of a few centimes.[16] I should think that your fate must have been similar.

Many thanks for writing to my wife.[17] She sends her kindest remembrances to you.

Ever yours | T. H. H.

RI MS JT/1/H/518
RI MS JT/1/TYP/9/2883–84
IC HP 8:29

---

1. *the establishment at N°̲ 14*: 14 Waverley Place, Huxley's house (which Tyndall and Hirst later rented in 1861, until Tyndall moved out to the RI after Faraday's death).

2. *Mont Blanc*: see letter 1320, n. 6.

3. *Allée Blanche*: a valley on the western side of Mont Blanc.

4. *the valley of Aosta*: a mountainous region in northwest Italy, with France to its west and Switzerland to the north.

5. *Brenva Glacier*: a glacier of the Aosta Valley east of Mont Blanc.

6. *You will be interested . . . together at length*: see letter 1436.

7. *Forbes has made a most egregious blunder*: see J. D. Forbes, *Travels through the Alps of Savoy* (Edinburgh: Adam and Charles Black; London: Longman, Brown, Green, and Longmans, 1843), pp. 200–207.

8. *La Brenva*: see n. 5.

9. *Aosta*: a city in the Italian region of the Alps.

10. *Great St Bernard*: an alpine pass in the Swiss canton of Valais near the the border with Italy, the lowest pass within the fifty miles between the two highest mountains of the Alps, Mont Blanc and Monte Rosa. The pass connects the Swiss canton of Martigny to Aosta in Italy.

11. *Col de Balme*: an alpine pass on the border of France and Switzerland, between the summits of Tête de Balme and Les Grandes Otanes.

12. *Old Simond*: Edouard Simond.

13. *'Je ne sais pas'*: I do not know (French).

14. *Neuchatel*: the capital of the Swiss canton of Neuchâtel.

15. *Pièrre a Bôt*: Pièrre-a-Bôt, a giant granite boulder situated on the Chamount, the hill above Neuchatel. See J. Murray, *A Handbook for Travellers in Switzerland, and the Alps of Savoy and Piedmont,* 7th edn (London: John Murray, 1856), pp. 138.

16. *centimes*: coins worth equal to one-hundredth of a franc (*OED*).

17. *writing to my wife*: letter missing; for Henrietta Huxley's reply to Tyndall, see letter 1426.

From Juliet Pollock   3 September [1857][1]   1430

2 Dorset Place Tunbridge Wells | Sept. 3[d].

My dear M[r]. Tyndall

We hope you may be able to come to us on Saturday week the 12.th inst. It will be a great pleasure to us to see you and I feel that we can make you comfortable here even if the weather should be less fine then we have hitherto had it.

Walter[2] sends you his love and begs to assure you that you <u>'need not have irons on your shoes for the rocks are the easiest things in the world'</u>!

It was amusing to see how the exaggerated view of them suggested by your letter[3] immediately reduced his own imagination on the subject.

My youngest[4] is just passing through the first troublesome event of life—vaccination, and my eldest[5] yesterday underwent that of a more advanced existence in parting with us to return to school. —

I trust that you have no friends in India for it is enough to <u>know</u> of the dark horrors of the scene there, without the additional pangs of anxious friendship. —

It is suggested, and the suggestion seems only too reasonable, that the peculiar atrocities of this war[6] have been committed by the advice of the Hindoo priests with a view to making a reconcilement between their people and ours, a total impossibility and of preserving by this means their own threatened religion and their own influence.

—For ourselves, we have the comfort of knowing my husband's Brother Richard[7] to be in the undisturbed district—the Punjaub[8]—but that security cannot take from us the calamitous sense of the suffering elsewhere. —

I respect the man who had the courage to shoot his wife and his child with his own hand to rescue them from torture.[9] —

Yours most truly | Juliet Pollock

1.   *[1857]*: the year is given by the reference to the Indian Rebellion of 1857.
2.   *Walter*: Walter Pollock.
3.   *your letter*: letter 1424.
4.   *my youngest*: Maurice Pollock.
5.   *my eldest*: Frederick Pollock.
6.   *the peculiar atrocities of this war*: the Indian Rebellion of 1857, when Indian soldiers (*sepoys*) of the British East India Company mutinied against the company's control of India. The rebellion led to the restructuring of Indian rule under the British Empire.
7.   *my husband's Brother Richard*: Frederick Richard Pollock (1827–99), William Frederick Pollock's younger brother and Major-General in the British Army who served in India. *Debrett's Peerage, Baronetage, Knightage, and Companionage, 1903* (London: Dean & Son, 1903), p. 492.
8.   *the Punjaub*: Punjab, a region in northern India (and now Pakistan) that was annexed by the British in 1849.
9.   *the man who had the courage . . . rescue them from torture*: during the rebellion, British men, women, and children were targeted and killed by the mutinous soldiers.

From Michael Faraday   5 September 1857   1431

Highgate | 5 Sep 1857

My dear Tyndall,

I have hesitated in writing to you to acknowledge yours,[1] for I thought you were coming home quickly—at least so I understood Anderson; but I hope you will enjoy the <u>quiet</u> country,[2] for I doubt whether Switzerland could be called <u>quiet</u> to you, occupied as you were. We shall be very glad to see you in due time, and then you and I will talk of the lectures and other things which you refer to. I do not feel as if my hand could write much distinctly, and indeed I doubt whether this will find you in Hampshire[3]; but wherever and ever yours

Very truly | M. Faraday

RI MS JT/1/TYP/12/4143
Faraday Correspondence, 5:3333
Typed Transcript Only

1.   *to acknowledge yours*: possibly letter 1419 of 15 August 1857.
2.   *enjoy the quiet country*: before settling back in London after traveling in the Alps, Tyndall was at Queenwood from 26 August to 9 September, then went to Highgate to see Faraday on 11 September, and to the Pollocks in Tunbridge Wells on 12 September. He returned to London on 14 September (Journal, RI MS JT/2/13c/1032–33).
3.   *Hampshire*: see letter 1424, n. 2.

From Frances Hooker   [6 September 1857][1]   1432

[Sept. 6th 1857]

My dear Dr Tyndall

Papa[2] has asked me to write to you for him (as he is extremely busy just now) to ask you if you will fulfil your promise of coming here next week? His autumnal horticultural show[3] is to come off on Wednesday the 16th and we should all be very glad if you could be present on the occasion. If you can come I think you will like to do so while we are still here. Willy[4] will be enchanted, he is constantly asking about you. You will perceive by my knowing your address that I have seen your letter to Joseph.[5] I think my thanks are due for it as much as his: for I was much interested in reading of your Swiss adventures[6] and very glad to find you have returned home strong and well after your narrow escapes.[7] When we meet I shall like to hear more about your stay in a place to which my thoughts often turn and especially about what you have been doing on the glaciers.

Joseph is at present at Kew[8] but comes down here on the 19th to spend Sunday here, going on to Manchester[9] the following day. On the 22d I hope to take my little ones home to Kew: by that time I trust I shall be quite strong and well and able to walk across Kew Green[10] which I certainly could not do now.

I am sorry to say that Mamma[11] has been so very unwell lately that her medical adviser has ordered her away from home and she is now staying with her brother[12] near Ashford.[13] This week she goes to Tunbridge Wells[14] and shortly afterwards to St Leonards.[15]

Believe me | Ever yours very sincerely | Frances H. Hooker.
Hitcham, Suffolk.

RI MS JT/1/TYP/8/2551
LT Typescript Only

---

1. *[6 September 1857]*: the date is given by relation to letter 1428 and LT's date on the typescript, which she probably took from the postmark.

2. *Papa*: John Henslow.

3. *autumnal horticultural show*: as part of his efforts to expand botanical education, Henslow oversaw an annual horticultural show at the Parish School in the village of Hitcham in county Suffolk, where he served as rector beginning in 1837. See P. Armstrong, *The English Parson-Naturalist: A Companionship Between Science and Religion* (Herefordshire, England: Gracewing Publishing, 2000), pp. 8–9.

4. *Willy*: William Hooker.

5. *your letter to Joseph*: letter 1428.

6.	*your Swiss adventures*: Tyndall spent most of July and August 1857 exploring in the Alps with Thomas Hirst and Thomas Huxley.

7.	*your narrow escapes*: in letter 1428, Tyndall described two moments of 'peril' during his climb of the Col du Géant, as well as a fall during his climb of Mont Blanc.

8.	*Kew*: see letter 1328, n. 8.

9.	*Manchester*: see letter 1311, n. 8.

10.	*Kew Green*: an open lawn just outside of the Royal Botanic Gardens, Kew on the northeast side, where the Hookers lived at 49 Kew Green.

11.	*Mamma*: Harriet Henslow (née Jenyns, 1797–1857), wife of Cambridge botanist John Stevens Henslow (1796–1861) and mother of Frances Hooker. She died on 20 November 1857.

12.	*her brother*: Leonard Blomefield formerly Jenyns (1800–93), an English clergyman and naturalist who was invited to accompany Captain Robert Fitzroy aboard HMS *Beagle*. He declined, and the position was offered to Charles Darwin (*ODNB*).

13.	*Ashford*: a town in Kent, England.

14.	*Tunbridge Wells*: see letter 1423, n. 1.

15.	*St Leonards*: St Leonards-on-Sea, a part of the town of Hastings, East Sussex, England.

## To Juliet Pollock				[8 September 1857][1]				1433

Queenwood Tuesday evg.

My dear M^{rs}. Pollock

Since I came here I have been endeavouring to throw my observations into an organic shape and the work has proceeded so slowly as to prevent me from returning to London as early as I intended. Tomorrow however I shall make my escape—or rather I shall deliberately put on town fetters[2] in exchange for the gentler chain which has tied me to my duties here. On Saturday I hope to have the pleasure of joining you and of inspecting with my own eyes the perils and dangers to which Walter[3] so rashly exposes himself in his excursions. I have no Bradshaw[4]—, otherwise I would name the train; but in the hope that you perpetuate at Tunbridge Wells[5] the duplex tea[6] of Montagu Square,[7] I shall aim to be with you between 6 and 7 oClock.

Most sincerely yours | J. Tyndall

RI MS JT/1/T/1193
RI MS JT/1/TYP/6/1926

1.	*[8 September 1857]*: the date is given based on relation to letter 1430, in which Juliet Pollock invited Tyndall to come to Tunbridge Wells. His journal shows he returned to London from Queenwood on 9 September 1857, and in his entry for 12 September, he wrote, 'To Tunbridge Wells and soon found myself at home with my friend Pollock, his excellent wife and little Walter at 2 Dorset Place' (Journal, RI MS JT/2/13c/1032–33).

2. *fetters*: a chain or shackle for the feet of a human being or animal (*OED*).

3. *Walter*: Walter Pollock.

4. *Bradshaw*: colloquial designation of 'Bradshaw's Railway Guide', a time-table of all railway trains running in Great Britain (*OED*).

5. *Tunbridge Wells*: see letter 1423, n. 1.

6. *duplex tea*: likely a reference to *tea* or *high-tea,* the term for the evening meal often associated with the working class.

7. *Montagu Square*: a square in Marylebone, London, where the Pollocks had their primary residence at no. 59.

From Thomas Archer Hirst   9 September 1857   1434

26 Rue de Lacépède | Sepr 9th 1857

My dear John.

As you are no doubt aware I received this morning a letter from Wright[1] requesting me to accept my old situation at Queenwood.[2] As I told you on our return from Chamouni[3]—foreseeing at the time the possibility of such a request being made—I feel the greatest disinclination to accept the offer. Disinclination indeed is too mild a term my feelings all prompt me to give at once a decided refusal. Nevertheless in every matter of this importance habit prompts me to seek your advice, and if possible your approval. Before giving Wright a decided answer therefore I should like to hear from you.

We were talking not long ago of the immense advantage in having found the work for which we were best fitted, and to which we were most devoted. I would sooner live in a garret[4] on bread and water with my mathematics for companion than live in a palace on the fat of the earth with other less congenial occupations. You know what the situation at Queenwood would be for me, you know the duties it involves and that in fulfilling those duties I can never reap that solid satisfaction and unenvious contentment with which my own pursuits are capable of rewarding me. I work slowly, too, and to do anything I must have time and leisure to abandon myself entirely to my subject; on this account I know from experience that at Queenwood my time would be so split up that I should be able to produce little or nothing of a mathematical kind. I shrink less at the personal inconveniences involved in the life, duties and anxieties at Queenwood than I do from the thought of having in a great measure to set aside my favourite study.

So much for this view of the question which would remain the same if Queenwood were a perfect institution of its kind or if I could come there under the best possible circumstances and have the entire management in my own hands. But there is another view of it. Is it advisable to tie myself to a tottering edifice which a miracle only can save from falling? The whole constitution of Queenwood is rotten to its core and fall it <u>must</u>. To attempt the task of restoring it *[to]* health what obstacles are there not in the way! The principal[5] is a mere zero or rather worse for he has a decidedly negative value. All your attempts at improvement are cramped first by his imbecility and again by all manner of pecuniary embarrassments, by your own undefined position and by the shortsightedness of another individual on the premises.[6]

The last public revelations in the 'Reasoner'[7] will very probably hasten Queenwood's ruin. Knowing this, and with the firm conviction that as long as Geo. Edmondson is at the head of affairs and stands in the same hollow relations to his partners as at present, and lastly feeling how incompatible the duties of the situation are with my own most loved objects of pursuit I must say I cannot persuade myself to entertain the idea of returning.

It is possible you might have something to say on the subject that would shake my resolution. If so tell me[8] and I will give it all the consideration which advice from you deserves in my eyes.

For the sake of Wright and Debus I would sacrifice much, but there are other ways of helping them than by tying myself to the wreck to which they still cling.

Write soon that I may give Wright an early reply.

Yours affectionately | <u>T. A Hirst</u>

P.S. How are you progressing with the Glacier subject? How are you? As for me I am working hard and as yet I enjoy excellent health.

'I desire no better lot on earth than health and strength to pursue with some success the pursuits which long experience has proved are most fitted to my tastes and abilities. My worthy, prudent ancestors have kindly provided me with food and clothing, and I would far rather continue to regulate my wants in accordance with their provisions than seek to increase my income and its attendant comforts by accepting any situation whatever. All things, therefore, conspire to induce me to work on patiently and perseveringly until some other position in life presents itself wherein without too great a sacrifice of pursuits so congenial to me and so adequate for presenting all the employment and self-discipline which a life requires, I can turn my powers more towards providing for the welfare of others.'

Extract from my journal.

RI MS JT/1/H/235

---

1. *Wright*: Richard Wright.
2. *my old situation at Queenwood*: see letter 1298, n. 13.
3. *Chamouni*: see letter 1353, n. 9.
4. *garret*: a turret projecting from the top of a tower or from the parapet of a fortification; a watch-tower (*OED*).

5.   *The principal*: George Edmondson.

6.   *another individual on the premises*: not identified.

7.   *last public revelations in the 'Reasoner'*: a mid-nineteenth century Secularist paper edited by George Holyoake (1817–1906). According to William Brock, 'after his wife's death in May 1857, and following the exposures in Holyoake's *Reasoner* that Queenwood College was still, in fact, owned by the socialists, Richard Wright tried unofficially to persuade Hirst to return and help arrest the declining enrolment' (Brock, 'Queenwood', p. 19).

8.   *If so tell me*: see letter 1435.

To Thomas Archer Hirst   12 September 1857   1435

12th <u>Sept</u>. 1857

My dear Tom.

I have just read your letter[1] to me and sit down at once to reply to it. Our thoughts on the subject were in a great measure similar: I knew that Wright[2] had written to you, but I had little idea of your closing with the offer: nevertheless as some unknown recess or influence might exist in your mind to determine you to accept the offer, I did not discountenance the act of writing to you. For my part my advice is clear and decided, that you ought to follow that line of study with all energy and devotion which your own experience of your constitution proves to be most congenial to you. I know you would never feel at peace in Queenwood[3]—and if you had any thoughts of coming there, I would advise it to be on the clear condition that you should have the management of the place entirely in your hands. If you liked the work, there is I believe room for noble activity at Queenwood, but if you have not made up your mind to follow out the life of an educator, in the sense in which it is understood in such an establishment, then I would say remain clear of the place while you are clear of it. With regard to other positions I have not the slightest doubt that such will present themselves whenever you make up your mind to seek them.

I, for my part, have ever looked forward, and look forward now with more undoubting confidence than ever, to the result of your studies. You are slow, but so is nature when she forms the oak. —It is a slowness which rises in a great measure from the comprehensiveness of your aims, and will in the end produce a result possessing a mass and value corresponding to the time spent in its achievement. It will come out of you someday as a great cannon roar which shall resound through the ages, while others spend their force in a succession of squibs which are soon forgotten. Be true then to your own aspirations, and may God grant you strength of will, and give you grace to preserve that strength of muscle upon which the accomplishment of intellectual ends so much depends.

I had a severe attack of cold at Queenwood, but it is gone and I am now strong. The glacier is a tangle, but I am getting gradually through it and hope soon to have it off hands. A brother of Prof. Thomson's[4] communicated a paper upon the subject[5] to the meeting in Dublin,[6] but I do not think he hits the mark. I saw Huxley the night before last and he talked much about you. I saw Faraday last night and as usual was delighted with the fine old fellow. I will gather some papers for you & send them to you in a few days.

Ever my dear Tom | affectionately yours | J. Tyndall

RI MS JT/1/T/641
RI MS JT/1/HTYP/502

1. *your letter*: letter 1434.
2. *Wright*: Richard Wright.
3. *Queenwood*: see letter 1298, n. 13.
4. *A brother of Prof. Thomson's*: James Thomson, brother of William Thomson.
5. *a paper upon the subject*: J. Thomson, 'On the Plasticity of Ice', in *Report of the Twenty-Seventh Meeting of the British Association for the Advancement of Science*; held at Dublin in August and September 1857 (London: John Murray, 1858), pp. 39–40. See letter 1396, n. 15.
6. *meeting in Dublin*: the twenty-seventh meeting of the BAAS, held in Dublin in August and September 1857.

## From Thomas Henry Huxley		14 September 1857		1436

THE | LONDON, EDINBURGH AND DUBLIN | PHILOSOPHICAL MAGAZINE | AND | JOURNAL OF SCIENCE. | [FOURTH SERIES.] | OCTOBER 1857.

XXIX. *Observations on the Structure of Glacier Ice.* | *By* T. H. HUXLEY, *F.R.S. &c.*
The Government School of Mines, | Jermyn Street, September 14, 1857.

MY DEAR TYNDALL,

IN the following pages I have given you some account of the experiments and observations upon the structure of glacier ice, which, at your suggestion, I commenced during our sojourn at the Montanvert[1] this autumn. No one knows better than yourself how much these subjects row under the hands of the inquirer, and how little claim my brief investigations have to the character of completeness. Nevertheless my conclusions, so far as they go, are based on such clear and decisive evidence, and are so totally opposed to the views entertained by the highest authorities, that I feel I shall be doing more good by publishing than by withholding them.

I will in the first place state what I have myself observed with regard to the structure and the permeability of glacier ice, and afterwards I will compare my results with those which have been arrived at by others.

Structure of Glacier Ice. —A mass of ice freshly extracted from any part of the Mer de Glace,[2] the Glacier du Géant,[3] or the glacier of La Brenva,[4] at a depth of 8 or 10 inches from the surface, always presented the following characters when examined either with the naked eye, or with a lens of a magnifying power of thirty or forty diameters.

It broke with a vitreous fracture; and when the surface was made even, either with a sharp knife, or by rubbing on a warm surface, it appeared perfectly smooth and glassy, exhibiting not the least trace of fissures. Minute shallow pits, however, were scattered over it, and became particularly obvious when a coloured fluid was poured on to the surface and then wiped away again, inasmuch as under these circumstances every pit retained a very small portion of the colour.

The mass was, as usual, traversed by a larger or smaller number of parallel blue veins (whose lenticular form was almost always very apparent, particularly in the Brenva); and when a thin section was made perpendicular to the plane of the veins and viewed by transmitted light, it became obvious that the ice formed one continuous mass, without fissures or interruptions of continuity of any kind. It contained, however, a multitude of small, closed, and perfectly distinct chambers, and it was to the absence or rarity of these in the course of the veins that the latter owed their transparency and blueness.

The form and contents of these chambers were exceedingly remarkable. In a blue vein, and in those parts of the intermediate "white ice" which were contiguous to a blue vein, they were always round or oval disks, with extremely flat and closely approximated sides; so that, viewed in one plane, they looked like circles; but in a plane at right angles to this, like narrow parallelograms. In the white ice midway between the blue veins, on the other hand, I very generally noticed an irregularity of form, which was in many instances so great that the cavities appeared to be ramified. The walls of the chambers very frequently appeared to be a little roughened, or, as it were, frosted.

Every chamber, without exception, which I carefully examined contained both water and air. The former was commonly present in larger quantities than the latter, which swam as a bubble in the water, and could very often be made to move about in the chamber like the bubble of a spirit-level.[5] It seemed to me, though I will not pretend to lay it down as a rule, that the air was more abundant in proportion to the water in the more irregular chambers. Where the air was in large proportion to the water, the bubble of course became more or less completely supported by the walls of the containing cavity, and to a certain extent assumed its form; but where, as in the majority of cases, the air-bubble was small in proportion to the water, its figure was spheroidal, and totally different from that of the containing cavity. I mention this particularly, because, as I shall show below, the chambers (which for distinction's sake I will term the "water-chambers") have been confounded with the air-bubbles, and the form which is characteristic of the one has been erroneously ascribed to the other.

I had no means of measuring the dimensions of the water-chambers, but at a guess I should say they varied from a tenth to a fiftieth or a sixtieth of an inch in diameter.

The line of contact of the water in the water-chambers with the ice was optically perfectly well defined, and easily distinguishable. Hence I have no hesitation in saying, that if canals or fissures of any appreciable size filled with water had existed in the ice, I must, with the magnifying power employed, have discovered some trace of them; but, I repeat, nothing of the kind was discernible in perfectly fresh ice.

If the existence of fluid water dispersed through its substance in closed chambers is shown by future observations to be a universal character of glacier ice (and I cannot imagine that a structure universally prevalent in the Mer de Glace, the Géant, the Brenva, and, as I shall show by-and-bye, from M. Agassiz's[6] figures, in the Aar glacier[7] also, is a mere local peculiarity), it appears to me to be a fact of primary importance. For what I have described is the structure of the unchanged ice of the glacier—of ice which has been protected from solar or atmospheric influences by that which covered it; and it must be remembered that the ice which is within a foot of the surface on the Mer de Glace opposite the Montanvert, must have formed a part of the very depths of the tributary glaciers. In other words, the ice which is at this moment, say a hundred feet below the surface in the Glacier du Géant, will, in consequence of ablation, form the superficial ice of some part of the Mer de Glace years hence. Consequently, unless it can be shown that the substance of a glacier, as it approaches the surface, is exposed to some influences capable of developing the water-chambers and their contents, it is to be presumed that the structure found near the surface in the lower part of a glacier is the structure which prevails throughout the thickness of the higher part; and hence that the structure described is that of unaltered glacier ice in general. This conclusion, as I shall immediately show, is directly confirmed by the boring experiments and by the figures of M. Agassiz.

M. Agassiz's deductions, however, are totally at variance with mine; and he is so generally quoted as an authority in these matters, that I feel compelled, however unwillingly, to enter into a detailed criticism of his views, which are contained in the following extracts from his 'Système Glaciaire',[8] numbered, for the sake of more convenient reference, in successive order.

(1) "At its origin, near the Névé,[9] the compact (or proper glacier ice) contains, like the ice of the Névé, a notable quantity of air. But there is this difference between the two, that in the compact ice the air, instead of being distributed through the whole mass, is united in small perfectly circumscribed bubbles, whilst the interspaces of these bubbles are perfectly transparent, so that without being as diaphanous[10] as ordinary water-ice, the compact ice has not the opacity of Névé ice. Moreover, it is more compact, and what is especially characteristic, it presents no trace of granular structure: a fragment exposed to the action of heat does not become resolved into grains of Névé, but breaks up into angular fragments.

"This difference of structure is accompanied by a greater impermeability; water no longer traverses the mass with the same ease and uniformity, but is seen to follow in preference certain angular routes which are the capillary fissures."—P. 151.

(2) "The means employed by nature to maintain this amount of plasticity and compressibility in glacier ice is the water which circulates throughout the mass, and which, while it lubricates it, contributes to maintain within it a constant temperature during the greater part of the year."—Pp. 152, 153.

(3) *"Superficial fissures which must not be confounded with the capillary fissures.*

"When during a fine summer day one travels over the upper regions of the compact ice (about the region of the Abschwung,[11] on the Aar glacier), a continual crepitation[12] is heard on all sides. It is caused by the bubbles of air which on approaching the surface escape through the ice, where they have been dilated by the effect of diathermanicity,[13] and cause the parietes of the ice to burst when they are no longer sufficiently strong to resist the dilatation of the air."—P. 153.

(4) "The air-bubbles undergo no less curious modifications. In the neighbourhood of the Névé, where they are most numerous, those which one sees at the surface are all spherical or ovoid; but by degrees they begin to be flattened, and near the end of the glacier there are some which are so flat, that they might be taken for fissures when seen in profile. The drawing, pl. 6, fig. 10, represents a bit of ice detached from the gallery of infiltration. All the bubbles are greatly flattened. But what is most extraordinary is, that, far from being uniform, the flattening is different in each fragment, so that the bubbles, according to the face which they offer, appear either very broad or very thin. I know of no more significant fact than this, since it demonstrates that each fragment of ice is capable of undergoing in the interior of the glacier a proper displacement independently of the movement of the whole."—P. 167.

(5) "The same flattening of the bubbles is found at a greater depth. While engaged in my boring experiments, I observed attentively the fragments of ice brought up to the surface by the borer. I found in them almost flat bubbles, perfectly similar to those of the fragment figured above, at all depths from 10 to 65 metres. I observed, besides, that in the fragments which proceeded from a great depth, all the bubbles without exception were strongly flattened, whilst at less depths there were some less compressed and even altogether round, as at the surface.

"It follows, hence, that a strong pressure is exercised in the interior of the glacier."—P. 167.

(6) "I ought also to mention a singular property of these air-bubbles, which at

first was very surprising, but afterwards admitted of very satisfactory explanation. When a fragment containing air-bubbles is exposed to the action of the sun, the bubbles insensibly enlarge. Soon, in proportion as they enlarge, a transparent drop shows itself on some point of the bubble. This drop in enlarging contributes its share to the enlargement of the cavity, and as it progresses it predominates over the air-bubble. The latter then swims in the midst of a zone of water, and incessantly tends to reach the most elevated point, at least if the flatness of the cavity does not hinder it."—Pp. 167, 168.

(7) In a note appended to this passage, M. Agassiz speaks of the irregularity of the walls of some of the bubbles, and adds,

"The same effect has been produced upon the bubbles of the fragment fig. 10. There also all the bubbles have enlarged by diathermanicity, and a little drop has developed in the middle of each. But as the cavities are very small, the drops do not yet move freely in their cavity."

It will be observed that in Nos. 1, 4, 5, 6, M. Agassiz confounds together the water-chambers and the air-bubbles under the common term of "bubbles," and he affirms (6) that the presence of water in the "air-bubbles" is the effect of exposure to the sun's rays, and of the different diathermanicity of air and ice*. A careful analysis of M. Agassiz's facts, however, is very instructive. In the first place, I recognize in his fig. 10. pl. 6, a fair, though rough and sketchy, representation of the general arrangement and form of the water-chambers with their contained air-bubbles. The chambers are as usual flattened, but the artist has rightly represented their contained air-bubbles as spheroidal. The strangest thing is, however, that M. Agassiz has taken the air-bubbles for drops of water, and the drops of water for air-bubbles, as any one who is familiar with the microscopic appearance of bubbles of air will see, on comparing the description in (7) with the figure 10. In the next place, I repeatedly exposed thin plates of ice to the sun, carefully watching the air-bubbles, without being able to observe the phaenomena detailed by M. Agassiz in (6); and I must frankly confess I do not understand how such changes as those described are reconcileable with the commonest properties of ice and air. How do the bubbles enlarge when exposed to the sun? M. Agassiz has already admitted that the chambers are closed (1), and we know that ice is not readily distensible; and therefore I hold it to be impossible that the bubbles should visibly dilate before the melting of the adjacent ice; and as to enlarging by the melting of the ice-wall, the fractional difference between the volume of water and the ice from which it proceeds, would be wholly imperceptible on such a scale. With regard to the explanation of the crackling noise given in (3), I can only say that I have repeatedly watched a thin lamina of ice melting, both by transmitted and reflected light, and that I have seen the walls of the chambers reduced to the thinnest pellicle[14] without being broken by pressure from within. The air-bubbles escape quite quietly as soon as their wall is perforated. Furthermore, the cavities left where the air-bubbles have been, are not fissures at all, but, as I have said above, rounded pits. Indeed, this is a necessary consequence from M. Agassiz's own statements with regard to the shape of the bubbles.

M. Agassiz affirms in (5), that ice brought up from a depth of 65 metres was perfectly similar in structure to that represented in his figure 10. The fact is important; but surely it alone affords sufficient evidence that "diathermanicity" has nothing to do with the formation of the cavities and their watery contents. And indeed in (4) this same piece of ice (fig. 10) is said to have come from the "gallery of infiltration," a cavity perfectly shaded, and bored many feet below the surface of the glacier. So that either this figure does not represent the structure of the glacier at this point, or the structure is unaltered, and diathermanicity has nothing to do with it.

It follows, therefore, that there is no evidence to show that the influence of solar radiation has anything to do with the structure; on the contrary, M. Agassiz's *facts* strengthen my case.

If it be the universal character of glacier ice to be full of closed cavities containing fluid water and air, it becomes a matter of extreme interest to ascertain how the air and the water come there; how it is that the water retains its fluidity, and how it is that the water-chambers are compressed. It may seem a common-place comparison, but the ice and its cavities containing water remind me of nothing so much as a Gruyère cheese, in which one so often meets with closed cavities containing fluid and air. Let the Nèvè represent moist curds and the glacier valley the cheese-press, and the analogy is perhaps closer than it looks. But these are questions for you to solve; and I will only venture on one other supposition, viz. that the water-chambers have the value of a register thermometer, indicating that the minimum temperature to which the mass of a glacier descends is never for long less than 32°; otherwise I cannot conceive how the water should remain fluid; and if it were once frozen, how could it melt again?

M. Agassiz makes a very important observation in (4), and one which I am glad to be able to confirm in the main. I took some pains to ascertain the general direction of the planes of the water-chambers, and I found that in the substance of the blue veins they were sometimes parallel to the plane of the latter, while in the white ice their planes were always more or less inclined to the veins, usually forming an acute angle, and never, so far as I have seen, a right angle with them. Furthermore, as Prof. Agassiz points out, the water-chambers are arranged in groups, all the members of the same group having parallel planes, while their direction is more or less inclined to that of neighbouring groups. It seems to me very probable that, as Prof. Agassiz suggests, the different directions of the planes of the cavities may indicate internal changes of place of segments of ice corresponding with the groups; but, as I have already said, no fissures separating these segments are to be found in the deep ice of a glacier, and hence we cannot with propriety speak of them as "fragments."

Such is the structure which I have found to obtain in all "deep" glacier ice, by which I mean, all ice situated more than a few inches below the surface. It is as solid as glass or marble, and as devoid of any but accidental and gross fissures. The glacier, however, where exposed to the atmosphere, presents what may be called a "superficial layer" of very different character. Every one who has had occasion to cut an escalier, must have been struck with the difference between the resistance to the ice-axe at the first flow and that at the fourth or fifth. At the first, the jar to the hand is slight, and fragments of ice fly in all directions; but, at the last, one might almost as well be hewing some hard though splintery wood. The reason of this at once becomes apparent on examining the superficial ice. It is composed of larger or smaller granules of exceedingly irregular form**, separated by very obvious fissures, but nevertheless so fitted into one another as to cohere with some firmness. The distance to which the fissures extend into the interior of the glacier (and hence the thickness of the superficial layer) varies a good deal; 7 or 8 inches is perhaps rather above than below their average depth; but however this may be, the important fact is, that whenever you clear away the superficial layer, you find beneath it what I have termed "deep" ice—that is, ice in which neither fissures nor granules are discernible; ice which tends to split parallel to the veins, and shows no disposition to break up into the angular fragments so characteristic of the superficial layer.

It has been said that mere optical examination is insufficient to disprove the existence of fissures in the deep ice, and that such fissures are present, though invisible in consequence of being filled with water. I have already shown that the line of contact of water and ice is optically well marked, and that there is every reason to believe that even the finest fissures would be visible under a sufficient magnifying power; but those who maintain the porosity of glacier ice, rest chiefly on the results of experiments made with coloured fluids. It is said that glacier ice becomes infiltrated throughout its substance with extreme readiness, the coloured liquid traversing fissures which are more particularly developed in the course of the blue veins. It became necessary, therefore, to repeat these infiltration experiments; and for this purpose, as you will recollect, I made use of the logwood[15] infusion which you had prepared, and which by its combined clearness and intensity of colour was excellently fitted for the object in view.

If a little of the infusion were poured upon the natural surface of the glacier, it immediately soaked in, spreading itself in all directions between the granules (but more rapidly, as I often observed, in directions parallel with the veins), and staining the whole thickness of the superficial layer. Whatever quantity might be poured on to the surface however, it penetrated no further than the superficial layer (unless there were some obvious crack in the deeper ice); and when the latter was cleared away with the axe, and the surface of the deep ice washed or even carefully rubbed with the hand, not a trace of the infusion could be found in it.

If a piece of the deep ice containing several blue veins were allowed to soak in the logwood infusion until it nearly melted away, it remained unstained, and either wiping it or passing it quickly through clean water rendered it perfectly clear and stainless.

But it is said that if cavities be made in the glacier and filled with a coloured infusion, the latter will soon, by means of the capillary fissures, infiltrate the surrounding mass. To determine this point, I selected a spot upon the north wall of a crevasse, just opposite the Montanvert, and between the centre and the west shore of the Mer de Glace, where the veins were well developed, their planes having a general north and south direction, but dipping at an angle of about 70° towards the centre of the glacier. On the northern aspect of the ice I cut away the superficial layer, so as to form two faces of a cube of about a foot in the side on the deep ice. One of these faces looked westward, and was consequently nearly parallel to the cleavage; the other looked northward, and was therefore nearly perpendicular to it. Perpendicular to the west face, and therefore to the structural planes, I bored a hole with an auger, about an inch in diameter and 9 inches long, and just sufficiently inclined to the horizon to hold the infusion of logwood, with which I filled it. I then thinned away the north face of the cube very carefully until the north wall of the whole was less than 2 inches in thickness—until, in fact, I could see the dark fluid through the substance of the several blue veins which it traversed with perfect distinctness.

For two hours not the slightest trace of leakage or infiltration into the substance of the ice forming the walls of this cavity could be observed; and the contour of the contained liquid remained perfectly sharp and well defined. It then began to leak at one point near its upper end through a small crack *in the white ice,* which led directly outwards. The liquid spread neither up nor down in the crack. Four hours afterwards no change whatever had taken place in the liquid contained in the lower part of the hole. At this time you joined me upon the ice, and you will recollect that I carefully thinned away the wall with a sharp knife until in some parts it was not more than ¼ of an inch thick. Still no infiltration occurred. The knife at length accidentally penetrated the wall, and the liquid at once flowed out. I then poured some clean water through the hole, and all trace of the coloured infusion was at once so thoroughly removed, that, on cutting away one wall, the other appeared perfectly clean and of its natural aspect.

I have given the details of this one experiment in order to show in what manner all were made; but it is unnecessary to be equally prolix with regard to the others. Suffice it to say, that, whether the holes were bored perpendicular to the structure or parallel with it, or at any intermediate angle, whether in white ice or in a blue vein, the result was precisely the same, not a particle of fluid making its way into the surrounding substance of the ice along the veins, nor in most cases in any other way. Occasionally a leakage would take place in the manner described above, but the fissures in these cases were gross and visible, and their direction had no reference whatsoever to that of the structure. Indeed, as the leakage always took place towards the surface, and not into the depth of the ice, I am inclined to think that these cracks were produced in cutting the ice to thin away the outer wall of the cavity.

I repeated these experiments in the neighbourhood of the Grand Moulin;[16] on the Moraine du Noire,[17] somewhat higher than the Couverele;[18] and on different parts of the Glacier du Géant, and everywhere with similar results. Furthermore, having carefully bored a vertical hole in the deep ice of the Mer de Glace, opposite the Montanvert, I filled it with the infusion, and having covered over the aperture with a roof of ice-blocks, I left it until the next morning. It rained hard during the night; and on revisiting the spot after an interval of about fifteen hours, I found that the covering blocks had slipped off, and that the liquid occupied only about the lower two-thirds of the cavity. No trace of infiltration could be discovered; but the lower part of the cavity had changed its figure from cylindrical to irregular and botryoidal. I conceive that the sinking of the fluid must be accounted for by the enlargement of the cavity consequent upon this botryoidal excavation of its walls; and I suppose that the ice-blocks proving an insufficient shelter, the rain poured into the hole, and keeping up a constant supply of comparatively warm liquid, eroded its walls in the way described. However this may be, the fact that the liquid had produced a fresh surface for itself, is important, as it shows that the absence of infiltration through the veins intersected by the cavities containing the coloured infusion is not dependent on a condensation of their walls by the auger.

To eliminate any error of this kind, however, I took a small block of the deep ice, and with a sharp knife fashioned it into a cup, whose walls varied in thickness from ¼ to ⅔rds of an inch. Filled with the infusion and surrounded with ice, this cup remained for two hours without showing a trace of infiltration along its structural planes.

I can only conclude from these experiments, that the chief substance of a glacier is as essentially impermeable as a mass of marble or slate; and that though it may be traversed here and there by fissures and cracks, these no more justify us in speaking of the glacier ice as "porous," than the joints and fissures in a slate quarry give us a right to term slate porous. We do not call iron porous because water runs out of a cracked kettle.

The extreme porosity of what I have termed the "superficial layer," however, is no less certain, and inasmuch as this layer is continually and rapidly wasting away at its surface, it must be as constantly reformed from the solid glacier ice beneath.

The fact observed by Professor Agassiz, that under a moraine[19] (that is, where covered and protected by stone and gravel) the superficial ice is of the same character as the deep, suggested the idea that the superficial layer is the result of the operation of atmospheric influences; and that just as a bed of impervious rock becomes broken up into fragments, separated by permeable interstices, down to a certain depth wherever it is exposed to the atmosphere, so the glacier ice when left unprotected undergoes a similar weathering and disintegration. I submitted this notion to the test of experiment in the following way:—Not far from the upper end of the Moraine du Noire, and on one bank of a stream which cuts its way down the Glacier du Géant, I cleared away the superficial layer and cut out a block of the deeper ice, which was then divided into two equal portions of irregular cuboidal form, and about 8 inches in the side.

The logwood infusion was poured on both of these, and was retained only by such portions of the superficial layer as had been allowed to remain. Water poured on to the blocks ran off them as it would run from marble or glass, sinking only into the remains of the superficial layer. I then placed the two blocks side by side, on an elevated ridge of the glacier, with their natural upper faces turned towards the sun, at this time (1·15 P.M.) shining brightly; the one block I left without protection, while the other was just covered by a stone of 4 or 5 inches in thickness, resting upon its upper face. At 1·40, that is to say in less than half an hour, I removed the block of stone and poured the infusion over both pieces of ice. The covered one could be as little infiltrated as before, while the face of the uncovered became at once beautifully injected, the fluid instantly running into a network of little superficial fissures which had developed themselves, and out of which the infusion could be only partially extracted by washing.

Both pieces of ice were well washed, and the stone was replaced on the one, while the other was left uncovered as before.

In the course of the ensuing half-hour I examined both blocks several times. The covered ice remained unchanged; but in the uncovered, the fissures extended further and further into the mass, which gradually assumed throughout the granular aspect of the superficial layer. Water poured on its surface soaked into it immediately, and a small quantity of the infusion spread out, the moment it reached the block, in the most beautifully ramified figure through the fissures. Particularly large and apparent fissures could thus be frequently observed traversing the middle of the blue veins. At length the fissures extended completely through the mass, which thus became truly sponge-like. Water poured on its surface, filling the interstices, gave the mass a clear and semitransparent aspect, though by no means to be compared to that of a blue vein. But as soon as the supply of water ceased, the fissures of the side uppermost immediately began to lose their water, which drained away below, and becoming filled with air, a whitish opake[20] hue succeeded. On reversing the block suddenly, what had been its under surface appeared at first clear, but the water soon deserting it, it rapidly whitened, while the previous upper and white surface became clear. Water poured upon the upper surface, traversed the mass and flowed out again below with the utmost ease. In fact it is impossible to conceive any more striking contrast in these respects, than that between the freshly extracted ice-block (or that which had remained under cover) and that which had been exposed.

So far as it may be permissible to draw a conclusion from the few experiments I made, I should say that direct exposure to the sun has much influence on the rapidity of this process of weathering; but it is by no means essential, for the northern faces of the walls of crevasses exhibit a well-developed superficial layer; and I have seen it even beneath huge boulders, where these were not in direct contact with the ice.

But one conclusion appears to me to be deducible from these experiments, and that is in perfect accordance with the results of ocular investigation. Glacier ice is essentially devoid of all pores, fissures and cavities, save the closed water-chambers; though of course, like all other brittle bodies, it is liable to become fissured and fractured by pressure from without. Fissures and cavities produced in this way, however, are accidental and not essential. But it is a remarkable feature of glacier ice, that it is liable to weather in a peculiar manner, becoming fissured and breaking up into irregular fragments to a certain depth. The superficial layer formed in this way is eminently porous, and absorbs fluids like a sponge.

In arriving at these results, however, I again regret to find myself in direct opposition to the current doctrine based on the statements of Prof. Agassiz, from whose 'Système Glaciaire' I continue my series of quotations.

(8) "Capillary fissures.—The true capillary fissures are very different from the superficial fissures which have just been described (3). They exist not merely at the surface, but are found on the walls of crevasses and in the interior of cavities where the rays of the sun never penetrate. They are larger than the little fissures which have just

been mentioned, and far less numerous, particularly in the regions in which the latter abound. Their distribution is not uniform in the interior of the compact ice," p. 154;

but (M. Agassiz goes on to explain) they are arranged in bands and zones, which, becoming more completely infiltrated with water than the intermediate ice, appear blue and transparent, and are the blue veins.

(9) "The quantity of bubbles with which the white ice is filled, is the reason why the fissures are more slowly propagated in it; the air, by its elasticity, being unfavourable to the formation of fissures (l'air qui est de sa nature élastique ne favorisant aucunement le crevassement[21]). By degrees, however, and in proportion as the infiltration perpetuates itself, the rigidity increases, and the fissures multiply in proportion. Every bubble that a fissure meets in its course loses its aëriform[22] contents. It becomes transparent, and the opacity of the mass is so far diminished. The consequence of this multiplication of fissures is continually to diminish the number of bubbles, and by this means to render the ice more and more transparent and blue.

"It will be evident to any one who has followed the progress of modern physics, that this phaenomenon is due solely to the diathermanicity of ice. The air first and then the water becomes heated through the ice. However minute may be the degree of heat which is thus transmitted to them, it is enough to melt a part of the ice which surrounds them, and thereby to increase the cavity in which they are imprisoned. I do not think, however, that any very great importance should be ascribed to this phaenomenon; and the fact that it is produced only when the ice is exposed directly to the rays of the sun, is in my eyes an indication that it exercises no notable influence on the mechanism of glaciers."—P. 157.

(10) "When the ice has acquired a certain degree of transparency, and the network of capillary fissures is fully established in it, water and air penetrate into the fissures with great facility. One may assure oneself of this in many ways, among others by the following experiment, which I have repeated many times. Let a cube of ice of a few decimetres on the side be detached from the bottom of a crevasse, in that part of the glacier where the ice is most transparent, and placed upon a rock. At first, a few fissures will appear on the surface, then these fissures will be gradually propagated into the interior, and the network becoming more and more complex, will by degrees reach the base. If, then, the block of ice be turned upside down, and water be poured upon it, all the fissures will disappear from above downwards, in the same order as they were formed. The block will remain perfectly transparent so long as it is saturated; but so soon as one leaves off watering, the fissures reappear where they last appeared when the block was reversed."—P. 161.

(11) "The angular fragments are the consequence and the product of the capillary fissures. When a morsel of compact ice is exposed for some time to the air, it becomes decomposed into a certain number of angular fragments, which are the smaller the more numerous the fissures. The same thing would happen to the glacier if its thickness were less, and if the external heat had access to it on all sides. Nevertheless its surface decomposes more or less, the fissures dilate in consequence of the circulation, and the fragments are so dislocated as to be moveable on one another without however becoming detached."—P. 163.

(12) "The angular fragments and the capillary fissures seem to disappear the

moment the ice is covered. Thus on sweeping clean a part of a moraine, or the side of a gravel cone, the ice beneath is found to be perfectly smooth, and apparently without a trace of a fracture. But it is sufficient to leave these same surfaces uncovered for some instants, and the capillary fissures immediately show themselves, and, in consequence, the angular fragments. They appear with such regularity, that one might be tempted to believe that they are formed spontaneously at the very moment of their appearance. But on examining them with a little attention, one becomes convinced that they are of older date.

"I by no means pretend to deny that heat, acting suddenly at the moment the moraine is uncovered, may not develope some cracks. I have myself seen such cracks form suddenly (par éclat[23]), but I conceive they are but few. If it were otherwise, and if the fissures were formed as they appear, it would be necessary to suppose that there are none in the ice of the moraine before it is uncovered, which would be contrary to all we know of the transformations of the ice."—P. 165.

(13) "Let us now make the opposite experiment, and cover with sand and gravel a portion of the surface of the glacier. However fissured and disaggregated it may be, the fissure and angular fragments will disappear at the end of some time so completely, that on removing the gravel the surface will be found as compact and transparent as that of a portion of moraine which has never been uncovered. And yet it is not probable that the fissures have reunited during the interval. It is, on the contrary, the gravel, which, intercepting the air and keeping the fissures full of water, renders them invisible, and gives to the whole mass a false appearance of compactness, which ceases the moment the air again has access to the fissure."—P. 166.

If the extract (8) were to be taken merely as a description of the superficial layer of a glacier, I should only have to object, that, so far as I have been able to observe, the colour of the disintegrated blue veins is not much affected by the water they contain, and that no amount of watery infiltration will confer on the white ice the beautiful transparency and colour of the blue veins. But Prof. Agassiz over and over again affirms that the whole substance of the glacier is traversed by capillary fissures, and his infiltration experiments are supposed by himself conclusively to demonstrate the fact. I must confess, however, that I have neither been able to observe what Prof. Agassiz supposes he has observed, nor, were our observations in unison, could I admit his explanation.

Take for instance the citation (9). How can the elasticity of the air-bubbles influence the formation of fissures in the continuous mass of rigid and eminently brittle ice which encloses them? How is the statement, that the ice becomes more rigid as the fissures are developed in it, these fissures being supposed to be filled with water, compatible with that made in (2), that this same water is the chief source of the plasticity and compressibility of ice?

Again, I am at a loss to understand the "diathermanicity" theory. Prof. Agassiz brings forward no experimental proof that air contained within ice is more heated by the sun's rays than the ice itself; and, *à priori*,[24] it seems improbable that the more diathermanous body should be more heated than the less. It is true, I cannot pretend to have "followed the progress of modern physics;" but I am emboldened to say this much by the fact, that you, who have, seem to find at least equal difficulty in adopting Prof. Agassiz's explanation.

With regard to the experiments detailed in (10), (12), and (13), it will be observed that my results in the main agree with those of Prof. Agassiz, if, as before, we confine ourselves to the superficial layer; but, as I have shown, it is an error to extend the conclusions drawn from the structure of the superficial layer to the deep ice. This, however, is what Prof. Agassiz has done; and it is curious to find him in (12) refusing to follow out a suggestion which would have led to the solution of his difficulty, because it is "contrary to all we know of the transformations of ice." What do we know at present of the transformations of the ice?

It is important to remark again, that *as regards matters of fact,* Prof. Agassiz's statements with respect even to the deep ice are, so far as they go, not essentially different from mine. He admits (12) that no fissures are at first visible in the deep ice;—had he taken the trouble to make the experiment, he would have found also that coloured liquids cannot be made to enter it;—and he admits that the establishment of a complete system of fissures through a block of ice, and its consequent permeability, are matters of time and exposure (10). See also citation (1), and p. 289 of the 'Système Glaciaire.'

I omitted to make the experiment detailed in (13). It is singular that in (12) Prof. Agassiz states that "the angular fragments and the capillary fissures seem to disappear *the moment* the ice is covered," while in (13) the operation is said to take some time; but, supposing the face to be as Prof. Agassiz says, it seems to me to be in the highest degree probable that the fissures *have* reunited during the interval. At any rate, I cannot admit Prof. Agassiz's explanation, for surely loose gravel is not exactly a substance calculated to "intercept air and keep fissures full of water."

It would take up too much space, and serve no useful purpose, to quote at length the account Prof. Agassiz gives of his infiltration experiments (Syst. Glaciaire, pp. 170–179). Those who will turn to the original, will find that they are all vitiated by the absence of any discrimination between the deep and the superficial ice, and between "capillary fissures" and accidental cracks. Not one of Prof. Agassiz's experiments affords the slightest evidence that capillary fissures are a primitive and essential constituent of the structure of the deep ice of a glacier.

The experiments of the Messrs. Schlagintweit[25] (l.c. p. 12) appear to me to be equally inconclusive; these gentlemen, like Prof. Agassiz, having omitted to take the precaution of clearing away the superficial layer from the mouth of the cavity to be filled with the infiltration of fluid. Unless this be done, the superficial layer sucks up the coloured liquid, which becomes diffused in the way they describe. And if the cavity (as may readily happen, especially with such large ones as those employed by these experimenters) communicates by an accidental fissure with some other part of the surface of the glacier (say the wall of a crevasse, or the roof of such a cavity as Prof. Agassiz's infiltration gallery), it should be well remembered that the fluid which drains through will not run out in a stream from the termination of the crack, unless the superficial layer has been cleared away; otherwise, it will fill the fissures of the superficial layer and appear as a great patch. The observer then, seeing nothing but fissures full of coloured infusion at each end of the course of the fluid, naturally enough imagines that in its intermediate course the fluid has traversed similar fissures. This conclusion would be at once dissipated by cutting away the superficial layer and laying open the infiltration cavity,—a precaution which does not seem to have occurred to either Prof. Agassiz or the Messrs. Schlagintweit.

I will conclude with a few words upon the relation of structure to the arrangement of dirt upon the surface of a glacier. The great "dirt-bands" have never been *proved* to be connected with any peculiar structure of the ice on which they lie, and it has been shown that they *may* be the mere result of the influence of the motion of the glacier upon the form of any patch of dirt scattered accidentally upon its surface; but besides these "dirt-bands," the dirt on a glacier frequently presents a definite arrangement upon a smaller scale, which *is* connected with the minute structure of the glacier. We have both observed, for instance, in those parts of the Mer de Glace in which the structure is vertical, that the superficial layer of the wall of a crevasse is weathered into granules of tolerably even size and similar form. Nevertheless, dirt (or a coloured infusion) accumulates in larger proportion in those fissures which are parallel with the cleavage, and thus, from a little distance, the surface of the ice appears as if striated or ruled with lines parallel to the structure. The lines are separated by the width of the granules, and there may be several interposed between two blue veins.

Why it is that those intergranular fissures which are parallel with the cleavage are the larger, is a question I will not for the present attempt to answer. It may be that the weathering takes place more rapidly in this direction, or it may be that these fissures being in the course of the flow of the water produced by the superficial waste of the ice, become enlarged more rapidly than the others.

These markings, and the similar ones frequently to be observed on the upper surface of a glacier, might be termed "dirt-lines," to distinguish them from the great "dirt-bands." There is a third mode of arrangement of dirt, which, like the "dirt-line," is dependent on the weathering of the ice, but the resulting striæ are broad streaks, and not mere lines. These may perhaps be termed "dirt-streaks."

I became acquainted with these quite recently, when, induced by Prof. Forbes's description and representation[26] of the "structure" of the glacier of La Brenva, I paid a visit to that glacier. Prof. Forbes states,—

"The alternation of bluish-green and greenish-white bands which compose this structure, gives to this glacier a most beautiful appearance as seen from the mule-road. An attempt has been made in plate 5 to give some idea of this most character-istic display, and which is better seen here than in any other glacier whatever with which I am acquainted. The sketch was taken by myself from the point marked *k* on the map in July 1842."—*Travels,* 2nd Ed., p. 203.

It must at once strike any one conversant with the ordinary character of the veined structure, that at the distance of the point on the mule-road from which Prof. Forbes's view is taken, any veins of the usual dimensions must be totally invisible; and I therefore approached La Brenva with the desire, if not the hope, of making the acquaintance of glacier structure on a new and gigantic scale.

Viewed from the mule-path, or from the old moraine at the commence-ment of the pine wood celebrated by De Saussure,[27] the lower part of the gla-cier of La Brenva exhibits numerous crevasses, which appear to run nearly parallel with its length, so that the icy mass is divided into a series of parallel crests or ridges. The lateral faces of these ridges form perpendicular cliffs of ice, and present dark stripes directed in a longitudinal and nearly horizontal direction; but where an end view of a ridge is obtained, the stripes run either horizontally and transversely (as in the more central parts of the glacier), or are curved up towards the sides (as in the more lateral parts).

These markings are evidently those described and faithfully figured, as the "structure" of the glacier, by Prof. Forbes; but I cannot say I should have called them bluish-green. They look to me simply dark and dirty. But I should state, that the weather, when I visited the glacier, was wet and cloudy.

Nevertheless, on descending on to the glacier itself, I found its structure, though very beautifully developed, to be in nowise remarkable for the size of its veins, which varied in length from an inch to eight or nine feet, and in breadth from a fraction of an inch to nine or ten inches. Veins of the latter dimensions, however, were rare; the majority having a thickness of less than an inch. The lenticular form was very well marked, and the veins were com-monly separated by less than their thickness of white ice. I need hardly say that these veins became indistinguishable at a very short distance.

The streaks so conspicuous a long way off, on the other hand, became less sharply defined as I approached, and at length showed themselves to be nothing more than accumulations of the fine dirt—spread more or less over the whole cliff-like wall of ice,—in streaks of four to ten inches in breadth, and of variable length. They ran parallel with one another and with the structure, at a distance varying from a few inches to six or seven feet; and they were entirely superficial, the dirt never extending deeper than the weathered superficial layer.

It became clear, therefore, that the markings were neither structure nor stratification, but a peculiar kind of dirt-marks; and the next point was to ascertain the conditions of their formation.

On close examination, the face of the ice-cliff exhibiting these markings appeared to be worn into a sort of wavy or rippled surface, the length of the ripples having a general direction downwards. The close-set veins, on the other hand, traverse the face of the ice, as has been said, nearly horizontally. The whole surface of the ice is more or less dirty, not half-a-dozen square inches being without its little grains of sand and minute gravel, brought down, as I imagine, by the water which continually trickles from above; but the greater part of this impurity is invisible from a small distance, unless where it is specially accumulated.

Such accumulation takes place in two localities: in the first place, on the little shelves afforded by the upper and more southerly aspects of the "ripples" above referred to. Here the dirt accumulates quite independently of the structure, and as a consequence either of the form or of the aspect of the part on which it lodges.

From a little distance these aggregations appear as spots and patches, but further off they cease to be visible, and the glacier between the horizontal streaks appears white.

These streaks mark the second locality in which the dirt aggregates. Now whenever I carefully examined the surface of the glacier at these points, I found it to be weathered into large granules, separated by coarse fissures which extended for a considerable depth into the substance of the glacier; while the parts intermediate between the streaks, and which appear white from a distance, presented very much smaller granules, with fissures proportionately finer, and extending inwards for but a very small distance. In short, where the dark streaks existed, the ice was deeply weathered and coarsely granular, affording lodgment for dirt to a depth of two or three inches; while the intermediate substance had undergone only superficial weathering, and its finely granular structure afforded but little facility for the intrusion of foreign matters.

The "dirt-streaks," then, are due to the unequal weathering of the ice; but why does the ice weather unequally? On seeking for an answer to this question, I found that every dirt-streak corresponded either with a very large blue vein, or with a closer aggregation than usual of smaller blue veins, while the intermediate substance contained a preponderance of the smallest blue veins: so that the coarse granules were the result of the weathering of parts of the glacier, composed either exclusively, or for the most part, of blue ice, while those in which the proportion of white ice was larger, weathered less deeply and into finer granules.

The markings of La Brenva, then, are neither ordinary dirt-bands, nor direct expressions of structure, nor direct evidence of stratification, but they are produced by the more ready lodgment of dirt in some parts of the superficial layer of the glacier than in others, in consequence of the more coarse and deep weathering of these parts; which, again, is the result of the predominance of blue ice over white in these localities.

Why blue ice should predominate at intervals in the substance of this glacier,—whether the like alternation of structure holds good in glaciers generally,—and whether it has any relation to a primitive stratification, are problems of great interest and well worthy of investigation.

With regard to the second, I will merely express a belief that some such alternation of structure does obtain in glaciers generally; for the appearances presented by good sectional views of glaciers, such as that exposed on the north side of the Allalein,[28] are so similar to those exhibited by La Brenva, that I cannot doubt the identity of their cause. I had been in the habit of regarding the appearances referred to as direct evidences of stratification; but if my supposition be correct, they will merely be evidences of an alternation of structure which may or may not depend on stratification.

Yours very faithfully, | T. H. HUXLEY.

* The Messrs. Schlagintweit (Untersuchungen, p. 17) adopt Prof. Agassiz's views on this point, and with him regard the presence of water as a local and partial phaenomenon.

* The superficial layer is particularly well described by the Messrs. Schlagintweit in their Untersuchungen über die physikalische Geographie der Alpen, 1850.

T. H. Huxley, 'Observations on the Structure of Glacial Ice', *Phil. Mag.*, 14 (1857), pp. 241–60

---

1. *Montanvert*: see letter 1415, n. 15.
2. *Mer de Glace*: see letter 1306, n. 12.
3. *Glacier du Géant*: a large glacier on the French side of Mont Blanc.
4. *La Brenva*: see letter 1429, n. 5.
5. *spirit-level*: a kind of levelling instrument for determining a horizontal line or surface, usually consisting of a hermetically-sealed glass tube filled with spirit and an air-bubble,

which, when the tube lies exactly horizontal, occupies a position midway in its length (*OED*).

6.   *M. Agassiz's*: Louis Agassiz.

7.   *Aar glacier*: see letter 1305, n. 2.

8.   *'Système Glaciaire'*: L. Agassiz, A. Guyot, and E. Desor, *Système glaciaire ou recherches sur les glaciers* (Paris: Victor Masson, 1847).

9.   *Névé*: compacted or hardened snow, as found on the upper part of a glacier (*OED*).

10. *diaphanous*: perfectly transparent (*OED*).

11. *Abschwung*: according to Tyndall, a rocky promontory 'where the Lauteraar and Finsteraar glaciers unite to form the Unteraar', which is part of the Aar glacier (*Glaciers of the Alps*, p. 388).

12. *crepitation*: a crackling noise (*OED*).

13. *diathermanicity*: diathermancy, the property of being diathermic or diathermanous; perviousness to radiant heat (*OED*).

14. *pellicle*: a small or thin skin; a fine sheet or layer covering a surface or enclosing a cavity (*OED*).

15. *logwood*: see letter 1353, n. 10.

16. *Grand Moulin*: on the Mer de Glace, a huge circular hole/cave in the glacier which attracted tourists and mountaineers.

17. *Moraine du Noire*: possibly the moraine of the Glacier du Tour Noire.

18. *Couverele*: Couvercle, the mountainside of Mont Blanc above the Glacier du Tacul, on which the current Couvercle Hut stands.

19. *moraine*: see letter 1415, n. 17.

20. *opake*: an alternative spelling of *opaque*.

21. *l'air qui est de sa nature élastique ne favorisant aucunement le crevassement*: 'the air, by its elasticity, being unfavourable to the formation of fissures' (French).

22. *aëriform*: of the form or nature of a gas; gaseous (*OED*).

23. *par éclat*: by breaking (French).

24. *à priori*: hence (*OED*).

25. *Messrs. Schlagintweit*: see letter 1417, n. 8.

26. *Prof. Forbes's description and representation*: James David Forbes, *Travels through the Alps of Savoy, and other parts of the Pennine chain, with observations on the phenomena of glaciers.* 2nd edn. (London: Longman, Brown, Green, and Longmans, 1845).

27. *De Saussure*: Horace-Bénédict de Saussure.

28. *Allalein*: the Allalin glacier in the valley of Saas in the Swiss canton of Valais, part of which formed the dam which created the Mattmark See (a lake). Today, the glacier is well above the current artificial dam (constructed in the mid-twentieth century).

From Walter Pollock   17 September 1857   1437

My dear Mr Tyndall,

I can write, you see;[1] I have got on well at, the riding school and I can trot—so I am not a little dot.

I am yours affectionately | the little demon.[2] | Please answer. | W. H. P. Tunbridge Wells[3] | xvij Sep: mdcccLvij.

RI MS JT/1/TYP/6/1917
LT Typescript Only

1. *I can write, you see*: a response to a remark Tyndall made in a letter to Walter's brother William; see letter 1339.
2. *the little demon*: Tyndall had climbed the Col du Géant with Edouard Balmat, whom he mentioned in letters 1419 and 1420, referring to him as 'my little demon' in 1420.
3. *Tunbridge Wells*: see letter 1423, n. 1.

From Gustav Wiedemann[1]   25 September 1857   1438

Lieber Herr College!

Erlauben Sie mir, durch diese Zeilen Herrn Dr. Esselbach bei Ihnen einzuführen, und Sie zu bitten, sich seiner freundlich bei seinem Besuche in England anzunehmen. Einen weiteren Empfehlung des Herren Esselbach an Sie wird es nicht bedürfen, da derselbe Ihnen durch seine schonen Untersuchungen über den ultravioletten Theil des Spectrums hinlänglich bekannt sein wird.

Ich hatte gehofft Sie in diesem *[Sommer]* bei uns in Basel *[oder]* *[dies]* auf der Naturforscherversammlung hier in Bonn wiederzusehen. Leider ist indeß *[meine]* Hoffnung getäuscht. Um so mehr wünsche ich, *[daß]* das nächste Jahr eine Gelegenheit bieten möchte, *[und]* Sie uns zuführen könnte.

Mit der Bitte, auch Ihrem verehrten Collegen Herrn Faraday auf das Angelegentlichste zu empfehlen

Ihr | *[Sehr]* ergebener | G. Wiedemann
Bonn. 25. Septbr 57.

Dear Colleague!

Permit me, through these lines, to introduce Dr. Esselbach[2] to you, and to ask you to kindly look after him on his visit to England. A further recommendation of Herr Esselbach to you will not be required, as he will be sufficiently known to you through his fine investigations on the ultraviolet part of the spectrum.[3]

I had hoped to see you here again this *[summer]* in Basel,[4] *[or]* *[this]* at the meeting of natural scientists here in Bonn.[5] Unfortunately, *[my]* hope has been deceived in the meantime. All the more do I wish *[that]* next year might provide an opportunity, *[and]* could bring you to us.

With the request to give my warmest compliments to your honoured colleague Herr Faraday as well

Your | *[Very]* humble | G. Wiedemann

Bonn. 25. September 57.

RI MS JT/1/W/39

---

1. *Gustav Wiedemann*: Gustav Heinrich Wiedemann (1826–99), a German physicist who studied under Gustav Magnus at the University of Berlin. He was professor of physics at the University of Basel from 1854–63 when he took a position at Braunschweig. Following Poggendorff's death in 1877, Wiedemann became editor of the *Annalen der Physik und Chemie* until his own death in 1899 (*CDSB*).

2. *Dr. Esselbach*: Ernst Esselbach (1832–1864), a German physicist who studied at Göttingen and at Königsberg under Hermann Helmholtz. In 1858 he began working for the telegraph works of Werner von Siemens at first in Berlin, then in London, and later as a Chief Superintendant of the telegraph to India. See B. Feddersen and A. von Oettingen (eds), J. C. *Poggendorff's Biographisch-literarisches Handwörterbuch der exakten Naturwissenschaften: 1858–83* (Leipzig: J. A. Barth, 1898), vol. 3, p. 419.

3. *fine investigations on the ultraviolet part of the spectrum*: E. Esselbach, 'Eine Wellenmessung im Spectrum jenseits des Violetts', *Poggend. Annal.*, 98 (1856), pp. 513–46.

4. *Basel*: a city in northwestern Switzerland, located on the Rhine river.

5. *meeting of natural scientists here in Bonn*: see letter 1414, n. 17.

From William Frederick Pollock   28 September 1857   1439

Tunbridge Wells | 28th. Sep. 1857

My dear Tyndal,[1]

I send you the Sonnet you admired[2]—together with my version of it, which I think is not so bad as when you saw it here—I, of course, understand the allusion at the end[3] to be to yourself when ascending Mont Blanc[4]—

I fear the Equinox[5] has put an end to our fine weather—It was raining yesterday & is worse today—My wife[6] & Walter[7] send their best remembrances—

I am | Yours truly | W. F. Pollock

RI MS JT/1/P/227
RI MS JT/1/TYP/6/1919

---

1.   *Tyndal*: William Frederick Pollock also misspelled Tyndall's name this way in letter 1474, as did Juliet Pollock in letter 1418. She also misspelled it as 'Tindal' in letter 1343.

2.   *the Sonnet you admired*: Tyndall had visited the Pollocks at Tunbridge Wells from 12–13 September. In his journal entry for 13 September 1857, he wrote of William Frederick Pollock: 'He showed me some Sonnets of Petrarch which he had translated—the man who could translate them as he did, or admire them so much as to feel a pleasure in translating them must have an element of nobleness in him' (Journal, RI MS JT/2/13c/1033).

3.   *the allusion at the end*: perhaps the sonnet to which Pollock referred to is Francesco Petrarch's (1304–74) poem 'Di pensier in pensier, di monte in monte', which discusses an ascent of mountains.

4.   *Mont Blanc*: see letter 1320, n. 6.

5.   *Equinox*: the autumnal equinox, which occurred on 22 September 1857.

6.   *My wife*: Juliet Pollock.

7.   *Walter*: Walter Pollock.

To Thomas Archer Hirst   [14 October][1] 1857   1440

London Wednesday morning | 1857

My dear Tom.

I have been thinking of writing to you almost every day during the last week, but I am also entangled in my subject[2] and find it hard for my thoughts to break away from it. This morning your note[3] came to me and was like a spark to tinder causing my previous resolution to ignite and take effect. I was going to have a walk before commencing work, but I will throw the walk back for half an hour and write to you in the mean time. On the whole I have been getting on well since my return. My knowledge of what I can bear, and when to cease working is increasing and altogether I find myself far more manageable. I wish I had greater working power; but it is a better wish to desire to make the most of that which I possess by abstaining from squandering it. I am hard at work upon the glacier question: since my return I have been experimenting upon ice and have obtained true—I think I might call them beautiful—results. I am now drawing up the paper[4] and believe me it requires a great amount of patient thought to get the spirit of the mass of facts which we collected rendered manifest. It will be a very heavy discharge of artillery against the viscous theory,[5] and will no doubt cause divers dissensions among scientific men. I had intelligence from Edinbro'[6] regarding Forbes some days ago, it is said that he is in a very disturbed state of mind about the subject.

This I am sorry to hear, and I shall take care in my next paper to give him the full measure of credit which I believe to be his due. Your feelings regarding that communication of Huxleys[7] are precisely my own. He had agreed that he should write me a letter upon the subject, but when the document came into my hand I was perfectly astonished. Nevertheless it is so clever that I read it all through with great interest. It is certainly a large paper to get out of two days observations. Huxley is swift and penetrating. I think however that it would be far safer if he had more inertia, and chewed the cud of thought more patiently. I confess I should have feared to publish all that he has published on the grounds which his observations furnish. The testimony of others is so distinct & circumstantial on this question of bubbles that before discarding this testimony I should have felt it necessary to look far deeper into the question.

People are beginning to pour into London and our active life will soon commence. I purpose talking to them upon heat[8] here before Easter. I have gone over the subject before and hence its preparation will not be so laborous. Nevertheless I hope to be able to put the question in new lights, and to expand portions which in my former course[9] were too meagrely dealt with. I have at earnest solicitation agreed to give a course of six lectures to the boys of Eton[10]—Last Thursday I gave my first lecture and my second comes on tomorrow. Next month my lectures at the London Institution[11] commence, so that I have plenty of work in front of me and around me. I very often dash off to the country on Fridays, and I rejoice to hear that you do the same. We must keep ourselves strong Tom. We must live as men, and not as puny weaklings to whom the joy of vigourous health is unknown. Carlyle says that real health is real holiness,[12] and I can conceive the truth of the statement, for as a moral being when my health is strong I feel myself better braver and nobler. Now that the thing is passed and gone I do love to think that we accomplished so much together upon the glaciers. I love to think on that day of toil which placed us beside each other on the top of Mont Blanc.[13] By the way I have had thoughts lately of devoting the hours of the evening which I dare not employ at taxing work in the writing of a book, in which I should connect the glorious objects which one encounters in the Alps with their principles. I think something might be done in this way which would have a wide circulation.

Debus was here a few days ago. Playfair & an American clergyman[14] called upon me asking whether I thought Debus a likely man for a post at Toronto[15] worth £350 a year. The result was that Debus came to town to see the clergyman, but it came to nothing. Heinrich's language is a great hinderniss[16] to him.

Frankland is now in London, he has left Owens College,[17] and a young fellow named Roscoe[18] succeeds him in Manchester.[19]

This is General Sabine's birthday, he came to me yesterday and engaged me to dine with him.

I will send you the papers very soon and now I want you to accomplish a commission for me. Go like a good fellow to Monsieur Duboscq, Rue de l'Odeon No 21. and request him to send me forthwith a box of coal points[20]—not a few merely but a good quantity. My stock is almost exhausted—indeed I had to borrow one or two a few days ago. Perhaps it would be safest if you would receive the points from Duboscq yourself and forward them to me. If they reach me within the coming fortnight it will answer.

And now Tom I will say good bye for the present & run out & take my walk. The morning was thick & foggy, but it is clearer now, and my walk will be all the pleasanter.

Affectionately yours | <u>John</u>

RI MS JT/1/T/964
RI MS JT/1/HTYP/504

---

1. *[Oct 14]*: the month and day are given based on Tyndall's mention of this day being Edward Sabine's birthday, which was 14 October 1788.

2. *entangled in my subject*: most likely his paper on ice, read at the RS on 17 December 1857 and published as J. Tyndall, 'On Some Physical Properties of Ice', *Phil. Trans.*, 148 (1858), pp. 211–29. Tyndall wrote in his journal on 18 October 1857 'Each day (available) I have been at my memoir. It is an entangled subject and required patient thought to link its parts together' (Journal, RI MS JT/2/13c/1035).

3. *your note*: letter missing.

4. *the paper*: J. Tyndall, 'On Some Physical Properties of Ice', *Phil. Trans.*, 148 (1858), pp. 211–29. Tyndall's lecture at the RI on this subject occurred on 22 January 1858.

5. *viscous theory*: see letter 1306, n. 8.

6. *intelligence from Edinbro'*: letter missing, apparently from someone in Edinburgh, Scotland, possibly James Coxe.

7. *that communication of Huxleys*: see letter 1436. Huxley's letter to Tyndall was published as 'Observations on the Structure of Glacial Ice', *Phil. Mag.*, 14 (1857), pp. 241–60.

8. *talking to them upon heat*: Tyndall presented 'Ten Lectures on Heat, Considered as a Mode of Motion' at the RI on Thursdays from 21 January to 25 March 1858. See RI Guard Book, vol. 2, p. 105.

9. *my former course*: Tyndall presented 'Twelve Lectures on Heat' at the RI on Tuesdays from 17 January to 4 April 1854.

10. *the boys of Eton*: students at Eton College, see letter 1296, n. 8. Tyndall had previously lectured there in 1856.

11. *the London Institution*: see letter 1319, n. 5. Tyndall gave 'Six Lectures on the Phenomena of Light' at the LI, beginning 14 December 1857 and each following week through 18 January 1858.

12. *Carlyle says that real health is real holiness*: in an inaugural address at Edinburgh on 2 April 1866, Thomas Carlyle stated, 'It is a curious thing, which I remarked long ago, and have

often turned in my head, that the old word for "holy" in the Teutonic languages, *heilig,* also means "healthy." Thus *Heilbronn* means indifferently "holy-well" or "health-well." We have in the Scotch, too, "hale," and its derivatives; and, I suppose, our English word "whole" (with a "w"), all of one piece, without any *hole* in it, is the same word. I find that you could not get any better definition of what "holy" really is than "healthy"' (T. Carlyle, *Critical and Miscellaneous Essays* 5 vols (London: Chapman and Hall, 1899), vol. 4, pp. 449–83, on p. 479). Perhaps Tyndall had heard this thought from conversation with Carlyle before it was expressed in the Edinburgh address.

13. *Mont Blanc*: in letters to Faraday (1419), Juliet Pollock (1420), and Hooker (1428), Tyndall described his and Hirst's ascent of Mont Blanc on 13 August 1857.

14. *an American clergyman*: not identified.

15. *Toronto*: the University of Toronto, where Tyndall was offered a position in 1855 but declined. See letter 1493, n. 13.

16. *hinderniss*: obstacle (German).

17. *Owens College*: founded in 1851, the college is now known as the University of Manchester.

18. *Roscoe*: Henry Roscoe (1833–1915), an English chemist who, like Tyndall, had studied under Robert Bunsen in Germany. Roscoe remained at Owens College until 1886 (*ODNB*).

19. *Manchester*: see letter 1311, n. 8.

20. *coal points*: or carbon point or charcoal point, used in electric apparatus.

To Stephen Hawtrey 15 October 1857   1441

My dear Sir,

My object in asking you last night not to write to me was to allow myself an opportunity of saying a word or two which might influence you in your proceedings with regard to the lectures.[1] I had been led to entertain the idea that the boys[2] would prefer having the lectures strictly instructive rather than amusing, and this led me to aim at a more accurate and consequently I fear to the mass of the audience less interesting exposition than I might otherwise have adopted. I confess that it would have pleased me to have interested the boys and parted from them in friendship; and it would be a matter of great regret to me to find that instead of the spontaneous enlistment of their attention, coercion had been thought necessary to keep them in order. I refuse a great many applications for lectures every year, being desirous rather of devoting my time to original research, and I went to Eton[3] partly influenced by your strong desire that I should do so, and partly by the natural liking which I feel for boys. It would therefore be a matter of great concern to me to find myself the indirect cause of the punishment of any one of them. Were it not that the sudden stoppage of the lectures might be taken as an evidence of anger or displeasure upon my part which I really do not feel, I would say let them stop now. But my impression is that it might be better to give one more and then cease. I would select a few of the best experiments in optics for the occasion and thus try to interest and gratify the boys. Or if you thought they would prefer some other subject I should be happy to meet their wishes as far as my ability will permit me to do.

I may say that it is not my intention to accept payment for these lectures. You will meet the expenses of the man whom I engage to accompany me,[4] and we are then clear.

Very truly yours | <u>John Tyndall</u>
Reverend Stephen Hawtrey

RI MS JT/2/13c/1036
RI MS JT/1/TYP/9/221–23
Transcript Only

1. *my lectures*: Tyndall had lectured for students at Eton College in Eton, Berkshire in 1856, and was asked to again in 1857. His series of six lectures began on 8 October 1857.
2. *the boys*: students at Eton College.
3. *Eton*: Eton College, see letter 1296, n. 8.
4. *the man whom I engage to accompany me*: not identified.

From Stephen Hawtrey   18 October 1857   1442

My dear Sir

Accept my best thanks for your most kind and valued letter.[1] I shewed it to the Head Master,[2] he said what a good thing it would be if the boys[3] could see it!

Fifty of them gave themselves up as implicated in making the noise and got their punishment before your letter came. To shew them how childish their conduct was, the Head Master gave them a lower school punishment, which was to write out 300 lines from the 'Propria quae maribus' & 'as in praesenti'[4] out of the Latin Grammar.

A great many boys have of their own accord said to me how disgraceful they thought it was, and they are most anxious the lectures[5] should continue. There is but one feeling among the elder boys. It was to the eldest of the boys that were at the lecture that the 50 noisy ones gave themselves up. I think it is very likely you will hear from them.

I will not say more now but that I hope you will continue the course. The Head Master has the same opinion of the value of the lectures that I have. I trust and fully hope that the result will be such that you will not regret having come to us. I feel how entirely it was owing to my earnest solicitations that you did so; this makes me all the more anxious that the boys should right themselves with you.

I am yours | very sincerely | Stephen Hawtrey

RI MS JT/2/13c/1037
RI MS JT/1/TYP/9/224–25
Transcript Only

1. *your most kind and valued letter*: letter 1441.
2. *Head Master*: Charles Goodford (1812–84), Head Master of Eton College from 1853–62 (*ODNB*).
3. *the boys*: students of Eton College, see letter 1296, n. 8.
4. *'Propria quae maribus' & 'As in praesenti'*: common exercises for writing in Latin in boy's schools. 'Propria quae maribus' refers to the rules of proper nouns, and 'As in praesenti' refers to the conjugation of verbs.
5. *the lectures*: see letter 1441, n. 1.

From William Frederick Pollock   19 October 1857   1443

Tunbridge Wells | 19ᵗʰ. Oct. 1857

My dear Tyndal

Thanks for your lines[1] on Richmond Park[2]—Barring the grander features of mountain & lake I believe there is as much beautiful country within an hour's reach of London, as in any part of England—and perhaps of all that is thus accessible, Richmond presents the choicest variety of charming scenery—Since your musings there, although not very many days since, Autumn must have made considerable havoc among the foliage—Here the leaves are in a remarkable degree persistent, both in colour & in attachment to their trees—but those on the limes and others of early decadence have already surrendered at discretion—and many a brown & yellow patch on the green of the distant woods—& many a rustling swirl of dry leaves at ones feet, remind one that the year is waning—& help to reconcile us to the coming hibernation in London—We are to leave the sweet nymphs[3] & naiads[4] of these Wells[5] on Friday—& the following day will see me at drudgery again for the old hag Themis[6]—

I thank you for the sight of the volume of Emerson's Orations,[7] which I have been reading with interest—It is impossible not to admire that on Emancipation[8] & next to it I like 'Man the Reformer'[9]—There are noble thoughts

& fine writing in all of them, but for the most part they seem to me too hazy & indefinite for the highest Truth & Beauty—They are grandiose & gigantic, but it is size without strength—more like the unsubstantial forms projected on the mist in the Brocken[10]—than like a living Hercules[11] able to kill the real dragons which infest the world—They fail to give me any feeling of effective power—& the same sense of general indistinctness prevents me from recognizing in them much of such beauty—of a very high kind—as might exist even in the absence of strength. As to some parts, of which I am bound to say that I do very entirely disapprove I will not trouble you with any remarks upon them—I will return you the volume from Montagu Square[12]—

Mrs Pollock[13] sends her kindest regards, & Walter[14] wishes particularly that you should know that he rides without a leading rein—& that he has had two out of three tumbles which are said to be necessary incidents on the road to good horsemanship—

We are going to offer ourselves to dine with the Moores[15] at Putney Heath[16] for Wednesday next—Perhaps we may meet then if not sooner—

Yours Ever | W. F. Pollock

RI MS JT/1/P/228
RI MS JT/1/TYP/6/1920

---

1. *your lines*: letter missing.
2. *Richmond Park*: the largest of London's Royal Parks established in the seventeenth century.
3. *nymphs*: in classical mythology, any of a class of semi-divine spirits, imagined as taking the form of a maiden inhabiting the sea, rivers, mountains, woods, trees, etc. (*OED*).
4. *naiads*: a nymph of fresh water, thought to inhabit a river, spring, etc., as its tutelary spirit (*OED*).
5. *Wells*: Tunbridge Wells (see letter 1423, n. 1).
6. *the old hag Themis*: Themis, the Greek Titaness of law, doubtless referring to Pollock's legal work.
7. *Emerson's Orations*: R. W. Emerson, *Orations, Lectures, and Addresses* (London: George Slater, 1849).
8. *that on Emancipation*: 'The Emancipation of the Negroes in the British West Indies' appeared on pp. 166–202 of the volume cited in n. 7.
9. *'Man the Reformer'*: 'Man the Reformer: a lecture on some of the prominent features of the present age' appeared on pp. 111–32 of the volume cited in n. 7.
10. *unsubstantial forms projected on the mist of the Brocken*: an atmospheric phenomena first observed at the Brocken, the highest peak of the Harz mountain range in northern Germany (which featured in Goethe's *Faust);* a magnified shadow of the spectator thrown on a bank of cloud in high mountains when the sun is low and often encircled by rainbow-like bands (*OED*). Tyndall wrote in his journal while exploring in the Alps with Huxley

in August 1856, 'On reaching a summit which overlooked the pass we found the valley beyond crammed with vapour, behind us however was the cloudless sun; suddenly each of us observed his shadow cast upon the pass in front of him, with a coloured halo round the head. Each raised his hand, the gigantic spectra before him did the same; and imitated all our further motions. We had in fact the spirit of the Brocken in all its splendour' (Journal, 24 August 1856, RI MS JT/2/13c/887).

11. *Hercules*: the Roman name for Heracles, son of Zeus in Greek mythology, known for his strength and defeating the dragon Ladon.
12. *Montagu Square*: see letter 1433, n. 7.
13. *Mrs Pollock*: Juliet Pollock.
14. *Walter*: Walter Pollock.
15. *the Moores*: James and Harriet Moore and their family Harriet Jane, Julia, John, and Graham.
16. *Putney Heath*: an open space in the London district of Putney.

From Edward Sabine   [19 October 1857][1]   1444

13. Ashley Place

My dear Sir—

In 1852 the declination[2] at Geneva[3] was 18°.2' W. The secular change[4] is not far from 6'.5 a year, westerly diminishing. Consequently in 1857, 5 years later it should be 32'.5 less than in 1852, or about 17°. 30'.

There is no doubt whatsoever that the Declination was West, not East, when Forbes's map[5] was drawn. And it is quite true that the line of Magnetic North[6] is set off erroneously in regard to the true Meridian[7] in his map. The error <u>may</u> not go further; the survey may be correct as regards the true or Astronomical bearings: and the draftsman's error in the compass rose[8] be the only mistake. If so it is an innocent one—but it may also be much more serious—of this I cannot judge, but you may be able to do so. If some parts have been laid down by astronomical bearings, and some by magnetic bearings with an erroneous declination allowed, the survey would be altogether vitiated—I hope that this is not the case; but that the error may be confined to the Rose.

You may allow 6' 1/2 as the secular change; it may be 1/2 a minute too little—If you should want greater exactness I can look further into the matter—

Sincerely yours | Edward Sabine. | Oct. 19th

RI MS JT/1/S/15
RI MS JT/1/TYP/4/1308

1. *[19 October 1857]*: the date is given based on mention of the year in the first paragraph, and
    the month and day are written in the signature line.
2. *declination*: of the magnetic needle: the deviation from the true north and south line, esp.
    the angular measure of this deviation (*OED*).
3. *Geneva*: see letter 1293, n. 3.
4. *secular change*: in scientific use, of processes of change: having a period of enormous length
    (*OED*).
5. *Forbes's map*: 'Map of the Mer de Glace of Chamouni. And of the Adjoining Mountains
    laid down from a detailed Survey in 1842 by Professor Forbes' in J. D. Forbes, *Travels
    through the Alps of Savoy* (see letter 1306, n. 8).
6. *Magnetic North*: the direction of the north magnetic pole (*OED*).
7. *true Meridian*: terrestrial meridian, or the great circle of the earth which lies in the plane
    of the celestial meridian of a place, and which passes through that place and the terrestrial
    poles; (also) that half of the latter circle which extends from pole to pole through the place,
    corresponding to a line of longitude (*OED*).
8. *compass rose*: a circle on a map showing the principal directions.

From Mary Anne Coxe   20 October [1857][1]   1445

9 Athol Crescent | Tuesday | October 20th.

Right glad indeed was I, dear Dr Tyndall, to see once more your well known
hand, and to find that I was still as *[1 word illeg]* say 'Tho' lost to sight, to
Memory dear'![2]

To think of little me having my health drank[3] on the top of the monarch
of mountains![4]

I shall be quite set up and conceited. I assure you I feel it no small compliment to have been remembered at such a place—and <u>by such a man</u>.

We tried for you with but faint hopes of success, on our way to Germany—
and again on our way back—and <u>very</u> disappointed I was not to see you.

I shall be delighted to get a pho:[5] it will be a real pleasure to me if good—
but a bad <u>likeness</u> I cannot bear (any thing totally <u>unlike</u> is much better) but
do pray let me have it, and I will tell you honestly, I can at any moment recall
you to my mind's eye as you really are, and I should not like <u>that</u> memory
mixed up with something unpleasant—and what monsters some are!

We spent 3 weeks at Homburg.[6] my brother[7] also with us—very enjoy-
ably—and with benefit to our health. I had a vague idea of making my fortune
there, in which laudable endeavour, I regularly lost a florin per diem,[8] but as I
never exceeded that sum, I can hardly come under the category of a 'professed
gambler'! What sums one does see there lost and won!

Your little Swiss box, do you remember it, is at this moment, as always, on my writing table—with its little key and all—quite a conceit!

James Coxe is gazetted Commissioner in Lunacy[9]—at a salary of £1000 or £1200—for it does not seem very clear which. He had two years intense labour—(I wonder if you would look at his 'report' if I sent it you?) gratuitously—and there will still be very hard work, and heavy responsibility. No more skeying[10] away to the Continent for us <u>now</u> I fear—we shall be tied by the leg for the future—and worse far, he will have to be a good deal away from home on tours of inspection of Asylums &c. but his heart is in the work—and sure I am he will do it well.

My brother left us yesterday to visit his master[11] at Inverness C.[12] but we hope to have him back presently, and to keep him till Parliament meets.

Now I hope you will just remember that you are—'Mortal'—and not do more than a mortal's work.

Please dont forget my photograph (be sure and pick me out <u>the best</u>) and with both our kindest regards, believe me now and always,

my dear Dr Tyndall | Yours very much | Mary Anne Coxe.

RI MS JT/1/TYP/1/274
LT Typescript Only

---

1. *[1857]*: the year is given based on mention of Tyndall having toasted Mary Coxe on top of Mont Blanc (see n. 3).

2. *'Tho' lost to sight, to Memory dear'*: from a song by composer and song writer George Linley (1798–1865), published *c.* 1848.

3. *having my health drank*: on the top of Mont Blanc on 13 August 1857, Tyndall and Hirst toasted to the health of Mary Coxe and other friends. See letter 1419, n. 10.

4. *the monarch of mountains*: see letter 1419, n. 10.

5. *pho*: a photograph. This portrait of John Tyndall—the frontispiece to this volume—was taken by the London-based Maull & Polyblank (later Maull & Fox), on 21 September 1857 (Journal, RI MS JT/2/13c/1034). Tyndall later wrote in his journal on 31 October 1857: 'Mrs. Coxe, and all other ladies that have seen it speak against my portrait' (Journal, RI MS JT/2/13c/1038); and on 19 March 1858: 'Joined, after a time, by Miss Drummond, when they all united in abusing my portrait, calling it the most horrid thing in the world, and affirming that they had never seen the Mr Tyndall therein represented' (Journal, RI MS JT/2/13c/1061).

6. *Homburg*: a town in Saarland, Germany.

7. *my brother*: William Cumming, see letter 1371, n. 6.

8. *florin per diem*: a daily allowance of one florin, a British coin worth two shillings. In the context of this letter, it would be the amount that Coxe was gambling per day.

9. *gazetted Commissioner in Lunacy*: Coxe was appointed the Commissioner of the General Board of Lunacy for Scotland in 1857.

10. *skeying*: to get clear, to sheer off (*OED*).
11. *his master*: possibly George Campbell (1823–1900), eighth Duke of Argyll (*ODNB*). William Cumming served as Campbell's physician and travelling companion, and became his private secretary, when Campbell served as Lord Privy Seal under two governments 1853–55 and 1859–60. See 'W. F. Cumming, M.D. Edin.', *Lancet,* 140 (1892), pp. 349–50.
12. *Inverness C*: Inveraray Castle, seat of the Duke of Argyll (Tyndall later advised on lightning conductors for the castle). It appears that LT mistranscribed 'Inveraray' as 'Inverness'.

To James David Forbes   21 October 1857   1446

21ˢᵗ. <u>Oct</u>. 1857. | Royal Institution of Great Britain [stamped]

My dear Sir

I regret much that the great length of time occupied in the printing of the accompanying paper[1] has prevented me from complying sooner with your request[2] to have a copy of it.

It was with much concern that I learned from Balmat[3] & others that you were prevented by the delicate state of your health from visiting the Mer de Glace[4] this year, and I should be truly rejoiced to learn that your strength is now restored. For six weeks last summer I was a sojourner at the Montanvert,[5] and thus augmented, in some measure, the scanty acquaintanceship with glaciers which I had previously possessed.

Will you allow me to say that the increased knowledge thus obtained, and the recent perusal of the works of others upon the same subject, have only served to enhance the high estimate which I had formed of your labours; and if I should finally differ from you on any point, I hope you will find it to be in a spirit not inconsistent with my conviction of the general excellence of your investigations.

Believe me dear Sir | most faithfully yours | <u>John Tyndall</u>

Dʳ Tyndall | Oc<u>t 1</u>857 | His visit to Mont <u>Anvert</u>[6]

StA JDF Incoming letters 1857, no. 85

1. *the accompanying paper*: J. Tyndall and T. H. Huxley, 'On the Structure and Motion of Glaciers', *Phil. Trans.*, 147 (1857), pp. 327–46.
2. *your request*: see letter 1349.
3. *Balmat*: Auguste Balmat.
4. *Mer de Glace*: see letter 1306, n. 12.
5. *Montanvert*: see letter 1415, n. 15.
6. *Dr Tyndall | Oct 1857 | His visit to Mont Anvert*: in Forbes's writing on the outside of the letter.

From James David Forbes   23 October 1857   1447

Edinburgh | 23 Oct 1857

My Dear Sir

I am obliged to you for sending me a copy of your Paper on Glaciers from the Philosophical Transactions[1] as well as for numerous former papers on different subjects with which you have favoured me.

I have also to thank you for the friendly opinion which you express in your accompanying Letter[2] of my Observations on Glaciers.

With respect to my health, (after which you were so good as to enquire) I am glad to say that it is now restored to very much the level at which it has stood for some years past. I did not hope in my projected tour to Switzerland to add any useful observations, but merely to revive pleasant recollections & perhaps to improve my health. My illness at Folkestone[3] (which probably had been some time coming on) entirely disappointed both these hopes. But I am thankful that it does not appear to have left permanent injury.

I shall of course be interested to receive some details of your newer observations and impressions.

I send by this post two small papers[4] & remain
dear Sir | Yours very truly | James D. Forbes
Dr Tyndall | &c &c &c

RI MS JT/1/F/32
RI MS JT/1/TYP/12/3942

---

1. *a copy of your Paper on Glaciers from the Philosophical Transactions*: J. Tyndall and T. H. Huxley, 'On the Structure and Motion of Glaciers', *Phil. Trans.*, 147 (1857), pp. 327–46.
2. *your accompanying Letter*: letter 1446.
3. *Folkestone*: a town in Kent, England, on the English Channel.
4. *two small papers*: not identified.

To Thomas Henry Huxley   [24][1] October 1857   1448

Saturday Morning

My dear Huxley,

Will come up to day at 6 o'clock but beg of you not to ask me to look at that paper upon glaciers[2]—I go for a dinner ʇand to see Mrs Huxley,[3] but for the present I have gone far enough on the ice—I enclose you a note received from Forbes[4] a few minutes ago. I hope you will be tender.

Ever Yours | J. Tyndall.

IC HP 8:30
Typed Transcript Only

———

1. *[24]*: the date is given based on relation to letter 1447.
2. *that paper upon glaciers*: T. H. Huxley, 'Observations on the Structure of Glacier Ice', *Phil. Mag.,* 14 (1857), pp. 241–60.
3. *Mrs Huxley*: Henrietta Huxley.
4. *note received from Forbes*: letter 1447.

From Samuel Haughton   31 October 1857   1449

Trin: Coll: Dub[1]: | 31 Oct. 1857.

My dear Sir,

I have just received and read with great pleasure your paper on the Structure and Motion of Glaciers.[2] We had an interesting discussion at the Dublin Meeting of the Br. Ass.[3] on Joints and Cleavage, in which I was left nearly alone in my support of the Mechanical Theory of Cleavage.[4] As I am, however, very obstinate by nature, I refused to be convinced by the vague allusions of my opponents to Polar Forces[5] &c.; and after the Meeting availed myself of some weeks residence in Dunmore E. Co Waterford,[6] to investigate the structure of the Joints and Cleavage of the old Red Conglomerate[7] in that locality—I have obtained several hundred accurate measurements, & I think I have discovered the law of the structure of the whole district.

I can reduce it to known mechanical principles, and feel satisfied that my idea as to the true nature of Slaty Cleavage is correct.

I believe that no English Geologist has any idea of the facilities for observation of such structure afforded by our Coast Sections in the South of Ireland, and that this country offers the key for the solution of many problems in Physical Geology.

I hope to send you for presentation to the Royal Society the detailed results of my summer's work before Christmas next—and think that you will agree with me that mechanical forces <u>only</u> can have produced the effects observable in the Old Red Conglomerate—I purposely chose this kind of rock as one in which the usual excuses of ignorance, such as Polar Force &c. —are of no avail.

I also think, that I can easily explain all the facts in Prof. Phillips[8] Report,[9] which at first sight, appeared at variance with the Mechanical Theory.

I hope to visit London in January next and to have an opportunity of talking with you on this subject.

I am, | Yours sincerely | Saml. Haughton

RI MS JT/1/TYP/2/485
LT Typescript Only

---

1. *Trin: Col: Dub*: Trinity College, Dublin.
2. *your paper on the Structure and Motion of Glaciers*: J. Tyndall and T. H. Huxley, 'On the Structure and Motion of Glaciers', *Phil. Trans.*, 147 (1857), pp. 327–46.
3. *Dublin Meeting of the Br. Ass*: the twenty-seventh meeting of the BAAS, held in Dublin in August and September 1857. The report of Haughton's paper is: S. Haughton, 'On a Model illustrative of Slaty Cleavage', in *Report of the Twenty-Seventh Meeting of the British Association for the Advancement of Science; held at Dublin in August and September 1857* (London: John Murray, 1858), p. 69.
4. *Mechanical Theory of Cleavage*: see letter 1295, n. 3.
5. *vague allusions of my opponents to Polar Forces*: the dispute over cleavage in slate was between those who argued for a crystalline model (polar forces) versus those who put forth a mechanical model (pressure). See letter 1295, n. 3.
6. *Dunmore E. Co Waterford*: Dunmore East, a tourist village in County Waterford on Ireland's southeastern coast.
7. *the old Red Conglomerate*: Haughton would soon publish on this, as communicated to the RS by Tyndall. See S. Haughton, 'On the Physical Structure of the Old Red Sandstone of the County of Waterford, Considered with Relation to Cleavage, Joint Surfaces, and Faults', *Phil. Trans.,* 148 (January 1858), pp. 333–48; and letters 1481 and 1484.
8. *Prof. Phillips*: John Phillips (1800–74), an English geologist who was the first to develop a global geologic time scale, and he coined the word Mesozoic. In 1857, Phillips gave at the RI 'Ten Lectures on Leading Questions in Geology'.
9. *Report*: J. Phillips, 'Report on cleavage and foliation in rocks, and on the theoretical explanations of these phænomena'. —Pt I', *Brit. Assoc. Rep. 1856*, pp. 369–96.

From Walter Pollock[1]   3 November 1857   1450

59 Montague Square | November 3rd 1857

My dear Mr Tyndall,

You must know that next year if we go to Tunbridge Wells[2] or any other place where there is a riding master I shall most likely have some more riding lessons, and I hope Fred[3] will have some too, and that we shall go to Tunbridge Wells and that I shall ride Seline and Fred Prince.[4] I was very much amused by your story,[5] though I don't believe it. If you think any of it true pray tell me so in your next letter. Fred writes this for me, and sends his regards to you. Pray write to me as soon as you can.

Yours affectionately | W. H. P. | (which I think very well written—| F. P. AMANUENSIS.[6])

Please come and see us some of these days | Yours for ever, | F. P.

RI MS JT/1/TYP/6/1921
LT Typescript Only

1. *From Walter Pollock*: this letter was written for Walter in the hand of Frederick Pollock, his brother.
2. *Tunbridge Wells*: see letter 1423, n. 1.
3. *Fred*: Frederick Pollock.
4. *Seline and Fred Prince*: Seline and Prince were the names of two horses, with Fred riding Prince.
5. *your story*: Tyndall had dined with the Pollocks on 29 October, so perhaps this refers to conversation then.
6. *AMANUENSIS*: one who copies or writes from the dictation of another (*OED*).

From Edward Sabine   6 November [1857][1]   1451

13. Ashley Place | Nov. 6.

Dear Tyndall.

There will be no Gov^t. Grant Com^c.[2] till after the Anniversary, and the last meeting of the present Council has I believe been held. The only way therefore that I can think of is that the Pres^t. & Officers should sanction an advance from the Donation fund[3] to be repaid from the Gov^t. Grant if they i.e. the G. G. Com^c. should approve of an apportionment of the Gov Grant to M^r. Gore.[4]

We were all in town together yesterday, & could have settled the matter at once if you had made the application a day or two earlier—now it will be a matter of Correspondence; Your best course I suppose would be to send the formal application, backed by your own and Faraday's opinion to Stokes, who acts as Secretary of the Gov^t. Grant Com^c. —and you might write to him to the same effect as you have to me in regard to keeping M^r. Gore in work when he is well disposed thereto. You shall have my voice, as far as it goes, for an advance from the donation fund.

We shall be at home tomorrow from 11 to 1 should you come this way. The medals were settled Chevreuil,[5] Copley[6] | Frankland[7]—& Lindley[8] | the Royals[9]

Sincerely yours | Edward Sabine

RI MS JT/1/S/42
RI MS JT/1/TYP/4/1310

1. *[1857]*: the year is given based on mention of RS medal recipients.

2. *Gov^t. Grant Com^e*: a RS committee established in 1849 for the purpose of granting funds
    to researchers. For more on the Government Grants, see R. M. MacLeod, 'The Royal
    Society and the Government Grant: Notes on the Administration of Scientific Research,
    1849–1914', *The Historical Journal*, 14 (1971), pp. 323–58; and M. Boas Hall, *All Scien-
    tists Now: The Royal Society in the Nineteenth Century* (New York: Cambridge University
    Press, 2002 (1984)), chapters 5 and 6.

3. *Donation fund*: a RS fund established in 1828 for the purpose of aiding experimen-
    tal research or rewarding researchers, or for other matters decided upon by the fund's
    committee.

4. *approve of an apportionment of the Gov Grant to M^r. Gore*: the *Roy. Soc. Proc.* show no
    record of Gore having received a Government Grant in 1857 nor in the years following.

5. *Chevreuil*: Michel Chevreul (1786–1889), a French chemist (*CDSB*). He received the
    1857 Copley Medal 'For his researches in organic chemistry, particularly on the composi-
    tion of the fats, and for his researches on the contrast of colours'. See 'Adjudication of the
    Medals of the Royal Society for the year 1857', *Phil. Trans.*, 147 (1857), p. 620.

6. *Copley*: the RS has awarded the Copley Medal annually since 1731, for 'outstanding
    achievements in research in any branch of science'.

7. *Frankland*: Edward Frankland. He received the Royal Medal 'For the isolation of the
    organic radicals of the alcohols, and for his researches on the metallic derivatives of alco-
    hol'. See 'Adjudication of the Medals of the Royal Society for the year 1857'.

8. *Lindley*: John Lindley (1799–1865), a botanist and professor at University College, Lon-
    don (*ODNB*). He received the Royal Medal 'For his numerous researches and works on
    all branches of scientific botany, and especially for his vegetable kingdom, and his genera
    & species of Orchideae'. See 'Adjudication of the Medals of the Royal Society for the year
    1857'.

9. *Royals*: the RS has awarded two Royal Medals each year since 1826, for 'the most import-
    ant contributions to the advancement of natural knowledge' (a third Royal Medal began
    being awarded each year in 1965).

To James Clerk Maxwell[1]   7 November 1857   1452

Nov. 7. 1857

My dear Sir

I am very much obliged to you for your kind thoughtfulness in sending
me your papers on the Dynamical Top & on the Perception of Colour,[2] as
also for your memoir on the Lines of Force[3] received some time ago. I never
doubted the possibility of giving Faraday's notions a mathematical form, and
you would probably be one of the last to deny the possibility of a totally different imagery by which the phenomena might be represented.[4]

Very sincerely yours | John Tyndall

Cambridge University Library MSS.Add.7655/II/13 and Add.7655/II/221

1. *James Clerk Maxwell*: James Clerk Maxwell (1831–79), a Scottish physicist who held professorships at King's College, London, and the University of Cambridge. He is best known for his work on electromagnetism, but also made contributions to thermodynamics and the study of colour vision (*ODNB*, *CDSB*).

2. *your papers on the Dynamical Top & on the Perception of Colour*: J. C. Maxwell, 'On a Dynamical for exhibiting the Phenomena of the Motion of a System of Invariable Form about a Fixed Point, with some Suggestions as to the Earth's Motion', *Transactions of the Royal Society of Edinburgh*, 21 (1857), pp. 559–70; and J. C. Maxwell, 'Account of Experiments on the Perception of Colour', *Phil. Mag.*, 14 (1857), pp. 40–47.

3. *your memoir on the Lines of Force*: J. C. Maxwell, 'On Faraday's Lines of Force', *Transactions of the Cambridge Philosophical Society*, 10 (1864), pp. 27–83. Maxwell read this paper in December 1855 and January 1856, and perhaps sent a copy to Tyndall before its later publication. Summaries were published in *Phil. Mag.* 11 (1856), pp. 404–5; and 12 (1856), pp. 316–20. Maxwell took care to publish in the *Phil. Mag.* 'because Faraday reads it and so does Tyndall', as he stated in a 22 February 1856 letter to William Thomson (Cambridge University Library MSS.Add.7342, M 95).

4. *I never doubted . . . might be represented*: here Tyndall contrasted his own material or particle-based view of polar substances acting by attraction and repulsion ('diamagnetic polarity') and Faraday's 'field theory' with its directive force only, which was put into mathematical form by Thomson and later Maxwell. See R. Jackson, 'John Tyndall and the Early History of Magnetism', *Annals of Science*, 72 (2015), pp. 435–89, on pp. 479–81.

To Rudolf Clausius   15 November 1857   1453

15th Nov. 1857.

My dear Clausius

The thought of writing to you occurred to me some time ago on reading your article 'on the motion which we call heat'[1] in which I felt great interest. It was translated by our friend Hirst for the Philosophical Magazine, and I hope your paper on the 'Electrolytes'[2] will soon appear also. I spent 6 weeks of the present summer beside the Mer de Glace[3] and was from 10 to 12 hours daily upon the ice. On the 13th of August, accompanied by Hirst and a single guide[4] I repeated Saussure's experiment of firing a shot on the summit of Mont Blanc.[5] I was regarded by the authorities of Chaumouni[6] as an exceptional case and was permitted to go where I pleased, and with as many & as few guides as I thought fit. Of course a considerable number of observations were collected during those 6 weeks, and I am now engaged in writing them out and discussing them. I dare say you saw a little paper of Huxley's in the Philosophical Magazine[7] some time ago. I have seen a good deal of the water and air-bubbles in ice since; the question is a very curious one for these water drops which are associated with the air-bubbles were assuredly once ice, and then the question arises how have they been melted? I think they have been melted by the quantity of heat which has been able to pass through the ice by <u>conduction</u>. The statement may appear paradoxical that ice can be melted by heat which has passed through another portion of the substance without melting it. But I think theory would lead us to expect a result of this kind. A particle at the surface of a mass of ice is in a very different mechanical condition from that of a particle within the mass, where it is surrounded and controlled on all sides by other particles. The quantity of heat necessary to give a particle at the surface the liberty of liquefaction, if I may use the term, would not I think be sufficient to give a particle within the mass the same liberty. But supposing a cavity—an airbubble, for example,—to exist within the ice, the particles surrounding the bubble would be in the same condition as those at the surface, and a quantity of heat may pass through the solid without melting it, which is capable of liberating the particles bounding the air cavity. I really see no other way of explaining the fact.[8] One may perhaps form a rough conception of the thing by thinking of the transmission of motion through a row of elastic balls, the last one of the row being liberated. The explanation given by Agassiz and the Schlagintweits,[9] is I think without basis; you know they suppose the air to be warmed by the heat which has passed through the ice, the air thus melting its surrounding walls.

While in Geneva[10] I thought of you, and made an excursion to the place where the Rhone enters the lake[11] to see the colour of its water: it was quite thick and muddy, but it was surprising how soon the suspended matter settled, the river losing its distinctive character and being absorbed in the general hue of the lake. Now the reason I thought of you was this. I am afraid we are too easily satisfied with the explanation usually given of the colour of the lake of Geneva; that it is merely the colour of the water itself. I know that pure water is blue, but the question still remains: is this blueness sufficient to account for the exceptional depth of hue of the lake of Geneva? Would a stratum of distilled water show an equal depth of colour? The river brings down an immense quantity of sediment, partly derived from the rubbing action of the glacier of the Rhone[12] upon its bed. Mud is extremely plentiful, and I think it must be the case that some of these mica[13] plates are so small as to remain suspended in the lake throughout its entire course. Faraday has obtained suspended gold so fine that it takes months to sink to the bottom of a bottle 6 inches high. But if the plates be in the lake, they must have an action like that which you attribute to the vesicles of water in the air, and hence, to their presence the colour of the water may be in some measure due. Do you not think this possible? I find the colour of snow depends materially upon its state of aggregation: on the summit of the Stelvio[14] last year where fine new snow was lying the hole made by my alpenstock[15] was filled with a delicate blue light. lower down the mountain, while the snow was more coarsely granular, this Effect almost wholly disappeared. I noticed the same upon the top of Mont Blanc this year: a hole made by the iron of my alpenstock was sufficiently deep to cause the colour to show itself in great beauty. You once mentioned to me I think that Brücke had made some experiments with the vapour of sulphur: can you tell me where they are described?[16] and tell me at the same time whether you retain your former opinion regarding the colour of the sky.

Have you formed any conception as to the manner in which a transverse undulatory motion, like that of radiant heat can build a tree in opposition to the force of gravity:[17] or how the chemical rays produce the tensions of vegetable & animal tissues? I was going to ask a few more questions, but I have given you enough of trouble for the present.

Believe me dear Clausius | Most sincerely yours | John Tyndall

RI MS JT/1/T/173

---

1. *your article 'on the motion which we call heat'*: R. Clausius, 'On the Nature of the Motion which we call Heat', *Phil. Mag.*, 14 (1857), pp. 108–27. Originally published as R. Clausius 'Ueber die Art der Bewegung, welche wir Wärme nenne', *Poggend. Annal.*, 100 (1857), pp. 353–80.

2. *hope your paper on the 'Electrolytes'*: R. Clausius, 'On the Conduction of Electricity in Electrolytes', *Phil. Mag.*, 15 (1858), pp. 94–109. Originally published as R. Clausius, 'Ueber die Elektricitätsleitung in Elektrolyten', *Poggend. Annal.*, 101 (1858), pp. 338–60.

3. *Mer de Glace*: see letter 1306, n. 12.

4. *a single guide*: Edouard Simond.

5. *Mont Blanc*: see letter 1320, n. 6.

6. *Chamouni*: see letter 1353, n. 9.

7. *a little paper of Huxley's in the Philosophical Magazine*: 'Observations on the Structure of Glacial Ice', *Phil. Mag.*, 14 (1857), pp. 241–60.

8. *I have seen a good deal of the water and air-bubbles in ice since . . . I really see no other way of explaining the fact*: this was perhaps one stimulus for Tyndall's 1859 work on the absorption of radiant heat by gases.

9. *the Schlagintweits*: two of the five Schlagintweit brothers (see letter 1416, n. 8), Adolf (1829–57) and Hermann (1826–82), wrote *Untersuchungen über die physikalische Geographie der Alpen* (Leipzig: Verlag von Johann Ambrosius Barth, 1850), updated as *Neue Untersuchungen über die physikalische Geographie und Geologie der Alpen* (Leipzig: Verlag von Johann Ambrosius Barth, 1854) which included their brother Robert as an author.

10. *Geneva*: see letter 1293, n. 3.

11. *the Rhone enters the lake*: near Le Bouveret, see letter 1293, n. 6.
12. *glacier of the Rhone*: see letter 1297, n. 12.
13. *mica*: a small glistening particle of talc, selenite, or other crystalline substance present in large numbers in the structure of a rock (*OED*).
14. *Stelvio*: Stelvio Pass in the Eastern Alps on the border of Switzerland and Italy, which Tyndall crossed on 2 September 1856. See *Glaciers of the Alps,* pp. 29–30.
15. *alpenstock*: see letter 1424, n. 6.
16. *where they are described?*: Clausius answered this question in his reply (see letter 1461).
17. *Have you formed . . . force of gravity*: this question is perhaps an early indication of Tyndall's view that the same forces operate in the organic as in the inorganic world.

From William Snow Harris   15 November 1857   1454

6 Windsor Villas | Plymouth | 15 November 1857

My dear Tyndall,

As we are to have a new Reform Bill[1] many in accordance with my own view are desirous to introduce what we term an 'Educational Franchise'—that is to say a Franchise based upon

Education and intelligence without regard to money qualification. By this we imagine the case of order and good Government would be greatly promoted—You will receive a pamphlet on this subject,[2] if upon consideration you find you can add your name to the enclosed list I am certain you would never regret it—The Memorial[3] will be supported by influential and intelligent men and I suspect not opposed by any.

I venture to put into your hands my last paper in the Phil. Magazine[4]— That perhaps may be superfluous, still I should wish you to possess a separate copy: I can not but believe that it contains some points on Statical Electricity of no small importance to our views of the probable nature and operation of Electrical Force. I am sure that there is no such thing as an insulated charged conductor—that is 'per se' The metal or other conducting surface is the mere coating or means of Electrical accumulation on some diaelectric in contact with it—that is the reason why a hollow conducting body appears to <u>contain</u> as they say—as much Electricity as a solid body of the same dimensions and form—See pages 8 and 9—

always most faithfully yours | W. Snow Harris

J. Tyndall Esq. Ph D F.R.S.

RI MS JT/1/TYP/2/474
LT Typescript Only

1. *new Reform Bill*: probably a reference to Palmerston's 7 May 1857 promise to introduce a bill to 'extend the franchise to classes of persons now unmeritedly excluded from that privilege' (HC Deb, 7 May 1857, vol. 145, cols 66–67). See also R. Saunders, *Democracy and the Vote in British Politics, 1848–1867: The Making of the Second Reform Act* (London: Routledge, 2011), ch. 4, pp. 101–30.

2. *a pamphlet on this subject*: *Parliamentary Reform: The Educational Franchise* (London: T. Hatchard, 1853). The pamphlet was initially published in 1853 and then reissued in 1857.

3. *The Memorial*: possibly the memorial in appendix A of H. Fortescue (Viscount Ebringon), *The Legal Profession and the Educational Franchise* (London: C. Roworth and Sons, 1858), p. 6. However, the signatories of the memorial are not shown.

4. *my last paper in the Phil. Magazine*: W. Snow Harris, 'Researches in Statical Electricity', *Phil. Mag.,* 14 (1857), pp. 81–100; and 14 (1857), pp. 176–83.

From Jules Lissajous 17 November 1857 1455

Paris 17 Novembre 1857

Mon cher Monsieur

Je viens de vous adresser par la poste un certain nombre d'exemplaires de mon mémoire destinés à quelques unes des personnes qui m'ont si cordialement accueilli à Londres. Je regrette de n'avoir pas pu disposer d'un plus grand nombre de ces mémoires, mais j'en ai eu à peine assez pour satisfaire à toutes les exigences du devoir et de la politesse.

Je vous prie de vouloir bien vous charger de leur distribution.

J'ai une foule de reproches à vous faire de ce que vous avez passé devant ma porte au mois de Septembre dernier sans vous arrêter, nous comptions bien sur vous et nous sommes même retournés en Bourgogne plus tôt afin de vous recevoir.

J'espère que vous nous rendrez cela plus tard, c'est une dette d'amitié dont je ne vous tiens pas quitte.

Veuillez, je vous prie, renouveller l'assurance de mon affectueux dévouement à votre excellent maître Mr Faraday pour lequel j'ai une véritable passion. Veuillez également me rappeler au souvenir de Mr Barlow et Mr Wheatstone que j'espérais voir à Paris.

Quant à vous continuez, je vous prie, à me conserver une place dans votre souvenir et dans votre amitié, et croyez bien que je n'oublierai ni votre cordiale hospitalité, ni votre concours dévoué dans la propagation de mes travaux scientifiques.

Tout à vous, de tout cœur. | J. Lissajous

Je ne sais si vous avez appris par les journaux que l'Empereur Napoléon par décret en date du 13 août dernier m'a nommé chevalier de l'ordre Impérial de la Légion d'honneur.

Paris, November 17th 1857

My dear Sir,

I just sent you by mail a few copies of my essay[1] intended for some of the people who welcomed me so warmly in London.[2] I regret not being able to send a larger number of these essays, but I barely had enough to meet the demands of duty and courtesy.

I would ask you to please take care of their distribution.

I hold a host of reproaches against you for passing by my door last September without stopping by;[3] we counted on your visit and we even returned to Burgundy earlier in order to welcome you.

I hope that you will indulge us later. It is a debt of friendship that I am unwilling to forget.

Please renew the assurance of my affectionate devotion to your excellent teacher Mr Faraday, for whom I hold great enthusiasm. Please send my regards to Mr Barlow and Mr Wheatstone whom I hope to see in Paris.

As for you, I beg of you to continue saving a place for me in your thoughts and in your friendship, and believe me that I will not forget your warm hospitality nor your devoted support in the circulation of my scientific work.

Yours very truly. | J. Lissajous

I do not know if you saw in the newspapers that the Emperor Napoleon,[4] by a decree on last August 13, appointed me Knight of the Imperial Order of the Legion of Honour.[5]

RI MS JT/1/L/26

---

1. *my essay*: possibly J. Lissajous, 'Étude optique des mouvements vibratoires', *Bulletin de la société d'encouragement*, 55 (1856), pp. 699–705; or J. Lissajous, 'Mémoire sur l'étude optique des mouvements vibratories', *Annales de chimie et de physique*, 51 (1857), pp. 147–231.

2. *in London*: see letter 1310, n. 2.

3. *passing by my door last September without stopping by*: Lissajous probably meant to write 'August'. On Tyndall's return to London from Switzerland in summer of 1857, he passed through Paris on 24 August, staying the night at l'Hôtel des Étrangers (Journal, 24 August 1857, RI MS JT/2/13c/1031).

4. *Emperor Napoleon*: Louis-Napoleon Bonaparte (1808–73), Emperor Napoleon III of the Second French Empire from 1852–70.

5. *appointed me Knight of the Imperial Order of the Legion of Honour*: a French order of merit (Chevalier) awarded for civil service.

From William Francis   20 November 1857   1456

Printing Office | Red Lion Court, Fleet Street E.C | Nov. 20th 1857.

Dear Tyndall

Thanks for Hirst's letter,[1] you shall have it back in a day or two; it will do <u>me</u> good to read it over and over again, so don't be angry at my not returning it at once. By all means let him send Poinsot;[2] his other requests shall also be attended to.

I send a duplicate proof of Ball's paper;[3] a proof has been forwarded by this evening's post to Dublin. I have not yet read it but have glanced at it, and like it much

Yours sincerely | William Francis
Prof. Tyndall

RI MS JT/1/HTYP/506
LT Typescript Only

---

1. *Hirst's letter*: letter missing.
2. *Poinsot*: presumably a paper by Louis Poinsot. Hirst often relied on Tyndall to supply him with recent issues of journals and scientific papers of interest while he lived on the Continent.
3. *Ball's paper*: J. Ball, 'Observations upon the Structure of Glaciers', *Phil. Mag.,* 14 (1857), pp. 481–504.

From Heinrich Debus   22 November 1857   1457

Queenwood am 22 Nov | 57

Mein lieber John!

Behalte das Buch so lange wie Du willst. Durch seine Zeilen zieht ein frischer, reiner ja himmlischer Geist der einen wie ein alter Freund anspricht und den man gern in seiner Nähe hat. Es verdient von vielen Menschen gekannt, gelesen und verstanden zu werden.

Das Object der Royal Society ist wie ich glaube sehr gut und daher möchte ich wohl gern ein Mitglied dieser Gesellschaft sein. Wenn Du daher glaubst dass ich zum Mitglied geeignet bin dann schlage mich vor. Aber ich wünsche auch dass mich die Gesellschaft gern nimmt und nicht aus Gunst und secundären Rücksichten. Für die Theilnahme an meinem Wohl und die freundliche Gesinnung die Du bei dieser und bei andern Gelegenheiten gegen mich gezeigt hast danke ich Dir recht herzlich. Ich freue mich auf die Ferien weil ich dann hoffen darf einmal wieder auf ein paar Stunden Deine Gegenwart zu geniessen. Es freut mich dass Frankland die Royal medale erhalten hat. Seine Arbeiten sind ebenso gut als die von einigen die diese Auszeichnung genossen haben. Über den innern Werth dieser Auszeichnung brauche ich nichts zu sagen; denn wir verstehn uns wie ich glaube vollkommen über diesen Punkt. Mit dem Wunsche dass es Dir auch ferner recht gut gehen möge

 Dein H. Debus.

Queenwood, 22 Nov | 57

My dear John!

Keep the book[1] as long as you want. Through its lines there runs a fresh, pure, indeed, heavenly spirit, which speaks to one like an old friend and which one is glad to have close by. It deserves to be known, read and understood by many people.

The objective of the Royal Society is, as I believe, very good and I therefore would very much like to be a member of this society.[2] If you therefore believe that I am suited to be a member, then propose me. But I also wish that the society is happy to accept me and not out of favour and secondary considerations. For your interest in my welfare and the friendly cast of mind that you have shown me on this and other occasions, I thank you most sincerely. I am looking forward to the holidays, because then I may hope to enjoy your presence once again[3] for a couple of hours. I am glad that Frankland has received the Royal Medal.[4] His works[5] are as good as those of others who have enjoyed this distinction. I do not need to say anything about the intrinsic value of this distinction, for we understand each other completely, as I believe, on this point. With the wish that you may continue to be very well

 Your H. Debus.

RI MS JT/1/D/23

---

1. *the book*: F. M. Müller, *Deutsche Liebe: Aus den Papieren eines Fremdlings* (Leipzig: F. A. Brockhaus, 1857). See Journal, 9 November 1857, RI MS JT/2/13c/1039. Tyndall subsequently lent this popular novel to Juliet Pollock (see letters 1470 and 1471).

2. *I therefore would very much like to be a member of this society*: Debus was not elected FRS until 1861, and would sit on the Society's council in the 1870s and 1880s (*ODNB*).

3. *I may hope to enjoy your presence once again*: from 26 August to 9 September 1857 Tyndall visited Queenwood where Debus taught chemistry. See Journal, RI MS JT/2/13c/1032.

4. *Frankland has received the Royal Medal*: see letter 1451, n. 7.

5. *His works*: see for example E. Frankland, 'On a New Series of Organic Bodies Containing Metals', *Phil. Trans.*, 142 (1852), pp. 417–44; E. Frankland, 'Researches on Organo-Metallic Bodies.—Second Memoir. Zincethyl', *Phil. Trans.*, 145 (1855), pp. 259–75; and E. Frankland, 'Researches on Organo-Metallic Bodies.—Third Memoir. On a New Series of Organic Acids Containing Nitrogen', *Phil. Trans.*, 147 (1857), pp. 59–78.

From Thomas Archer Hirst   26 November 1857   1458

26 Rue de Lacépède | Paris <u>Nov</u>^r 26/57

My dear John.

In a few days I expect John Martin will call upon you on his way through London to Paris. I have asked him to leave with you the sum of £22. 6p. which I believe to be the amount I now owe to you. He was going to bring me money but as I have more than sufficient for my own use I shall never have a better opportunity of paying my debts. Some time ago I sent him a letter to give you when paying the money but the letter contained simply my version of the account between us. I am afraid he has lost it but if so I will send you another. If you can so arrange it you will have a good opportunity of sending me, by him, any memoirs you may have set aside for me.[1] I also write by this post to Wright of Broughton[2] requesting him to forward to the Royal Institution a parcel of books which I wish John Martin also to bring to me. I hope this parcel will arrive before John Martin but if not please tell him you expect it daily and then he will perhaps wait a day longer in London should the same be necessary.

I am sorry to have to tell you that I have not been able to work for several days. I am suffering from an obstinate attack of my old complaint—the bowels. I have just returned from a couple of days walk in the Forest of Fontainebleau[3] and although much better I am not yet at all well. With a little more patience however I hope to be as good as ever again.

How are you? Send me a line when you have the time and inclination

Yours affectionately | <u>T. A. Hirst</u>

RI MS JT/1/H/236

---

1. *any memoirs you may have set aside for me*: see letter 1456, n. 2.
2. *Wright of Broughton*: Richard Wright.
3. *Forest of Fontainebleau*: a historic forested area about forty miles southeast of Paris, a popular holiday destination for Parisians.

From William Mathews[1]   27 November 1857   1459

MY DEAR SIR,

I am sorry that Birmingham Institute[2] will not have the advantage of a lecture from you, but thank you very much for your kind offer of your paper[3] which I am very anxious to see.

I write to-day on a different subject. You will see from the enclosed prospectus[4] that several active mountaineers are about to form "an alpine club".[5] Do you feel disposed to join us?[6] We should be extremely glad of your co-operation and I have no doubt of the success of the project—Rule VII[7] appears to me questionable. I am not answerable for it.

Did Mr. Huxley accompany you in the ascent of Mont Blanc?[8]

I am, My dear Sir, | Yours very faithfully, | WILLIAM MATHEWS Junr.

Eve and Creasey, p. 386

---

1. *William Mathews*: William Mathews (1828–1901), an English botanist, surveyor, mountaineer, and founding member of the Alpine Club in 1857 (ODNB).

2. *Birmingham Institute*: the Birmingham and Midland Institute was founded in 1852 'for the Diffusion and Advancement of Science, Literature and Art amongst all Classes of Persons resident in Birmingham and the Midland Counties'.

3. *your paper*: J. Tyndall and T. H. Huxley, 'On the Structure and Motion of Glaciers', *Phil. Trans.*, 147 (1857), pp. 327–46.

4. *enclosed prospectus*: Mathews likely sent to Tyndall the first circular (of three) describing the proposed formation of the club. The text of the first circular, and changes to the second and third, are included in D. F. O. Dangar and T. S. Blakeney, 'The Rise of Modern Mountaineering 1854–65', *Alpine Journal*, 62 (1957), pp. 16–38, on pp. 26–29.

5. *"an alpine club"*: the Alpine Club was founded at a dinner meeting at Ashley's Hotel in London on 22 December 1857. Tyndall was elected a member on 27 November 1858.

6. *to join us?*: Tyndall did not join as a founding member of the Alpine Club.

7. *Rule VII*: in the enclosed prospectus, Rule VII stated, 'A candidate shall not be eligible unless he shall have ascended to the top of a mountain 13,000 feet in height'. Rule VII was removed from the third circular.

8. *Did Mr. Huxley accompany you in the ascent of Mont Blanc?*: Thomas Huxley had joined Hirst and Tyndall in the Alps to summit its tallest mountain, Mont Blanc, and further explore the region from 10–22 August. They left Chamonix on 12 August, spent the night in a cabin on the Grand Mulets, and reached the summit, without an exhausted Huxley, on 13 August (Journal, RI MS JT/2/13c/1002, 1030). Huxley returned to London by 3 September, see *Life and Letters of Thomas Henry Huxley*, ed. L. Huxley, 2 vols (London: Macmillan and Co., 1900), vol. 1, p. 146.

To Thomas Archer Hirst   [28][1] November 1857   1460

Nov. 1857

My dear Tom

I have been spending the end of the day with Frankland who has removed from Manchester[2] and is now Lecturer at Saint Bartholomew's Hospital.[3] It is nearly 11 oC. but I think a line or two to you before I go to bed will not make my sleep less sound. Your last letter[4] was written to me in one of those dewy moments when the heart gets rid of its bondage and the feelings pour themselves like fluids into expression. It did me good to read the letter, and it did old Francis good also—I send you his note[5] which also contains his reply to your request on behalf of D[r]. Simpson.[6] I would not make the highest demands on the friendship of Francis: I do not think him quite capable of that half romantic self-sacrifice which one friend is sometimes ready to make for the sake of another: but there is a bottom of pure human good nature in the man; a tremulous sympathy with every act of feeling and of kindness which is highly refreshing to observe in a man plunged as he is in the midst of London business. Did I not see this fountain of good feeling in him I think I should have quarrelled with him long ago. I am rejoiced to learn that you have got another stroke of work off hands, and to find that you have made Liouville's acquaintance. It is I am firmly persuaded only the beginning of things. I might have got you a mathematical tutorship in the family of the late King of France[7] some days ago, but I thought, and think it better, that for the present you should not be disturbed by any thing of the kind. With regard to our comparative ways of living my dear Tom rest assured of it that there are times when my spirit thirsts for that tranquility which you habitually enjoy. A man's life in London is often peddled into driblets by the multiplicity of the demands upon his time, and this to him who longs to pour the entire strength which God has given him in one direction, is often distressing There is a wholeness of purpose possible to solitude which is not attainable in London. I do not complain—our lives must be taken in their entirety, and from this point of view I have reason to be thankful that I am what I am.

I am still working at the ice paper,[8] but hope soon to send it in to the Royal Society—Yesterday I made an important experiment with perfectly dry ice, which I had cooled down by solid carbonic acid to a temperature 1200Faht. below the freezing point. Under the hydraulic press it was squeezed into a mass as white as the whitest table salt. If a mass of quartz were crushed in the same manner there could not be more white & less transparence. This has an important bearing upon the conversion of the Névé[9] into ice. I have also made a multitude of experiments on the action of heat upon ice, and on the manner in which it disintegrates itself sometimes. These experiments have thus far kept the paper on hands.

It has gone 11 Tom and I fear if I continue writing that the wheel thus set going will continue to revolve. I was on the Royal Society Council last year, and proposed Frankland for a Royal Medal which I am happy to say was awarded him.[10] On Thursday last I was obliged to give a brief account of our trip up Mont Blanc[11] at the Philosophical Club[12] dinner.[13] But I am getting into gossip again, and will now lay my paper aside—Haply tomorrow will carry me to another sheet.

'Tomorrow' is come and passed: rolled on like a water-born pebble into the ocean of eternity. I worked all day at my ice and at my paper: shook a number of minor duties from my shoulders, and at 8 o'clock this evening began to feel cloudy. I knew that the heaviness would soon be compacted into pain if I persisted, So I shut all books and went off to a theatre. The brain sobered down and I walked home with the vigour of Driglington[14] in my limbs: It is now long past the noon of night. sound and deep be your slumbers my dear boy. There is no restorer like sleep.

I am a curious oscillating machine—I suppose we are all so, and that it is simply human to be so. Sometimes physical strength is here, and the will is defective. Sometimes the will is paramount, and we wish that the brain was iron to carry out our aspirations. Thank God however for the strength he gives—our wisdom and duty it is to use it aright; to apply but not to squander it. Still one would fain sometimes condense one's force to a cannon ball; it seems as if difficulties might thus be shattered which defy an equal amount of strength more widely distributed. But I suppose it is the impatient individual nature that thus expresses itself.

This letter was begun on the 20th—it is now the 28th and your note[15] has come to spur me on. I thought Tom that you would be calling upon me for money instead of sending it to me—However I suppose you can afford it and hence I am content. John Martin has not yet made his appearance. I will send you the papers on the stereoscope[16] by him. Ever & always | your affectionate John

RI MS JT/1/T/936
RI MS JT/1/HTYP/505–6

---

1. *[28]*: Hirst began writing this letter on 20 November and finished it on 28 November.
2. *Manchester*: see letter 1311, n. 8
3. *Saint Bartholomew's Hospital*: in 1857, Frankland left his position as professor of chemistry at Owens College in Manchester to be the lecturer in chemistry at Saint Bartholomew's Hospital in London.
4. *Your last letter*: letter missing. See letter 1456, n. 1.
5. *his note*: letter 1456.
6. *his reply to your request on behalf of D^r. Simpson*: Francis had written to Tyndall, 'his other requests shall also be attended to', and this likely referred to sending scientific papers or journal issues to Hirst. Maxwell Simpson (1815–1902), an Irish chemist who studied chemistry at the University of Marburg in the 1850s and married Anna Hirst's sister Mary Martin in 1845 (*ODNB*).

7. *the late King of France*: probably Louis Philippe I (1773–1850), who was King of France from 1830 until he was forced to abdicate during the French Revolution of 1848.

8. *the ice paper*: J. Tyndall, 'On Some Physical Properties of Ice', *Phil. Trans.*, 148 (1858), pp. 211–29.

9. *Névé*: see letter 1436, n. 9.

10. *proposed Frankland for a Royal Medal which I am happy to say was awarded him*: see letter 1457, n. 4.

11. *our trip up Mont Blanc*: in letters to Faraday (1419), Juliet Pollock (1421), and Hooker (1428), Tyndall described his and Hirst's ascent of Mont Blanc on 13 August 1857.

12. *Philosophical Club*: see letter 1299, n. 7.

13. *dinner*: Tyndall recounted his August 1857 ascent of Mont Blanc to the Philosophical Club on 19 November 1857. See T. G. Bonney, *Annals of the Philosophical Club of the Royal Society, Written from Its Minute Books* (London: Macmillan and Co., 1919), p. 136.

14. *Driglington*: Drighlington, a village in West Yorkshire, in northern central England, that was the site of the 1643 Battle of Adwalton Moor in the First English Civil War.

15. *your note*: letter 1458.

16. *papers on the stereoscope*: Tyndall's two articles on the stereoscope: 'On Binocular Vision and the Stereoscope', *Journal of the Photographic Society*, 3 (1856), pp. 96–102 and pp. 116–21.

From Rudolf Clausius   8 December 1857   1461

Zürich, 8 Dec. 1857.

Lieber Tyndall,

Sie haben mich durch Ihren Brief sehr erfreut. Ich würde ihn eher beantwortet haben, aber ich war, als ich ihn empfing, gerade damit beschäftigt Auszüge aus Ihrer und Huxley's Abhandlung in den Phil. Trans. und aus dem Briefe von Huxley im Phil. Mag. anzufertigen, um sie der hiesigen naturforschenden Gesellschaft vorzutragen und dann in der von dieser Gesellschaft herausgegebenen Vierteljahresschrift drucken zu lassen, und da dachte ich dann, ich wolle mit meiner Antwort bis zur Vollendung dieser Arbeit warten, weil sich vielleicht im Verlauf derselben noch einige interessante Fragen darböten. Ich habe den Vortrag gestern gehalten, und er hat grosses Interesse erregt. Da wegen eines vorausgegangenen anderen Vortrages die Zeit nicht mehr ausreichte, um auch noch eine Discussion über den Gegenstand zu eröffnen, so hat sich die Gesellschaft vorbehalten, in der nächsten Sitzung über 14 Tage noch einmal auf den Gegenstand einzutreten. Sie können sich denken, dass die Gesellschaft, vor welcher ich den Vortrag hielt, sehr competent war, über die Sache zu urtheilen. Fast alle kennen die Gletscher aus vielfacher eigener Anschauung; ferner befanden sich darunter Mousson, der das Buch geschrieben hat, Escher von der Linth, welcher mit Agassiz zusammen viele Versuche, unter andern die von Huxley bekämpften Infiltrationsversuche, gemacht hat, Heer, der bei seinen Wanderungen zur Untersuchung lebender und fossiler Pflanzen ebenfalls den so oft gesehenen Gletschern sein Interesse zugewandt hat, u.a. Alle haben sich sehr anerkennend über Ihre Untersuchungen ausgesprochen, besonders Escher v.d.L. der doch selbst bei der früheren Theorie betheiligt ist. In Bezug auf Ihre Experimente, Eis durch Druck zu formen, wurde bemerkt, dass, obwohl das Zerbrechen und Wiederzusammenfrieren ein wesentlich neues Moment sei, welches sowohl für die Physik im Allgemeinen, als auch speciell für die Gletscher von Wichtigkeit sei, doch ein gewisser Grad von Plasticität dabei mitwirken müsse, weil daohne sich unregelmässig zerbrochene Stücke, wenn sie auch noch so klein wären, nicht zu einem continuirlichen Ganzen zu ammenfügen könnten. Dass eine geringe Plasticität des Eises in der Nähe des Schmelzpunctes stattfindet, wird auch aus einer anderen Thatsache wahrscheinlich, auf welche ich Sie aufmerksam machen möchte, obwohl sie Ihnen wahrscheinlich auch bekannt ist. Bei Wachs ist bekanntlich das, was man gewöhnlich specifische Wärme nennt, bei den Temperaturen, wo es anfängt weich zu werden, grösser als bei sehr tiefen Temperaturen, oder bei so hohen Temperaturen, wo es ganz geschmolzen ist. Person hat gefunden Compt. rend. XXIII 626:

Temperaturgrenzen	-20° bis +2°;	+6° bis 26°;	26° bis 42°;	42° bis 58°;	66 bis 102°
Lat. Wärme	0,39	0,52	0,79	1,72	0,54

Dieses beruht natürlich darauf, dass während des Erweichens schon eine gewisse Quantität Wärme <u>latent</u> wird; dass also bei diesen Temperaturen das, was man gewöhnlich specifische Wärme nennt, schon die Summe der wirklichen specifischen W. und und eines Theiles der latenten Wärme ist. Will man also beim Wachs die latente Wärme vollständig bestimmen, so muss man nicht blos den Moment des wirklichen Zerfliessens betrachten, sondern auch das ganze Temperaturintervall des Weichwerdens. Dasselbe nun, was beim Wachs im grossen Maßstab stattfindet, hat Person im kleinen Maßstab auch beim Eis gefunden, Compt. rend. XXX 526. Er findet, dass die specifische Wärme bei niedern Temp. etwas kleiner ist als dicht bei 0°, und dass man die latente Wärme etwas grösser erhält, wenn man die Anfangstemperatur des zu schmelzenden Eises nicht zu nahe unter 0°, sondern wenigstens 2° unter 0 nimmt. Daraus erklärt er es, dass er die Zahl 80 fand, während de la Provostaye u. Dessain und Regnault nur 79 oder etwas darüber angeben. Person meint daher, <u>dass auch das Eis bei einer Temp. etwas unter 0°, etwa bei -1.5° oder -2° schon anfange weich zu werden</u>. Vielleicht lassen sich die Erscheinungen des Gletschereises noch vollständiger erklären, wenn Sie beide Erklärungsgründe vereinigen bei grösseren Formveränderungen ein Zerbrechen des Eises und nachheriges Wiederzusammenfrieren; bei sehr kleinen Formveränderungen dagegen ein plastisches Nachgeben. Sollten sich bei der Discussion über 14 Tage wesentlich neue Puncte herausstellen, so werde ich sie Ihnen schreiben. Der Auszug aus der ersten Abhandlung (Phil. Trans) wird im nächsten Heft unserer Vierteljahresschrift gedruckt. Sobald der Druck vollendet ist, werde ich Ihnen ein Exemplar zuschicken. Ich hoffe Ihre Gedanken der Hauptsache nach richtig wiedergegeben zu haben. An einer Stelle (in der Anmerkung über die durch seitliche Verschiebung entstehenden Zug- und Druckkräfte,) habe ich mir erlaubt, eine etwas andere Figur zu geben als Sie, weil ich glaubte, dass dadurch die Sache noch etwas anschaulicher würde. Der Auszug der zweiten Abh. erscheint in dem darauf folgenden Hefte.

Ihre Annahme über die Entstehung der Wassertropfen in den kleinen Höhlungen des Eises scheint mir ganz zulässig zu sein. Sie selbst haben durch Ihre Versuche hinlänglich bewiesen, dass eine Wasserhaut, die an <u>beiden</u> Seiten von Eis begrenzt ist, sich ganz anders verhält, als die Oberfläche, indem sie friert, während das Eis an der Oberfl. thaut; und dasselbe kann unzweifelhaft auch auf die inneren Oberflächen der kleinen Höhlungen angewandt werden. Man kann sich die Möglichkeit der Entstehung von flüssigem Wasser vielleicht auch so klar machen. Es ist bekannt, dass das Eis als fester Körper verdampfen kann. Wenn es nun richtig ist, was ich in meiner Abhandlung angenommen habe, dass der Zustand, welcher eintritt wenn ein Raum mit Wasserdampf gesättigt ist, nicht darin besteht, dass gar keine Verdampfung mehr stattfindet, sondern darin, dass Verdampfung und Niederschlag gleich stark sind, so kann man das auch auf jene kleinen Höhlungen anwenden. Ist nun die Temp. 0°, oder ein Minimum über 0° so schlägt sich vielleicht der Dampf, welcher sich von der Eiswand entwickelt hat, nicht wieder als Eis, sondern als Wasser nieder. Dadurch könnte eine kleine Quantität Eis in Wasser verwandelt werden, bis die innern Wände der Höhlung ganz mit Wasser benetzt wären, und nun das Verdampfende ebenfalls Wasser wäre, so dass eine wirkliche Compensation zwischen Verdampfung und Niederschlag einträte. Vielleicht könnte sich auch von der benetztenden Wasserhaut ein Theil am tiefsten Puncte der Höhlung sammeln. Ich habe dieses nur als eine mögliche Vorstellungsweise angeführt; es ist aber wohl nicht einmal nöthig, die Verdampfung zu Hülfe zu nehmen, um sich zu erklären, dass bei einer Temperatur, die gerade an der Grenze zwischen Frieren und Thauen steht, eine Masse, die nur an <u>einer</u> Seite den Cohäsions oder Crystallisationskräften der festen Theile des Körpers ausgesetzt, leichter flüssig wird, als eine solche, die diese Kräfte von <u>beiden</u> Seiten oder rund herum erleidet. Was das Eindringen der Wärme in das Eis anbetrifft so geschieht dieses wohl in grösserer Tiefe, wie Sie es annehmen durch Leitung, und nur in der Nähe der Oberfläche bis zu einer gewissen Tiefe kommt die directe Strahlung in Betracht.

Was Sie mir über die Farbe des Genfer Sees schreiben, hat mich sehr interessirt, indessen halte ich die Sache für so schwierig, dass ich nicht wage ein bestimmtes Urtheil darüber abzugeben. Ich glaube dass der Genfer See mit seiner blauen Farbe nicht allein dasteht; so soll z.B. auch das Wasser des Mittelländischen Meeres sehr blau sein, was besonders in der blauen Grotte von Capri auffällig sein soll. Ausserdem finde ich bei ein und demselben Gewässer eine grosse Abhängigkeit der Farbe von Nebenumständen. Ob das Wasser tief oder flach ist, ob es ruhig oder bewegt ist, ob die Atmosphäre heiter oder trübe ist. Der Züricher See zeigt zuweilen, wenn er ruhig und das Wetter klar ist, auf dem grössten Theile seiner Fläche eine tief blaue Farbe, und die Stellen, wo er flach ist, unterscheiden sich davon ganz deutlich durch eine fast hellgrüne Farbe. An anderen Tagen erscheint der ganze See mehr grünlich, und die seichten Stellen unterscheiden sich sehr wenig von den tiefen. Ich glaube daher dass bei der Erklärung der Farbe, diese Nebenumstände ebenfalls eine grosse Berücksichtigung verdienen. Dass die Farbe des Wassers in ihrer Entstehungsart mit der Farbe der Atmosphäre verwandt ist, wie Sie am Genfer See vermuthen, scheint allerdings auch aus anderen Beobachtungen hervorzugehen. So führt schon Newton (Opt. B.I P.II Prop 10) die Beobachtung von Halley an, welcher beim Tauchen tief im Meer an einem sonnenhellen Tage fand, dass die obere Fläche seiner Hand, welche von directen Sonnenlichte durch das Wasser hindurch beschienen war, roth erschien, die andere Fläche dagegen, (im reflectirten Lichte,) grün. Dass Brücke Experimente mit Schwefeldämpfen gemacht hat, ist mir nicht bekannt; die Versuche, von denen ich gesprochen habe, sind mit einem feinen Mastix-Niederschlage in Wasser gemacht. (Es ist möglich, dass ich damals gleichzeitig über Versuche von Brame mit Schwefel gesprochen habe, die sich aber nicht auf die Farbe beziehen.) Die recht beachtenswerthe Abhandlung von Brücke, welche zum grossen Theil gegen mich gerichtet ist, befindet sich: Sitzungsberichte der Wiener Ac: Juli 1852 und Pogg. Ann. LXXXVIII 363. Ich habe darauf geantwortet in demselben Bande von Pogg. Ann. S 543, indem ich zu meinen frühern Gründen, welche dafür sprechen, dass das niedergeschlagene Wasser der Atmosphäre die Form von Bläschen hat, noch neue hinzufüge, und dann ausspreche inwiefern ich an meiner früheren Erklärung der blauen Farbe des Himmels festhalte. Wenn man nämlich zugiebt, dass die reflectirenden Theilchen der Atmosphäre den gewöhnlichen Gesetzen der Reflexion und Brechung folgen (und dafür scheint die bei der Reflexion eintretende Polarisation zu sprechen,) so glaube ich an meinen Schlüssen festhalten zu müssen. Sagt man aber, die reflectirenden Theilchen der Atmosphäre sind so klein, dass die Reflexions- und Brechungsgesetze auf sie gar keine Anwendung mehr finden, so finden auf solche Theilchen auch meine Schlüsse allerdings keine Anwendung, aber dann darf man auch nicht mehr, wie Brücke, von Interferenz des an beiden Flächen reflectirten Lichtes sprechen, sondern dann befindet sich die ganze Sache noch auf einem vollständig unbekannten Gebiet.

Ihre Bemerkung, dass ein mit dem Alpenstock gemachtes Loch in dem Hochschnee ein schönes Blau zeigt, erregte in mir den Gedanken, den Sie gewiss auch schon gehabt haben, dass der vollkommen weisse Schnee das Licht mit derselben Farbe reflectirt, mit welcher es auffällt. Da nun in eine solche Höhlung nur Licht vom blauen Himmel eindringen kann, so dachte ich, dass sie darum blau erschiene. Ich sprach mit Escher v.d.L. darüber. Der hatte die Beobachtung dieser blauen Farbe auch oft gemacht, und meinte, er habe sie auch bei trübem Wetter gesehen. Ich fragte ihn, ob die Höhlung blauer erschienen sei, als der Himmel? Das könnte er nicht mit Sicherheit bejahen, aber er meinte, er glaube es: Für diesen Fall weiss ich für jetzt keinen Grund für die Entstehung dieses Blau anzugeben.

Ihre Frage, wie die strahlende Wärme Bäume construiren könne, kann ich noch weniger beantworten. Mit dieser Kühnheit habe ich mir die Frage noch nie gestellt, sondern habe mich nur gefragt, wie die strahlende Wärme an der Oberfläche der Blätter die Kohlensäure der Luft in Kohlenstoff und Sauerstoff zerlegen könne. Aber selbst diese vereinfachte Frage kann ich nicht beantworten, und muss mich für jetzt damit begnügen, dass die strahlende Wärme lebendige Kraft besitzt, und dass daher die <u>Möglichkeit</u> vorhanden ist, dass sie die Anziehungskraft der Atome überwinden, und somit eine Arbeit thun kann.

Für die Aufnahme meiner Abhandlung "über die Art der Bewegung etc" in das Phil. Mag, danke ich Ihnen sehr, und bitte auch T.A. Hirst meinen herzlichen Dank für die Anfertigung der Übersetzung zu sagen. Für den Fall dass Sie wie Sie in Ihrem Schreiben erwähnen, auch eine Übersetzung meiner Abh. "über die Electrolyten" aufnehmen, möchte ich Sie bitten einen kleinen Redactionsfehler, welcher sich darin eingeschlichen hat, zu verbessern. Auf der vorletzten Seite heisst es:

$$\frac{1}{41.\ c^2}$$

Grad C.
<u>zu erwärmen, wenn c die specifische Wärme der Flüssigkeit ist</u>.
Statt dessen muss es heissen:

$$\frac{1}{41.\ c.\ 5}$$

Grad C.
<u>zu erwärmen, wenn c die specifische Wärme und 5 das specifische Gewicht der Flüssigkeit ist</u>.

Ebenso muss die darauf folgende Formel statt,

$$\frac{p}{41.\,c}$$

Grad C.

heissen:

$$\frac{p}{41.\,c.\,5}$$

Grad C.

Als ich in diesem Sommer hörte dass Sie in der Schweiz seien, habe ich immer geglaubt Sie würden auch nach Zürich kommen. Ich würde mich sehr darüber gefreut haben. Es war damals der Prof. Beetz, welchen Sie von Berlin her kennen, und welcher jetzt in Bern ist, eine Zeitlang bei mir zum Besuch. Da haben wir oft davon gesprochen, ob Sie kommen würden, und ich habe jeden Tag das Fremdenblatt nachgelesen, aber immer vergebens.

Auf Ihre neue Abhandlung über die Gletscher, bin ich sehr gespannt. Wenn Sie sie mir nach dem Druck ebenfalls zuschicken wollen, wie die frühere, werden Sie mich sehr erfreuen.

Mit freundlichen Grüssen an Sie und die Hrn Huxley und Hirst, | Ihr | R. Clausius.

Zurich, 8 Dec. 1857

Dear Tyndall,

You have given me much pleasure through your letter. I would have answered it sooner, but when I received it I was busy preparing excerpts from your and Huxley's paper in the Phil. Trans.[1] and from the letter by Huxley in the Phil. Mag.,[2] in order to give a lecture on them to the natural history society[3] here and then to have them printed in the quarterly journal published by this society,[4] and then I thought that I would wait with my reply until I had completed this task, because some further interesting questions might perhaps present themselves in the course of this. I gave the lecture yesterday,[5] and it excited great interest. As there was no longer enough time to start a discussion on the subject as well due to another lecture[6] *[that had preceded mine]*, the society has left open the possibility of entering into the subject again in the next session in a fortnight's time. You can imagine that the society to which I gave the lecture was very competent to judge the matter. Almost everyone is familiar with the glaciers from frequent personal experience; furthermore, there were among them Mousson, who has written the book,[7] Escher von der Linth, who, together with Agassiz, has done numerous experiments, among others the infiltration experiments that were opposed by Huxley,[8] Heer,[9] who, on his walking tours to investigate living and fossil plants, has also turned his interest to the glaciers he had seen so often, among others. All have spoken very appreciatively of your investigations, especially Escher v.d.L., who is himself party to the earlier theory.[10] With regard to your experiments in forming ice through pressure,[11] it was remarked that, although breaking apart and freezing together again are a fundamentally new factor which is of importance both for physics in general as well as for glaciers in particular, a certain degree of plasticity must nevertheless also be involved in it, because without it, irregularly broken pieces, even if they were still so small, could not fit together to make a continuous whole. That a slight plasticity of ice occurs near melting point is probable from another fact to which I would like to draw your attention, although it is probably known to you already. With wax it is well-known that what one usually calls specific heat is greater at the temperatures where it begins to soften than at very low temperatures, or at such high temperatures where it has completely melted. Person[12] has found, Compt. rend. XXIII 626:[13]

Temperature limits	-20° to +2°;	+6° to 26°;	26° to 42°;	42° to 58°;	66 to 102°
Lat. heat	0,39	0,52	0,79	1,72	0,54

This of course is due to the fact that during the softening process, a certain quantity of heat is already becoming <u>latent</u>; that is, that at these temperatures, what one usually calls specific heat is already the sum of the actual specific heat and a part of the latent heat. If, with wax, one thus wants to fully determine the latent heat, then one must consider not merely the moment it actually melts away completely, but also the entire temperature interval of the process of softening. The same process that occurs with wax on a large scale, Person has now found on a small scale with ice as well, Compt. rend. XXX 526.[14] He finds that the specific heat at lower temperatures is somewhat less than close to 0°, and that one keeps the latent heat somewhat greater if one takes the initial temperature of the ice to be melted not too close to below 0°, but rather at least 2° below 0. It is from this that he explains the fact that he found the figure of 80, while de la Provostaye[15] and Dessain[16] and Regnault only state 79 or somewhat above this. Person therefore believes <u>that ice too already begins to soften at a temperature somewhat below 0°, roughly at -1.5° or -2°</u>. Perhaps the phenomena of glacial ice can be even more fully explained, if you combine both explanations: with major changes in form /there is/ a shattering of the ice and a subsequent refreezing of it; with very small changes in form, however, /there is/ a plastic yielding. Should some fundamentally new points emerge at the discussion in a fortnight's time, I shall write to you. The excerpt from the first paper (Phil. Trans) will be printed in the next issue of our quarterly journal.[17] As soon as printing is finished, I shall send you a copy. I hope I have correctly conveyed your thoughts in the main. In one place (in the note on tractive and compressive forces occurring through lateral displacement) I have taken the liberty of providing a somewhat different figure than you do, because I thought that the matter would thereby become somewhat clearer still. The summary of the second paper will appear in the following issue.[18]

Your assumption about the formation of water drops in the small cavities of ice seems to me to be quite acceptable. You yourself have sufficiently proven through your experiments that a film of water that is restricted on <u>both</u> sides by ice behaves quite differently from the surface, in that it freezes while the ice on the surface thaws; and the same can undoubtedly be applied to the inner surfaces of the small cavities as well. One can perhaps make the possibility of the formation of liquid water plain to oneself in this way too. It is known that ice as a solid body can vaporise. If what I have assumed in my paper[19] is now correct, that the state which occurs when a space is saturated with steam does not lie in the fact that no vaporisation at all is occurring any longer, but rather in the fact that vaporisation and condensation are equally strong, then one can also apply *[this]* to those small cavities. If the temperature is now 0°, or a minimum above 0°, then the vapour which has developed from the ice wall will possibly not condense again as ice, but instead as water. In this way, a small quantity of ice could be transformed into water, until the inner walls of the cavity were completely moistened with water, and now the evaporate was also water, so that a real compensation between vaporisation and condensation would occur. Perhaps part of the moistening water film could also collect at the deepest point of the cavity. I have offered this only as a possible way of imagining it; but it is probably not even necessary to use vaporisation to explain to oneself that at a temperature that is just on the boundary between freezing and thawing, a mass that is exposed on only <u>one</u> side to the cohesive or crystallisation forces of the solid parts of the body becomes liquid more easily than one that experiences these forces on <u>both</u> sides or all around. As far as the penetration of heat into ice is concerned, this probably happens at a greater depth, as you assume, through conduction, and only near the surface up to a certain depth does direct radiation come into consideration.

What you write to me about the colour of Lake Geneva[20] has interested me very much, however, I consider the matter to be so difficult that I dare not give a specific opinion on it. I believe that Lake Geneva is not alone with its blue colour; the water of the Mediterranean Sea, for instance, is also supposed to be very blue, which is supposed to be striking in the Blue Grotto of Capri[21] in particular. In addition, I find that with one and the same body of water, colour is greatly dependent on accompanying circumstances. Whether the water is deep or shallow, whether it is calm or choppy, whether the atmosphere is bright or dull. From time to time, Lake Zurich[22] displays, when it is calm and the weather is clear, a deep blue colour on most of its surface, and the places where it is shallow differ quite clearly from this through an almost light green colour. On other days, the whole lake appears more greenish, and the shallow places differ very little from the deep ones. I therefore believe that in explaining colour, these accompanying circumstances also deserve major consideration. The fact that the colour of the water is, in the way it comes about, related to the colour of the atmosphere, as you suspect from Lake Geneva, certainly seems to emerge from other observations as well. Even Newton (Opt. B.I P.II Prop 10)[23] cites the observation by Halley,[24] who, when diving deep in the sea on a sunny day, found that the upper surface of his hand, which was lit by direct sunlight through the water, appeared red, the other surface, however (in the reflected light) green. I did not know that Brücke has done experiments with sulphur vapours;[25] the experiments of which I spoke are done with a fine mastic precipitate in water. (It is possible that I spoke at the same time then about experiments by Brame with sulphur,[26] which, however, do not have any reference to colour). The quite noteworthy paper by Brücke, which is largely directed against me, can be found in: Sitzungsberichte der Wiener Ac: July 1852 and Pogg. Ann. LXXXVIII 363.[27] I have responded to it in the same volume of Pogg. Ann. p. 543,[28] by adding some more new reasons to my earlier ones supporting condensed water in the atmosphere having the form of tiny bubbles, and then expressing the extent to which I abide by my earlier explanation of the blue colour of the sky. If, you see, one admits that the reflecting particles of the atmosphere follow the usual laws of reflection and refraction (and the polarisation that occurs with reflection seems to support this), then I believe that I have to abide by my conclusions. If, however, one says that the reflective particles of the atmosphere are so small that the laws of reflection and refraction do not apply to them at all, then my conclusions would certainly not apply to such particles either, but then one may no longer speak, like Brücke, of the interference of the light reflecting on both surfaces, but then the whole matter will still be in a completely unknown territory.

Your remark that a hole made in snow high up in the mountains with an alpenstock[29] displays a beautiful blue[30] gave me the idea, which you will certainly have had already too, that completely white snow reflects light with the same colour with which the light falls. Now as only light from the blue sky can penetrate into such a cavity, I thought that the latter would appear blue for that reason. I spoke to Escher v.d.L. about it. He had often observed this blue colour too, and thought that he had also seen it in dull weather. I asked him whether the cavity had appeared bluer than the sky? He could not affirm this with certainty, /he said/, but he thought he believed it to be so. In this case, I cannot for now give any explanation for the origin of this blue.

I can give even less of an answer to your question of how radiant heat could build trees. [31] I have never asked myself this question with that kind of temerity, but rather have only wondered how radiant heat on the surface of the leaves could break down the carbonic acid of the air into carbon and oxygen. But I could not answer even this simplified question, and must satisfy myself for now with the fact that radiant heat possesses living force, and the <u>possibility</u> is therefore present that it can overcome the attractive force of atoms, and thereby perform work.

I thank you very much for including my paper 'über die Art der Bewegung' in the Phil. Mag[32] and ask that you also give T.A. Hirst my sincere thanks for preparing the translation. In the event that you also include a translation of my paper 'über die Electrolyten,'[33] as you mention in your letter,[34] I would like to ask you to correct a small editorial error which has crept into it. On the second-to-last page it says:

to be heated [by][35]

$$\frac{1}{41.\ c}$$

degrees C.

where c is the specific heat of the liquid.
Instead of this, it must read:

to be heated [by]

$$\frac{1}{41.\ c.\ 5}$$

degrees C.

where c is the specific heat of the liquid and 5 is the specific weight of the liquid.

Likewise, instead of

$$\frac{p}{41.\ c}$$

degrees C.
the following formula must read:

$$\frac{p}{41.\ c.\ 5}$$

degrees C.

When I heard this summer that you were in Switzerland,[36] I always believed you would also come to Zürich. I would have very much enjoyed this. At the time, Prof. Beetz,[37] whom you know from Berlin and who is now in Berne,[38] was visiting me for a while. We often spoke about whether you would come, and I read the newspaper every day, but always in vain.

I am looking forward very much to your new paper on glaciers.[39] If you are also going to send it to me after it has been printed, like the earlier one, you will please me very much.

With kind regards to you and Herr Huxley and Herr Hirst, | Your | R. Clausius.

RI MS JT/1/TYP/7/2226–30
LT Typescript Only

<hr>

1. *your and Huxley's paper in the Phil. Trans.*: J. Tyndall and T. H. Huxley, 'On the Structure and Motion of Glaciers', *Phil. Trans.*, 147 (1857), pp. 327–46.

2. *the letter by Huxley in the Phil. Mag.*: T. H. Huxley, 'Observations on the Structure of Glacial Ice', *Phil. Mag.*, 14 (1857), pp. 241–60.

3. *give a lecture on them to the natural history society*: the Naturforschende Gesellschaft in Zürich, founded in 1746.

4. *printed in the quarterly journal published by this society*: R. Clausius, 'Ueber die Structur un Bewegung der Gletscher; von John Tyndall und Thomas H. Huxley', *Vierteljahrsschrift der Naturforschenden Gesellschaft in Zürich,* 3 (1858), pp. 36–61.

5. *I gave the lecture yesterday*: 7 December 1857.

6. *another lecture*: not identified.

7. *the book*: A. Mousson, *Die Gletscher der Jetztzeit* (Zürich: Fr. Schulthess, 1854).

8. *infiltration experiments that were opposed by Huxley*: Huxley, in his 'Observations on the Structure of Glacial Ice' (see n. 2), discusses Agassiz's experiments found in L. Agassiz, *Nouvelles études et expériences sur les glaciers actuels: leur structure, leur progression et leur action physique sur le sol* (Paris: Victor Masson, 1847), pp. 170–79. Agassiz's *Nouvelles études* was the first and only part of a planned multi-volume work edited by Agassiz, Arnold Guyot, and Édouard Desor, titled *Système glaciaire ou recherches sur les glaciers, leur mécanisme, leur ancienne extension et le rôle qu'ils ont joué dans l'histoire de la terre.*

9. *Heer*: Oswald Heer (1809–83), a Swiss biologist and paleontologist who was a professor of botany and entomology at the University of Zurich (*CDSB*).

10. *the earlier theory*: see letter 1306, n. 8 and n. 10.

11. *your experiments in forming ice through pressure*: see J. Tyndall, 'Observations on Glaciers', *Roy. Inst. Proc.,* 2 (1854–58), pp. 320–27.

12. *Person*: Charles Person (see letter 1312, n. 5).

13. *Compt. rend. XXIII 626*: C. C. Person, 'Solution d'un problème sur las fusion des alliages', *Paris, Comptes Rendus,* 23 (1845), pp. 626–29.

14. *Compt. rend. XXX 526*: C. C. Person, 'Sur la chaleur latente de fusion de la glace', *Paris, Comptes Rendus,* 30 (1850), pp. 526–28.

15. *de la Provostaye*: Frédéric Hervé de la Provostaye (1812–63), a French physicist who held positions at the académie de Paris, the académie de la Seine, and worked as an inspector of scientific education in secondary schools. See E. Desains, 'Notice sur la vie et les travaux de M. Hervé de la Provostaye', *L'Institut, journal universel des sciences et des sociétés savantes en France et à l'étranger,* 32 (1849), pp. 89–91.

16. *Dessain*: François Edouard Desains (1812–63), a French physicist and professor at the lycée Napoléon who studied electricity, light, and the properties of ice. See J. Vieille, 'Desains (François-Edouard)', in *Mémorial de l'Association des anciens élèves de l'École normale 1846–1876* (Versailles: Cerf et Fils, 1877), pp. 197–200. For Desains's values of

the specific heat of ice, see E. Desains, 'Mémoire sur la chaleur spécifique de la glace', *Paris, Comptes Rendus*, 20 (1845), pp. 1345–47.

17.　*the first paper (Phil. Trans) will be printed in the next issue of our quarterly journal*: see n. 4.

18.　*the second paper will appear in the following issue*: T. Huxley, 'Beobachtungen über die Struktur des Gletschereises', *Vierteljahrsschrift der Naturforschenden Gesellschaft in Zürich*, 4 (1859), pp. 1–12.

19.　*my paper*: R. Clausius, 'Ueber die bewegende Kraft der Wärme und die Gesetze, welche sich daraus für die Wärmelehre selbst ableiten lassen', *Poggend. Annal.*, 79 (1850), pp. 368–97.

20.　*What you write to me about the colour of Lake Geneva*: see letter 1453.

21.　*Blue Grotto of Capri*: a sea cave found on the island of Capri, Italy. When sunlight enters an underwater opening the sea water turns bright blue in colour and illuminates the cave.

22.　*Lake Zurich*: a lake in Switzerland located southeast of Zürich.

23.　*(Opt. B.I P.II Prop 10)*: I. Newton, 'Prop. X. Prob. V. By the discovered Properties of Light to explain the permanent Colours of Natural Bodies', *Opticks*, Book I, Part II (London: S. Smith and B. Walford, 1704), pp. 135–41, with the specific mention of Halley on p. 139.

24.　*Halley*: Edmund Halley (1656–1742), an English astronomer and second Astronomer Royal at the Greenwich Observatory (*ODNB*).

25.　*experiments with sulphur vapours*: not identified

26.　*experiments by Brame with sulphur*: Charles Brame (1813–88), a French physician and professor of chemistry and pharmacy at the École de médecine de Tours. He conducted many investigations of sulphur many of which were published in the *Comptes rendus hebdomadaires de séances de l'Académie des sciences*. See C. Viel, 'Charles Brame (1813–1888), un Pionnier de recherche à l'École de médecine et pharmacie de Tours', *Mémoires de l'Académie des Sciences, Arts et Belles-Lettres de Touraine*, 24 (2011), pp. 105–26, esp. p. 120.

27.　*noteworthy paper by Brücke, which is largely directed against me, can be found in*: Sitzungsberichte der Wiener Ac: July 1852 and Pogg. Ann. LXXXVIII 363: E. Brücke, 'Ueber die Farben, welche trübe Medien im auffallenden und durchfallenden Lichte zeigen', *Poggend. Annal.*, 88 (1853), pp. 363–85. Brücke's paper critiqued R. Clausius, 'Ueber die blaue Farbe des Himmels und die Morgen- und Abendröthe', *Poggend. Annal.*, 76 (1849), pp. 188–95.

28.　*I have responded to it in the same volume of Pogg. Ann. p. 543*: R. Clausius, 'Ueber das Vorhandenseyn von Dampfbläschen in der Atmosphäre und ihren Einfluss auf die Lichtereflexion und die Farben derselben', *Poggend. Annal.*, 88 (1853), pp. 543–56.

29.　*alpenstock*: see letter 1424, n. 6.

30.　*Your remark that a hole made in snow . . . displays a beautiful blue*: see letter 1453.

31.　*your question of how radiant heat could build trees*: see letter 1453.

32.　*for including my paper 'über die Art der Bewegung' in the Phil. Mag*: R. Clausius, 'On the Nature of the Motion which we call Heat', *Phil. Mag.*, 14 (1857), pp. 108–27. Originally published as R. Clausius 'Ueber die Art der Bewegung, welche wir Wärme nenne', *Poggend. Annal.*, 100 (1857), pp. 353–80.

33. *a translation of my paper 'über die Electrolyten'*: R. Clausius, 'On the Conduction of Electricity in Electrolytes', *Phil. Mag.*, 15 (1858), pp. 94–109. Originally published as R. Clausius, 'Ueber die Elektricitätsleitung in Elektrolyten', *Poggend. Annal.*, 101 (1858), pp. 338–60.

34. *your letter*: see letter 1453.

35. *[by]*: although Clausius did not include it in this letter, 'um', the German equivalent of 'by', occurred immediately before the misprinted formula in the original article (see n. 33).

36. *you were in Switzerland*: Tyndall, accompanied by Thomas Hirst for a good portion of the trip, travelled in the Alps from 7 July to 23 August 1857.

37. *Prof. Beetz*: Wilhelm von Beetz (1822–86), a German physicist who worked as an assistant to Gustav Magnus and held professorships in Berlin, Bern, and Erlangen (*ADB*).

38. *Berne*: see letter 1297, n. 10.

39. *your new paper on glaciers*: J. Tyndall, 'On some Physical Properties of Ice', *Phil. Trans.*, 148 (1858), pp. 211–29.

From Michael Faraday 9 December 1857 1462

Brighton | 9 Dec 1857

My dear Tyndall,

I cannot resist the pleasure of saying I have very much enjoyed your paper.[1] Every part has given me delight. It goes on from point to point beautifully. You will find many pencil marks, for I made them as I read. I let them stand, for though many of them receive their answer as the story proceeds, yet they shew how the wording impresses a mind fresh to the subject, and perhaps here and there you may like to alter it slightly, if you wish the full idea, i.e. not an inaccurate one, to be suggested at first; and yet after all I believe it is not your phrase, but the natural jumping to a conclusion, that affects, or has affected, my pencil.

We return on Friday, when I will return you the paper.

Ever truly yours | M. Faraday

RI MS JT/1/TYP/12/4144
Faraday Correspondence, 5:3363
Typed Transcript Only

---

1. *your paper*: most likely a draft of the paper Tyndall would read before the RS on 17 December. Published as J. Tyndall, 'On Some Physical Properties of Ice', *Phil. Trans.*, 148 (1858), pp. 211–29.

From Samuel Haughton   14 December 1857   1463

Geological Society of Dublin, | Trinity College. | 14 Dec. 1857.

My dear Sir,

I have prepared four good diagrams and a map to illustrate my paper on the 'Cleavage and Joint Planes of the Old Red Limestone conglomerate of the Co. Waterford'.[1]

My paper itself is nearly ready but I fear will not be finished before Christmas; I must be in London myself before 18th Jan., for the Artillery Examination;[2] would it answer to bring my paper and diagrams to you at that time myself.

I have many Tables in my paper, and some easy Mechanical work based on 345 accurate observations. As you know more of the Royal Society than I do, perhaps you would kindly give me a hint as to the length of my paper; I mean as to whether I should condense greatly; or draw my inferences freely.

I am, | Yours very sincerely | Saml. Haughton.

John Tyndall Esq.

RI MS JT/1/TYP/2/486
LT Typescript Only

1. *my paper on the 'Cleavage and Joint Planes of the Old Red Limestone conglomerate of the Co. Waterford'*: see letter 1449, n. 7.

2. *Artillery Examination*: at the Royal Military Academy at Woolwich, Haughton had taught classes in scientific subjects to prepare students for becoming commissioned officers of the Royal Artillery and Royal Engineers. Haughton, with Joseph Allen Galbraith (1819–90), professor of natural and experimental philosophy at Trinity College, Dublin, also wrote a series of scientific and mathematical textbooks (the *Galbraith and Haughton Scientific Manuals*) which were widely employed in schools training young men for military service. See N. D. McMillan, 'Samuel Haughton, 1821–1897', in Mark McCartney and Andrew Whitaker (eds), *Physicists of Ireland: Passion and Precision* (Philadelphia: Institute of Physics Publishing, 2003), pp. 106–15, on p. 107–8; and M. De Arce, P. Wyse-Jackson, and N. McMillan, 'Patriotism, pedagogy and profit: Galbraith and Haughton's Mathematical Series (1851–91)', *History Ireland*, 23:2 (2015), pp. 26–29.

From Frances Hooker   [15 December 1857]¹   1464

Kew | Tuesday Eve.

My D.ʳ J.T.

Shall we be so fortunate as to find you disengaged on Friday² Papa³ is with us, & the Busks⁴ have promised to come to us on that day, you will join us, it will give us much pleasure. Our dinner hour be 6 p.m. . . . . come as much sooner as you can.

I have not answered your kind letter to me at Hitcham.⁵ I felt very grateful to you for y[ou].ʳ sympathy in my sorrow,⁶ but I was so disinterested to write anyone, that even you were no exception.

I am very glad you will come out next week⁷—it is an age since we have had more than the merest glimpse of you.

RI MS JT/1/TYP/8/2556

---

1. *[15 December 1857]*: the date is given by reference to Tyndall's dining with the Hookers on 18 December (see n. 2), the recent death of Frances Hooker's mother (see n. 6), and Tyndall's spending Christmas Day 1857 with the Hooker's at Kew (see n. 7).

2. *on Friday*: Tyndall noted in his journal entry beginning on 17 December 1857 that he dined at the Hookers on Friday, 18 December and 'met pleasant people there', perhaps the Busks that Frances mentions (Journal, RI MS JT/2/13c/1041).

3. *Papa*: John Henslow.

4. *the Busks*: George Busk (1807–86), a British naval surgeon and zoologist, and his wife Ellen Busk (1816–90) (*ODNB*).

5. *your kind letter at Hitcham*: letter missing. Frances likely was staying at her father John Henslow's home in the village of Hitcham in county Suffolk, where he served as rector of the Parish School beginning in 1837. See P. Armstrong, *The English Parson-Naturalist: A Companionship Between Science and Religion* (Herefordshire, England: Gracewing Publishing, 2000), pp. 8–9.

6. *my sorrow*: most likely because of her mother Harriet Henslow's (1797–1857) recent death on 20 November. On 6 September 1857, Hooker had written to Tyndall that her mother was 'very unwell' (see letter 1432).

7. *you will come out next week*: Tyndall spent Christmas Day 1857 with the Hookers at Kew (Journal, RI MS JT/2/13c/1042).

From Edward Sabine   15 December 1857   1465

Dear Tyndall—

Lord Wrottesley made so short a stay in London about the time when we were to have had a meeting of the Gov. Grant Com^{c1}: that he was compelled to defer its meeting till January, (towards the middle) when he will be in town again. I fear this will delay M^r. Gore's supply of means. With the recommendations you name his application, I should think, could scarcely fail of success.

I received this morning the enclosed[2] from Hansteen[3] of Christiania.[4] I am in some difficulty how to reply to it—Does Faraday lecture now in the mornings in the laboratory as formerly? If so, my reply would be that the course commences (when?) and terminates (when?)—and that subscription to the course is (what?)—and further that M^r. Hansteen's introduction could ensure his being formally known to M^r. Faraday and Anderson. —But I could say no more—Can you enable me to reply as above? or can you suggest anything better?

Sincerely yours | Edward Sabine | Dec. 15. 57

RI MS JT/1/S/16
RI MS JT/1/TYP/4/1309

———————

1. *Gov. Grant Com^c*: see letter 1451, n. 2.
2. *the enclosed*: not identified.
3. *Hansteen*: Christopher Hansteen (1784–1873), a Norwegian astronomer and physicist (*CDSB*).
4. *Christiania*: a previous name for Oslo, Norway.

To Edward Kennedy[1]   16 December 1857   1466

16$^{\underline{th}}$ Dec. 1857.

My dear Sir

It is very kind of you to send me your circular[2] though I do not belong to the Alpine Club.[3] The reasons which prevent me I have given in my note to M^r. Mathews.[4] I ask to be excused simply for the sake of my pursuits, and I hope that the scientific side of the Alpine question will not suffer by this arrangement

I shall take an interest in your proceedings and if permitted shall be glad to join you from time to time.

Do you not think that one black ball in five gives a great preponderance to the dissentient? Sometimes men in a club do things through a mere spurt of ill temper. This rule would always place a very great power in the hands of a very small number of members—Would you not think one black ball in <u>three</u> a sufficient security?

If rule VII[5] is to stand might it not be modified thus 'ascended a mountain to the height of 13000 feet'? It is the <u>elevation</u> I suppose that gives <u>the claim</u> not the reaching of the 'top' of the Mountain. M[r]. Coleman,[6] for example, once lost his baton on the grand plateau[7] and was obliged to return without reaching the summit of Mont Blanc.[8] Would not this be a sufficient claim upon his part even though he had not afterwards reached the top of the mountain?

This bit of information I gleaned from a document in the hut of the Grands Mulets.[9]

Yours very truly | <u>John Tyndall</u>

Alpine Club

---

1. *Edward Kennedy*: Edward Shirley Kennedy (1817–98), an English mountaineer and founding member of the Alpine Club. See C. E. M., 'In Memoriam. Edward Shirley Kennedy', *Alpine Journal*, 19 (1898), pp. 152–56.

2. *your circular*: see letter 1459, n. 4.

3. *I do not belong to the Alpine Club*: for the Alpine Club, see letter 1459, n. 5. While Tyndall wrote that he did not belong to the club, the club in fact had not been formed. In letter 1459, William Mathews informed Tyndall 'that several active mountaineers are about to form "an alpine club"'.

4. *my note to M[r]. Mathews*: letter missing (a reply to letter 1459). William Mathews, see letter 1459, n. 1.

5. *rule VII*: see letter 1459, n. 7.

6. *M[r]. Coleman*: Edmund Thomas Coleman (1824–92), an alpine mountaineer and artist (see 'Founders of the Alpine Club', in *ODNB*). He published *Scenes from the Snow-fields; being Illustrations of the Upper Ice-world of Mont Blanc, from Sketches Made on the Spot in the Years 1855, 1856, 1857, 1858* (London: Longman, Brown, 1859). He would later climb in the Cascade Range in the Pacific Northwest of the United States.

7. *the grand plateau*: a relatively flat region on the north side of Mont Blanc, beyond the Grand Mulets.

8. *Mont Blanc*: see letter 1320, n. 6.

9. *Grand Mulets*: see letter 1419, n. 9.

From George Edmondson   16 December 1857   1467

12m0.16—1857.

My dear John Tyndall,

Our concert and gala night[1] come off on the 21st. It would make it complete, if thou could join us. I am well aware we have nothing to offer thee equal to the attractions thou art surrounded with,—a hearty welcome and repose for thy weary head, are the inducements we have to offer.

I am | thy sincere Friend | Geo. Edmondson.

RI MS JT/5/14/-
LT Typescript Only

———————

1.   *concert and gala night*: the end of term Christmas event at Queenwood College.

From William Hopkins   16 December 1857   1468

Cambridge | Dec. 16 57

My dear Sir,

I returned your MSS[1] by post about two days ago. You will find various comments in pencil. Some of them might perhaps have been modified had I been able to read successive portions of your paper in the order in which you intend to arrange them. I had not time to read them as they came and the <u>discontinuity</u> in the paging made it impossible for me to place them afterwards in their proper order.

I quite believe that ice has very little extensibility but I think you have carried your notions respecting it too far. I recommended some revision of that part of your paper. Forbes's notion of the almost <u>indefinite</u> extensibility of ice, was always to me one of the great difficulties of his theory.[2] Even without considering it all but <u>in</u>extensible, your views approximate much nearer to those which I entertained when I investigated the motion and internal action of glaciers, than they do to Forbes's notion. Only after I had broken the ice by tension, I did not know well how to <u>join</u> it again. According to your notions a glacier <u>must break</u> in consequence of its motion. According to Forbes the breaking was an accidental rather than a necessary consequence of the motion.

I should like to have seen your paper on the Physical properties of ice,[3] but as soon as I feel well enough (I have lately been laid up with bronchial affection) I am off to Manchester,[4] and should have no time at present to read your paper.

I am satisfied that no <u>partial</u> observations will now be of any essential service as to glacial theories. A number of points on the surface of a glacier ought to be deter[mine]^d by accurate ad- measurements, and these points watched and re-measured from time to time for a good many years. It is only in this manner we can really know how a glacier <u>conducts </u>itself during a length of time.

Yours truly | W. Hopkins.

RI MS JT/1/TYP/2/612
LT Typescript Only

---

1. *your MSS*: based on the discussion of 'breaking', this must be a draft of Tyndall's paper received and read at the RS on 20 May 1858 and published as 'On the Physical Phenomena of Glaciers–Part I. Observations on the Mer de Glace', *Phil. Trans.*, 148 (1859), pp. 261–78. An abstract, received on 10 June 1858, was published as 'Observations on the Mer de Glace–Part I', *Roy. Soc. Proc.*, 9 (1857–59), pp. 245–47.
2. *his theory*: see letter 1306, n. 8.
3. *your paper on the Physical properties of ice*: J. Tyndall, 'On Some Physical Properties of Ice', *Phil. Trans.*, 148 (1858), pp. 211–29, which Tyndall read to the RS on 17 December 1857, the day after Hopkins wrote this letter.
4. *Manchester*: see letter 1311, n. 8.

From Sarah Faraday 17 December [1857][1] 1469

Royal Institution

Dear Dr. Tyndall

Can you not come up to us for an hour or two (now we are all pretty well) and give us the pleasure of your company? to day will suit us, or to morrow if you like it better—come to tea or at tea time, and you shall have your choice of a slice of beef or a mutton chop with it. I would say come at our dinner time 2.30. but that does not suit you so well and we do not get such a nice time for chat, but at any rate come, and believe me

Very sincerely yours | S. Faraday | Thursday Dec 17th.

RI MS JT/1/TYP/12/4167
Faraday Correspondence, 6:4732u
Typed Transcript Only

---

1. *[1857]*: in the *Faraday Correspondence*, this letter is given as either 1857 or 1863 (in those years, 17 December fell on a Thursday). LT's note in the typescript suggests 1857 would be more likely: 'before 1860 (when they were at Hampton Court)'.

From Juliet Pollock   [17 December 1857][1]   1470

59 Montague Square | Thursday night.

My dear Mr Tyndall,

Will you dine with us on Monday next at 7? I return the Deutsche Liebe[2] with many thanks: it has really interested me and it contains some passages that seem to me beautiful both in feeling and expression 'Hilf den Menschen, wo du kannst, liebe sie und danke Gott das du ein solches Menschen Herz wie das ihrige auf Erden gesehen, gekannt, geliebt, und verloren hast.'[3] is one of these.

We go to St. Julians[4] on the 24th for a few days.

Believe me my dear Mr Tyndall | Most truly Yours | Juliet Pollock.

Frederick's[5] regards to you. He is reading Wilhelm Meister!!'[6] Carlyle's doing into English.[7]

RI MS JT/1/TYP/6/1904
LT Typescript Only

1. *[17 December 1857]*: the date is given by the reference to *Deutsche Liebe, Wilhelm Meister,* and the invitation to dinner. For Tyndall's reply the following day see letter 1471, dated Friday 18 December 1857.

2. *Deutsche Liebe*: see letter 1457, n. 1.

3. *'Hilf den Menschen, wo du kannst, liebe sie und danke Gott das du ein solches Menschen Herz wie das ihrige auf Erden gesehen, gekannt, geliebt, und verloren hast'*: F. M. Müller, *Deutsche Liebe: Aus den Papieren eines Fremdlings* (Leipzig: F. A. Brockhaus, 1857), p. 174. The translation is, 'Help your fellow-men wherever you have the opportunity; love them, and thank God that it has been your lot on earth to see, to know, to love—and to lose such a human heart as hers'. See F. M. Müller, *German Love: From the Papers of an Alien*, trans. S. Winkworth (London: Chapman and Hall, 1858), pp. 151–52.

4. *St. Julians*: the family home of Juliet Pollock's cousins, the Herries, in the town of Sevenoaks in Kent, southeast of London.

5. *Frederick's*: William Frederick Pollock. Although both father and son went by 'Frederick', in his reply to Juliet (letter 1471), Tyndall understood her to have been referring to her husband, William Frederick Pollock.

6. *Wilhelm Meister*: *Wilhelm Meister's Apprenticeship,* the second novel of Johann Wolfgang von Goethe, first published in 1795–96.

7. *Carlyle's doing into English*: J. W. von Goethe, *Wilhelm Meister's Apprenticeship: A Novel from the German of Goethe*, 3 vols, trans. T. Carlyle (London: G. & W.B. Whittaker, 1824).

To Juliet Pollock   18 December 1857   1471

18<u>th</u> <u>Dec. 1857</u>

My dear M^{rs}. Pollock,

That black, bloated venomous spider <u>Fate</u>,[1] who spinneth from circumstances the fibres of his inexorable net, and whom at present I abhor with the combined vigour of the Celt & Saxon in my constitution,[2] will tie me to the lecture table of the London Institution[3] on Monday.

I knew you would like <u>that</u> paragraph in the Deutsche Liebe.[4]

Wilhelm Meister[5] has left on me a curiously composite impression—it is crammed full of practical wisdom, and still there is a mixture of mud with the gold which chills my admiration of the work. With M^r. Pollock's[6] prepossessions the very honey of the book will I doubt not be converted into acid. He will hate the old intellectual Titan[7] worse than ever.

Will you give my love to the boys,[8] & believe me
ever yours sincerely | <u>John Tyndall</u>

RI MS JT/1/T/1130
RI MS JT/1/TYP/6/1927

---

1. *spider <u>Fate</u>*: in Greek mythology, Arachne, a talented mortal weaver who challenged the goddess Athena to a weaving contest. Athena transformed Arachne into a spider for her hubris.

2. *the Celt & Saxon in my constitution*: ancestors of Tyndall's on his grandfather William Tyndall's (1766–1823) side had emigrated from Gloucestershire, England to Leighlinbridge, County Carlow, Ireland around 1670, where Tyndall was later born.

3. *London Institution*: see letter 1319, n. 5.

4. *<u>that</u> paragraph in the Deutsche Liebe*: see letter 1457, n. 1.

5. *Wilhelm Meister*: see letter 1470, n. 6.

6. *M^r. Pollock's*: William Frederick Pollock.

7. *old intellectual Titan*: Thomas Carlyle, who translated Goethe's *Wilhelm Meister's Apprenticeship* from German into English. See letter 1470, n. 7.

8. *the boys*: Frederick, Walter, and Maurice Pollock.

From Thomas Archer Hirst   19 December 1857   1472

Paris. Dec[r] 19th 1857

My dear John

John Martin has arrived with my books and a few of your memoirs.[1] He tells me he has lost a small letter[2] I sent to him long ago to deliver to you; in it was the following, (as far as I know) correct account of our money transactions since we last struck a balance about three years ago at Vine Cottage[3]

	£ s
I owe you	
for sundry small sums rec[d] at Broughton[4]	2. .5
for one very large sum ' ' '	20. .0
for money paid to Pridie	14. .13
' ' ' ' Wright	25. .0
' Bill drawn on you at Pau[5]	10. .0
' fares from Chamouni[6] (50 francs)	2. .0
Total	73. .18

	£ s
To set against this I sent you a draft from Paris	50–0
paid Duboscq for Char. points[7]	1–12
Total	51. .12

which deducted from	73. .18
leaves the amount I sent by John Martin	22. .6

So much for money matters

Your articles on the Stereoscope[8] pleased me very much. They are clearly and well written. You will republish them some day and if you do I would advise you to enter more into detail with respect to the <u>lenticular stereo-scope</u>.[9] It struck me on reading it that though perfectly lucid as far as it goes the lenticular

not so completely exhausted and finished off as the reflecting telescope. As there is no fear of 'your wearying the reader' too you might turn the pseudoscope inside out as it does its *[hat]*. It would serve to familiarize your readers, especially young ones with the true way of viewing this and related subjects. As I read it over (only once) I detected but one error and that a very small one; in case of republication however it might as well be corrected as not. Near the top of the second column of page 101 you say 'every pair of corresponding points of the base are <u>more</u> distant from each other &c' . . . you mean <u>less</u>

At present I am tolerably well and am working accordingly my time is a little more broken than it was for I attend a course of lectures by Liouville[10] at the College de France and <u>work them out</u> in my usual way which you may guess takes time. The subject (Theory of Numbers) is one I have hitherto too much neglected

I have at last made Poinsots acquaintance the finest old French gentleman I have ever met He is very old and feeble but his faculties still preserve their characteristic clearness. I visited him at his own house the other day and we had a long talk of two hours, on leaving him he insisted upon me taking from him copies of <u>all he has ever written</u> which to me was a valued and valuable gift. How are you? Write some day to

Your affectionate Son | <u>T A Hirst</u>

I have not included in our account the £9 sent to H. Hill[11] since I expect he will soon repay you the £4 of it which is still owing.

RI MS JT/1/H/237

---

1. *my books and a few of your memoirs*: see letter 1458.

2. *a small letter*: see letter 1458.

3. *Vine Cottage*: Hirst's residence in Broughton. He had previously mentioned to Tyndall that he had 'moved to my new residence and have commenced my initiation into the mysteries of household matters. I have already started in good earnest to make the place a neat and habitable one, by the time you next visit Queenwood I shall have a roof to shelter to you and you may dwell under it in peace as long as you wish and thank nobody for your lodging' (see letter 0977, Thomas Hirst to John Tyndall, 5 October 1854, *Tyndall Correspondence,* vol. 4).

4. *Broughton*: a village in Hampshire, England near to Queenwood College and Stockbridge, and location of Broughton Hill which Tyndall mentioned in letter 1306. Hirst moved to a new residence there in the fall of 1854 (see letter 0977, Thomas Hirst to John Tyndall, 5 October 1854, *Tyndall Correspondence,* vol. 4).

5. *Pau*: see letter 1298, n. 4.

6. *Chamouni*: see letter 1353, n. 9.

7. *Char. Points*: see letter 1440, n. 20.

8. *articles on the Stereoscope*: see letter 1460, n. 16.

9. *lenticular stereoscope*: the lenticular stereoscope was developed by Brewster to include lenses to make viewing the double image easier. It enjoyed great popularity around the time of the Great Exhibition in London in 1851.

10. *a course of lectures by Liouville*: Liouville gave eleven lecture courses on the theory of numbers while teaching at the College de France from 1851–82. See K. S. Williams, *Number Theory in the Spirit of Liouville* (New York: Cambridge University Press, 2011), p. 9.

11. *H. Hill*: Henry Hill, a friend of Hirst's.

To Auguste de la Rive 22 December 1857 1473

Royal Institution, Dec, 22nd, 1857.

My dear Friend

I send you a short abstract[1] of a paper[2] on some physical properties of ice which was read before the Royal Society a few days ago.

The observations upon glaciers in which the paper originated will I hope be communicated to the Royal Society at the end of January, or early in the month of February.[3] As soon as it is read I shall be happy to give you a short account of what it contains.

The proofs[4] are coming in from Mr Walkers regularly: I have already received as far as page 603. The work will prove invaluable to English men of science as a book of reference. To the general English public it would I think be more acceptable if the style were more purely English: but the probability is that you did not write for the general public, and therefore regard the question of style to be of small importance.

Your friend Faraday is very well and is preparing to give the young people a course of 6 lectures at Xmas.[5] Sabine, Grove, Wheatstone, Gassiot &c are all well.

I had a glorious season upon the ice, and clung faithfully to the Montanvert[6] for 6 weeks—I descended from it but once to Chamouni[7] and that was for the purpose of getting the laws regarding guides abolished in my case.[8] On the 13th of August I walked with a friend[9] and a single guide[10] to the top of Mont Blanc[11] and learned a good deal.

Will you kindly remember me to Madame DelaRive[12] and to your son and daughter.[13] I have not forgotten the pleasant day I spent among you

Believe me | always Yours Sincerely | J. Tyndall

RI MS JT/1/TYP/1/353
LT Typescript Only

---

1. *short abstract*: J. Tyndall, 'On Some Physical Properties of Ice', *Roy. Soc. Proc.*, 9 (1857), pp. 76–80.

2. *paper*: J. Tyndall, 'On Some Physical Properties of Ice', *Phil. Trans.*, 148 (1858), pp. 211–29.

3. *The observations upon glaciers . . . month of February*: Tyndall's paper on these observations was not received and read at the RS until 20 May 1858 and published as 'On the Physical Phenomena of Glaciers–Part I. Observations on the Mer de Glace', *Phil. Trans.*, 148 (1859), pp. 261–78

4. *proofs*: see letter 1293, n. 16.

5. *give the young people a course of 6 lectures at Xmas*: the RI's annual Christmas Lectures for children began in 1827 and still continue. Over his career, Faraday gave nineteen Christmas Lectures, with 'Static Electricity' as the topic in 1857. Tyndall would give twelve beginning in 1861. See F. A. J. L. James (ed.), *Christmas at the Royal Institution: An Anthology of Lectures* (Hackensack, NJ: World Scientific, 2007).

6. *Montanvert*: see letter 1415, n. 15.

7. *Chamouni*: see letter 1353, n. 9.

8. *getting the laws regarding guides abolished in my case*: see letter 1419, where Tyndall related to Faraday that he was able to bypass climbing rules regarding having guides because his pursuit was scientific in nature rather than recreational. The following summer, Tyndall would have difficulty with this (see letter 1551).

9. *a friend*: Thomas Hirst.

10. *a single guide*: Edouard Simond.

11. *Mont Blanc*: see letter 1459, n. 8.

12. *Madame DelaRive*: Louise-Victoire-Marie de la Rive (née Fatio, 1808–74), whom he married in 1855 after his first wife passed away in 1850.

13. *your son and daughter*: De la Rive had two sons and three daughters with his first wife, the writer Jeanne-Mathilde Duppa (1808–50), whom he married in 1826. One of his sons, Lucien de la Rive (1834–1924), was also a physicist; the other, William de la Rive (1827–1900), was a writer and politician.

From William Frederick Pollock   26 December 1857   1474

S[t]. Julian's | Sevenoaks | 26th. December 1857

My dear Tyndal[1]

You were so kind as to call a few nights since[2] to ask about us—We were out, driving with Forster,[3] and we are sorry to have missed seeing you—<u>Now I can tell you authentically where we are—and that is—here</u>—a sentence which would be thought full of a deep transcendental meaning, if it was found in the writings of a German metaphysicker[4]—but which to you, as chiefly a physicker, is only intended to mean what it does mean—& from me, as neither, perhaps if possible means rather less—The category of place—if you believe in place, that is—& can conceive an absolute here, more easily than some philosophers can an absolute—being thus disposed of—I may now proceed to that other perplexity—Time—which whether you may take it to be a function of Space—or that of Time—in a differential series to be taken as integrated between limits from a central zero to a plus and minus infinity as truly both ways—yet does assert itself in gross and palpable forms in clocks, calendars, histories, astronomical cycles, and other tangibilities and substantialities—and cannot wholly be disregarded in these mundane existences of ours—and so tell you that we came here the day before yesterday—& shall remain as I suppose until Thursday or Friday next—So that on Saturday, we may perchance see you meeting as juveniles round Faraday's Lecture Table—Meanwhile here & now we are—far more cheerful than I could have expected—and in the midst of a country looking perhaps more green than it should do in Christmas week—The boys[5] well & enjoying the outdoor & indoor recreations of the place—Walter the other day came to his mother in much indignation at a Scientific Glossary[6]—in which he had found 'Solid—see Fluid' and then turning to the F's—'Fluid not Solid'—which as he justly complained left him not much wiser than before—Then he framed quite by himself his own definition of a fluid as <u>stuff that you can't make a hole in</u>'—which is really as good a one as could be given in so few words—as stating one of the most obvious consequences of perfect mobility among the particles—The said Walter sends his love—Mrs Pollock desires her kindest regards—and I will wish you a merry Christmas & a happy new year—& so farewell—

Yours truly | W. F. Pollock

RI MS JT/1/P/230
RI MS JT/1/TYP/6/1928

---

1. *Tyndal*: William Frederick Pollock also misspelled Tyndall's name this way in letter 1439, as did Juliet Pollock in letter 1418. She also misspelled it as 'Tindal' in letter 1343.

2. *to call a few nights since*: letter missing.

3. *Forster*: John Forster (1812–76), an English biographer, critic, and editor of various journals (*ODNB*).

4. *metaphysicker*: German for *metaphysicist,* a person concerned with questions of a metaphysical, non-empirical, or hypothetical nature (*OED*).

5. *The boys*: Frederick, Walter, and Maurice Pollock.

6. *a Scientific Glossary*: not identified.

From Heinrich Debus   27 December 1857   1475

Broughton Decbr | 27th 57—

Lieber John!

Deinen Brief erhielt ich am Donnerstag morgen. Edmondson war in London und ich wusste nicht von wem er die Burners gekauft hatte; Daher konnte ich Dir nicht sogleich Antwort geben, sondern sehe mich erst heute im Stande Dir die gewünschte Auskunft mitzutheilen. Edmondson sagt mir dass Du die Burners von Simpson & Maule bekommen kannst. Seit Mittwoch wohne ich hier in Broughton bei Wright. Ich habe Queenwood ungern verlassen und meine Wohnung an einem andern Ort bezogen. Doch nach den Erfahrungen die ich Christmass 1855 und Christmass 1856 mit Edmonson gemacht habe, konnte ich nicht anders handeln. Mr Wright hat keine Knaben und hat daher genug Platz für mich. Mrs Wright ist sehr freundlich und so gefällt es mir recht gut hier. Jeden Tag gehe ich nach Queenwood und arbeite dort im Laboratorium. Queenwood ist jetzt ein einsamer Ort; keine Seele bewegt sich in der Nachbarschaft und es sieht aus wie eine Leiche.—Ich weiss nicht warum mir der Tom nicht schreibt. Wright erhält <u>sehr oft</u> Briefe von Tom ich aber habe dies Privilegium seit den 30th August nicht genossen. Da ich aber höre dass er wohl, thätig und zufrieden ist so habe ich mich mit Geduld in das Warten finden müssen ohne sonst beunruhigt zu sein. Heute gehe ich zu Tribe um dort zu essen. Tribe war anfangs der vergagenen Woche in Farnham wo er von Seiner Lordship of Winchester mit Priests Orders belegt wurde. Fox ist stolz auf seinen Sohn William; wie ich höre soll der letztere ein recht gutes Examen gemacht haben.

Mit herzlichen Grüssen | Dein | H. Debus.

Broughton Decbr | 27th 57—

Dear John!

I received your letter[1] on Thursday morning. Edmondson was in London and I did not know whom he had purchased the burners from; for this reason I could not give you an answer right away, but am only today in a position to give you the desired information. Edmondson tells me that you can get the burners from Simpson & Maule.[2] I have been living here in Broughton[3] at Wright's since Wednesday. I have reluctantly left Queenwood[4] and moved into lodgings at another place. But after the experiences I went through in Christmas 1855 and Christmas 1856 with Edmondson,[5] I could not do otherwise. Mr Wright does not have any boys and therefore has enough space for me. Mrs Wright is very kind and thus I like it here very much. I go to Queenwood every day and work there in the laboratory. Queenwood is a lonely place now; not a soul stirs in the neighbourhood and it looks like a corpse.—I do not know why Tom[6] does not write to me. Wright receives letters from Tom <u>very often</u> but I have not enjoyed this privilege since the <u>30th</u> of August. But as I hear that he is well, active and content, I have had to resign myself to waiting patiently without being otherwise worried. Today I am going to Tribe's[7] to dine there. Early last[8] week, Tribe was in Farnham,[9] where he was invested with Priest's Orders by His Lordship of Winchester.[10] Fox[11] is proud of his son William;[12] as I hear, the latter is supposed to have done a very good exam.[13]

With warm greetings | Your | H. Debus.

1. *your letter*: letter missing.
2. *Simpson & Maule*: the London firm Simpson, Maule, and Nicholson, founded in 1853 by George Simpson, George Maule, and Edward Nicholson. The firm sold scientific apparatus, and produced a variety of chemicals. During the 1860s it became one of the largest manufacturers of dyes in Britain (*ODNB*).
3. *Broughton*: see letter 1472, n. 4.
4. *Queenwood*: Queenwood College (see letter 1298, n. 13), where Debus had taught chemistry.
5. *experiences I went through in Christmas 1855 and Christmas 1856 with Edmondson*: not identified. While Debus held a generally unfavourable view of Edmondson, earlier Debus letters (in *Tyndall Correspondence,* vol. 5) do not reference anything specific about what Debus alludes to here.
6. *Tom*: Thomas Hirst.
7. *Tribe's*: Walter Harry Tribe (1832–1909), an English clergyman ordained, by the Bishop of Winchester, in 1857 at which time he became curate of Broughton, Hampshire. He became Rector of nearby Stockbridge between 1860–67, after which he served in several locations in India, before becoming the Archdeacon of Lahore from 1885–92. See 'The Rev. W. H. Tribe', *Times,* 19 May 1909, p. 13.
8. *last*: in the German, Debus spelled 'vergangenen' incorrectly.
9. *Farnham*: a town in Surrey, England, approximately 35 miles from London, and bordering Hampshire. Farnham Castle is one of the historic homes of the Bishop of Winchester.
10. *His Lordship of Winchester*: Charles Richard Sumner (1790–1874), an English clergyman who in 1827 became Bishop of Winchester (*ODNB*).
11. *Fox*: Luther Owen Fox (1811/12–79), a physician and surgeon who lived in Broughton, Hampshire. Fox studied at University College, London and St Bartholomew's Hospital. He obtained his license from the Society of Apothecaries in 1833, and became a Fellow of the Royal College of Surgeons in 1852, before receiving an MD from St Andrews. He attended the pupils and staff at Queenwood College and was on close terms with Tyndall, who often visited and dined with him. See 'Luther Owen Fox, M.D.', *Lancet,* 113 (1879), pp. 900–901.
12. *William*: William Tilbury Fox (1836–79), a physician and dermatologist, and son of Luther Owen Fox. Tilbury Fox attended Queenwood College, and later studied medicine at University College Hospital. He received his MD in 1858 (*ODNB*).
13. *a very good exam*: probably the second 1857 exam for Bachelor of Medicine at the University of London. Tilbury Fox received a Gold Medal and University Medical Scholarship in 1857. See 'Medical News', *British Medical Journal,* 1 (1857), pp. 1015–16.

From Henry Moseley   27 December [1857][1]   1476

32 Cumberland Terrace | Regents Park | 27 Dec.

My dear Sir,

I will endeavour to alter the days appointed for your examination and will communicate with you further on the subject next week.

The scheme of our examination was drawn out by a Committee composed chiefly of officers of the Engineers and Artillery and have received the official sanction of the Ordnance I apprehend we have no alternative but to conform to it however much we may dissent.

Yours truly | Henry Moseley.

RI MS JT/1/TYP/3/900
LT Typescript Only

---

1. *[1857]*: the year is given by the reference to Tyndall becoming an examiner for commissioned officers of the Royal Artillery and Royal Engineers at the Royal Military Academies at Woolwich and Sandhurst for the newly-established Council of Military Education in January 1858. See letter 1306, n. 2. Tyndall wrote in his journal on 17 December 1857 that 'Colonel Portlock called on me last Wednesday, enquiring why I was not on the list of the Woolwich examiners—I told him the reason. He proposed to me to become one of the examiners of the Commissioners, which I agreed to' (Journal, RI MS JT/2/13c/1041).

1858

From Carlo Matteucci   2 January 1858   1477

Pisa 2 jan. 1858

Mon cher M. Tyndall

J'ai bien des remerciements à vous faire de l'extrait que vous m'avez communiqué de votre mémoire sur la glace: aussitôt que je pourrai, je le ferai traduire pour le <u>Cimento</u>.

J'espère que M. Francis aura reçu tout le <u>Cimento</u> de cette année et qu'il voudra bien continuer à m'envoyer le Philosophical Magazine: ce n'est pas pour l'argent, mais je n'aurais pas votre journal aussitôt que je l'ai maintenant. Veuillez m'arranger cette affaire. J'ai une bonne nouvelle à vous donner: j'ai fini ou plutôt un jeune Physicien d'ici a fini pour moi une belle série d'expériences avec l'appareil de Weber. Elles ont été parfaitement bien faites et il n'y a plus rien à dire: un cylindre diamagnétique isolant dans une spirale électromagnétique a ses deux extrémités dans un état magnétique contraire, et il se comporte comme s'il avait des pôles opposés à ceux d'un cylindre magnétique. Soyez sûr que les actions du cylindre diamagnétique augmentent avec la force du courant mais elles ne varient pas comme font les corps magnétiques avec la longueur des cylindres. Je vais retourner pour quelques jours encore sur l'influence de la division sur le pouvoir diamagnétique.

En attendant je vous dirai que j'ai passé deux mois sur un sujet tout nouveau et qui m'a fait bien du plaisir. Aujourd'hui même j'ai envoyé l'extrait de ce travail à M. de Senarmont. C'est un phénomène nouveau d'induction magnétique qui met en évidence une propriété très singulière de conductibilité électrique et de structure développée dans le fer par la torsion. Excusez moi si je ne puis pas vous entretenir longuement sur ce sujet: je me borne à vous décrire en deux mots, peut-être la plus simple des expériences que je vous prie de dire à Faraday et de montrer si vous voulez à la Société Royale.

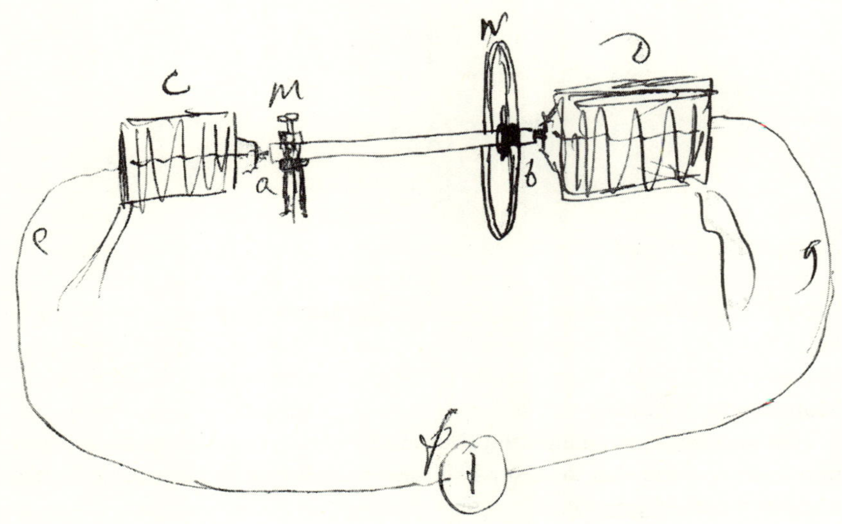

ab—cylindre de fer

cd—deux électro-aimants avec le *[trou]* dans l'axe comme pour l'expérience de Faraday.

aefg fil de cuivre sur le galvanomètre

m. mâchoire pour fermer le cylindre

N—roue en bois pour tordre le cylindre

Le cylindre de fer doux ab est fixé dans un appareil de torsion; m c'est une mâchoire de laiton pour fixer un cylindre; n la roue au centre de laquelle est fixée l'autre extrémité du cylindre, et à l'aide de laquelle on peut le tordre; a e f g b un fil de cuivre soudé à ses extrémités aux extremités du cylindre de fer: *[dans]* ce circuit entre un galvanomètre & à fil court. Le cylindre est aimanté par les deux électro-aimants c d. On peut aussi bien aimanter le cylindre avec une spirale. Voici l'expérience: en tordant et en détordant le cylindre on a des courants induits très forts de sens opposés. Le sens se renverse en tordant dans le sens opposé ou en renversant les pôles. J'ai donné les lois de ces phénomènes dont l'explication consiste à imaginer que le cylindre de fer est composé d'un faisceau de fibres qui par la torsion se forment en spirale autour de l'axe magnétique qui reste invariable par l'invariabilité de la force magnétisante et par la propriété du fer doux. Vous verrez dans le mémoire tous les détails et cette explication n'est en défaut dans aucun cas. Cela explique très facilement les courants induits sur une seconde bobine &c &c

Mille félicitations à M. Faraday et à de la Rive

C. Matteucci

Pisa 2 January 1858

My dear Mr Tyndall,

Many thanks for the excerpt from your memoir on the ice[1] you communicated with me; I will have it translated for the <u>Cimento</u>[2] as soon as I can.

I hope that Mr Francis would have received all the <u>Cimento</u> from this year and that he will continue to send me the Philosophical Magazine: it is not for the money, but I would not have your journal as early as I have it now. Please sort out this matter for me. I have good news to give you: I have finished or rather a young physicist from here[3] has finished for me a great series of experiments with Weber's[4] device.[5] They have been perfectly done and there is nothing more to say: a diamagnetic cylinder isolated in an electromagnetic spiral has its two extremities in an opposite magnetic state, and it behaves like it had poles opposite to the ones of a magnetic cylinder. Rest assured that the actions of the diamagnetic cylinder increase with the strength of the current but they do not vary as the magnetic objects do, with the length of the cylinders. I will come back for a few more days on the influence of the division on the diamagnetic power.

In the meantime, I will tell you that I spent two months on a brand new subject and that I enjoyed it a lot. Today, I sent the excerpt of this work to Mr. de Senarmont.[6] It is a new phenomenon of magnetic induction that highlights a very unique property of electric conductibility and of a structure developed in the iron by the torsion. Excuse me if I cannot say more on this subject: I restrict myself to describing in two words, perhaps the simplest experiment, which I beg you to tell Faraday and show if you wish to the Royal Society.

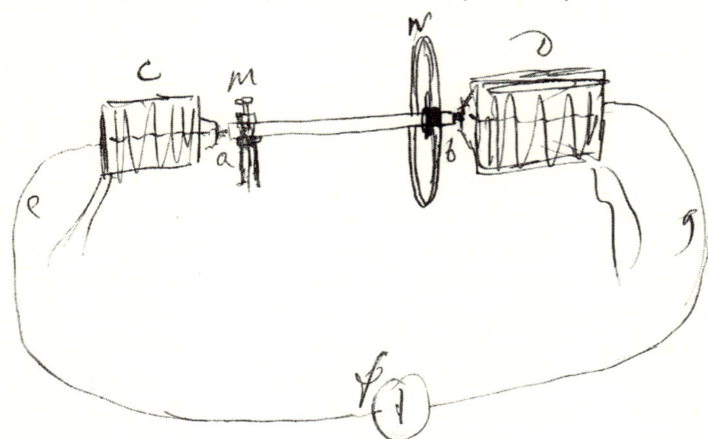

ab—iron cylinder

cd—two electro-magnets with the hole in the axis like for Faraday's experiment.

aefg copper wire on the galvanometer[7]

<u>m</u>. jaw to close the cylinder

<u>N</u>—wooden wheel to twist the cylinder

The soft iron cylinder <u>ab</u> is fixed into a torsion device, <u>m</u> is a jaw made of brass to fix a cylinder; <u>n</u> the wheel at the centre of which the other extremity of the cylinder is fixed, and with the help of which we can twist it, <u>a</u> <u>e</u> f g <u>b</u> a copper wire soldered to the extremities of the iron cylinder: *[in]* this circuit between a galvanometer & with short wire. The cylinder is magnetized by the two electro-magnets <u>c</u> <u>d</u>. We can as well magnetize the cylinder with a spiral. Here is the experiment: by twisting and untwisting the cylinder we obtain very high induced currents from opposite directions. The direction is reversed by twisting in the opposite direction or by reversing the poles. I gave the laws of these phenomena whose explanation consists in imagining that the iron cylinder is composed of a beam of fibres which under the effect of torsion are shaped in a spiral around the magnetic axis that stays invariable by the invariability of the magnetising force and by the property of the soft iron. You will see in the essay all the details and this explanation is not faulted in any case. It explains very easily the induced currents on a second bobbin of &c &c

A thousand congratulations to Mr Faraday and Mr. de la Rive

C. Matteucci

RI MS JT/1/M/63

---

1. *excerpt from your memoir on the ice*: J. Tyndall and T. H. Huxley, 'On the Structure and Motion of Glaciers', *Phil. Trans.*, 147 (1857), pp. 327–46.

2. *translated for the Cimento*: it does not appear that Tyndall's paper was ever translated and published in *Il Nuovo Cimento,* a physics journal established in 1855 by Matteucci and Raffaele Piria (1814–65), an Italian chemist, as a continuation of the journal *Il Cimento* (est. 1844).

3. *a young physicist from here*: not identified.

4. *Weber's*: Wilhelm Eduard Weber (1804–91), a German physicist and professor of physics at Göttingen and Leipzig, noted for his work on electrodynamics. In 1833, he and German mathematician Carl Gauss (1777–1855) invented the electromagnetic telegraph (*CDSB*).

5. *device*: Weber's apparatus to explore diamagnetism, which was a bone of contention between Matteucci and Tyndall. Matteucci appears to be comparing diamagnetics and magnetics. Tyndall had used apparatus for his diamagnetism experiments built by Leyser and specified by Weber.

6. *Mr. de Senarmont*: Henri Hureau de Sénarmont (1808–62), a French physicist and mineralogist of the École des Mines in Paris (*CDSB*).

7. *galvanometer*: an apparatus for detecting the existence and determining the direction and intensity of a galvanic current (*OED*).

To Thomas Archer Hirst   8 January 1858   1478

8th Jan. 1858[1]

My dear Tom

Your little note[2] was like a bunch of violets to my moral sense. A happy new year to you my dear boy and dozens of them till both the heads of yourself & your father grow hoar.[3] I am rather plunged in duties. Faraday[4] is now lecturing—the Prince of Wales[5] and his addentants[6] come to every lecture, and the throng is immense. half a dozen French princes are also there. The Prince comes 10 minutes before the time & it sometimes falls to my lot to amuse him during this interval. Some days ago, I half choked him by the fumes of explosive antimony—Yesterday I showed him Chladni's figures,[7] and caused him to use the fiddle bow[8] himself. He is a mild amicable boy. The glacier paper[9] is finished but not yet sent in—I hope to be able to forward it in a few days. I thought I had sent you a copy of last years paper.[10] This post will repair the omission. It gives me sincere satisfaction to see the ties of amity getting closer between you & Wertheim: from the first moment of my acquaintance with him I liked to him. He is a first rate <u>moral</u> workman, That is he works with sincerity and not for shew, and his results are worthy of all confidence. Pray present my very kindest remembrances to him when you see him next. And now as regards the bowels. For three years I was a martyr to constipation. I had almost given up all hope of being cured, but thank God it has completely vanished. I feel the strongest confidence that observation of the effects of food, <u>smoking</u> and exercise would lead you to a cure. Coffee I found very detrimental to me—taken after dinner it was within an inch of killing Goethe[11]—He said it paralyzed his bowels. Chocolate is good & gentle—A small thing may have very great influence; hence <u>watch the small things.</u>

I dine today with my friend Pollock.[12] The more I see of him the stronger my affection for him becomes. His wife[13] is altogether admirable. Huxley often asks after you—Whenever you come to London you will find a circle of ready made friends. I go to an excellent lady[14] tomorrow evening to read for her & her daughters the account of our trip up Mont Blanc.[15]

I told you that I had given up the examination of the Artillery & Engineers,[16] A Military Commission has been lately formed for extending & arranging the system of examination; and a member of it Gen. Portlock,[17] called upon me a few days ago wishing me to become an examiner.[18] I thought it might lead to something important and agreed to try it. There are to be 10 examinations a year & for this the sum of £150 is allowed. It is too small, but the labour I am told will not be great. This morning I received the notification that his Royal Highness the Com[r]. in Chief[19] with the concurrence of the Secretary at War[20] had appointed me examiner for the present year.

What do you intend to do with yourself next Summer. I have never seen the glaciers about Zermatt[21] and they are essential to the completion of my knowledge. Shall we march there together and climb to the top of Monté Rosa?[22] There is an Alpine Club established in London which I was asked to join but declined,[23] one of their conditions is that a candidate must have reached a height of 13000 feet.

Now goodbye Tom. | Ever yours affectionately | John Tyndall.

RI MS JT/1/T/635
RI MS JT/1/HTYP/507

1. *1858*: Tyndall originally wrote 1857 but the 7 is crossed out and replaced by an 8.

2. *Your little note*: letter missing.

3. *hoar*: grey-haired with age (*OED*).

4. *Faraday is now lecturing*: Michael Faraday gave the sixth and final Christmas lecture in the series 'Distinctive Properties of Common Metals' at the RI on 7 January 1858. The Prince of Wales and his brother Prince Albert attended the entire series; their father Prince Albert attended the first lecture only. See *Faraday Correspondence*, 5:3068.

5. *Prince of Wales*: Albert Edward (1841–1910), eldest son of Queen Victoria and Prince Albert, who succeeded to the throne of the United Kingdom as King Edward VII when Queen Victoria died in 1901 (*ODNB*).

6. *addentants*: attendents.

7. *Chladni's figures*: Ernst Chladni (1756–1827), a German physicist and musician, is best known for his work in acoustics (*CDSB*). In his lectures on sound and the later published version, Tyndall discussed and demonstrated Chladni's discovery that when a fiddle bow is drawn across a side of a square glass plate with sand on top of it, the sand will move and create predictable patterns consisting of two or more discernible nodal lines and sometimes with more complex configurations depending on the placement of the bow. See J. Tyndall, *Sound. A Course of Eight Lectures* (London: Longmans, Green, and Co., 1867), pp. 139–48.

8. *use the fiddle bow*: the stringed bow with which a fiddle is played (*OED*).

9. *The glacier paper*: J. Tyndall, 'On Some Physical Properties of Ice', *Phil. Trans.*, 148 (1858), pp. 211–19.

10. *last years paper*: J. Tyndall and T. H. Huxley, 'On the Structure and Motion of Glaciers', *Phil. Trans.*, 147 (1857), pp. 327–46.

11. *Goethe*: see letter 1406, n. 3.

12. *my friend Pollock*: William Frederick Pollock.

13. *His wife*: Juliet Pollock.

14. *an excellent lady*: Maria Drummond and her daughters Mary, Emily, and Fanny. In letter 1367, Tyndall shared with Hirst his positive feelings toward the family.

15. *the account of our trip up Mont Blanc*: in letters to Faraday (1419), Juliet Pollock (1420), and Hooker (1428), Tyndall described his and Hirst's ascent of Mont Blanc on 13 August 1857.
16. *examination of the Artillery & Engineers*: see letter 1476, n. 1.
17. *Gen. Portlock*: Joseph Portlock (1794–1864), a surveyor and geologist who was a member of the Council of Military Education and served as President of the Geological Society of London from 1857–58 (*ODNB*).
18. *to become an examiner*: see letter 1476, n. 1.
19. *his Royal Highness the Comr. in Chief*: George William Frederick Charles (1819–1904), Prince George, second Duke of Cambridge and cousin of Queen Victoria, was the Commander-in-Chief of the Forces from 1856–95 (*ODNB*).
20. *the Secretary at War*: Lord Panmure (Fox Maule-Ramsay).
21. *Zermatt*: see letter 1297, n. 7.
22. *Monté Rosa*: the second highest mountain in the Alps located between Switzerland and Italy near the town of Zermatt.
23. *I was asked to join but declined*: see letters 1459 and 1466.

From Tommy Grieve[1]   11 January 1858   1479

56 Cambridge Terrace | Hyde Park | January 11th 1858.

My dear Dr. Tyndall,

It is a long time since I wrote to you and I suppose you think I have quite forgotten you. On Friday as we were returning from the Tower[2] we past[3] the Institution[4] but it was late so I did not go in to see you, but I hope I will see you some day. My eldest brother[5] is in Italy just now he is in rather bad health. I am going to leave Brighton[6] at Midsummer then I think I will go to school at Edinburgh for a year or two I am going out to Ceylon[7] about eighteen to be a coffee planter. On Saturday when we went down to see the Leveathan[8] I caught a very bad cold I don't think it will go away for some time I must say good bye for I have no more to write.

With kind love | I remain | your affectionate Friend | Tommy Grieve.

RI MS JT/1/TYP/1/467
LT Typescript Only

---

1. *Tommy Grieve*: a student at Queenwood who corresponded with Tyndall after he left his position as a teacher there in 1853.
2. *the Tower*: the Tower of London, an historic castle situated on the north side of the River Thames.
3. *past*: Grieve wrote 'past' instead of 'passed'.

4.   *the Institution*: the RI.

5.   *My eldest brother*: not identified.

6.   *Brighton*: see letter 1365, n. 7.

7.   *Ceylon*: the name of the island nation Sri Lanka under British colonial rule, to the south-east of India, which achieved independence in 1948.

8.   *Leveathan*: Leviathan, officially named *SS Great Eastern,* was at the time the largest ship ever built. Designed by Isambard Kingdom Brunel (1806–59), its launch on 31 January 1857 on the River Thames was a public spectacle.

From Frances Hooker   [13 January 1858][1]   1480

Kew

My dear Dr Tyndall

I thank you heartily for your pleasant letter,[2] which came to help to cheer me and charm my thoughts from the sorrowful tidings they were dwelling on.[3] Have you not remarked that Death seldom strikes a single blow? and now I am mourning the loss of a very dear cousin[4] whom I loved as a sister. More sorrow is gathering in the horizon too: in truth this New Year has opened very gloomily to me.

Many thanks for your kind enquiries. The children[5] are all well again I am happy to say, and myself also, for I too was amongst the invalids. Joseph[6] is still coughing and I do not think he is looking very well. Still he is better. This is his busy week of examination at the India House;[7] he was there yesterday and will be again to-morrow and Friday.

I am very grateful to you for the promised tickets for your lectures. I quite hope to be present both on the Friday evening[8] and also at as many of your morning lectures[9] as I can possibly manage to get to. Had I wished it I could not have been at Mr Faraday's, having scarcely left the house since Christmas, and certainly not been able to get to town till yesterday when I had a busy day. I thought of you as I passed down Albemarle St.[10]

Why do you undertake examinations[11] that you don't like and that don't pay much? Dr Sinclair[12] sails for New Zealand next week.[13] I had the amusement of helping him to choose silk dresses to take to his nieces: you know ladies always like shopping! I hope you will soon be able to carry out your intention of coming here, you know you are always sure of a welcome. When am I to be honoured with your proof sheets? This morning I have before me a long piece of writing of a very different description, relative to Dr Livingstone's expedition.[14] You asked me how I liked Newman's[15] book on the Soul.[16] Now that I have finished it I answer 'very much,' better than his 'Phases of Faith'[17] as there is less in it to which I hesitate to assent. His books are all most agreeable to read; he writes so well and puts his arguments so clearly. It is certainly a very great treat to get hold of a well-written book, be the subject matter what it may.

Farewell, | I am ever yours most sincerely | Frances H. Hooker

RI MS JT/1/TYP/8/2557
LT Typescript Only

1. *[13 January 1858]*: the date is given by the reference to Andrew Sinclair sailing for New Zealand, which occurred after 17 January 1858 (see n. 12), the mention of Tyndall's Friday evening lecture (see n. 8), and his morning lectures (see n. 9). LT noted on the typescript letter that it was on a Wednesday, and as this letter predates the next one from Frances Hooker (1482, she mentioned receiving the lecture tickets), the only Wednesday in January that fits is 13 January.

2. *your pleasant letter*: letter missing.

3. *dwelling on*: Frances Henslow's great-aunt Ann Henslow died in 1857.

4. *dear cousin*: Jemima Maria Hicks Jenyns (1829–58), maternal cousin of Frances Hooker, who died at Ryde on the Isle of Wight on 9 January 1858. See *Isle of Wight Observer,* 16 January 1858, p. 4.

5. *children*: by this time, Joseph and Frances Hooker had four children: William Henslow Hooker (1853–1942), Harriet Anne Hooker (1854–1945), Charles Paget Hooker (1855–1933), and Maria Elizabeth Hooker (1857–63). They would go on to have three more.

6. *Joseph*: Joseph Hooker.

7. *examination at the India House*: since 1854, Hooker was a botanical examiner for candidates for the Indian Medical Service under the East India Company, which occurred at its headquarters at the East India House in London, *Life and Letters of Thomas Henry Huxley,* ed. L. Huxley, 2 vols (London: Macmillan, 1900), vol. 1, p. 385).

8. *Friday evening*: Tyndall's Friday evening lecture at the RI on 22 January 1858, 'On Some Physical Properties of Ice'.

9. *morning lectures*: see letter 1440, n. 8.

10. *Albemarle St.*: the location of the RI.

11. *undertake examinations*: see letter 1476, n. 1.

12. *Dr. Sinclair*: Andrew Sinclair (1794–1861), a Scottish surgeon and botanist who collected plants with Joseph Hooker in Australia in the 1840s. He served as New Zealand's Colonial Secretary from 1845–56, and then returned to Scotland and made acquaintances with other scientific men in London, including Tyndall (*ODNB*).

13. *sails for New Zealand next week*: after his time in Scotland and England, Sinclair returned to New Zealand in January 1858 to collect plants, but drowned trying to cross a river in 1861. In a February 1858 letter to Charles Darwin, Joseph Hooker mentioned that 'Sinclair sailed a month ago' (Joseph Hooker to Charles Darwin, (25) February 1858, in F. Burkhardt, et al. (eds), *The Correspondence of Charles Darwin,* 24 vols (New York:

Cambridge University Press, 1991), vol. 7 [1858–59], p. 34). While the exact date of his departure is not known, it would have been after 17 January, as Tyndall recorded in his journal for that date: 'To Kew. Took Hooker and Dr Sinclair with us and promenaded the gardens and pleasure grounds' (Journal, RI MS JT/2/13c/1047). Thomas Huxley was among the party.

14. *Dr. Livingstone's expedition*: David Livingstone (1813–73), Scottish missionary and explorer, travelled in central and southern Africa. He returned to Britain in 1856 to be hailed as a hero. His *Missionary travels and researches in South Africa* was published in 1857 (*ODNB*). Livingstone corresponded with Joseph Hooker (see Livingstone Online).

15. *Newman's*: Francis William Newman (1805–97), an English scholar and brother of the well-known theologian Cardinal John Henry Newman (1801–90) (*ODNB*)

16. *book on the Soul*: F. W. Newman, *The Soul, her sorrows and her aspirations: an essay towards the natural history of the soul, as the true basis of theology* (London: J. Chapman, 1849).

17. *'Phases of Faith'*: F. W. Newman, *Phases of Faith; or, passages from the history of my creed* (London, J. Chapman, 1850).

From George Gabriel Stokes   13 January 1858   1481

69 Albert S̲t̲. Regents Park | Jan. 13th 1858

My dear D[r]. Tyndall,

You have not remembered quite rightly the views I expressed about the dark bands, or fixed lines, of the solar spectrum and the bright lines of electric spectra.[1] As to the former you were right. I said that I thought it most probable that they were a phenomenon of gaseous absorption produced by the atmosphere partly of the sun partly of the Earth. But as to the <u>bright</u> lines you have substituted for my views, from imperfect recollection, something utterly different. In the former case light is the rule, a dark line the exception; but here light is the exception, and we have <u>isolated</u> bright lines. There is nothing whatsoever to lead us to suppose that the generally wanting light ever existed. Everything tends to show that in the voltaic <u>arc</u> (not the incandescent poles) ponderable matter sets the ether in vibration in a manner analogous to a finite number of notes of particular pitch.

It is well known that alloys exhibit simultaneously the bright lines due to the several metals of which they are composed; and it is just what one would expect in accordance with the view I have explained.

What put the explanation which you supposed to be mine of the bright lines into your head was probably an indistinct recollection of a very remarkable experiment of Foucault's[2] which I mentioned to you, in which he shewed that the voltaic arc between charcoal points[3] was at the same time a source of light giving out the bright line (darker line) D and an absorbing medium extinguishing light of that particular refrangibility[4] (<u>not</u>, observe, all light <u>except</u> that agreeing in refrangibility with D). I mentioned I believe what I regarded as an illustration of this experiment of Foucault's derived from the analogy of sound.

My wife[5] desires me to say that she is much obliged to you for the ticket.[6] We hope to be present at the lecture.

Yours very truly | G. G. Stokes

P.S. I presume you would like the 21ˢᵗ for your day at the R.S.[7] and believe it is the same subject that you are to bring forward at the R.I.

RI MS JT/1/S/219
RI MS JT/1/TYP/4/1361–62

1. *bright lines of electric spectra*: possibly a reference to a lecture Stokes gave at the RI on 18 February 1853, and published as 'On the Change of Refrangibility of Light, and the exhibition thereby of the Chemical Rays', *Roy. Inst. Proc.,* 1 (1851–54), pp. 259–64. Stokes's paper on the same subject 'On the Change of Refrangibility of Light', *Phil. Trans.,* 142 (1852), pp. 463–562, had earned him the RS's Rumford Medal in 1852.

2. *very remarkable experiment of Foucault's*: Jean Bernard Léon Foucault (1819–68), a French physicist, published a note on his experimental observation in the 7 February 1849 number of *L'Institut, journal universel des sciences et des sociétés savantes en France et à l'étranger,* 17 (1849), pp. 44–46. Reprinted as L. Foucault, 'Note sur la lumière de l'arc voltaique', *Annal. Chim. et Phys.,* 58 (1860), pp. 476–78. Also see G. G. Stokes, 'On the Simultaneous Emission and Absorption of Rays of the same definite Refrangibility; being a translation of a portion of a paper by M. Léon Foucault, and of a paper by Professor Kirchhoff', *Phil. Mag.,* 19 (1860), pp. 193–97.

3. *charcoal points*: see letter 1440, n. 20.

4. *refrangibility*: the ability of light (or other rays) to be refracted (*OED*).

5. *My wife*: Mary Stokes (née Robinson, 1823/24–99), married on 4 July 1857. Together they had five children.

6. *the ticket*: on 22 January 1858, Tyndall lectured at the RI, 'On Some Physical Properties of Ice'.

7. *your day at the R.S.*: presumably Stokes was mistaken in stating that he believed Tyndall's subject for the RS on 21 January was the same as his subject at the RI on 22 January. Tyndall read Samuel Haughton's paper on rock cleavage at the RS on 21 January 1858. See letter 1449, n. 7.

From Frances Hooker   [16 January 1858][1]   1482

My dear Dr Tyndall,

Are you disengaged on Tuesday next? (the 19th inst) and will you dine with us on that day at 6 p-m. We hope to see Dr Sinclair[2] who sails for New Zealand[3] a few days later and we have also asked Dr Lindley and Capt and Mrs Sullivan.[4] He was formerly Governor I think at the Falklands.[5]

Thank you very much for kindly sending me the two tickets.[6] I look forward with much pleasure to making use of them.

Ever believe me | Yours most sincerely | Frances H. Hooker.

RI MS JT/1/TYP/8/2558
LT Typescript Only

---

1. *[16 January 1858]*: the date is given by its relation to letter 1480, and that 19 January 1858 fell on a Tuesday. It is possible this letter could have been written on 14 or 15 January as well.

2. *Dr Sinclair*: see letter 1480, n. 12.

3. *sails for New Zealand*: see letter 1480, n. 13.

4. *Capt and Mrs Sullivan*: Bartholomew James Sulivan (1810–90) and his wife Sophia (née Young, 1809–90). Sulivan was a British naval officer and hydrographer (*ODNB*). He served as a lieutenant under Captain Robert Fitzroy on the second voyage of HMS *Beagle,* from 1831–36 on which Darwin was aboard.

5. *the Falklands*: while Sulivan was not a governor in the Falkland Islands (a disputed British territory off the eastern coast of southern Argentina), he did survey the islands while commanding HMS *Arrow* (1838–39) and HMS *Philomel* (1842–45). Hooker had also visited the islands on the Ross expedition (1839–43).

6. *two tickets*: Frances had previously thanked Tyndall 'for the promised tickets for your lectures' (see letter 1480, n. 8 and n. 9).

From Auguste de la Rive   18 January 1858   1483

Genève, le 18 jan. 1858.

Mon cher Monsieur,

J'ai bien reçu votre bonne lettre du 22 décembre, ainsi que l'extrait de votre mémoire sur la glace. Je vous remercie infiniment de l'un & de l'autre. Votre extrait est imprimé dans le numéro des <u>Archives de la Bibl. Univ.</u> qui paraît après demain; il nous a tous beaucoup intéressés, & nous sommes impatients de connaitre l'autre travail que vous annoncez sur les glaciers, & dont vous nous promettez également un extrait, sur lequel nous comptons. J'ai été heureux d'apprendre que votre séjour sur les glaciers a si bien réussi: il est vrai que vous avez été favorisé par une saison comme on a rarement & que vous avez su mettre à profit ce temps magnifique.

Je ne saurais trop vous exprimer ma reconnaissance pour la peine que vous avez prise de corriger les épreuves de mon troisième volume; j'espère être dans le cas de vous prouver cette reconnaissance autrement que par des mots. Je regrette ce que vous me dites des défauts de la traduction; cela m'étonne d'autant plus que M. Walker sait très peu le Français, & qu'il me paraissait, soit par ce motif, soit parce qu'il aimait bien le sujet de mon ouvrage, bien qualifié pour le travail dont je l'ai chargé. Monsieur Marcet qui lit en anglais mon Traité, est bien frappé aussi jusqu'à un certain point des gallicismes de Monsieur Walker, mais il trouve que cet inconvénient est compensé par la grande clarté du style, qualité, il faut la connaître, qu'on ne rencontre pas toujours dans les ouvrages scientifiques anglais; personne ne doit mieux le sentir que vous qui êtes si remarquable pour la clarté & la netteté de votre exposition.

J'espère que maintenant le 3è volume a paru & que Mr Longman en a envoyé, comme je le lui avais demandé, un exemplaire à la Société Royale—à laquelle je vous prie d'avoir la bonté d'en faire hommage en mon nom. Je l'ai aussi chargé d'en envoyer un exemplaire, d'abord à vous, ensuite à MM. Faraday, Grove, Wheatstone, &c. Veuillez me rappeler au bon souvenir de ces excellents amis quand vous les verrez; merci des bonnes nouvelles que vous me donnez d'eux tous, et en particulier de notre chef Faraday. Dites-lui tout particulièrement mille choses de ma part.

Voilà donc Mr Matteucci qui a finalement reconnu la polarité diamagnétique, j'ai une lettre de lui où il m'annonce cette nouvelle. Que dites-vous de Mr Despretz qui donne comme nouveau le dépôt du peroxyde de plomb à l'électrode positif ? C'est une nouvelle sottise à ajouter à tant d'autres. Le <u>Phil. Mag.</u> ne devrait pas reproduire de semblables inepties, lui qui ne dit rien bien souvent de choses plus importantes. J'espère ne pas tarder à vous communiquer les résultats de quelques recherches que je fais cet hiver, mais qui ne sont pas encore terminées. Maintenant que je suis débarrassé du travail considérable que m'occasionnait, comme vous devez le comprendre, la rédaction de mon ouvrage, je puis me livrer d'avantage que par le passé aux recherches de laboratoire—, et j'en jouis beaucoup. De votre côté ne m'oubliez pas, cher Monsieur, donnez moi souvent de vos nouvelles & croyez moi votre dévoué & bien affectionné

A . de la Rive.

Ne sachant pas l'adresse de Mr Grove, je me permets d'insérer sous ce pli à votre adresse, une lettre que je vous prie d'avoir la bonté de lui faire parvenir.

Geneva, 18 Jan. 1858.

My dear Sir,

I have indeed received your letter from 22 December,[1] as well as the abstract of your essay on the ice.[2] I am infinitely thankful for both. Your abstract is printed in the edition of the <u>Archives de la Bibl. Univ.</u> which will be published the day after tomorrow;[3] it was very interesting for all of us, & we are impatient to know your other work on the glaciers, of which you also promised an abstract that we are counting on. I was happy to learn that your stay in the glaciers was so successful: it is true that you have been favored by a season we rarely have, and you knew how to make good use of this wonderful weather.

I cannot stress enough how grateful I am for the trouble you took to correct the proofs of my third volume;[4] I hope I will find the occasion to demonstrate my gratitude by means other than words. I am sorry about what you said of flaws in the translation; it is even more surprising since Mr Walker knows little French, and that he seemed, either for this reason, or because he liked the matter of my work, well qualified for the work I assigned him to do.[5] Monsieur Marcet,[6] who reads my Traité[7] in English, is also struck by the Gallicisms of Mister Walker,[8] but he finds that this inconvenience is compensated by the clarity of the style, quality, we must recognize it, that we do not see in many of the British scientific books; nobody can feel it better than you who is so remarkable for the clarity and the sharpness of your writing.

I hope that by now the 3rd volume was published & that Mr Longman[9] has sent, as I requested, a copy to the Royal Society that I beg you to gift them in my name. I have also asked him to send a copy, first to you, then to Messrs Faraday, Dove, Wheatstone, etc. Please send my regards to these excellent friends when you see them; thank you for all the excellent news you give of them, and particularly our mentor Faraday. Tell him in particular thousands of wishes from me.

So it is Mr Matteucci finally recognized the diamagnetic polarity,[10] I have a letter from him where he announces it. What do you think about Mr Despretz who gives for novelty the peroxide deposit with positive electrodes?[11] It is a new stupidity to add to many others. The <u>Phil. Mag.</u> should not publish such ineptitudes, this magazine which so often says nothing about more important things. I hope to communicate with you without further ado the results of some research that I did this winter, but which are not yet finished. Now that I am done with the considerable task of writing my book, as you surely know yourself, I can engage in lab research more than in the past—, and I enjoy it a lot. On your end, do not forget me, dear Sir, send news often, & remember me as your very devoted and affectionate

  A. de la Rive

Not knowing the address of Mr Grove, I allow myself to enclose with this cover addressed to you, a letter[12] that I beg you to be so kind as to forward to him.

RI MS JT/1/D/93
RI MS JT/1/TYP/1/354–55

---

1. *your letter from 22 December*: letter 1473.
2. *abstract of your essay on the ice*: J. Tyndall, 'On Some Physical Properties of Ice', *Roy. Soc. Proc.,* 9 (1857), pp. 76–80.
3. *Your abstract is printed . . . the day after tomorrow*: J. Tyndall, 'Sur quelques propriétés physiques de la glace', *Archives des sciences physiques et naturelles,* 1 (1858), pp. 5–10.
4. *correct the proofs of my third volume*: see letters 1293 and 1299.
5. *assigned him to do*: Tyndall mentioned the poor translation in letters 1299 and 1353.
6. *Monsieur Marcet*: François Marcet (1803–83), a Swiss physicist who worked with de la Rive on the specific heat of gases. See 'Notes', *Nature,* 27 (1883), p. 587.
7. *Traité*: treatise (French); see n. 4.
8. *Gallicisms of Mister Walker*: see letter 1378, n. 2.
9. *Mr Longman*: probably either Thomas (1804–79) or William (1813–77), members of the Longman publishing family (*ODNB*).
10. *So it is Mr Matteucci finally recognized the diamagnetic polarity*: see letters 1340 and 1353.
11. *gives for novelty the peroxide deposit with positive electrodes*: this is perhaps in reference to C. Despretz, 'Note on the Decomposition of Certain Salts, Particularly Lead-Salts, by the Action of the Voltaic Current', *Phil. Mag.,* 15 (1858), pp. 78–80.
12. *a letter*: letter missing.

From Samuel Haughton   18 January 1858   1484

9 Norfolk St, Strand | 18 Jan. 1858

My dear Sir,

I leave with this note my paper and a map of the part of the Co. Waterford examined[1]—scale 1 inch to the mile, just published by the Geological Survey.[2]

Please obtain admission for Mr Galbraith[3] and myself on Thursday evening.[4]

I have given on the last page a short description of my diagrams—In case the Royal Society should see fit to publish my paper,[5] I should like the printer to send me the 'copy' with the 'revise', as I have no record of my figures except in my note book.

It will scarcely be necessary to illustrate the paper as it is quite intelligible without the diagrams—or if you wish, I could have them drawn to any scale

I am, | Yours very sincerely | Saml. Haughton

John Tyndall Esq.

RI MS JT/1/TYP/2/487
LT Typescript Only

---

1. *my paper and a map of the part of the Co. Waterford examined*: Tyndall read Haughton's paper at the RS on 21 January 1858. See letter 1449, n. 7.
2. *Geological Survey*: Geological Survey of Ireland, founded in 1845.
3. *Mr Galbraith*: Joseph Galbraith (1819–90), professor of natural and experimental philosophy at Trinity College, Dublin from 1854–70. With Samuel Haughton he wrote a series of scientific and mathematical textbooks (the *Galbraith and Haughton Scientific Manuals*) which were widely employed in schools training young men for military service. See N. D. McMillan, 'Samuel Haughton, 1821–1897', in M. McCartney and A. Whitaker (eds), *Physicists of Ireland*: *Passion and Precision* (Philadelphia: Institute of Physics Publishing, 2003), pp. 106–15, on p. 107–8.
4. *Thursday evening*: see n. 1.
5. *to publish my paper*: see n. 1.

From John Barlow   [22 January 1858][1]   1485

My dear Tyndall

Earnest congratulations on your success to day.[2] You made your philosophy as transparent as you rendered your ice by judicious squeezing. —There are several points which you made, which the Princes[3] will remember as long as they live.

. . .

But that is not the object of this letter—

Would you look in on us on <u>Friday</u> evening. Perhaps Pollock will do the same & Mrs Pollock would be but too welcome if she ventures out.[4]

You would meet Babbage Sir C. Fellows[5] & L^y Shelly[6] my very old friend, to whom I want to introduce you

Ever yours | John <u>Barlow</u>

RI MS JT/1/B/22

---

1. *[22 January 1858]*: the date is given by the reference to a lecture Tyndall gave at the RI (see n. 2). The invitation apparently refers to the following week, since on 29 January, Tyndall visited Barlow as recorded in his Journal: '[a] short time at Barlows after the lecture'. The lecture he refers to was William Grove's 'On Molecular Impressions by Light and Electricity' at the RI (Journal, RI MS JT/2/13c/1049).
2. *your success today*: on 22 January 1858, Tyndall lectured at the RI, 'On Some Physical Properties of Ice'.

3.   *the Princes*: probably Princes Albert Edward (1841–1910) and Alfred Ernest (1844–1900), the two oldest sons of Queen Victoria and Prince Albert. They had attended Michael Faraday's recent lecture series. See letter 1478, n. 4.

4.   *if she ventures out*: Juliet Pollock did attend, see letter 1486.

5.   *Sir C. Fellows*: Charles Fellows (1799–1860), a British archaeologist, mountaineer, and member of the BAAS. In 1827, he discovered the modern route to the top of Mont Blanc (*ODNB*).

6.   *Ly Shelly*: possibly Lady Frances Shelley (née Winckley, 1787–1873), daughter of Thomas Winckley and a member of the Duke of Wellington's circle with a talent for harp and piano. In an 1844 letter from William Frederick Pollock in *Personal Remembrances of Sir Frederick Pollock: Second Baronet, Sometime Queen's Remembrancer,* 2 vols (London: MacMillan and Co., 1887), it is mentioned 'This morning we have been at Lady Shelley's. She gives *matinées* at Fulham, which are much the fashion, on Saturdays in June' (vol. 1, p. 214).

From Juliet Pollock   23 January 1858   1486

My dear M^r. Tyndall,

I must congratulate you at once upon the complete success of your lecture last night.[1] It was admirably clear and interesting throughout, and triumphantly overturned one theory[2] while it substituted another[3] which seems to me a very beautiful one, though I cannot pretend to be a competent judge of its merits.

I may venture to speak more confidently of your manner of treating your subject and I do so with unqualified praise.

A year had passed last Thursday since I had heard you lecture[4] and I found that you had gained more than a year upon your audience. You have as it appears to me lost that sense of their presence which had at moments to impede the flow of your thoughts when you addressed them.

It gives me a very high degree of pleasure to look forward to your whole course on heat[5]—

In writing to you I feel sure that you will acquit me of presumption and that you will quite understand that the only value I suppose attached to my praise is that of perfect sincerity. You have reason to know that I am sincere; the expression I am now giving you of the gratification I felt last night falls very short of the actual feeling. Frederick[6] was as much delighted as I was with your lecture and this assurance is more worth giving you as his understanding of the subject is of course beyond mine—

I do not wait to say all this till Monday,[7] when I am to have the pleasure of seeing you, for I believe you would not like to hear it said while there were others present to listen to it.

Walter's[8] love—he thanks you for your letter[9] and he has a particular question to ask you on Monday—

Yours most truly | Juliet Pollock | 59 Mont. Sq. | Saturday 23.ᵈ Jan. | 1858

RI MS JT/1/P/167

---

1. *your lecture last night*: on 22 January Tyndall lectured at the RI, 'On Some Physical Properties of Ice'.
2. *overturned one theory*: see letter 1306, n. 8.
3. *substituted another*: see letter 1306, n. 10.
4. *since I had heard you lecture*: on 23 January 1857, Tyndall lectured at the RI on 'Observations on Glaciers'.
5. *your whole course on heat*: see letter 1440, n. 8.
6. *Frederick*: William Frederick Pollock.
7. *till Monday*: 25 January 1858, when Tyndall had dinner with the Pollocks and Debus (Journal, 25 January 1858, RI MS JT/2/13c/1049).
8. *Walter's*: Walter Pollock.
9. *your letter*: letter missing.

To Juliet Pollock   23 January 1858   1487

23ʳᵈ Jan. 1858

My dear Mʳˢ. Pollock

How kind it is of you to write to me[1] thus. I do not think that I am a candidate for popular applause but the warm, truthful commendation of a friend commends itself with inexpressible sweetness to my heart. I thank you for your note. It is precious to me—not so much for the assurance it conveys that the lecture[2] was a success (which I still doubt) as for the spirit of goodness which prompted you to write it. Of course I know you are sincere—I defy you to be otherwise with me or any body else. The gods made you sincere, and it is to them I return thanks for the happiness which the friendship of you & yours has ever given me.

most sincerely yours | John Tyndall

RI MS JT/1/T/1131
RI MS JT/1/TYP/6/1930

---

1. *to write to me*: letter 1486.
2. *the lecture*: on 22 January Tyndall lectured at the RI, 'On Some Physical Properties of Ice'.

From Henry Clifton Sorby   28 January 1858   1488

Broomfield, Sheffield. | Jan. 28/58.

My dear Tyndall,

You will remember we tried to freeze the water in the fluid cavity in quartz, but could not.[1] I have now made out the true facts of the case, but cannot just see why they should be so. In tubes of moderate diam., water does not freeze till the tem., is about 23°F, unless a bit of ice is in contact, when it does easily at 31°. It therefore as it were objects to <u>commence</u> crystallizing until much below the thawing point. This of course is well known. However if kept perfectly still in moderately large tubes it freezes at 20°, without any hesitation. What however is remarkable is the fact that in very minute tubes it will not freeze till the tem. is about 6°F unless in contact with ice, when it does at 31°. At 10° it certainly will not, even when knocked about. I have proved whether it was frozen or not by making use of polarized light—Ice has the power of depolarizing, but water not. The presence of a small quantity of salt lowers the tem. at which the water freezes, and there is apparently some dissolved in the water in the fluid cavity we experimented on, and therefore it did not freeze even at 0°. A large fluid cavity in another specimen froze just like water, and thawed exactly at 32°.

I cannot think of any good explanation of the above facts. Can you suggest any? If you can, I should like to have your opinion: and, if you think well, ask Faraday what he thinks about it.

Since you are so much thrown in the way of information, could you put me in the way of some lectures,[2] suitable for our Lit. & Phil. Society?[3] We require first rate, popular, scientific or literary lectures. The sort we have had from Hunt[4] and Lancaster[5] were very good for us. We should like some of a literary or art character or first rate ones on astronomy. If you could tell me of any likely people, it would be a great kindness, since all we had calculated on appear to be failing us.

Yours very truly | H. C. Sorby.

RI MS JT/1/TYP/4/1353
LT Typescript Only

---

1. *You will remember we tried to freeze the water in the fluid cavity in quartz, but could not*: for Sorby's account of this observation, see H. C. Sorby, 'On the Microscopical Structure of Crystals, indicating the Origin of Minerals and Rocks', *Quarterly Journal of the Geological Society,* 14 (1858), pp. 453–500, on pp. 471–75.

2. *could you put me in the way of some lectures*: for Tyndall's response, see letter 1492.

3. *Lit. & Phil. Society*: the Sheffield Literary and Philosophical Society was founded in 1822 and served as a meeting place for individuals to share and debate a variety of topics of educational and cultural interest. Sorby organized lectures for the Society and served as its president five times.
4. *Hunt*: see letter 1311, n. 9.
5. *Lancaster*: Edwin Lankester (1814–74), an English naturalist who served as President of the BAAS for a quarter-century, was friends with Darwin and Thomas Huxley, and through studying water through a microscope was instrumental in curbing cholera outbreak in London (*ODNB*). Lankester lectured for the Sheffield Literary and Philosophical Society in 1848 on 'The Natural History of Plants Yielding Food', and in 1851 on 'Plants and Vegetable Substances Used in the Arts and Manufactures'.

From Thomas Archer Hirst   6 February 1858   1489

Paris. Feb. 6th 1858

My dear John.

It is late and high time that I put aside my pen for the night. Before doing so however I will say a word if it be but to tell you that I am alive and well. In fact my health is far better than it was when I last wrote to you.[1] For the last three weeks I have been able to work very well though I have taken great care never to push my forces too far. I often regret my small capacity for work or rather I wish for a greater but I begin to see that to gain what I desire I must commence by caring for my health.

How are you? Tell me some day soon. Little as I deserve a letter from you I long for one often. I read hastily in the R.S. Proceedings the account of your experiments on Ice.[2] They appeared to me good and important. By the bye Wertheim told me to night that Moncel[3] had given the Geological Society[4] an account of your researches on Ice and Glaciers and made some remarks thereon which would have interested me. In the 'Institut' or in 'Cosmos'[5] you may <u>perhaps</u> see some notice of the same. I will look out & let you know. The memoir of Brücke you have given in the Phil. Mag.[6] interested me much. In many respects it is exceedingly good but he has not yet handled the subject as I could wish to see it handled. I have a great desire to look into the matter myself. Perhaps I may. Nous verrons.[7]

Yours affectionately | T.A. Hirst

P.S. | My Aunt Mrs Battersby of Bristol[8] has been and still is seriously ill. It is just possible that events might call me to England (for a short time) sooner than I intended. I hope not, but God's will be done. Poinsot too is dangerously ill.

RI MS JT/1/H/239

1. *last wrote to you*: the letter is probably missing. Letter 1472 (dated 19 December 1857) is the previous letter in this volume from Hirst to Tyndall.

2. *in the R.S. Proceedings the account of your experiments on Ice*: J. Tyndall, 'On Some Physical Properties of Ice,' *Roy. Soc. Proc.*, 9 (1858), pp. 76–80.

3. *Moncel*: Théodose Achille Louis Du Moncel (1821–84), a French physicist who promoted the use of electricity (*CDSB*)

4. *Geological Society*: Société géologique de France.

5. *'Institut' or in 'Cosmos'*: *L'Institut, journal universel des sciences et des sociétés savantes en France et à l'étranger*, and *Cosmos, Revue encyclopédique hebdomadaire des progrès des sciences.*

6. *memoir of Brücke you have given in the Phil. Mag.*: Tyndall translated papers for publication in the *Phil. Mag.*, such as the mentioned paper by German physiologist Ernst Brücke (1819–92): E. Brücke, 'On Gravitation and the Conservation of Force,' *Phil. Mag.*, 15 (1858), pp. 81–90.

7. *Nous verrons*: see letter 1359, n. 7.

8. *Aunt Mrs Battersby of Bristol*: Eliza Battersby (*bc.* 1806), sister of Thomas Hirst's mother, was married to an Anglican clergyman.

To Thomas Archer Hirst 8 February 1858 1490

8th. February 1858

My dear Tom.

I cannot say that it is late, and still I am sleepy—very sleepy. I had no sleep last night, but thank the gods I have a prospect of it tonight. What need I more than those little notes of yours[1] to assure me that the old fire still burns. The consciousness of this is sufficient for me revived as it is by the little missives you send me from time to time. I have not written to you since I gave my Friday evening.[2] On that evening I was a weary broken down man, still the gods gave me power & utterance. I thought the lecture was not far from being a failure, but I received a flight of notes from my friends[3] next day congratulating me upon its wonderful success. My great enemy is want of sleep, but I still hope to discover the conditions which shall secure it to me. I know that want of sleep is itself the result of a more remote cause, which I shall find out and annihilate some day. I have just got over three military examinations,[4] and have sent today a short report to the generals constituting the committee of military education which will probably make them wish they had a more pliant examiner in my place. I saw Col. Wynne[5] today—They are thinking of sending him out in command of an expedition to China.[6] I hope the gods will frustrate this, it would be a terrible blow to poor M^{rs}. Wynne.[7] Eddy[8] was with me last night and enquired very affectionately after you. I think you will like my little paper[9] when you see it in full, it is tidy, and sends at least one gross blunder to the devil. I had Debus with me a short time ago—he appears to be unhappy about your not writing to him. He is a great difficulty with me—He is so good—so worthy of being assisted, and still so difficult to assist. His want of language, and forgetfulness of the claims of the world, which in some circumstances would be charming, but which in London looks like a defect of character, make against him. Francis was with me today and brought me your little preparatory remarks to Poinsot's paper.[10] If you let me I will run my eye over them and perhaps alter a word here and there. My ordinary lectures[11] are going on satisfactorily, I make a little investigation of each and thus obtain considerable mastery over the subject. Every body is wonderfully kind to me. perhaps I am generalizing too quick—I mean really that I have a circle of friends who are good at heart towards me. Lying outside of these I have a number of friendly acquaintances. In short the world never treated a chap much better than it treats me. And still I feel that my strength lies, perhaps my very effect upon my friends lies, on the reliance upon inner principle rather than on outward support. This very kindness has caused me quite recently to make the resolution not to dine out any where for a month. I want to try the effect of this upon my health, and if I find it good I will extend the principle. Tomorrow I dine with a namesake of John Martin's, namely Baron Martin,[12] and afterwards until the 10th of March my dining out is at an end. It is a terrible invasion of a workings man's time in London. On Friday last I was invited to meet Livingstone the African traveller.[13] I found him a man full of sound practical sense. I sat beside a Mr Buckle, who has recently published a book that has caused very great sensations in London. A history of civilization in England.[14] He is the literary lion of the day. His conversation is wonderfully energetic, and his voice at table generally masks all other voices. Soon after sitting down we had a little sparring as to the influence of moral agents in the culture and advancement of the world. After the ladies left the table our differences broke out into open war. It was thoroughly good humoured, but energetic in the extreme. He talked loud and I responded with corresponding vigour. All other voices at the table sank, and I could see from the earnest eyes and fixed facial muscles of the rest of the guests how much they felt interested in the issue. Once or twice I observed a smile break over the countenance of Livingstone when I charged my gallant antagonist, and this encouraged me a little. I saw both his position and my own with perfect clearness and thus on the main points I kept my footing secure, while my mind was sufficiently free to avail itself of the accidents that arose out of the discussion. He swept at me sometimes like a billow;[15] he lifted me, but did not overwhelm me. I never lost my balance for an instant. After a time we both withdrew our forces with feelings of mutual respect. I trusted throughout to my practical acquaintance with the operations of my own mind, while he brought his vast store of acquired knowledge like battering rams into the field against me. This is the second great conflict that we have had, and from each I had the good fortune to escape with a whole skin. Now my dear Tom there is a sheet of gossip to amuse you: I think I have paid you very well for your little note.

Ever affectionately yours | <u>John Tyndall</u>

RI MS JT/1/T/642
RI MS JT/1/HTYP/508–9

---

1. *little notes of yours*: letter 1489.
2. *my Friday evening*: on 22 January, Tyndall lectured at the RI, 'On Some Physical Properties of Ice'.
3. *a flight of notes from my friends*: letter 1485 from John Barlow and 1486 from Juliet Pollock. Tyndall mentioned in his journal entry for 23 January 1858 having received a letter from a 'Miss Moore' (either Harriet Jane Moore or Julia Moore), but this letter is missing (Journal, RI MS JT/2/13c/1049).
4. *military examinations*: under the newly formed Military Council on Education, Tyndall examined candidates for the Royal Military College, Sandhurst on 'Experimental sciences' on 30 January 1858. See *Sandhurst Royal Military Academy. Copy of the rules and regulations under which candidates are admitted to compete for admissions into the Royal Military College at Sandhurst,* HC 195 (1857–58), xxxvii, p. 537. For more on Tyndall as an examiner, see letter 1306, n. 2.
5. *Col. Wynne*: George Wynne.
6. *an expedition to China*: Wynne served and fought in several battles during the summer of 1858 in the second Opium War, a conflict between China and Britain over trade. He left for China in early August (see letter 1504).
7. *M^rs. Wynne*: Anne Wynne (née Osborne).
8. *Eddy*: Edward Wynne (1837–89), George and Anne Wynne's son who took the entrance examinations to the Royal Military Academy at Woolwich in 1856 and again successfully in 1857 (see letter 1412).
9. *my little paper*: J. Tyndall, 'On Some Physical Properties of Ice', *Phil. Trans.,* 148 (1858), pp. 211–29.
10. *Poinsot's paper*: L. Poinsot, 'On the Percussion of Bodies', *Phil. Mag.,* 15 (1858), pp. 161–80; 15 (1858), pp. 349–59; and 18 (1859), pp. 241–59.
11. *My ordinary lectures*: see letter 1440, n. 8.
12. *Baron Martin*: Samuel Martin (1801–83), an Irish judge and politician who was appointed as a Baron of the Exchequer in 1850. He married William Frederick Pollock's sister Frances in 1838 (*ODNB*).
13. *Livingstone the African traveller*: see letter 1480, n. 14.
14. *A history of civilization in England*: H. T. Buckle, *History of Civilization in England,* vol. 1 (London: John W. Parker and Son, 1857). Buckle published a second volume in 1861, but the rest of the anticipated 14 volumes were never completed.
15. *billow*: a body of men sweeping onward (*OED*).

To Thomas Henry Huxley [17 February 1858][1] 1491

My dear Hal,

By Jove I am very glad. A few days ago I was speculating on the long phiz[2] you were likely to draw, if <u>it</u>[3] proved a boy. But you are as nicely balanced as John Gilpin[4] when placed in equilibrio by his wine bottles. 'Pon my soul[5] (—I don't often swear thus) the gods are very kind to you! Things could not have happened more nicely if you yourself had been the patentee of the molecular architecture. I wonder when an equal blessedness is likely to fall upon me. You ought to feel thankful to heaven and especially thankful to your wife[6] for managing these things in so excellent a manner. I shall be rejoiced to learn that the young life thus thrown upon the seedfield of time develops itself to its father's satisfaction, and that the noble matronly stem from which it sprung has become thoroughly reinvigorated.

Always yours | John Tyndall.

RI MS JT/1/TYP/9/2887
IC HP 8:33
Typed Transcript Only

---

1. *[17 February 1858]*: the date is given by reference to the birth of Huxley's daughter Jessie on 16 February 1858.
2. *phiz*: a face or facial expression, slang derived from the word 'physiognomy' (*OED*).
3. *it*: the birth of Huxley's daughter Jessie on 16 February 1858 (d. 1927) (*ODNB*).
4. *John Gilpin*: a reference to the comedic poem 'The Diverting History of John Gilpin' (1782) by William Cowper (1731–1800). Tyndall means that Huxley now had one son (Noel, 1856–60) and one daughter.
5. *'Pon my soul*: an English expression meaning 'I swear', which appears in Shakespeare's *Othello* (1622) and several times in C. Dickens, *The Old Curiosity Shop* (London: Chapman & Hall, 1841).
6. *your wife*: Henrietta Huxley.

To Henry Clifton Sorby 17 February 1858 1492

17th Feb. 1858

My dear Sorby

Your note[1] came to me when I was in the midst of a series of military examinations,[2] and my health has been so shaky since as to take all the spirit of writing out of me. The facts to which you refer can only be referred in a general way to the influence of the attraction of the sides of the quartz upon the liquid within it—which attraction opposes itself to that of crystallisation: There is a specific attraction between ice & water, as between every crystal & the liquid in which it is formed. You know that crystals suspended in a liquid may be nursed so as to become of enormous size while no particle of solid matter is deposited on the walls of the containing vessel. With regard to the lectures,[3] I am a bad source of information at present, because I have been compelled to keep so much at home—for astronomy I dont suppose you could have a better man than Grant of the Astronomical Society;[4] I have never heard him myself, but he has lectured at the London Institution[5] & I believe given satisfaction.

Very sincerely yours | John Tyndall

Sheffield Archives SLPS/51/107–17 Feb 1858

---

1. *Your note*: letter 1488.
2. *military examinations*: see letter 1476, n. 1, and letter 1490, n. 4.
3. *With regard to the lectures*: in letter 1488, Sorby asked if Tyndall could 'put me in the way of some lectures, suitable for our Lit. & Phil. Society?', especially in astronomy.
4. *Grant of the Astronomical Society*: Robert Grant (1814–92), an astronomer and editor of the Royal Astronomical Society's *Monthly Notices* from 1852–60 (*ODNB*).
5. *London Institution*: see letter 1319, n. 5.

To Michael Faraday   19 February 1858   1493

Friday, 19 Feb 1858

My dear Mr. Faraday,

Will you grant me your patience while I endeavour to lay before you a few reflections which arise out of my recent conversation with you regarding the shortening of my course of lectures.[1] At the time of our conversation I was so deeply sensible of the kindly feelings which, as you informed me, the Managers[2] had manifested towards me, that I thought of nothing else. Perhaps I now labour under a misconception, and if so, I am sure that you will have the goodness to set me right.

The matter, as I understand it, stands thus:—I have agreed to perform certain duties, for which the Institution grants me a certain annual salary, and which the Managers have the right to require of me. My friend Dr. Bence Jones is to give his opinion that my health is impaired by these duties, and the Managers, acting on his report, propose to continue my salary at its present amount, while they diminish the quantity of work which I have to perform.

I feel that your own natural disposition enables you to solve mine so thoroughly that I have no fear of your misunderstanding what I have to say upon this proposition. I ask you, therefore, would you not in my place consider that such an arrangement would place you in a doubtful position as regards the Institution? The acceptance of the change, under the proposed conditions, would, I fear, materially hamper a feeling of independence which, throughout my life, I have endeavoured to maintain. If an alteration is to be made at all I certainly should not like to see it made on the grounds of my being unable to discharge the duties which the Managers regard as merely a reasonable return for the terms which they grant me.

Will you in your kindness permit me to draw your attention to the exact circumstances connected with my joining the Institution?[3] I had three offers before me at the time. The R.I. offered me £200 a year for 19 lectures. Another institution[4] would have given me the same sum, an assistant, a laboratory and funds for experiments, for 6 lectures annually; and I was afterwards given to understand that this offer would be made still better.—At the conclusion of one of my first lectures at the R.I. a gentleman[5] connected officially with a Government establishment,[6] who had heard of my probable appointment at the R.I. took me aside and advised me strongly not to accept the professorship, telling me that I should regret it if I did, for another and a better post was open to me.

Subsequent to this, as you are aware, the late Sir Henry DelaBeche[7] sought to induce me to accept an appointment in the Government School of Mines. Before me lies a letter[8] from poor Edward Forbes,[9] in which he communicated to me the strong desire of himself and his colleagues to have me among them.[10] There was nothing in my agreement with the R.I. to prevent me from holding this post in connexion with my professorship, but the fear that such a connexion might interfere with my duties at the Royal Institution caused me to refuse a position which, in many respects, would have been extremely agreeable to me.

This however occurred subsequent to my appointment at the R.I. and therefore could not influence my decision when the professorship of physics[11] was offered to me. I might refer to other cases which occurred still later—to offers from Woolwich and elsewhere,[12] one of which referred to a position, not however in England,[13] worth £600 a year. I refer to these instances on account of their bearing up on a passage of a letter which I shall cite presently, and to show that since I came to the Institution I have exhibited no disposition to be unfaithful towards it.

But when I made my choice in 1853 the two offers first mentioned were before me, and side by side with these was the offer from the Royal Institution. A short time before the official communication reached me I received a letter[14] from which I extract as follows:—'Mr. Faraday proposed you as

Professor of Physics, to receive £200 a year and to give 19 lectures. To show what we might, and ought to do, he read a statement of what the Managers had done for Davy, and he said he saw no reason why we should not do the same now. Davy had £100 and rooms the first year, the second year £200; the third £300; the fourth £400; the fifth an extra hundred for an excursion; the sixth it was repeated, and he was allowed to take a travelling assistant and have his expenses paid; the seventh he had a multitude of other appointments.... I have no doubt that a scale sliding upwards will be the result, and you will be allowed any sum however great, for experiments'.

On Thursday the 26th of May 1853 a letter[15] reached me from which I will make one or two extracts:—'The Managers met to day, and I am requested to communicate to you officially, that in consequence of a recommendation from Professor Faraday the Managers are desirous of proposing you for election as Professor of Natural philosophy, with £200 a year'.... 'The £200 may and will be increased <u>in a year</u>, and there is no reason why after a time you should not have £400 (or more) if you devote yourself to the R.I.—Davy had, and Mr. Faraday has quoted him as a precedent for this proceeding. We had a very full meeting, and all were for you, and some for offering you more at first, while all agreed that it was very poor pay'.

These are the inducements which I had to weigh against the two other offers to which I have referred. I accepted the professorship at the R.I., but I think it will be granted that some discrepancy exists between the above extracts and my actual position after the lapse of nearly five years. I confess that this has often been a source of dissatisfaction to me.—Dissatisfaction perhaps as much with myself as with others. For had I not reason to infer that I had not come up to the Managers' expectations regarding me, and that the discrepancy to which I have referred was to be attributed, in part at least, to failure of my own?

I have been repeatedly urged by two or three friends who are acquainted with the above circumstances to bring them under the notice of the Managers. But my repugnance to such a step has been hitherto insurmountable. Urged by a friend, I wrote a letter upon the subject last autumn; but when finished I put it aside, and it has lain in my drawer ever since. At present, however, I think perfect frankness on my part is better than remaining silent as I have hitherto done.

Thus far I have looked at my connexion with the R.I. from a commercial point of view merely. Need I say that from first to last this is the consideration which has had least influence with me. The extracts which I have quoted indicated a feeling towards me which I regarded as a thousand times more precious than any pecuniary emolument that the Royal Institution could bestow. This was my feeling at the time, as my letters if consulted, would shew. It is my feeling now. The Institution has surrounded me with relations which to me are above all price. Indeed my feelings towards the Institution are such as would induce me, were I sufficiently independent, to give it my best services without any pecuniary return whatever; for no return of the kind which it could make is to be put in comparison with the happiness which I have derived from the personal relationships, for which I have to thank my connexion with the Institution.

If I were permitted to give utterance to my feelings on the point in question I would express the hope that should the Managers legislate upon the subject, they will not suffer themselves to be influenced in any degree by considerations regarding my health; but be simply guided by what they consider to be just towards me and advantageous to the Institution and to science. Were I relieved to some extent from lecturing I should not become an idler. Other, and perhaps higher work, would be before me. This, and not my health, is with me the point of chief importance, but as things stand at present I find my power of following up such work far more limited than I could wish it to be. This is the point of view from which I should like the question to be regarded. For my health I have no fears, for it is always in my power to repair its injuries by giving myself rest, and I know the reliance which I can place upon a constitution naturally tough and unshaken by physical intemperance.

Believe me dear Mr. Faraday | Ever yours most faithfully | John Tyndall

To be opened when Mr. Faraday has abundance of leisure, but not before. | Professor Faraday.[16]

RI MS JT/1/TYP/12/4077–81
Faraday Correspondence, 5:3397
Typed Transcript Only

---

1. *my course of lectures*: on 18 April 1857, Tyndall wrote to Faraday asking if his load of 19 lectures per year could be decreased (see letter 1377). On 1 March 1858, the Managers approved a change to Tyndall's position that he would deliver no more than a dozen lectures per year. See *Faraday Correspondence,* 5:3397, n. 2.

2. *The Managers*: see letter 1377, n. 5.

3. *my joining the Institution*: the RI.

4. *Another institution*: the LI, see letter 1319, n. 5. Tyndall wrote about this offer in his journal entry for 20 February 1853 (Journal, RI MS JT/2/13b/601).

5. *a gentleman*: possibly John Percy (1817–99), Professor of Metallurgy at the Royal School of Mines from 1851–79. In his journal entry of 6 July 1853, Tyndall described meeting with many people at a Royal Society meeting, and that 'Percy warns me to bind myself to the RI. I may regret it he says' (Journal, RI MS JT/2/13b/608). It appears than that in this letter, he misremembered this conversation, writing that it occurred 'At the conclusion of one of my first lectures at the R.I.'.

6. *a Government establishment*: the Government School of Mines and Science Applied to the Arts, established in 1851 (from 1863 known as the Royal School of Mines).

7. *Sir Henry DelaBeche*: Henry De la Beche (1796–1855), a geologist and first director of the Ordnance Geological Survey (*ODNB*).

8. *a letter*: see letter 0861, Edward Forbes to John Tyndall, 27 December 1853, *Tyndall Correspondence*, vol. 4.

9. *Edward Forbes*: Edward Forbes (1815–54), a highly productive naturalist who published on a wide range of geological and zoological topics. He held a chair of botany at King's College London, and shortly before his early death was appointed to a professorship of natural history at the University of Edinburgh (*ODNB*).

10. *among them*: at the School of Mines.

11. *professorship of physics*: Tyndall's official title at the RI was Professor of Natural Philosophy.

12. *offers from Woolwich and elsewhere*: Tyndall wrote in his journal on 23 November 1854 that he was informed by Charles Wheatstone of an open position at the Royal Military Academy at Woolwich for which he would later work as an examiner for commissioned officers (see letter 1306, n. 2 and Journal, RI MS JT/2/13b/699).

13. *a position, not however in England*: in the spring of 1855 Tyndall was offered a chair at the University of Toronto in Canada and director of its Observatory, but he declined because the position would have him doing scientific observation rather than experimentation. Being far from England was also not desirable for him, despite the much higher pay the positon would have afforded than at the RI. See letter 1074, Edward Sabine to John Tyndall, 24 April 1855, and letter 1075, John Tyndall to Edward Sabine, 26 April 1855, both in *Tyndall Correspondence*, vol. 5.

14. *a letter*: see letter 0756, Henry Bence Jones to John Tyndall, 9 May 1853, *Tyndall Correspondence*, vol. 4.

15. *a letter*: see letter 0762, Henry Bence Jones to John Tyndall, 23 May 1853, *Tyndall Correspondence*, vol. 4.

16. *To be opened . . . Professor Faraday*: written on the envelope.

To unidentified[1] 19 February 1858 1494

19th Feb. 1858

Dear Sir.

The enquiry to which you refer[2] will I hope be printed in fall in the Philosophical Transactions—Next week however I hope to have in my possession an abstract of the lecture referred to in the Athenaeum.[3] It will give me pleasure to send you a copy as soon as I am in a position to do so.

Very sinc[ere]ly yours | John Tyndall | Thanks for your paper.[4]

Royal College of Physicians

1. *unidentified*: possibly William Brinton (1823–67), Robert Liveing (1834–1919), Francis Sibson (1814–76), or Hermann Weber (1823–1918), physicians, Fellows of the Royal College of Physicians, avid mountaineers, and members of the Alpine Club (*ODNB*). See also G. H. Brown, *Lives of the Fellows of the Royal College of Physicians of London,* 10 vols (London: Royal College of Physicians, 1955–2000), vol. 4 (1825–1925).
2. *enquiry to which you refer*: probably J. Tyndall, 'On Some Physical Properties of Ice', *Phil. Trans.,* 148 (1858), pp. 211–29.
3. *an abstract of the lecture referred to in the Athenaeum*: on 30 January 1858 the *Athenaeum* printed an account of Tyndall's 22 January lecture at the RI, 'On Some Physical Properties of Ice'. See 'Royal Institution', *Athenaeum,* 1579 (30 January 1858), p. 150. The lecture was printed as J. Tyndall, 'On Some Physical Properties of Ice', *Roy. Inst. Proc.,* 2 (1854–58), pp. 454–77.
4. *your paper*: not identified.

To [George Gabriel Stokes][1]   20 February 1858   1495

20th. Feb. 1858

My dear Sir

I return one of Mr Matthiessen's[2] papers.[3] It contains the results of experiments on the electric conductivity of a great number of bodies. These results are undoubtedly valuable, and the Philosophical Transactions would, in my opinion be the proper place for recording them

very truly yours | John Tyndall

I hope to send on the second paper[4] very soon.

RS RR/3/197 Tyndall 20 February 1858

1. *[George Gabriel Stokes]*: this letter is attributed to George Gabriel Stokes by its relation to letter 1498.
2. *Mr Matthiessen's*: Augustus Matthiessen (1831–70), a British physicist and chemist, educated in Germany, who worked with Robert Bunsen (*ODNB*).
3. *paper*: A. Matthiessen, 'On the Electric Conducting Power of the Metals', *Phil. Trans.,* 148 (1858), pp. 383–87.
4. *second paper*: A. Matthiessen, 'On the Thermo-electric Series', *Phil. Trans.,* 148 (1858), pp. 369–81.

From Michael Faraday   20 February 1858   1496

Royal Institution | 20 Feb 1858

Dear Tyndall,

Your letter of yesterday[1] has taken me by surprise in this respect:—that it seems to imply that you had reason to believe that I was conscious of an understood engagement, rising year by year £100, according to the terms of the first quotations from a letter by a third person[2] which you have sent me:— or of an engagement at £200, which was to be increased in a year, and should rise up to £400 or more after a time, according to your second quotation.[3] I was not conscious of any such understood engagement, and you may be quite sure that if I had felt myself answerable in any way, either by expression or implication, for such an understanding, I would have seen it carried out or else the whole matter rearranged. At the same time it is also true that I quoted Davy's case to the Managers[4] as a good precedent and an example for their departing from the course which, through circumstances, they had been following for many years. I should have been very glad indeed if I could have felt at the time that they had the means of doing it; and you must not suppose that I would not in that case have urged some such course between that time and this. I have often considered whether I should not aid such facility, and do more good to the Institution by retiring from the place I hold, than by keeping it. Whilst waiting, however, I hoped that other engagements would rise up, that could be held by you conjointly with the one here, and give you that return on the part of science which you so thoroughly deserve for your labours in its cause.

Your letter to me of last April[5] I brought (according to your request) before the Managers at the time, and again last Monday. They shewed the kindest feeling. They could know nothing of the sense of the extracts which yours to me of the 19th inst. contains, but only of the case of Mr. Brande,[6] to which you refer in that of April, and they seemed quite ready to go beyond that, and will meet on the 1st of next month to settle that matter. Your last letter[7] puts the subject altogether on new grounds, and will take them unawares perhaps more than it did me, for though they may know your worth to the Institution and to science, they cannot be aware of the latter point to the same extent as I am. They could not be conscious of the understood engagement, and they have not the power to be liberal patrons of science, or to permit themselves to reward it in proportion to their admiration of it.

And now, my dear Tyndall, let me say that my chief care in this note had been to write only that which would be necessary to clear away misapprehensions and remove the implied charge of injustice or forgetfulness from either myself or the Managers or any other person. We know your value to the cause of science, and therefore to us; but we also know our incompetency to express the proper sense of it. However on that I must not speak too hastily for others. I know you agree with me in the matter of open dealing, and therefore am sure that if you have entertained the impressions above referred to you will let me make the true state of the case known to the Board of Managers at its next meeting.

Ever, my dear Tyndall, | Truly yours, | M. Faraday.

RI MS JT/1/TYP/12/4145–46
Faraday Correspondence, 5:3398
Typed Transcript Only

1. *Your letter of yesterday*: letter 1493.
2. *a letter by a third person*: see letter 0756, Henry Bence Jones to John Tyndall, 9 May 1853, *Tyndall Correspondence*, vol. 4.
3. *your second quotation*: see letter 0762, Henry Bence Jones to John Tyndall, 23 May 1853, *Tyndall Correspondence*, vol. 4.
4. *the Managers*: see letter 1377, n. 5.
5. *Your letter to me of last April*: letter 1377.
6. *Mr. Brande*: William Thomas Brande (1788–1866), an English chemist, who in 1812 succeeded Humphry Davy as the chair of chemistry at the RI (*ODNB*).
7. *Your last letter*: letter 1493.

To Michael Faraday   22 February 1858   1497

22 Feb 1858

My dear Mr. Faraday,

Though deeply sensible of the kindness which suggested the arrangement, I did not quite like the idea of having my duties lessened on the ground that I was incompetent through ill health, to perform them, my salary at the same time remaining intact. I therefore sent to you, on the 19th, a statement of facts[1] which I thought calculated to shew, that I might reasonably expect to see these duties diminished on grounds which should relieve me from a feeling of a weighty personal obligation. My conversation with you on Saturday removes all uncertainty from my mind as to the real origin of the extracts which I laid before you. They doubtless expressed the private views of that member of their body[2] whom the Managers had deputed to communicate with me, and did not commit the Managers generally. Still, standing as they did, in such close juxtaposition with an avowedly official communication, it was natural that I should place, as I did place, perfect reliance upon them. At the time referred to I was a comparative stranger to the writer of those letters, but he is now an intimate friend, the generosity of whose character, and the kindness of whose heart, I appreciate so highly that it would give me pain to see him reminded, in a manner which must be unpleasant to him, of the inducements which he held out to me. It is solely through a desire to spare him this unpleasantness that I now ask you not to communicate my letter of the 19th. to the managers. That <u>you</u> were 'conscious of an understood engagement',[3] and had neglected to see it carried out, is an idea which I never entertained.

With regard to the precise matter in hand I would say, that if the Managers are disposed to regard me purely in the light of scientific lecturer to the Institution, the terms which they now give me do no credit to their liberality. I call, for example, the terms of the London Institution[4] liberal, and they are about the same, per lecture, as I receive from the Royal Institution. True the labour in both cases, is widely different, but that is beyond the question. I have therefore not the slightest fault to find with the terms as a lecturer—They are good terms—and if the Managers would have the kindness to shorten my course,[5] and reduce my salary in the same proportion I shall retain a position which I consider it an honour to be called upon to fill. But it is my misfortune to desire to do something beyond lecturing. This desire, existing side by side with the demand for lectures, often, it is true, produces weariness of brain; but for this I am myself accountable, and I should shrink from the idea of becoming on this account, a burden to the Royal Institution. I am sure you will enter into my feelings, and will not ascribe to it any want of appreciation, on my part, of the great kindness of the Managers, when I say, that any change which diminishes my duties, while it leaves me in possession of my present salary, could only be agreeable to me under the condition that it is made upon scientific grounds alone, and without any reference to the state of my health.

Believe me dear Mr. Faraday | Ever yours most faithfully | John Tyndall

RI MS JT/1/TYP/12/4082–83
Faraday Correspondence, 5:3399
Typed Transcript Only

---

1. *a statement of facts*: see letter 1493.

2. *that member of their body*: Henry Bence Jones. See letter 0756, Henry Bence Jones to John Tyndall, 9 May 1853, and letter 0762, Henry Bence Jones to John Tyndall, 23 May 1853, both in *Tyndall Correspondence*, vol. 4.

3. *'conscious of an understood engagement'*: see letter 1496.

4. *London Institution*: see setter 1319, n. 5.

5. *shorten my course*: see letter 1493, n. 1.

To George Gabriel Stokes   5 March 1858   1498

5th <u>March</u> 1858

My dear Sir.

M[r]. Matthiesen's[1] paper on the Thermoelectric series[2] appears to me to be a highly valuable one, and well worthy of a place in the Philosophical Transactions.[3] The following points are, I think, worthy of M[r]. Matthiesen's attention:—When a statement is made from which important deductions are to be drawn, or upon which calculations are to be based, the statement, if new, ought to be <u>proved</u>, and if old, the <u>reference</u> to the proof of it ought to be given. It ought also to be borne in mind that there are terms, arising and originally out of theoretic considerations, which though familiar, and determinate in Germany are not so in England; and it would be well to avoid a cause of obscurity so easily received. An explanatory phrase here and there would add much to the perscuity[4] of the paper. The introduction I think is capable of considerable improvement in the above respect, and it would be a pity to see the interest of a paper on which so much skill and labour have been bestowed diminished, in any degree, by a defect of style or arrangement.

Very truly yours | John Tyndall
Prof. Stokes | <u>Sec. R.S.</u>

RS RR/3/198 Tyndall 5 March 1858

---

1. *Mr. Matthiesen's*: see letter 1495, n. 2.
2. *paper on the Thermoelectric series*: A. Matthiessen, 'On the Thermo-electric Series', *Phil. Trans.*, 148 (1858), pp. 369–81.
3. *well worthy of a place in the Philosophical Transactions*: Tyndall acted as a referee for the *Phil. Trans.*, reviewing submissions and sending reports to Stokes, the journal's editor.
4. <u>perscuity</u>: Tyndall most likely meant to write 'perspicuity', meaning 'lucidity' (*OED*).

From William Hopkins   16 March 1858   1499

Cambridge | March 16 1858

My dear Sir,

I sent by last evening's post a copy of my memoir on the temperature of the earth and planets.[1] I may mention a point which may be liable to be mistaken. You will find two kinds of heat spoken of and designated respectively by H_o and H_i and radiating from a spherical envelope, which represents the external sources of radiant heat falling on the earth. H_i is supposed to have the power of radiating completely thro' a certain imaginary shell (which may represent the earth's atmosphere) so that the shell is supposed completely <u>dia-thermanous</u>[2] for this kind of heat; but it is not diathermanous for H_o which represents in fact all the heat from whatever source which is absorbed by the imaginary shell. In the actual case it represents the portion of the solar and stellar heat which is absorbed by the atmosphere. My reasoning and conclusions do not depend on the ratio of H_i to H_o so far at least as regards the nature of these conclusions. It is of no consequence whether it be 100: 5 or 100: 25.

Yours very truly | W. Hopkins.

RI MS JT/1/TYP/2/613
LT Typescript Only

---

1. *a copy of my memoir on the temperature of the earth and planets*: W. Hopkins, 'On the External Temperature of the Earth, and the Other Planets of the Solar System', *Transactions of the Cambridge Philosophical Society*, 9 (1855), pp. 628–72.
2. <u>*diathermanous*</u>: having the property of freely transmitting radiant heat (*OED*).

To Michael Faraday   19 March 1858   1500

19th <u>March</u> 1858

My dear M^r. Faraday

The change recently made in the number of my lectures,[1] and the manner in which it has been made, are both so gratifying to me, that I feel very desirous to express to the Managers[2] how highly I appreciate this proof of their confidence. Would you have the goodness to convey to them my warmest thanks for the consideration they have shewn me? I know not, of course, what the future may bring forth; but I cherish the hope that the alteration which has been made in such a kindly spirit, will eventually be found to harmonize both with the interests of the Institution[3] of[4] Science.

Believe me | dear Mr Faraday | most truly yours | <u>John Tyndall</u>

RI MS AD/03/D/04/04
Faraday Correspondence, 5:3404

---

1. *change recently made in the number of my lectures*: see letter 1493, n. 1.
2. *the Managers*: see letter 1377, n. 5 for a list of the Managers of the RI from May 1857 to April 1858.
3. *the Institution*: the RI.
4. *of*: it appears that Tyndall neglected to write 'and' before 'of'.

From Thomas Romney Robinson[1]   24 March 1858   1501

ON FLUORESCENCE PRODUCED BY THE AURORA.
To John Tyndall, Esq. | Observatory, Armagh, | March 24, 1858.

MY DEAR SIR,

I DO not know whether the fact mentioned in the enclosed[2] has been noticed before. If not, perhaps you may think it worth inserting in the Philosophical Magazine.

Yours ever, | T. R. ROBINSON.

On the 14th instant an aurora was visible here of more than the average brightness. At 11 P.M it showed an arch extending from W. to N.E. by E., which emitted a few yellow streamers; and the sky above it was covered with diffused light, over which brighter portions flickered like waves, extending several degrees beyond the zenith.[3] I availed myself of the opportunity to try whether this light was rich in those highly refrangible[4] rays which produce fluorescence, and which are so abundant in the light of electric discharges; and I found it to be so. A drop of disulphate of quinine on a porcelain tablet seemed like a luminous patch on a faint ground; and crystals of platinocyanide of potassium were so bright, that the label on the tube which contained them (and which by lamplight could not be distinguished from the salt at a little distance) seemed almost black by contrast.

These effects were so strong in relation to the actual intensity of the light, that they appear to afford an additional evidence of the electric origin of this phænomenon, and as such I hope they may interest some of your readers.

Phil. Mag., 15 (1858), pp. 326–27

---

1.   *Thomas Romney Robinson*: Thomas Romney Robinson (1793–1882), an Irish astronomer, physicist, and director of the Armagh Observatory from 1823 until his death (*ODNB*).

2.   *The enclosed*: personal observations of Thomas Robinson made at Armagh observatory on 14 March 1858.

3.   *zenith*: see letter 1415, n. 14.

4.   *refrangible*: see letter 1481, n. 4.

From Juliet Pollock   27 March 1858   1502

59 Montagu Squ. W. | March 27th. | <u>1858</u>

My dear M[r]. Tyndall,

I have directed Parker[1] to forward to you a small vol. of my writing called <u>New Friends</u>.[2] It is a work dedicated to Walter.[3]

Fred[k4] will put into your hands on Monday a copy of the book destined for M[rs]. Alfred Tennyson.[5] If you will take it to her with a note from me it will serve as an introduction for you—or perhaps I had better write to her beforehand to prepare her mind.

I have an answer this morning from Macready[6]—<u>he loses his temper</u> (I quote him) over this slander of Horatio[7] and he is certain that the person who speaks in such terms of such a man must be <u>a priss</u>[8] and a <u>pedantic coxcomb</u>,[9] but it happens that he is unwell and too physically feeble to express with all the strength he could wish what he feels on the subject. —

I consider that for a sick man he has done pretty well—I am glad to be assured by him that his view of Horatio's character coincides with my own for his conclusions are the result of my careful study assisted by a peculiarly vivid apprehension.

Yours alw[ays]—most truly | Juliet Pollock

RI MS JT/1/P/168

---

1. *Parker*: John William Parker (1792–1870), an English publisher and printer (*ODNB*). He published works by many well-known authors, including John Stuart Mill, Henry Thomas Buckle, G. H. Lewes, William Whewell, and Charles Kingsley.

2. *small vol. of my writing called New Friends*: in W. Pollock, *Personal Remembrances of Sir Frederick Pollock: Second Baronet, Sometime Queen's Remembrancer*, 2 vols (London: MacMillan and Co., 1887), it is mentioned 'In this year was published New Friends, written by my wife for our son W[alter], as *Julian and his Playfellows* had been for F[rederick]' (vol. 2, p. 77).

3. *Walter*: Walter Pollock.

4. *Fred[k]*: William Frederick Pollock.

5. *M[rs]. Alfred Tennyson*: Emily Tennyson. Tyndall was planning a visit to the Isle of Wight where the Tennysons lived. See letter 1506.

6. *Macready*: William Charles Macready (1793–1873), a famous English actor who performed in many Shakespearean plays. He was retired and unwell at this point (*ODNB*). Juliet Pollock was a close friend who later wrote a memoir about him, *Macready as I Knew Him* (London, 1884).

7.   *slander of Horatio*: possibly a reference to a review of an actor playing Horatio in a recent performance of *Hamlet*. Macready played the role of Hamlet regularly from the 1820s through 1851. The review and performance have not been identified.

8.   *priss*: a prissy person, precise and over-particular (*OED*).

9.   <u>coxcomb</u>: a foolish, conceited, showy person, vain of his accomplishments, appearance, or dress (*OED*).

From Thomas Archer Hirst   [28][1] March 1858   1503

Paris. <u>March.</u> 1858

My dear John.

It is Sunday morning and the bright sun shining upon me in bed awoke me earlier than usual. I do not know by what association the spring sun and yourself presented themselves, but it is a fact that no sooner had I rubbed my eyes and exclaimed what a glorious morning! than I added 'Up you lazy dog and write to John before your breakfast'. It must be a long time since I wrote to you,[2] I think it was mid-winter and now it is spring. Nothing very remarkable has occurred in the interim or you would certainly have heard of it. My health has been steadily improving since Xmas. About that time I was very poorly, so ill that I went one day to consult a physician about myself. He was a sensitive fellow and after I had explained to him all the circumstances I must confess he gave me good advice and little medicine. The medicine was merely to pull me over the immediate difficulty. The rest of his advice was for permanent observance. Amongst other things he advised me to take Cod Liver oil regularly, and to take more energetic exercise than my usual walks which, besides swallowing up a good deal of time, were apt to be come too mechanical and thus to lose their efficacy. Accordingly I purchased a pair of dumb-bells and ever since I have made a practice of swinging them rigorously for a quarter of an hour every day. To this and the cod liver oil, which I have found to be exceedingly beneficial, I attribute chiefly the improved state of health I have ever since enjoyed. By your advice too I gave up coffee which has possibly had a good effect also for it is quite true that now, after having ceased to take it habitually, I find that a cup of coffee has the same effect upon me as you describe it to have upon yourself. Lastly I cut off two or three cigars a day and that has done me no harm. As a natural consequence of improved health I have worked better. I have attended two courses of lectures and derived considerable profit from them. By means of Liouvilles lectures on the Theory of Numbers[3] I have filled up a gap in my mathematical education, and by means of Chasles lectures on Higher Geometry[4] I have increased my power of coping with my own researches. The latter proceed slowly but still not unsatisfactorily. The subject of equally attracting bodies which I am completing is extremely rich in geometrical consequences. It continually forces me to step aside to examine purely geometrical questions, and I allow it to do so because the latter appear to me just as important as the former, and without the one it is certain the other would be incomplete. Thus for a long time I was engaged with curves of the fourth degree, whose attracting properties bear a close relation to those of the conic sections. After devoting much time to them and obtaining some results I thought to be both new and interesting I made the acquaintance of M. Chasles (of which more by and bye) and found that much of what I had done he had already published. My capacity for reading is so limited that I have ceased to be surprised at such events; neither am I so much discouraged at re-finding what is already known, it diminishes the available results of my time and labour it is true, but it increases my power to prosecute other researches. Besides it seldom happens that two independent researches on the same subject are so coincident as to render the publication of both unadvisable, and I shall be able to collect together as much material—of which Chasles has not robbed me—as will make a small memoir.[5] I am anxious before the time for my summer holiday arrives to arrange my writing work for publication. I must make every other project give way to this, and on this account I am afraid I shall not have finished in time to accompany you to Switzerland; much as I should enjoy another trip like our last I see little prospect of taking it, not only for the above reason but because it is more than probable that I must go to England this summer, and my income will not allow me to make the two journeys, even if I could spare the time.

I have mentioned my acquaintance with Chasles, and as it has formed one of the pleasantest incidents of the winter I must tell you about its commencement. I had attended his lectures regularly and often had the intention of stepping up to him and introducing myself, but my habitual reserve or hesitation accompanied by bad health caused me to postpone doing so from day to day until the lectures were nearly finished. At last on one of my brighter days I had the courage to break the ice, and we had a very pleasant conversation. He regretted much that I had not introduced . . . [6]

RI MS JT/1/H/240·

1. *[28]*: the day is given by the reference to a 'Sunday morning' in March. Tyndall's reply (letter 1507) was dated Saturday, 10 April 1858. The previous Sunday that fell in March was 28 March.
2. *since I wrote to you*: letter 1489.
3. *lectures on the Theory of Numbers*: see letter 1472, n. 10.

4. *lectures on Higher Geometry*: Chasles lectured at the Collège de Sorbonne of the University of Paris, where he held the chair in higher geometry since 1846.
5. *as will make a small memoir*: possibly T. A. Hirst, 'On the quadric inversion of plane curves', *Roy. Soc. Proc.*, 14 (1865), pp. 91–106.
6. *I had not introduced . . .* : fragment only, letter is incomplete.

From George Wynne   [5 April 1858]¹   1504

Dear Tyndall.

Good bye, I embark for China tomorrow, I am so sorry to miss you.

RI MS JT/2/13c/1065
RI MS JT/2/9/314

---

1. *[5 April 1858]*: in his journal for 5 April 1858, Tyndall wrote, 'a card from Colonel Wynne bearing the following to me most sorrowful inscription "Dear Tyndall" Good bye, I embark for China tomorrow, I am so sorry to miss you". It smote my heart like the tidings of the death of a dear friend' (Journal, RI MS JT/2/13c/1065).

From Heinrich Gustav Magnus   6 April 1858   1505

Berlin 6 April 58

Mein theurer Tyndall

Seit Ewigkeiten habe ich nicht von Ihnen gehört, und fast glaube ich daß es meine Schuld ist, und daß ich einen Ihrer Briefe unbeantwortet gelassen. Ich freue mich um so mehr dass Hʳ. Robert, mit dem wir, wie Sie sich vielleicht erinnern, in Gesellschaft seiner liebenswürdigen Frau, eine Fahrt nach Potsdam gemacht haben, es übernehmen will Ihnen mündlich Nachricht von mir zu geben. Ich benutze diese Gelegenheit zu folgender Bitte: Sie waren so gut meinem Bruder Eduard eine Anzahl von Photograghien zu besorgen, die mir große Freude gewähren, und die allgemein hier sehr gefallen. Könnten Sie mir davon senden

Faraday   2 Exemplare
Tyndall   1   ”
Stokes   1   ”

Das Portrait von Ihnen und von Stokes ist vortrefflich, wie die meisten andern die Sie so gut waren für mich auszuwählen; das von Faraday gefällt mir weniger. Ich sage dies nur damit wenn etwa verschiedene Aufnahmen gemacht worden sind, ich nichts dagegen hätte von ihm eine andere zu erhalten, aber nicht von Ihnen und von Stokes. Sollte ein gutes Portrait von Wheatstone sich in dieser Samlung finden, so bitte fügen Sie es bei.

Herr Robert hält sich nur 2 Tage in London auf, da diese Zeit wahrscheinlich nicht ausreicht um ihm die Blätter mitzugeben, so sind Sie wohl so gut sie mir durch Nutt zu senden.

Gegen meine Electrolytischen Untersuchungen sind 3 Angriffe erschienen, aber nichts als Dummheit. Ich werde diesmal nicht unterlassen zu antworten.

Lassen Sie mich doch mündlich durch Hr. R. wissen wie es Ihnen geht; oder was mir noch mehr Freude machen würde, schreiben Sie mir einmal wieder.

Grüssen Sie alle Freunde die sich meiner erinnern; vor Allen aber M[r]. Faraday

Bei Ihren hiesigen Bekannten ist alles unverändert!

Von Herzen | Ihr | G Magnus.

Frau und Kinder grüßen bestens. | D[r]. Tyndall.

Berlin 6 April 58

My dear Tyndall

I have not heard from you for an eternity, and I almost believe that it is my fault, and that I have left one of your letters[1] unanswered. I am all the more delighted that Herr Robert,[2] with whom, as you will perhaps remember, in the company of his amiable wife,[3] we went on a trip to Potsdam,[4] wants to undertake to give you news of myself in person. I am taking advantage of this opportunity for the following request: You were so good as to provide my brother Eduard[5] with a number of photographs,[6] which give me great delight and which everyone here likes very much. Could you send me of those

Faraday   2 copies
Tyndall   1   ”
Stokes   1   ”

The portrait of you and of Stokes is excellent, as are most of the others that you were so good as to select for me; that of Faraday I like less.[7] I say this only so that, if, by any chance, some different photographs have been taken, I would have no objection to receiving another one of him, but not of you and of Stokes. Should there be a good portrait of Wheatstone in this collection, then please include it.

Herr Robert will be staying in London only for 2 days, and as this time will probably not be sufficient to give him the prints, then be so good as to send them to me through Nutt.[8]

Three attacks have appeared against my electrolytic investigations,[9] but nothing other than stupidity. This time I shall not refrain from replying.

Do let me know verbally through Herr R. how you are doing; or, what would bring me even more pleasure, write to me again some time.

Give my regards to all friends who remember me; but above all Mr Faraday. Everything remains the same with your acquaintances here!

Sincerely | Your | G Magnus.

My wife and children[10] send their best regards. | D^r. Tyndall.

RI MS JT/1/M/22

1. *one of your letters*: letter missing.

2. *Herr Robert*: not identified.

3. *his amiable wife*: not identified.

4. *Potsdam*: the capital city of the German state Brandenburg, close by Berlin.

5. *my brother Eduard*: Eduard Magnus (1799–1872), a portrait and genre painter, and elder brother of Gustav Magnus (*ADB*).

6. *photographs*: in the German, Magnus spelled 'photographien' incorrectly.

7. *I like less*: on 21 September 1857, Tyndall and Eduard Magnus went to Maull & Polyblank to have their individual portraits taken (Journal, RI MS JT/2/13c/1034). Tyndall's portrait is reproduced as the frontispiece to this volume.

8. *Nutt*: David Nutt (1810–63), a publisher and bookseller located at 270 Strand, London. See 'Deaths', *Times,* 30 November 1863, p. 1.

9. *Three attacks have appeared against my electrolytic investigations*: G. Magnus, 'Elektrolytische Untersuchengen', *Poggend. Annal.,* 102 (1857), pp. 1–54. His reply to his critics is G. Magnus, 'Ueber directe und indirecte Zersetzung durch den galvanischen Strom', *Poggend. Annal.,* 104 (1858), pp. 553–80.

10. *My wife and children*: Bertha, Anna, Christine, and Paul Magnus.

To Juliet Pollock   [7 April 1858][1]   1506

<u>Wednesday</u>.

My dear Mrs Pollock

Are we not unfortunate? There goes the driving wind which one would readily encounter were it not charged with rain—and thus it has poured down upon us for the last two days. Yesterday I called upon M^rs. Tennyson[2] & delivered up your book.[3] Her husband[4] came in and I spent half an hour with them promising to return to dinner at 5 o'C. I had a note from M^rs. Tennyson[5] on Monday asking me to come on that day, but I craved liberty to transfer the thing to Tuesday. In fact, how I know not, some letters which had been addressed to me at the post office found their way to their house, so that they were apprised of my coming before I informed them of it myself.

I found M^rs. Tennyson just what you described her. Well I found them at dinner, Mr Juett[6] (I am not sure of the spelling, but he was pulled up some time ago at Oxford touching orthodoxy). was also there. We had 'honest talk and wholesome wine',[7] and two fine little boys joined us for a time. After M^rs. Tennyson went away we continued our talk, and after that again I ascended to the upper story, into the poet's own holy place; here he filled a pipe for me, lighted it himself and transferred it to my lips. and we smoked and talked for another hour. We talked of Maud[8] and its critics of poetry, of Mr Buckle's lecture,[9] which he and Juett had glanced at and thought empty—of Christianity and the influence of the imagination. Tennyson does not dazzle, but there is that about him which pulled me like the force of gravity—a thorough candour and brotherliness, if I may use the expression, an absence of all artificial fences, so that there is no hindrance to the play of natural affinities. I left him as if I had known him for years, promising to call again before quitting the Island.[10]

I quit this place tomorrow & turn my steps homeward. On Saturday at the furthest I hope to reach London

always yours most sincerely | John <u>Tyndall</u>

RI MS JT/1/T/1202
RI MS JT/1/TYP/6/1995

---

1. *[7 April 1858]*: the date is given by the reference to visiting Emily Tennyson, wife of Alfred Tennyson, at their home Farringford House, Freshwater, on the Isle of Wight (see n. 10). Tyndall wrote in his journal entry for 6 April 1858: 'At 12 o'clock Harry and myself went to call on Mrs Tennyson, found her at home, she was such as might be inferred from the description of Mrs Pollock—"charming" is the word which would best express her character' (Journal, RI MS JT/2/13c/1066).
2. *M^rs. Tennyson*: Emily Tennyson
3. *your book*: see letter 1502, n. 2.
4. *Her husband*: Alfred Tennyson.
5. *a note from M^rs. Tennyson*: letter missing.
6. *Mr Juett*: Benjamin Jowett (1817–93), a theologian who held the Regius Professorship of Greek at Oxford (*ODNB*).
7. *'honest talk and wholesome wine'*: words from Tennyson's 1852 poem (and invitation) 'To The Rev. F. D. Maurice', ii.18: 'You'll have no scandal while you dine | But honest talk and wholesome wine, | And only hear the magpie gossip | Garrulous under a roof of pine:'.
8. *Maud*: Tennyson's poem 'Maud' (1855).
9. *Mr Buckle's lecture*: on 19 March 1858, Henry Buckle delivered a Friday Evening Discourse at the RI entitled 'The Influence of Women on the Progress of Knowledge', which was later printed in *Fraser's Magazine* for April 1858. This was Buckle's only public lecture.

10. *quitting the Island*: Tyndall vacationed on the Isle of Wight, England's largest island and popular destination in the English Channel, with Henry Bence Jones from 3–8 April. He visited the Tennysons at their home on the island at Freshwater. Tyndall noted in his journal that they left the island on 8 April: 'No walking along the cliff to day, so we go by Yarmouth and Lymington' (Journal, RI MS JT/2/13c/1069).

To Thomas Archer Hirst   10 April 1858   1507

Saturday 10th. April 1858.

My dear Tom.

The very day that I received your note[1] I sat down to write to you, but something stopped my hand, and what I wrote remains still in my drawer. It gladdened me very much to hear of the improved state of your health. I sometimes think that if a man lived a life 'according to Reason' as Fichte would say,[2] that his health might be preserved much more vigorous than is usually the case. I wonder would cod liver oil be good for me!

Since I heard from you I have been down in the Isle of Wight,[3] but was peculiarly unfortunate as regards the weather. I reflected on your proposal about Fontainebleau,[4] but found it out of the question, though the temptation was indeed strong. The only part of your note that does not please me is that where you extinguish my hope of having you along with me in the summer. I had fully calculated upon another campaign with you. By the way Tom do you intend to make Paris your residence always? I am afraid you live a far too lonely life there, and are perhaps shirking a contact with the world which is sure to be demanded of you at last. On last Tuesday you would not guess where I found myself:—sitting in a little room, at the top of a house within reach of the sea roar, with a gentleman with black hair & black beard at my side, watching him fill his own pipe, then fill one for me, then put my pipe in his mouth, light it well, and transfer it to my lips. The man was Alfred Tennyson! Eight years ago I heard you read some of his poetry in some corner of St Martin's Court.[5] I little thought at that time what 8 years would bring forth.

I am very glad to find that you have made the acquaintance of Chasles.[6] You will be a great Mathematician Tom. If you were here I should be strongly tempted to read the Differential & Integral Calculus[7] again under your guidance. And now good bye. Many claims are pulling me. Col. Wynne[8] I deeply regret to say is sent off to China[9]

ever yours affectionately, John

RI MS JT/1/T/643
RI MS JT/1/HTYP/510

----

1. *your note*: letter 1503.

2. *'according to Reason' as Fichte would say*: a reference to a lecture from the German philos-
    opher Johann Gottlieb Fichte (1762–1814) entitled 'Life According to Reason'. Fichte is
    considered one of the founding figures of post-Kantian German Idealism.

3. *Isle of Wight*: Tyndall visited Alfred Tennyson several times at his island home Farringford
    House (see letter 1506 and Journal, RI MS JT/2/13c/1064–69).

4. *Fontainebleau*: see letter 1458, n. 3.

5. *St Martin's Court*: a pedestrian alley in central London.

6. *made the acquaintance of Chasles*: in letter 1503, Hirst recounted to Tyndall that he met
    the French mathematician Michel Chasles while attending a course of his lectures in Paris.

7. *Differential & Integral Calculus*: probably A. de Morgan, *The Differential and Integral Cal-
    culus* (London: Baldwin and Cradock, 1842); or perhaps S. F. Lacroix, *Elementary Treatise
    on the Differential and Integral Calculus* (Cambridge: J. Deighton and Sons, 1816).

8. *Col. Wynne*: George Wynne.

9. *sent off to China*: see letter 1490, n. 6. In his journal entry for 5 April 1858, Tyndall wrote:
    'a card from Colonel Wynne bearing the following to me most sorrowful inscription "Dear
    Tyndall" Good bye, I embark for China tomorrow, I am so sorry to miss you". It smote my
    heart like the tidings of the death of a dear friend' (Journal, RI MS JT/2/13c/1065; this is
    also letter 1504).

From Thomas Archer Hirst   17 April 1858   1508

Paris. <u>April</u> 17<u>th</u> 1858

My dear John

Although I was a little disappointed you did not accept my invitation[1] and
come to spend your Easter holidays with me I was glad to learn that you had
done the next best thing—been to the Isle of Wight[2] and made the acquaintance of Tennyson.[3] You tell me little or nothing about him simply because
you write your letter in a hurry I hope when you have a little more leisure you
will give me more particulars of your interview.

A few days ago I went to hear a popular lecture by Jamin on the Theory
of Glaciers.[4] It was delivered at a kind of club 'Cercle du sociéte des Chemins de Fer'[5] to an amateur audience and consequently essentially popular—
It was a good lecture nevertheless and interested me amongst other things I
was conscious of a pardonable throb of pride when I heard him refer to the
experiment of Tyndall '<u>un physicien Anglais trés distinguê</u>'[6] at the same time
I thought he might have made more use of the researches of that same Tyndall. He only mentioned your experiments on the capability of ice to take any
form under pressure. He did not enter into the subject of regelation even but
accepted Forbes Theory of plasticity[7] together with that of the infiltration

and subsequent freezing of water in the body of the glacier. As a consequence his lecture whilst it pleased me as far as its style was concerned disappointed me. I expected to hear more of what you had done.

You ask me if I intend to make Paris my residence always. No. My intention was to spend the winter in Paris and on the whole I am well satisfied with the result. I have made myself tolerably familiar with French and at the same time I have made more progress with my own work than if I had been living elsewhere and avoiding less that contact with the world which simply disturbs. I trust you will find that I shall not <u>shirk</u> any other contact whenever the same is demanded from me.

My present intentions are not to stir from Paris until my winters work is put into order. After that, unless I should be called to England by more important affairs, I think of turning my steps towards Italy and there spending next winter as I have here spent the last. I have entertained this project for many years and in all probability I shall never have again so good an opportunity of carrying it out.

I do not lead so lonely a life as you appear to fear in fact I am becoming almost a diner out. Last night I was at a party at Chasles' again.

I am sorry to hear of Col. Wynne's departure for China[8] it must be a source of great regret to M[rs] Wynne.[9] Let us hope however that he will not be long absent. I was a little amused at Thompson's[10] letter in the last number of the 'Proceedings'.[11] He receives the account of your experiments one day & writes off the next to shew how those experiments confirm his brothers theory[12] and 'to add in his own part a physical explanation of the blue veins in glaciers' Fortunately for himself, perhaps, he 'has not time to execute the latter intention'. What he has to say on the former subject is to me very obscure and unsatisfactory. When will the full account of your experiments in question be published?[13]

Yours affectionately—<u>T A Hirst</u>

RI MS JT/1/H/241

---

1. *my invitation*: Easter Sunday in 1858 fell on 4 April. Hirst's letter of 28 March (1503) may have included the invitation (the letter is incomplete).

2. *Isle of Wight*: see letter 1506, n. 10.

3. *Tennyson*: Alfred Tennyson.

4. *popular lecture by Jamin on the Theory of Glaciers*: Jules Jamin (1818–86), a French physicist and professor at École Polytechnique in Paris from 1852–81 (see 'Discours prononcés aux obsèques de M. Jamin', *Paris, Comptes Rendus*, 102 (1886), pp. 337–43. In his journal for 18 April 1858, Hirst wrote: 'The incidents of the week have been a lecture on the Theory of Glaciers by Jamin delivered at the "Cercle du Société des Chemins de Fer". Wertheim procured me a card of invitation. The lecture was popular but goo[d]. Tyndall's experiments

were but very briefly referred to and the lecturer appeared to accept Forbes theory entirely'
(Brock & MacLeod, *Hirst Journals,* p. 1351).

5. *'Cercle du société des Chemins de Fer'*: originally founded as the Conférence des chemins
de fer (1844), and later renamed the Cercle des chemins de fer, the railway society was an
organization for those involved in the development of railways in France. See Cercle des
chemins de fer, *Annuaire de 1862* (Paris: Paul Dupont, 1862), p. 75.

6. *'un physicien Anglais trés distingué'*: a very distinguished English physicist (French).

7. *Theory of plasticity*: see letter 1306, n. 8.

8. *Col. Wynne's departure for China*: see letter 1507, n. 9.

9. *Mrs Wynne*: Anne Wynne

10. *Thompson's*: William Thomson.

11. *letter in the last number of the 'Proceedings'*: W. Thomson, 'Remarks on the interior Melting
of Ice', *Roy. Soc. Proc.,* 9 (1858), pp. 141–43.

12. *his brothers theory*: James Thomson. See J. Thomson, 'On the Plasticity of Ice, as manifested in Glaciers', *Roy. Soc. Proc.,* 8 (1857), pp. 455–58. See letter 1435.

13. *When will the full account of your experiments in question be published?*: see letter 1468, n. 1.

From Thomas Henry Huxley 20 April 1858 1509

April 20th 1858.

My dear Tyndall,

At the meeting at Reeks[1] where you were <u>not</u>, there was some talk of
working up the public in science through the Saturday Review.[2] That was
before that delicate flower the 'Scientific Review'[3] budded only to fade.

Maskelyne,[4] the proposer of this scheme, has now seen the 'Saturday's'
Editor,[5] and wants half a dozen of us to join together to work out the plan.

Saturday's Editor offers space and pay for a scientific article, not a review,
once a fortnight, such articles to be inserted without alteration or modification. He will also be glad to insert Reviews, if we like to do them. It is suggested that Sylvester, Maskelyne, you, Frankland, myself, Hooker,[6] Ramsay
and Smyth,[7] should form a corps (d'elite!)[8] each member of which should
pledge himself to supply at least one article in three months. The subject to be
at the pleasure of the individual in his own branch of science.

The plan seems to me so feasible that I promised at once to write to you
and Hooker. It seems to me to offer the advantages without the trouble and
responsibility of our 'Scientific Review'—and it is not too much of a pull
upon one's time.

Let me know if you agree, because we must have a preliminary meeting to
discuss our plan of operations.

Ever yours very faithfully | T. H. Huxley.

IC HP 8:35
Typed Transcript Only

1. *meeting at Reeks*: Trenham Reeks (1823–79), registrar at the Government School of Mines and Science Applied to the Arts in London. See 'Trenham Reeks', *Nature,* 20 (1879), pp. 38–39

2. *Saturday Review*: *The Saturday Review of Politics, Literature, Science, and Art* was a weekly London newspaper established in 1855.

3. *'Scientific Review'*: Huxley, Joseph Hooker, and Tyndall had discussed establishing a 'Scientific Review' newspaper to communicate science to the public. The idea was abandoned and as this letter indicates efforts would go toward writing science articles for the already established *Saturday Review,* although Hooker declined. On the 'Scientific Review', see R. Barton, '"Huxley, Lubbock, and Half a Dozen Others": Professionals and Gentlemen in the Formation of the X Club, 1851–1864', *Isis,* 89 (1998), pp. 410–44, on pp. 428–29.

4. *Maskelyne*: Mervyn Herbert Maskelyne (1823–1911), a mineralogist and grandson of Astronomer Royal Nevil Maskelyne (*ODNB*).

5. *the 'Saturday's' Editor*: John Cook, from 1855–68. See letters 1528 and 1574.

6. *Hooker*: Joseph Hooker.

7. *Smyth*: Warington Smyth (1817–90), an English geologist and professor at the Government School of Mines and Science Applied to the Arts in London, and older brother of Charles Piazzi Smyth (1819–1900), the astronomer royal for Scotland (*ODNB*).

8. *corps (d'elite!)*: a select group of people (French).

To Thomas Archer Hirst   21 April 1858   1510

Wednesday 21st. April / 58.

My dear Tom.

I have just read your letter,[1] and laughed at the last paragraph—laughed at the perfect similarity of your thoughts to my own. I rather think it is 'fortunate for himself' that he did not attempt what he[2] proposed. But perhaps I am wrong, for the man who is unintelligible is in a measure unassailable.

I was greatly pleased with my visit to Tennyson.[3] I found him a brotherly soul, the very sound of whose voice told you that a man was behind the organ. I dined with him, and discussed many matters with him—His perfect simplicity of character delighted me. He has a very superior woman for a wife; and two sons who appear to me to be a pair of the sweetest lyrics that ever sprung from Tennyson. We parted from each other as if we had been friends for years.

Last night I formed a fraction of a small but distinguished party at Lord Ashburton's.[4] He is a man of fine earnest soul. I had thoughts of declining his invitation as my head was not over well, but I am glad I did not do so, as my brain brightened, and I took my own share of the conversation—I think they were all Lords. He introduced me to two or three of them, and to the old Marquis of Lansdowne,[5] in fact nobody could be kinder. At dinner Lord Stanhope[6] was at my right; and would you guess who was on my left? Thomas Carlyle. We were the most joyful pair at the table, though nobody lacked glad cheerfulness. It was a regular, thorough face to face bit of talk, and I enjoyed it mightily. We spoke of old Leslie,[7] who was Forbes' predecessor; of Forbes & Agassiz, of Kepler,[8] of the habit of mind to which nature will reveal herself, and of those habits from which she withdraws. We talked of some books which are now causing a stir in London. At intervals my noble right hand neighbour talked about the Glaciers; he had spent some time among them last year. Carlyle appeared to be in choice condition, strong and fervent. His countenance is a wonderful one, and his eyes light up sometimes with almost superhuman radiance. It seemed to me that I quite got a hold of the old chap, and that we are likely to be permanent friends. He purposes coming to my lecture tomorrow.[9]

Now Tom dont be angry with me for the 'shirk'—though to say the truth I used the word to 'rile' you. Thank you for the bit of news about Jamin.[10] I am glad to hear of the dining—and still <u>it</u> has its healthy limits as well as other things. I am obliged to be very sparing in it. But Tom I know such a splendid family[11] here, Rich, cultivated, beautiful & good—good to excellence.—a mother and three daughters, who rank among the choicest bits of humanity that I ever met, or hoped to meet. Earth is like heaven tonight, the air is so bland & the moon & stars so beautiful goodnight Tom

always yours affectionately | <u>John</u>
Many thanks for the Memoir.

RI MS JT/1/T/892
RI MS JT/1/HTYP/511

---

1. *your letter*: letter 1508.

2. *he*: William Thomson.

3. *my visit to Tennyson*: see letter 1507.

4. *Lord Ashburton's*: Bingham Baring, 2nd Baron Ashburton (1799–1864), an Oxford-educated British politician who switched from Whig to Tory, and was elected FRS in 1854 (*ODNB*). The meeting was at The Grange, Ashburton's country residence.

5. *old Marquis of Lansdowne*: see letter 1404, n. 9.

6. *Lord Stanhope*: Philip Henry Stanhope (1805–75), fifth Earl Stanhope, a British politician and historian who took a leading role in the establishment of the National Portrait Gallery in 1856. He was titled Viscount Mahon from 1816 until he succeeded to the earldom in 1855 (*ODNB*).

7.   *old Leslie*: John Leslie (1766–1832), a Scottish mathematician, physicist, and professor of natural philosophy at Edinburgh University (*ODNB*).

8.   *Kepler*: Johannes Kepler (1571–1630), a German mathematician and astronomer best known for his laws governing the motion of planets (*CDSB*).

9.   *my lecture tomorrow*: on 15 April 1858, Tyndall commenced an additional three lectures (through 29 April) which continued his 'Ten Lectures on Heat, Considered as a Mode of Motion' given on Thursdays from 21 January to 25 March 1858 at the RI. See RI Guard Book, vol. 2, p. 105, and 'Royal Institution of Great Britain', *Morning Post*, 8 April 1858, p. 1.

10. *news of Jamin*: see letter 1508, n. 4.

11. *a splendid family*: Maria Drummond and her daughters Mary, Emily, and Fanny. Tyndall had mentioned to Hirst in January that he was going to visit the family and read to them 'the account of our trip up Mont Blanc' (see letter 1478). In letter 1367, Tyndall had shared with Hirst his positive feelings toward the family.

To Thomas Henry Huxley   [21 April 1858][1] Wed   1511

My dear Huxley,

The thing seems very feasible in my eyes,[2] and I think I could promise to perform my share of the business. Sound your men and let us have a meeting about it.

I think it may become a very important agency and exercise a salutary influence in this quackridden country.

I was at Lord Ashburton's[3] last night and had some talk about you and Hooker. He wishes to know you both very much. He is a fine earnest fellow and you will like him. I wish he was a member of the Phil. Club.[4]

I sat beside Carlyle at dinner. He appeared to be in choice trim, and we talked most congenially together. Of Buckle he says 'He's a weak watery unfruitful creature out of whom no good can come'!

Ever Yours | John Tyndall.

Kindest regards to Mrs Huxley.[5] I will soon call to see her <u>if permitted</u>.

IC HP 8:34
Typed Transcript Only

---

1.   *[21 April 1858]*: the date is given by relation to Tyndall's journal entry of 20 April 1858: 'I know not what I did during the day: dined at Lord Ashburton's' (Journal, RI MS JT/2/13c/1071).

2.   *The thing seems very feasible in my eyes*: see letter 1509.

3.   *Lord Ashburton's*: see letter 1510, n. 4.

4.   *Phil. Club*: see letter 1299, n. 7.

5.   *Mrs Huxley*: Henrietta Huxley.

From Thomas Henry Huxley   [22 April 1858]¹   1512

14 Waverley Place | Thursday.

My dear Tyndall,

I met John Murray last night. He talked about you, said he wanted to call upon you and begged that I would let you know he would be glad to make an arrangement about the publication of your Lectures.² He will be very glad if you will look in upon him.

I fear you could not have had a very pleasant trip.³ Here the weather has been diabolic.

Imagine my astonishment just now at getting an invitation to dinner from Buckle the Great:⁴ I shall decline, as I have too much of the Arab about me to eat a man's salt⁵ and then pitch into him. And I am afraid I shall be tempted to do the latter some day or the other.

I made an awful muddle à propos of you the other day. Plücker⁶ came to the Museum⁷ with Wheatstone who introduced us. You know the awkward pause that ensues when one has made one's bow under those circumstances. I wanted to say something and knew I had heard <u>something</u> about Plücker from you, so I must needs tell him that you were a great ally of mine and that I had often heard you speak of him!!

It was not till he was gone that I came to a sense of what I had done.
Ever yours faithfully | T. H. Huxley.

RI MS JT/1/TYP/9/2899
LT Typescript Only

———————

1. *[22 April 1858]*: the date is given based on relation to letter 1511 (written on a Wednesday, 21 April, in which Tyndall made a comment about Henry Buckle that Huxley here replied to).

2. *publication of your Lectures*: Tyndall presented 'Ten Lectures on Heat, Considered as a Mode of Motion' at the RI on in early 1858 (see letter 1440, n. 8). While the idea to publish these lectures with John Murray was proposed, Tyndall did not publish with John Murray until 1860 when he decided to publish a book about his mountaineering trips (see letters 1530 and 1563), titled *Glaciers of the Alps*. Tyndall gave a series of lectures on heat again in 1862, and these were published the following year as *Heat Considered as a Mode of Motion* (London: Longman, Green, Longman, Roberts, & Green, 1863).

3. *could not have had a very pleasant trip*: see letters 1506, 1507, and 1510.

4. *Imagine my astonishment . . . Buckle the Great*: in letter 1510, Tyndall wrote to Huxley about Carlyle's thoughts about Henry Buckle (see letter 1506, n. 9), and here it appears that Huxley is remarking on the coincidence.

5. *too much of the Arab about me to eat a man's salt*: 'to partake of his hospitality. Among the Arabs to eat a man's salt was a sacred bond between the host and guest. No one who has

eaten of another's salt should speak ill of him or do him an ill turn' (in E. C. Brewer, *Dictionary of Phrase and Fable* (1870), 2 vols (London: Cassell & Co., 1895), vol. 2, p. 1069).

6. *Plücker*: on Plücker and Tyndall's relationship, see letters 1299 and 1360. Tyndall met with Plücker on 10 April at August Hofmann's. From his journal entry: 'Hofmann arranged to bring Plücker and myself together. He laid aside his coldness and we talked together for a long time in a very friendly manner' (Journal, RI MS JT/2/13c/1069).

7. *the Museum*: The Museum of Practical Geology and the Government School of Mines and Science Applied to the Arts (Royal School of Mines as of 1863), was founded in 1851 at Jermyn Street in London. Huxley held the position of Professor of Natural History there from 1854–72, and continued with the School of Mines when it moved to South Kensington in 1872 until 1885. For an overview of the role of the School of Mines in science education, see R. Macleod, '"Instructed men" and Mining Engineers: The associates of the Royal School of Mines and British Imperial Science, 1851–1920', *Minerva* 32 (1994), pp. 422–39.

From William Hopkins   24 April 1858   1513

Cambridge | April 24 '58

My dear Sir,

A great current has always been described by Rennell[1] and others (as I have understood) as originating in the Indian Ocean and usually attributed to the trade winds. It runs south westerly along the S.E. coast of Africa, doubles the Cape takes a N. Westerly course and then a more westerly one along the Equator (where it known as the Equational current) to the coast of Brazil and the Caribbean Sea. It seems there to be in some degree lost as a distinct current, but I have never doubted but that there must be a movement in that sea which supplies the water poured out from the Gulf of Mexico as the proper Gulf stream continuing its course N. Easterly to our own shores. Hence it is that I have spoken of the Gulf stream as being <u>reflected</u> from the coasts of America.

I trust we shall ere long know more about ocean currents than we do at present. I have never speculated on the <u>causes</u> of them and in fact have been disposed to avoid doing so on account of our imperfect knowledge of the facts. I have always regarded the trade winds a <u>vera causa</u>[2] of currents but have thought little as to other causes which may exist. I have not considered with any care the cause you mention,[3] but supposing it to produce an <u>easterly</u> current in the northern Atlantic, how shall we account for the <u>westerly</u> currents to the South of the Cape of Good Hope and Cape Horn, both of which I imagine are ascertained beyond doubt. However if you pursue the subject I shall be glad to hear your speculations upon it.[4]

Yours truly | (in haste) | W. Hopkins.

RI MS JT/1/TYP/2/614
LT Typescript Only

---

1. *Rennell*: James Rennell (1742–1830), an English cartographer and historian who studied the currents of the Atlantic and Indian oceans (*ODNB*).
2. *vera causa*: a true cause (*OED*).
3. *the cause you mention*: not identified (presumably in a letter from Tyndall to Hopkins that is missing).
4. *if you pursue the subject I shall be glad to hear your speculations upon it*: at the time of this letter, Tyndall was about to give the last of the three extended lectures on heat (April 29, see letter 1510, n. 9). He was interested in heat as it related to wind and ocean currents, as this reply to a missing letter to Hopkins presumably shows. In Tyndall's published 1862 heat lectures, *Heat as a Mode of Motion* (London: Longman, Green, Longman, Roberts & Green, 1863), he discussed wind and ocean currents in chapter 6 (pp. 167–93).

From Thomas Romney Robinson[1]   26 April 1858   1514

April 26. 58 | Observatory | Armagh

Dear Sir

In the Astron Nachrichten Nº 1138 is a paper[2] by Steinheil[3] on Specula of Silvered Glass which I think may be worth inserting in the Phil Magazine.[4] As to his claim of priority over Foucault,[5] I cannot offer an opinion, as I have no access to the Algemeine Zeitung.[6] Some years since he proposed a very crude plan, the deposition of Silver of Gold by the Electrotype on polished mould.

The photometric results agree closely with those obtained by Mr Grubb[7] and myself, and show clearly that a Reflecting telescope with specula of silver, may have more light than a good achromatic of equal aperture.

He seems however to have contemplated only a spherical figure for the concave speculum, and proposes to correct it by a small achromatic which is over corrected. This would lose light and introduce complicated adjustment; and he seems not to be aware that the parabolic figure can be given to glass as well as speculum metal. I think also that the plans which he proposes for improving meridian and Equatorial instruments will be found impracticable from the very great difficulty of forming the large plane specula which they require.

Believe me | yours ever | T. R. Robinson
Dr Tyndall

RI MS JT/1/R/32
RI MS JT/1/TYP/3/1019

1. *Thomas Romney Robinson*: see letter 1501, n. 1.
2. *a paper*: see Steinheil, Dr., 'On the Advantages to be derived from the Use of Silver Mirrors for Reflecting Telescopes, and on a Novel Method of Mounting such Instruments (Translation of a letter from C.A. Steinheil of Munich, to Professor C.A.F. Peters of Altona, published in No. 1138 of the Astronomische Nachrichten)', *Monthly Notices of the Royal Astronomical Society*, 19 (1858), pp. 56–60.
3. *Steinheil*: Karl August Steinheil (1801–70), a German physicist and astronomer who was professor of mathematics and physics at the University of Munich, 1832–70. In 1854 he opened an optical workshop in Munich that became renowned for its instruments and telescopes for astronomy and telegraphy (*CDSB*).
4. *which I think may be worth inserting in the Phil Magazine*: Steinheil's paper was not published in the *Phil. Mag.*
5. *Foucault*: Jean Bernard Léon Foucault (1819–68), a French physicist known for the invention of the Foucault pendulum, an instrument he used to demonstrate the rotation of the Earth in 1851. Tyndall translated and abridged Foucault's work for the *Phil. Mag. (CDSB)*.
6. *Algemeine Zeitung*: a daily political journal in Germany, founded in 1798.
7. *Mr Grubb*: Thomas Grubb (1800–78), an Irish optical engineer (*CDSB*).

To Harriet Jane Moore or Julia Moore   [30 April 1858][1]   1515

Friday evg

My dear Miss Moore

You are a true friend & comforter. Yesterday's lecture[2] was to me a horrible failure.—no matter. I must cast the thought of such things behind me and go on to something better. I shall be rejoiced to come to you sometimes at Petersham[3] & am ready to undergo any amount of penance!

Ever yours J. Tyndall

RI MS JT/1/T/1090
RI MS JT/1/TYP/3/862/1

1. *[30 April 1858]*: the date is given by relation to Tyndall's journal entry of 29 April 1858: 'Lectured: but did not by any means fill my own ideal: no matter I must cast defeat behind me and go on to better things' (Journal, RI MS JT/2/13c/1072).
2. *Yesterday's lecture*: Tyndall continued his 'Ten Lectures on Heat, Considered as a Mode of Motion' at the RI. See letter 1510, n. 9.
3. *Petersham*: an area of southwest London on the south side of the River Thames, where the Moore family had their residence. The wealthy Moores had close connections with the RI and frequently entertained Tyndall.

From Frances Hooker [2 May 1858][1] 1516

8 Norfolk Road | Hove. | Brighton | Sunday Evg.

My dear Tyndall

My conscience misgives me that I have not half expressed to you the real gratitude I feel for all the instruction and pleasure you have given me in your lectures; and so you must allow me to write and thank you for all the pains you have taken to make your subject intelligible to me. This I do most heartily, and I feel that if I do not know so much about the matter as I ought, the fault is mine and not yours. At any rate, all I do know is due to you. Probably none of your audience were more thoroughly ignorant of such subjects than I, till you commenced my enlightenment.

I was sorry you could not come to us on Friday.[2] I had rather set my heart on seeing you that day, as it was my birthday,[3] which I always like to feel a pleasant day.

It is bitterly cold and raw here, and I shall not be sorry to find myself at home again next Saturday, unless the weather changes. Joseph[4] came down with me yesterday and is now on his way back to Kew,[5] a miserable evening for a journey. No fewer than three ladies said nothing but Dr Tyndall's lecture[6] would have tempted them out on Thursday! So I am afraid you were answerable for a good many headaches!

Lebe wohl.[7] I hold myself in readiness for any work you may deem me capable of executing,[8] and very much honoured by being allowed to do it.

Yours ever | Frances Hooker.

RI MS JT/1/TYP/8/2560
LT Typescript Only

---

1. *[2 May 1858]*: the date is given by the reference to Frances Hooker's birthday, 30 April, which fell on a Friday in 1858 (but not again until 1869). 2 May was the following Sunday.

2. *you could not come to us on Friday*: Tyndall likely was unable to get away from the RI since on the Thursday he gave the last of an additional three lectures (29 April) which continued his 'Ten Lectures on Heat, Considered as a Mode of Motion' (see letter 1510, n. 9) , and on 30 April there was a Friday Evening Discourse delivered by Andrew Ramsay, 'On the Geological Causes That Have Influenced the Scenery of Canada and the North-East Provinces of the United States', which Tyndall attended. In his journal entry for 30 April, he wrote 'Heard Ramsay' (Journal, RI MS JT/2/13c/1072).

3. *as it was my birthday*: Frances Hooker was born on 30 April 1825. See S. M. Walters and E. A. Stow, *Darwin's Mentor: John Stevens Henslow, 1796–1861* (New York: Cambridge University Press, 2001), p. 108.

4. *Joseph*: Joseph Hooker.
5. *Kew*: where Frances and Joseph Hooker lived. See letter 1328, n. 8.
6. *Dr Tyndall's lecture*: see n. 2.
7. *Lebe wohl*: farewell (German).
8. *I hold myself in readiness for any work you may deem me capable of executing*: Frances Hooker read manuscripts for Tyndall and provided feedback.

From George Biddell Airy   3 May 1858   1517

Royal Observatory Greenwich | S. E. | 1858 May 3

My dear Sir

I have been reading with great interest your paper about Glaciers.[1] I would beg leave to ask you one question upon it: not I hope quite idly, as I have had some castle-in-the-air ideas of visiting the Grimsel[2] this year: to which place I have not been for nearly 30 years.

Philos. Mag. 1858 May.

Page 380. From this reasoning I conclude that in the ordinary conditions of a glacier and excepting the near approach to a cataract, the laminae at the middle are transverse to the valley.

Page 383, diagram & p. 384: it would seem from these that the laminae at the middle are longitudinal, not only a little below A where the squeeze of the interfering glaciers is sufficiently obvious, but also towards RS[3] where the glacier must have long since, assumed its normal conditions as a single ice-stream.

Am I correct in either of these interpretations? and if in both, how are they reconciled?

I am, my dear Sir, | Yours very faithfully | G B Airy
Professor Tyndall

RI MS JT/1/A/33
RI MS JT/1/TYP/1/23

---

1. *your paper about Glaciers*: J. Tyndall and T. H. Huxley, 'On the Structure and Motion of Glaciers', *Phil. Mag.*, 15 (1858), pp. 365–88. This is a reprint of the original article in *Phil. Trans.*, which was published in the May edition.
2. *Grimsel*: the mountain pass separating the upper valley of the river Aare from the upper valley of the Rhône.
3. *RS*: on a diagram in 'On the Structure and Motion of Glaciers' (see n. 1), the line between points R and S on a glacier experiment Tyndall had performed that represent a point past when two merging glaciers have formed a single glacier.

From George Biddell Airy   7 May 1858   1518

Royal Observatory Greenwich | S. E. | 1858 May 7

Dear Sir

I am much obliged to your explanation about the Aar Glacier[1] and the trouble which you have taken in sending it. The matter is now quite clear.

In the subjects of regelation and viscosity or plasticity,[2] it seems to me that there is no essential difference between you and Prof. Forbes: for I can hardly imagine viscosity except as a destruction and restoration of organization: so that, in my view, the regelation is not an opposition to Prof. Forbes' viscosity, but is an explanation of its details, or of its modus operandi. But in the formation of the laminae of ice, your theories are entirely opposed.

I cannot tell very precisely the probable time of my journey to Switzerland: it depends partly on the time of my sons' return from College.[3] And I cannot well tell my course; it will depend partly on their tastes and partly on weather. But I will keep a watch for you, and should be delighted to fall in with you.

Probably I may start soon after the middle of June, entering I think by Geneva.[4]

I am, dear Sir, | Very faithfully yours | <u>G B Airy</u>
Professor Tyndall

RI MS JT/1/A/34
RI MS JT/1/TYP/1/23

---

1. *your explanation about the Aar Glacier*: letter missing, but surely Tyndall's reply to letter 1517. For the Aar Glacier, see letter 1305, n. 2.

2. *subjects of regelation and viscosity or plasticity*: see letter 1306, n. 8 and n. 10.

3. *my sons' return from College*: Wilfrid Airy (1836–1925), and Hubert Airy (1838–1903), both attended Trinity College at the University of Cambridge. The Easter (summer) term usually ended in early June. See J. A. Venn (ed.), *Alumni Cantabridgienses Part II: From 1752 to 1900*, 6 vols (Cambridge: Cambridge University Press, 1940), vol. 1, pp. 21–22.

4. *Geneva*: see letter 1293, n. 3.

To Thomas Archer Hirst   [14 May 1858][1]   1519

<u>Friday</u>.

My dear Tom

I am sure you will be glad to hear that your friend M. Poinsot was yesterday elected a Foreign Fellow of the Royal Society. I lose no time in letting you know this, so that you, if you like, may be the first to communicate the intelligence to him

affectionately yours | J. Tyndall

RI MS JT/1/T/644
RI MS JT/1/HTYP/512

———

1.   *[14 May 1858]*: the date is given by the addition of 'May 14 / 58.' on the letter, presumably written by Hirst.

From Johann Heinrich Jakob Müller[1]   20 May 1858   1520

Hochgeehrtester Herr Profeßor

Vor Kurzem erfuhr ich, daß die naturforschende Gesellschaft in Basel Ihr werthvolles „Philosophical Magazine" durch Tausch erhält und es wurde dadurch der Wunsch rege auf ähnlichem Wege diese mir so werthvolle Zeitschrift erwerben zu können. Ich bin deßhalb so frei mich in dieser Angelegenheit an Sie zu wenden. Was ich dagegen bieten könnte wäre

1) Alle von mir erscheinenden Bücher, zunächst also die eben erst vollendete 5te Auflage meines Lehrbuchs der Physik in zwei Bänden (Pouillet Müllers Lehrbuch der Physik und Meteorologie) in welcher ich auch ihre ausgezeichneten Arbeiten über Diamagnetismus besprochen habe

2) Alle von mir verfaßten Abhandlungen, von denen ich bereit bin eine Abschrift mitzutheilen gleichzeitig mit der Absendung an Poggendorf oder ein andres Journal.

3) Die Verhandlungen der naturforschenden Gesellschaft zu Freiburg i B, welche in ähnlicher Weise erscheinen, wie die der Basler Gesellschaft, nur daß sie mehr naturhistorisches bringen, obgleich sie auch viel physicalische, chemische und physiologische Aufsätze enthalten. So steht z B in einem fruheren Hefte ein Aufsatz von mir über Fluorescenz und gegenwärtig ist wieder eine Mittheilung über diesen Gegenstand im Druck.

Ich bin eben mit einer Experimentaluntersuchung über die Vertheilung der strahlenden Wärme im Sonnenspectrum beschäftigt, welche mir bereits interessante Resultate geliefert hat, namentlich glaube ich durch sie auch einen Erklärungsgrund für die Erscheinungen der Fluorescenz gefunden zu haben.

Da Sie längere Zeit in Deutschland sich aufhielten so kann ich wohl voraussetzen, daß Sie deutsch lesen und desshalb erlaubte ich mir auch deutsch a Sie zu schreiben, denn obgleich ich englisch ziemlich lesen kann, so würde mir doch die Abfassung eines englischen Briefes große Schwierigkeiten machen. Durch eine baldige Beantwortung meines Briefes, die ich Sie natürlich englisch abzufassen bitte /würden/ Sie mich in der That /sehr/ verbinden.

Indem ich bitte mir nicht übel nehmen zu wollen, daß ich Sie mit dieser Angelegenheit belästigt habe verbleibe ich hochachtungsvoll

Ihr ergebenster | Dr. J Müller | Professor der Physik

Freiburg <u>im Breisgau</u> | den 20ten May 1858.

Mr. | <u>John Tynall</u>. F.R.S. | Professor of Natural Philosophy | in the royal Institution | London

Most esteemed Herr Professor

I recently learned that the natural history society in Basel[2] receives your valuable "Philosophical Magazine" by exchange and I thereby became interested in being able to acquire this, to me, so valuable journal in a similar manner. I therefore take the liberty of turning to you in this matter. What I could offer in exchange would be

1) All books by me that will be appearing, starting with the just completed 5th edition of my textbook on physics in two volumes (Pouillet Müllers Lehrbuch der Physik und Meteorologie)[3] in which I have also discussed your excellent works on diamagnetism.

2) All papers authored by me, of which I am willing to submit a copy at the same time as I send them to Poggendorf[4] or another journal.

3) The Verhandlungen der Naturforschenden Gesellschaft zu Freiburg i. B.,[5] which appear in a similar fashion to those of the Basel Society,[6] only that they offer more on natural history, although they also contain many essays on physics, chemistry and physiology. There is, for instance, in an earlier[7] issue, an essay by me on fluorescence,[8] and at the present time there is again a communication on this subject at press.[9]

I am currently busy with an experimental investigation of the distribution of radiant heat in the solar spectrum,[10] which has already provided me with some interesting results; in particular, I believe that I have found through it an explanation for the phenomena of fluorescence.

As you stayed in Germany for a lengthy time, I can probably assume that you read German and for this reason I allowed myself to write to you in German as well, for although I can read English reasonably, writing a letter in English would, however, cause me great difficulties. I would indeed be *[very]* much obliged to you by a prompt answering of my letter, which I of course ask you to write in English.

In asking that you will not hold it against me for having bothered you with this matter, I remain most respectfully

Your very humble servant | Dr. J Müller | Professor of Physics

Freiburg <u>im Breisgau</u>[11] | 20th May 1858.

Mr. | <u>John Tynall</u>.[12] F.R.S. | Professor of Natural Philosophy | in the royal Institution | London[13]

RI MS JT/1/M/153

1.	*Johann Heinrich Jakob Müller*: Johann Heinrich Jakob Müller (1809–75), a German physicist, mathematician, and professor at the University of Freiburg whose research interests included optics and radiant heat (*CDSB*).

2.	*the natural history society in Basel*: the Naturforschende Gesellschaft in Basel, founded in 1817.

3.	*Pouillet Müllers Lehrbuch der Physik und Meteorologie*: C. Pouillet and J. Müller, *Lehrbuch der Physik und Meteorologie*, 5th edn, 2 vols (Braunschweig: Vieweg, 1858).

4.	*Poggendorf*: the *Annalen der Physik und Chemie,* while under the editorship of Johann Poggendorff from 1824 to 1876 was referred to as *Poggendorff's Annalen,* abbreviated as *Poggend. Annal.* in this volume.

5.	*Verhandlungen der Naturforschenden Gesellschaft zu Freiburg i. B.*: Berichte über die Verhandlungen der Naturforschenden Geselleschaft zu Freiburg im Breisgau.

6.	*those of the Basel Society*: Verhandlungen der Naturforschenden Gesellschaft in Basel.

7.	*earlier*: in the German, Muller did not include an Umlaut over the 'u' in 'fruheren'.

8.	*an essay by me on fluorescence*: J. Müller, 'Ueber Fluorescenz', *Verhandlungen der Naturforschenden Geselleschaft zu Freiburg,* 1 (1853–58), pp. 49–64, 97–99.

9.	*a communication on this subject at press*: J. Müller, 'Thermische Fluorescenz', *Verhandlungen der Naturforschenden Geselleschaft zu Freiburg,* 1 (1853–58), pp. 510–12; J. Müller, 'Intermittirende Fluorescenz', *Verhandlungen der Naturforschenden Gesellschaft zu Freiburg,* 1 (1853–58), pp. 513–14.

10.	*radiant heat in the solar spectrum*: J. Müller, 'Untersuchungen über die thermischen Wirkungen des Sonnenspectrums', *Poggend. Annal.,* 105 (1858), pp. 337–59.

11.	*Freiburg im Breisgau*: a city located in Baden-Württemberg, in the southwest of Germany.

12.	*Tynall*: misspelled by Muller.

13.	*Mr. | John Tynall. F.R.S. | Professor of Natural Philosophy | in the royal Institution | London*: written on the envelope, Tyndall's name misspelled in the original.

From Lyon Playfair   22 May 1858   1521

34 Cleveland Square | Hyde Park 22 May/58

My dear Tyndall

From the tone of your mind I dare say that you abominate the system of testimonials by which places are now bought & won—So do I & I do not intend, if I can possibly help, to print one in my canvass for the Edinburgh Chair of Chemistry.[1] But my Edinburgh supporters insist upon my having them in reserve, lest it should be alledged that I could not procure any.

Can you therefore in conscience oblige me so far as to write me a letter[2] stating that you have on <u>recent</u> occasions (for my rivals alledge I have not lectured since 1853) heard me lecture on Chemical subjects at the Royal Institution[3] & your general opinion as to my qualifications as a lecturer. Without professing to give an opinion on Chemical Evidence, perhaps you will kindly allude to my position in London, as being President of the Chemical Society,[4] to show that I take active interest in the pursuit of my science.

Yours truly | Lyon Playfair

RI MS JT/1/P/116
RI MS JT/1/TYP/3/984

---

1. *my canvass for the Edinburgh Chair of Chemistry*: Playfair was successful in obtaining this position, and only resigned when he moved into politics in 1869.

2. *write me a letter*: letter missing.

3. *heard me lecture on Chemical subjects at the Royal Institution*: Playfair's lectures at the RI on chemical subjects include 'On Three Important Chemical Discoveries From the Exhibition of 1851' on 27 February 1852, and 'On the Chemical Principles involved in Agricultural Experiments' on 30 May 1856.

4. *being President of the Chemical Society*: Playfair served as President of the Chemical Society of London from 1857–59.

From Rudolf Clausius   23 May 1858   1522

Zürich, 23 Mai, 1858.

Lieber Tyndall,

Sie haben Sich durch Ihre schönen Arbeiten über die Gletscher um die Schweiz so verdient gemacht, dass ich mir erlaubt habe, der hiesigen natur-forschenden Gesellschaft den Vorschlag zu machen, Sie zum Ehrenmitgliede zu ernennen. Nachdem der Präsident (welcher gegenwärtig Hr. Mousson ist) diesen Vorschlag der Gesellschaft in ihrer letzten Generalversammlung vor-gelegt hat, ist dieselbe mit der grössten Bereitwilligkeit darauf eingegangen, und es wird Ihnen von dem Secretär der Gesellschaft das Diplom bald zuge-sandt werden. Ich freue mich sehr Sie in dieser Beziehung als meinen College begrüssen zu können, und wünsche, dass diese Ernennung dazu beiträgt, Ihr Interesse für die Schweiz vielleicht noch etwas zu erhöhen.

Ich habe noch immer gezögert, den Auszug der Abhandlung von Huxley anzufertigen, weil ich hoffte, Ihre zweite Abhandlung, welche Sie im Decem-ber der Royal Society vorgelegt haben, und wovon ein kurzer Auszug in den Proceedings mitgetheilt ist, würde bald vollständig erscheinen, so dass ich bei meiner Bearbeitung beide zusammen berücksichtigen könnte. Sie können mir vielleicht gelegentlich schreiben, wie bald Ihre Abhandlung erscheinen wird.

Ich schicke Ihnen mit diesem Briefe zugleich ein Exemplar einer kleinen Abhandlung 'über die Natur des Ozon', und ein zweites Exemplar für Herrn Hirst, dessen Wohnung ich nicht weiss, und welches Sie wohl so gut sind, ihm bei Gelegenheit zukommen zu lassen. Es wird mir lieb sein, wenn Sie mir einmal schreiben, ob Ihnen Abhandlungen, welche ich so wie diese unter Band mit der Post schicke, ganz portofrei abgeliefert werden. Auf dem hiesi-gen Postbureau hat man mir die Versicherung gegeben, dass dieses der Fall sei.

Mit freundlichen Grüssen | der Ihrige | R. Clausius.

Zurich, 23 May, 1858.

Dear Tyndall,

You have rendered such outstanding services to Switzerland through your fine works on glaciers, that I have taken the liberty of making the proposal[1] to the natural history society here[2] that it appoints you an honorary member.[3] After the president (which is currently Herr Mousson), presented this proposal to the soci-ety in its last general meeting, it went into the matter with the greatest alacrity, and you will soon be sent the certificate by the secretary of the society.[4] I am very pleased to be able to welcome you in this respect as my colleague, and wish that this appointment will contribute to raising your interest in Switzerland perhaps even more.

I have still been holding off preparing the summary of the paper by Huxley,[5] because I was hoping that your second paper,[6] which you presented to the Royal Society in December, and of which a short summary is communicated in the Proceedings,[7] would soon appear in full, so that I could consider both together in my adaptation. Perhaps, when you get the opportunity, you can write to me to say how soon your paper will be appearing.

I am sending you with this letter a copy of a short paper 'über die Natur des Ozon',[8] and a second copy for Herr Hirst, whose residence I do not know, and which you will perhaps be so good as to send to him when you have the opportunity. I shall be glad if you write to me at some time as to whether papers which I send by post, like these ones, in a postal wrapper are delivered to you completely post-free. At the post office here, one gave me the assurance that this was the case.

With kind regards | Your | R. Clausius.

RI MS JT/1/TYP/7/2231
LT Typescript Only

1. *I have taken the liberty of making the proposal*: in letter 1524, Tyndall thanked Clausius for this.

2. *the natural history society here*: see letter 1461, n. 3.

3. *appoints you an honorary member*: Tyndall became an honorary member, or 'Ehrenmitglied' (German), of the the Naturforschende Gesellschaft in Zürich in 1858. See 'Verzeichniss der Mitglieder der naturforschenden Gesellschaft in Zürich', *Vierteljahrsschrift der Naturforschenden Gesellschaft in Zürich*, 4 (1859), p. 4.

4. *the secretary of the society*: Hermann Pestalozzi Bodmer (1826–1903) was secretary of the Naturforschende Gesellschaft from 1857–60. See E. Rübel, 'Geschichte der Naturforschenden Gesellschaft in Zürich', *Neujahrsblatt, herausgegeben von der Naturforschenden Gesellschaft in Zürich*, 149 (1947), p. 73.

5. *the paper by Huxley*: T. H. Huxley, 'Observations on the Structure of Glacial Ice', *Phil. Mag.*, 14 (1857), pp. 241–60; it eventually appeared as T. Huxley, 'Beobachtungen über die Struktur des Gletschereises', *Vierteljahrsschrift der Naturforschenden Gesellschaft in Zürich*, 4 (1859), pp. 1–12.

6. *your second paper*: J. Tyndall, 'On some Physical Properties of Ice', *Phil. Trans.*, 148 (1858), pp. 211–29.

7. *short summary is communicated in the Proceedings*: J. Tyndall, 'On some Physical Properties of Ice', *Roy. Soc. Proc.*, 9 (1857–59), pp 76–80.

8. *'über die Natur des Ozon'*: R. Clausius, *'über die Natur des Ozon'*, *Poggend. Annal.*, 103 (1858), pp. 644–52, which appeared in English as R. Clausius, 'On the Nature of Ozone', *Phil. Mag.*, 16 (1858), pp. 43–51.

To George Gabriel Stokes   25 May 1858   1523

25th. May 1858

My dear Sir

M[r]. Gassiots paper[1] appears to me to be well worthy of a place in the Philosophical Transactions.[2] In it a great number of experiments, which have invoked the application of considerable skill and labour, are described. Their explanation must be included in any complete theory of the electric discharge, and from their number and variety, and the critical circumstances under which some of them have been made, they will probably render efficient help to the founding of a theory. These are the circumstances which induce me to recommend the publication of the paper.

Very sincere[l]y yours | John Tyndall
Prof. Stokes &c &c.

RS RR/3/110 Tyndall 25 May 1858

---

1.   *M[r]. Gassiots paper*: J. P. Gassiot, 'The Bakerian Lecture: On the Stratifications and Dark Band in Electrical Discharges as Observed in Torricellian Vacua', *Phil. Trans.*, 148 (1858), pp. 1–16.

2.   *well worthy of a place in the Philosophical Transactions*: Tyndall acted as a referee for the *Phil. Trans.*, reviewing submissions and sending reports to Stokes, the journal's editor.

To Rudolf Clausius   28 May 1858   1524

28th May 1858

My dear Clausius

I feel very much gratified by the proof of your friendship which you have just given,[1] and greatly honoured by the distinction conferred upon me by the Naturforschenden Gesellschaft[2] of Zurich.[3] I hope you will endeavour to express this my feeling to the society in better words than I can make use of.

To M. Mousson personally I would beg of you to convey my best thanks for the trouble which he has so kindly taken. I hope soon to prove both to you and the Society[4] that my interest in the glacier question is unabated, and I hope the facts which I have to bring forward, whether they lead to any modification of our theoretic views or not, will still be regarded as a contribution to our knowledge.

The paper to which you refer was more upon the physical properties of Ice than upon the glaciers.—As soon as it is printed[5] I will send it to you, but the Royal Society is very slow in printing its memoirs. I communicated the first part of the observations upon the Mer de Glace[6] to the Society a few days ago,[7] and of this paper[8] I hope to be able to send you a pretty full abstract within the coming fortnight.

All your papers reach me with the greatest regularity, and free of all charge. This is not the case with others. An abhandlung[9] has sometimes cost me more than two Thalers:[10] but I do not now take them in. Your paper on Ozone[11] reached me yesterday, but I have not yet had time to look at it. It shall appear in due time in the Philosophical Magazine.

Suppose a verdampfende[12] drop of water to be placed in the centre of an indefinitely large space. Then, according to your theory, the quantity of vapour in a given volume of this space would be inversely proportional to the square of the distance. Do you think facts support this inference?

We are about to lose the present president of the Royal Society,[13] and, to the honour of the council be it said, a deputation, headed by Lord Wrottesley, waited upon Faraday a few days ago asking him to accept the presidency.[14] I wish he felt himself strong enough to do so—it would be an immense gain to science in England. But he will live and die plain Michael Faraday. He is one of the most glorious spirits that God ever sent upon the earth.—noble, true, and simple as a child.

I hope the day will come when I shall be able to return the compliment you have conferred on me. At present the claims of older men weigh with the council of the R.S. very much: but the day must come sooner or later, when we shall be colleagues in a double sense.

Believe me dear Clausius | Ever sincerely yours | John Tyndall

RI MS JT/1/T/174

---

1. *proof of your friendship which you have just given*: see letter 1522.

2. *Naturforschenden Gesellschaft*: a Swiss scientific society founded in 1746 to promote the study of the natural sciences, of which Tyndall was recently elected an Honorary Member. See letter 1522.

3. *Zurich*: see letter 1333, n. 11.

4. *the Society*: see n. 2.

5. *As soon as it is printed*: J. Tyndall, 'On Some Physical Properties of Ice', *Phil. Trans.*, 148 (1858), pp. 211–29.

6. *Mer de Glace*: see letter 1306, n. 12.

7. *a few days ago*: 20 May 1858.

8. *this paper*: see letter 1468, n. 1.

9. *abhandlung*: a treatise (German)

10. *Thalers*: German silver coins.

11. *Your paper on Ozone*: see letter 1522, n. 8.

12. *verdampfende*: vaporizing (German).

13. *the present president of the Royal Society*: John Wrottesley, from 1854–58.

14. *to accept the presidency*: Faraday did not accept the request, and Wrottesley was succeeded by Benjamin Brodie, see letter 1353, n. 21.

To Joseph Dalton Hooker   2 June 1858   1525

2nd. <u>June 1858</u>

My dear Hooker,

D[r]. Seemann[1] has been kind enough to send me that Diploma.[2] Will you have the goodness to say to me what sum you sent to Breslau[3] on the receipt of yours—It is left to my generosity to pay certain fees; how much ought I to pay? Of course the paying of these fees is a totally different thing from paying for the <u>distinction</u>. Do instruct me like a good fellow. I had written to D[r]. Seemann[4] but will detain the note until I hear from you.

Ever yours | <u>John Tyndall</u>

Kew JDH, vol. 4, 1260

---

1. *D[r]. Seemann*: Berthold Seemann, see letter 1427, n. 2.

2. *Diploma*: presumably Tyndall's certificate of membership in the Leopoldina. See letters 1365 and 1427.

3. *Breslau*: the Leopoldina was administered at Breslau during the later part of Christian Nees von Esenbeck's (1776–1858) tenure as president from 1818–58 (*CDSB, ADB*).

4. *I had written to Dr. Seemann*: letter missing.

From Juliet Pollock   [5 June 1858][1]   1526

59 Montagu Squ. W | Saturday <u>night</u>

What a man you are!

When one has supposed that at your last appearance[2] your climax was reached you come out again[3] surpassing yourself. What height will you reach at last.

I was not well—but you made me better. I had w. me a beautiful young Lady Augusta Ritchie[4] w. whom I wanted to introduce you but your table was deserted when I reached it.

So it must be another time. To day I am going w. people to see the horses at Holborn.[5]

The performance is said to be first rate.
Your friend | Juliet Pollock

RI MS JT/1/P/218
RI MS JT/1/TYP/6/1929

---

1. *[5 June 1858]*: the date, though uncertain, is given by the reference to Tyndall's previous
   two Friday Evening Discourses at the RI.
2. *your last appearance*: Tyndall's previous Friday Evening Discourse at the RI, 'On Some
   Physical Properties of Ice', occurred on 22 January 1858.
3. *come out again*: on 4 June 1858, Tyndall gave another Friday Evening Discourse at the RI
   titled 'On the Mer-de-Glace'.
4. *Lady Augusta Ritchie*: probably Charlotte Augusta Ritchie (1820–78), a cousin of the
   English writer William Thackeray. See C. Ritchie, *A Memoir* (London: Spottiswoode &
   Co., 1879).
5. *the horses at Holborn*: possibly a performance at Astley's Amphitheatre, famous for its
   equestrian acts. It was located in Lambeth in central London.

To Juliet Pollock [23 June 1858][1] 1527

<u>Wednesday.</u>

My dear M^rs. Pollock

As far as my vision reaches I see no barrier between me and the enjoyment
of the trip to St Albans.[2] I shall be most glad to join you and I am sure I shall
return morally & physically sounder for the trip.

I spent a delightful day at Petersham[3] on Monday—M^r. & M^rs. Faraday &
Miss Barnard[4] were there. I acted the part of a dutiful donkey in drawing M^rs.
Faraday round the grounds. Miss Moore[5] took two photographs[6] of us all in a
group. The negative looks very well, which is, of course, the first step towards
a good positive.

From them I heard, and was sorry to hear it, that you had recently lost a
friend,[7] and that your friend Mr Macready[8] had lost a daughter.[9]

I am busy clearing off the coil of small matters which had aggregated
during the session. I look forward with longing eyes to the Alps. Two military
examinations[10] which I have to attend occur in July, and this holds me back
some days later than I expected—nevertheless when the time comes I shall
make amends for all.

I intend to write today to a guide[11] in Lauterbrunnen[12] who has the reputation of being 'the boldest man in the Oberland'.[13] If I can secure him I shall
be all right, as he is not likely to insist upon a retinue of guides to keep him

in countenance. I am assured by some who know him that he will go with me anywhere.

What splendid weather! When are <u>you</u> going out of town?—How is Walter[14]—how is Mr Pollock[15]—I have just had a cup of coffee, & it has made me quite <u>gossipy</u>. Can you inform me how an agent so purely physical & material can produce a result so purely spiritual? I wish you could—it would save me some bewilderment. Perhaps however it is better that the thing should be hid—It leaves room for <u>wonder</u>, which is a wonderful moral agent in its way.

I must contrive to see you before Saturday week, and to learn all about the time of starting.

After I came in from Petersham on Monday M^r. Faraday took me with him to see Ristori[16]—I had never seen her previously—She is a wonderful creature.

When are you going to pay Walter & myself that ice which you owe us. I think we have a right to charge you interest upon it. According to the laws of compound interest you owe us two instead of one.

Ever yours | J. Tyndall

Have you seen 'Thorndale or the conflict of opinions?'[17]

RI MS JT/1/T/1199
RI MS JT/1/TYP/6/2067–68

---

1. *[23 June 1858]*: the date is given by the reference to the death of Lydia Macready on 20 June 1858 (see n. 7) and Tyndall's mention of two military examinations coming up in July. Wednesdays occurring after 20 June 1858 and before July were 23 and 30 June. Given the news of Lydia Macready's death, 23 June seems more likely.

2. *St Albans*: a city in England, northwest of London.

3. *Petersham*: the home of the Moore family. See letter 1515, n. 3.

4. *Miss Barnard*: Jane Barnard.

5. *Miss Moore*: either Harriet Jane Moore or Julia Moore.

6. *two photographs*: the photographs, in the collection of the RI, show Michael and Sarah Faraday, Tyndall, Faraday's niece Jane Barnard, and either Harriet Jane Moore or Julia Moore positioned around a bench. The photographs were taken by either Harriet Jane Moore or Julia Moore on 21 June (given the date of this letter), and one is reproduced as the frontispiece to *Tyndall Correspondence*, vol. 5.

7. *a friend*: Lydia Jane Macready (1842–58), who died on 20 June 1858.

8. *Mr Macready*: William Macready, see letter 1502, n. 6.

9. *a daughter*: Lydia Macready.

10. *military examinations*: in 1858 Tyndall was an examiner in 'Experimental sciences' for the Council of Military Education (see letter 1476, n. 1). He gave an admission examination for the Sandhurst Military College on 5 July at King's College, and an examination for direct commissions on 15 July at Burlington House. See Council of Military Education, *Report,* Command Paper 2603 (London: Her Majesty's Stationery Office, 1860); *Examination Papers used at the Examinations for Direct Commissions, and for Admission to the Royal Military College, in July, 1858* (London: Harrison, 1858); and Journal, 15 July 1858, RI MS JT/2/13c/1076. For more on Tyndall as an examiner, see letter 1306, n. 2.
11. *a guide*: Christian Lauener.
12. *Lauterbrunnen*: a village in the Swiss canton of Bern.
13. *Oberland*: the geographically higher region of the Swiss canton of Bern.
14. *Walter*: Walter Pollock.
15. *Mr Pollock*: William Frederick Pollock.
16. *Ristori*: Adelaide Ristori (1822–1906), an Italian stage actress of tragic roles, performed in London in late June–July 1858. See H. Morley, *The Journal of a London Playgoer from 1851 to 1866* (London: George Routledge & Sons, 1866), p. 204–18.
17. *'Thorndale or the conflict of opinions?'*: W. Smith, *Thorndale or the conflict of opinions* (London: William Blackwood and Sons, 1857).

From John Douglas Cook 7 July 1858 1528

My dear Sir

I much like your idea,[1] and shall feel obliged for any articles on the subjects you have had the kindness to suggest with which you may favour me.

Perhaps, if you are passing this way,[2] you will have the goodness to give me a call before you leave.

Ever most truly yours | John D. Cook. | The Albany | July 7/58

RI MS JT/1/C/40
RI MS JT/1/TYP/1/225

---

1. *your idea*: the proposal by Tyndall and fellow men of science to publish scientific articles in the *Saturday Review,* of which John Cook was the editor (1855–68). See letters 1509 and 1511.
2. *this way*: John Cook resided in the Albany, an apartment building in Piccadilly, London.

To Edward William Cooke[1]   12 July 1858   1529

12th July 1858

My dear M[r] Cooke.

The Rhone Glacier[2] is a very fine one. Looked down upon from the top of the Mayen Wand,[3] at the Grimsel[4] side, it is a noble mass. It has also deep crevasses enough. The Under Grindelwald[5] glacier is also fine and accessible. Ascend a sufficient distance toward the so-called Eismeer,[6] you will find crevasses and shafts which display the colour beautifully. For colour the end of Rosenlaui Glacier[7] is also superb. The <u>lower ends</u> of most glaciers afford you the means of studying the colour and the ice generally. They do not enable you to study the <u>structure</u> of <u>glacier</u> ice but that is a thing you probably don't need. The ice minarets on the Unter Grindelwald Gl., looked down upon as you descend the glacier, are splendid when the Sun shines upon them. The Mer de Glace[8] and the Glacier de Bossons[9] offer you ample material for study. They are within the valley of Chamouni.[10] The <u>seracs</u>[11] of the Glacier du Géant[12] (one of the tributaries to the Mer de Glace) are fine & picturesque. Scarcely anything in the ice-world more so. The ice cascade of the Talèfre[13] would also furnish you with the materials for study. & there are such crevasses which show the veined <u>structures</u> beautifully near Trelaporte.[14] As far as my experience goes I think the Mer de Glace would suit your purpose as well as any. but I have not yet seen the Monte Rosa[15] district. Thither I go this year. The only glacier map with which I am acquainted is that of Forbes[16] of the Mer de Glace which I have found very useful.

Ever yours | John Tyndall

Jeff Weber Rare Books S1539

---

1. *Edward Cooke*: the catalogue from Jeff Weber Rare Books describing this letter states it is from Tyndall to a 'W. Cooke'. However, the 'W' was a misreading for 'Mr', and the likely recipient was Edward William Cooke (1811–80), a painter and alpinist, with whom Tyndall was sharing alpine locations for painting subjects (*ODNB*).

2. *Rhone Glacier*: see letter 1297, n. 12.

3. *Mayen Wand*: a very steep, vertical rock wall near the Grimsel.

4. *Grimsel*: see letter 1517, n. 2.

5. *Under Grindelwald*: see letter 1312, n. 3.

6. *Eismeer*: the glacial area above the Grindelwald glaciers.

7. *Rosenlaui Glacier*: a glacier in the Swiss Bernese Alps.

8. *Mer de Glace*: see letter 1306, n. 12.

9. *Glacier de Bossons*: a large glacier on the north side of Mont Blanc.

10. *Chamouni*: see letter 1353, n. 9.
11. *seracs*: see letter 1420, n. 20.
12. *Glacier du Géant*: see letter 1436, n. 3.
13. *Talèfre*: see letter 1416, n. 6.
14. *Trelaporte*: the easternmost of the Chamonix needles of the Mont Blanc massif, overlooking the Mer de Glace.
15. *Monte Rosa*: see letter 1478, n. 22.
16. *only glacier map with which I am acquainted is that of Forbes*: see letter 1444, n. 5.

To John Murray   12 July 1858   1530

<u>12th July 1858</u>

My dear Sir

I have been so unfortunate as to miss seeing you three times. Today I should have waited had not the time at my disposal been very brief. —What I intended to say to you however can be easily said in a note.

I have not been able to write out the lectures on Heat:[1] nor shall I be able to do so this year. I go to Switzerland this week and purpose remaining there for a couple of months. While there I intend to throw my observations & reflections into the form of a book[2] which I think might be made very interesting to the public generally. Besides embodying my glacier work it would contain the description of the more striking natural phenomena which I have encountered during the last 2 or 3 years, and would connect in a strict, yet simple and lively manner, these phenomena with their physical causes. I should write it in a very free spirit and should devote a chapter now and then to subjects of present & general interest, which however have not as yet at all made their way into the mass of society. M[r] Longman[3] asked me to write such a book, but I think it right in the first place to bring it under your notice. What do you think of it?

Very sincerely yours | <u>John Tyndall</u>

NLS

---

1. *lectures on Heat*: Murray had proposed publishing Tyndall's lectures. See letter 1512.
2. *I intend to throw my observations & reflections into the form of a book*: see letter 1512, n. 2.
3. *Mr Longman*: either Thomas Longman (1804–79) or William Longman (1813–77) of the London publishers Longman, Green, Longman, Roberts, & Green (*ODNB*). William Longman is more likely as he was also a mountaineer. See letter 1535.

From Auguste de la Rive   12 July 1858   1531

Genève, le 12 Juillet, 1858.

Mon cher Monsieur,

J'espérais toujours vous voir arriver au milieu de nous; du moins on nous en avait donné l'espoir; c'est ce qui faisait que je ne vous écrivais pas. Mais vous ne paraissez point, & en effet avec le temps affreux que nous avons depuis près de quinze jours, il n'y aurait rien de bon à faire pour vous sur les glaciers. Toutefois j'espère bien que l'été ne se passera pas sans que nous ayons le plaisir de vous voir arriver. Vous seriez très aimable en particulier si vous pouviez vous arranger de façon à être le 1ᵉʳ Août à Berne. C'est le lendemain matin 2 Août que s'ouvre la séance de la Société Helvétique des Sciences Naturelles à laquelle assisteront plusieurs naturalistes & physiciens de mérite dont vous connaissez, je crois plusieurs. Nous avons entre autres M. R. Clausius, *<1 word missing>*, Wiedemann, Merian, Studer, Dufour, de Candolles, Pictet, Plantamour, Soret &c. M. Matteucci & quelques autres savants étrangers nous honorent de leur présence; il y aura aussi bien des géologues, notre M. Studer, M. M. Escher, Favre, Masson, *<1 word missing>* &c, y seront. Tâchez donc d'être des nôtres; ce sera un bien grand plaisir que vous nous ferez à tous & à moi en particulier.

J'ai sur la conscience de ne vous avoir pas assez témoigné toute ma reconnaissance pour la peine que vous avez prise de corriger les épreuves de mon troisième volume; mais si je ne vous l'ai pas assez dit, je l'ai du moins bien senti. Me permettriez-vous de vous offrir comme un faible témoignage de ma gratitude une collection des Archives de l'électricité que j'ai publiées pendant cinq années & qui sont assez utiles à avoir dans sa bibliothèque pour les consulter. J'en ai préparé un exemplaire que je vous remettrai à votre passage à Genève, si comme je l'espère, vous y passez. Je serai aussi heureux de vous montrer mes dernières expériences sur l'influence de l'aimant sur la lumière électrique; elles sont assez curieuses & belles à voir. Il y a quelques points dans ces expériences sur lesquels j'aimerais appeler votre attention & m'entretenir avec vous.

Votre dernier travail sur les glaciers nous a tous ici vivement intéressé. Je vous remercie d'avoir pensé à m'en envoyer un extrait que j'ai reçu. M. Soret a imprimé dans la <u>Bibl. Univer.</u> la traduction entière du mémoire qui a paru dans le <u>Phil Mag</u>.

J'espère que M. Faraday continue à être bien portant; rappelez moi à son bon souvenir & croyez moi, vous-même, votre tout dévoué & bien affectionné | Aug. de la Rive.

Geneva, 12 July 1858.

My dear Sir,

I was still expecting to see you turn up amongst us. At least we were given the hope, and that was the reason why I did not write to you. But you are not showing up, & in fact with the dreadful weather that we have been experiencing for almost a fortnight, there would be nothing good for you to do up on the glaciers. However, I hope summer will not go without the pleasure of seeing you arrive. You would be especially kind if you could make arrangements to be in Berne[1] on the 1st of August. The next morning August 2nd is the opening session of the Société Helvetique des Sciences Naturelles[2] that will be attended by many great naturalists & physicians, many of whom I believe you know. We will receive Messrs. R. Clausius, *<1 word missing>*, Wiedemann,[3] Merian,[4] Studer, Dufour,[5] de Candolles,[6] Pictet,[7] Plantamour,[8] Soret[9] &c. Mr. Matteucci & some other foreign savants who will grace us with their presence; there will also be many geologists; our Mr. Studer, Messrs. M. Escher, Favre,[10] Masson,[11] *<1 word missing>* &c, will be there. Please try to join us;[12] it will be a great pleasure for us all & especially for me.

It weighs on my heart that I have not demonstrated all my gratitude to you for the trouble you took to correct the proofs of my third volume.[13] But if I have not told you enough, at least I really felt it. Would you allow me to offer you as a humble token of my gratitude, a collection of the Archives de l'éléctricité, which I published for five years[14] & are quite useful to own in one's library for reference. I prepared a copy that I will hand you when you visit Geneva,[15] as I am hoping you will. I will also be happy to show you my latest experiments on the influence of the magnet on electric light; they are really curious and beautiful to look at. There are a few points in these experiments on which I would like to draw your attention & discuss with you.

Your latest work on glaciers[16] was of great interest to us all here. I thank you for remembering to send the excerpt that I received. Mr Soret printed in the <u>Bibl. Univer.</u> the entire translation of the essay that appeared in the <u>Phil. Mag.</u>[17]

I hope that Mr. Faraday is still in good health; remind him of me & believe that I am your devoted & truly affectionate | Aug. de la Rive.

RI MS JT/1/TYP/1/338

---

1. *Berne*: see letter 1297, n. 10.
2. *Société Helvetique des Sciences Naturelles*: the meeting of the Société Helvetique des Sciences Naturelles (Swiss Society of Natural Sciences, now Swiss Academy of Natural Sciences, founded in 1815) was held in Bern, at the beginning of August 1858.

3. *Wiedemann*: Gustav Heinrich Wiedemann , see letter 1438, n. 1.

4. *Merian*: Peter Merian (1795–1883), a Swiss geologist who held professorships in physics, chemistry, and geology at the Univeristy of Basel, in addition to serving as the director of the Natural History Museum of Basel (*HLS*).

5. *Dufour*: Guillaume Henri Dufour, see letter 1378, n. 8.

6. *de Candolles*: Alphonse de Candolle (1806–93) and Casimir Pyrame de Candolle (1836–1918), French-Swiss botanists and sons of Augustin Pyramus de Candolle (1778–1841), another Swiss botanist (*HLS*).

7. *Pictet*: Francois Jules Pictet de la Rive (1809–72), a Swiss zoologist and paleontologist (*HLS*).

8. *Plantamour*: Emile Plantamour (1815–82), a Swiss astronomer and director of the Geneva Observatory from 1839–82 (*HLS*).

9. *Soret*: Jacques-Louis Soret, see letter 1417, n. 5.

10. *Favre*: Alphonse Favre (1815–90), a Swiss geologist and alpinist (*HLS*).

11. *Masson*: Antoine Masson (1806–60), a French physicist and professor of physics at the Royal College of Louis-le-Grand. See B. W. Feddersen and A. J. von Oettingen (eds), *J. C. Poggendorff's biographisch-literarisches Handwörterbuch zur Geschichte der exacten Wissenschaften* (Leipzig: J. A. Barth, 1898), vol. 3, p. 881.

12. *Please try to join us*: in his reply (letter 1534), Tyndall wrote 'With regard to the meeting at Berne it would give me great pleasure to go there, and to meet the gentleman you mention,' but that 'the demands upon me this year are so imperative, and the quantity of work which I must accomplish is so great that I fear to encourage myself with the hope of being able to join you'. No other letters nor journal entries indicate that he attended the August meeting or visited with de la Rive. When Tyndall was about to return from the Continent to London in mid-September, he wrote to de la Rive that he would not be able to see him (letter 1558).

13. *correct the proofs of my third volume*: see letter 1293, n. 16.

14. *for five years*: 1841–45.

15. *Geneva*: see letter 1293, n. 3.

16. *Your latest work on glaciers*: see letter 1468, n. 1.

17. *printed in the <u>Bibl. Univer.</u> the entire translation of the essay that appeared in the <u>Phil. Mag.</u>*: the paper J. Tyndall and T. H. Huxley, 'On the Structure and Motion of Glaciers', *Phil. Mag.*, 15 (1858), pp. 365–88, was translated and published as J. Tyndall et T. Huxley, 'Sur la Structure et le Mouvement des Glaciers', *Archives des sciences physiques et naturelles*, 2 (1858), pp. 200–31. The Swiss journal *Archives des sciences physiques et naturelles* was previously known as *Bibliothèque universelle de Genève*.

From Frances Hooker   [12 July 1858]¹   1532

Monday Even.

My dear Mʳ. Tyndall,

    Trusting that 'this week' does not absolutely mean tomorrow, I will write & thank you for your note,² & wish you farewell, & a pleasant journey.

    What should you say if we peeped in on you, in your solitude at the Distel Alp?³ More extraordinary things have happened!—Seriously, our thoughts are turning lovingly towards the mountains, though rather indefinitely at present—& *if* we go, & <u>where</u> we go, are points at present undecided. When we have quite made up our minds, I will write & tell you; as we might possibly meet for awhile, which would be very pleasant—

    At any rate, we cannot get away for another month—

    I suppose if I direct to you, to the care of M. le Curé Imseng,⁴ it will find you?

    Lebe wohl, und gute Reise.⁵ | Yours always sincerely | F H Hooker—

RI MS JT/1/H/497
RI MS JT/1/TYP/8/2554

---

1.   *[12 July 1858]*: the date is given based on a note from LT on the typescript letter stating that 13 July 1858, a Tuesday, was written on the envelope (which is now missing), making the day this letter was written, 'Monday Eve', 12 July. Tyndall presumably wrote it on that Monday and posted it the following day.

2.   *your note*: letter missing.

3.   *Distel Alp*: part of the mountainside beyond the southern end of the Mattmark See in the Saas valley.

4.   *M. le Curé Imseng*: Johann Josef Imseng (1806–69), the curé (priest) of Saas Fee, the upper village above the main village of Saas in the Swiss canton of Valais. Imseng, who accompanied Tyndall over several days in late August 1858, was also the first alpine skier in Switzerland (*HLS*). Tyndall had planned to ascend the Dom with Imseng, but they were prevented by bad weather.

5.   *Lebe wohl, und gute Reise*: Farewell, and good travel (German).

To Thomas Archer Hirst   [14 July 1858][1]   1533

Wednesday

My dear Tom.

I am at length able to say that I start from here on Friday morning, and reach Paris at 11.55 on the same day. I go by Dieppe.[2]

I will take with me the books & tobacco.

I will put up at the Hotel Vivienne,[3] and spend Saturday with you. On Sunday I proceed to Strasburg.[4]

Yours affectionately | John Tyndall

RI MS JT/1/T/645
RI MS JT/1/HTYP/512

1. *[14 July 1858]*: the date is given by the addition of 'July 14. 1858' on the letter, presumably written by Hirst.

2. *Dieppe*: a channel port in the Normandy region of France and popular entry point by ferry from Britain.

3. *Hotel Vivienne*: a hotel in Paris, France.

4. *Strasburg*: Strasbourg, a city in eastern France. Travelers to Switzerland from Paris would take a railway to Strasbourg, and then another railway from Strasbourg south to Basel in Switzerland.

To Auguste de la Rive   21 July 1858   1534

21st July 1858

My dear Sir

It is very kind of you to write to me, for indeed I do not deserve it at your hands. Your last letter[1] reached me as I was on the point of starting from London, and until now, when I find myself confined by the rain to a little Inn at Alpnacht[2] by the lake of Lucerne,[3] I have not had a quiet half hour to reply to it.

With regard to the meeting at Berne[4] it would give me great pleasure to go there, and to meet the gentleman you mention. Many of them I have never seen. M. M. Matteucci, Studer & Escher are among this number, and I have long wished to see them. If I can contrive to join you I will, but the demands upon me this year are so imperative, and the quantity of work which I <u>must</u> accomplish is so great that I fear to encourage myself with the hope of being able to join you. Should you see M. Matteucci would you kindly become my mediator and ask him not to think harshly of me for not answering his last letter.[5] I deferred doing so from day to day, the chief cause of my delay being ill health which compelled me to cut away almost all my correspondence.[6]

Let me now say that I cannot help regretting that you should think it necessary to make me any return for the little labour which I expended in the reading of your proofs.[7] It was a pleasure to me to do so; and I only wish that I could render you ten times the service. If you have not sent those books away, then I beg of you to keep them. I frankly confess to you that a friendly thought on your part towards me would be of more value in my estimation than your whole library were it transported bodily to London.

If I can at all accomplish it, be assured that I shall feel great delight in paying you a visit and in witnessing your experiment. My only fear is that the large demands for <u>work</u> made upon me this year will compel me to forego <u>pleasure</u> altogether. However if I find that I can make the work and a visit to Geneva[8] harmonise I shall be truly rejoiced to do so.

Believe me my dear Sir | Ever most sincerely yours | John Tyndall

I expect to be at Zermatt[9] at the time the Berne meeting[10] commences. Excuse this bad paper.

RI MS JT/1/T/372
RI MS JT/1/TYP/1/356

---

1. *your last letter*: letter 1531.
2. *Alpnacht*: a village in the canton of Unterwalden (now Obwalden) in Switzerland, on the west side of the Lake of Lucerne.
3. *lake of Lucerne*: Lake of Lucerne, or Luzerner See, in central Switzerland.
4. *the meeting at Berne*: see letter 1531, n. 2.
5. *his last letter*: possibly letter 1477.
6. *cut away almost all my correspondence*: there is a noticeable decrease in correspondence by Tyndall for the months of June and July 1858.
7. *reading of your proofs*: see letter 1293, n. 16 and n. 17.
8. *Geneva*: see letter 1293, n. 3.
9. *Zermatt*: see letter 1297, n. 7.
10. *Berne meeting*: see letter 1531, n. 2.

From Joseph Hooker   [late July or early August 1858][1]   1535

My dear Tyndall

I cannot bear the idea of your going to that screw Longman[2] with your wares[3] when John Murray[4] is in the same street. I hope I have not taken too great a liberty in simply asking John whether he thinks such works as yours would suit him.

Everyone I know who has dealt with Longmans are disgusted with them and scientific men especially, and if your lectures on Sound and Heat[5] should lead to a series of such works on Physics they would prove most valuable property and could not be in better hands than Murrays both for the author's and public's sake.

I am going to Ireland[6] on Thursday night for a week or ten days.
Ever yours | J. D. Hooker.

RI MS JT/1/TYP/8/2552
LT Typescript Only

1. *[late July or early August 1858]*: the year is given based on the mention of Longman and Murray with regard to Tyndall's planned book about glaciers (see n. 3), and it surely falls between letter 1530 (12 July) and letter 1563 (September). The month is given based on Hooker's mention of taking a trip to Ireland, which appears to have occurred in August 1858. He was still writing letters from Kew at the end of July, and away from Kew for three weeks in August 1858 (see L. Huxley, *Life and Letters of Sir Joseph Dalton Hooker,* 2 vols (London: John Murray, 1918), vol. 1, p. 362). In a letter to Hooker dated 30 July 1858, Charles Darwin wrote, 'I hope you & M[rs] Hooker will have a very very pleasant tour' (Charles Darwin to Joseph Hooker, 30 July 1858, in F. Burkhardt, et al. (eds), *The Correspondence of Charles Darwin,* 24 vols (New York: Cambridge University Press, 1991), vol. 7 (1858–1859), p. 140–41).

2. *that screw Longman*: probably William Longman of the London publishers Longman, Green, Longman, Roberts, & Green. See letter 1530, n. 3. Tyndall would use this publisher for almost all of his subsequent books.

3. *your wares*: J. Tyndall, *Glaciers of the Alps* (London: John Murray, 1860).

4. *Murray*: John Murray, who published *Glaciers of the Alps,* implying that Tyndall took Hooker's advice. He later explained why he transferred to Longman for his subsequent works, as Longman offered two thirds of the profits and paid him on the day of publication, while Murray had claimed part of the profits from American sales of an edition he had not published. See John Tyndall to John Murray, 4 September 1868, National Library of Scotland.

5. *lectures on Sound and Heat*: for Tyndall's lectures on sound, see letter 1308, n. 7. For his lectures on heat, see letter 1440, n. 8 and letter 1510, n. 9.

6. *going to Ireland*: see n. 1.

To Heinrich Debus   4 August 1858   1536

4th. <u>August</u> 1858.

My dear Heinrich,

I promised to write to you. I will now try to redeem that promise, though if my letter be as drowsy as I am myself it will send you to the land of dreams before you have finished it. My route[1] hitherto has been thus:—To Paris, to Zürich,[2] to Lucerne[3] across the three lakes, to Meyringen[4] over the Brüning[5] to Rosenlaui,[6] where I stayed a night and found next morning that a traveller had stolen my thoroughbred English boots, and left a pair of vile broken-winded continental <u>souliers</u>[7] in their place. A strong man was sent after him: he was overtaken upon the Scheideck[8] and deprived of his booty.[9] After examining the glacier we went on to Grindelwald.[10] Here we spent two days upon the glaciers.[11] Afterwards we crossed the Strahleck.[12] The normal allowance here is two guides for one traveller, but we reversed the regulation by being two travellers[13] to one guide. The clouds were sulky & threatening when we set out, but we got quite above them, and had magnificent glimpses from the summit of the pass. It amused me to see the guide relaxing in his demands—first <u>two</u> guides were absolutely essential to cross the pass with safety: then he thought a herd taken part of the way from the pastures which fringe the glacier, might do. Afterwards, of his own free will, he gave up the herd, and saw that we were competent to do the thing ourselves. We spent two nights at the Grimsel,[14] and afterwards a day upon the Rhone glacier.[15] Thence down the Valley to Viesch,[16] whence we ascended to a hotel[17] which stands midway up the Eggishorn.[18] The view from the summit of the Eggishorn is magnificent, and the grandest glacier in Switzerland winds it way for miles within sight among the mountains. I have traced the glacier to its origin. Five valleys, each flanked by noble mountains send down their frozen contents to the same place of convergence and forms there the trunk of the great Aletsch glacier.[19] This scene is most wonderful and most beautiful. The mountain masses are grand: the Jungfrau,[20] the Monk,[21] the Eiger,[22] the Aletsch horn[23] and others of similar magnitude send down the snows from their shining shoulders. In the upper regions the snow appears as smooth as polished marble and clothing as it does mountain forms in themselves beautiful, renders the scene lovely as well as majestic.

My guide[24] met me at Meyringen, he is a capital fellow, but I fear not quite enterprising enough. He is married, and has already had a brother[25] smashed to pieces upon the mountains, perhaps these circumstances augment his prudence. But there is a remarkable guide attached to this hotel. A low sized man of 33, who gives me the impression of great strength and great decision. The landlord spoke of him to me as 'the celebrated Bennen.'[26] 'Your man' said he 'is taller and larger, but he is not so strong as Bennen; and Bennen knows no fear: what man can accomplish he will do. He is proud of his calling and will shun no danger. He is conscientious, and depend upon it if you lose your life in his company he will lose his in yours, for he will die to save the man he leads.' The day before yesterday I sent for M[r] Bennen. I wished to make some observations upon the highest attainable mountain summit, and the Finsteraarhorn[27] is the highest mountain of the Oberland.[28] The mountain had been ascended but never with a single guide. I laid the following proposal & question before Bennen. 'Wherever you go I will follow and I do not think that I shall need any help from you—will you undertake to accompany me to the top of the Finsteraarhorn?'—xxxxxx—[29]

The night before last[30] I found myself stretched in a mountain cavern in the wildest place imaginable. A single mountaineer[31] was beside me and oh! how I envied his power of <u>snoring</u>. Once or twice I endeavored to get up an artificial snore in order to drown his operations, but I had finally to desist and listen to him for nearly five hours! However I had laid in a stock of sleep the night previous & could afford to lie awake. A cloudless moon was in the heavens and the stars twinkled all colours. At 3oC. we were on foot. The Jungfrau was before us shining in the moonlight. It was most tempting—'shall we try the Jungfrau?' I demanded 'just as you like' was the reply—I paused and weighed matters, I suffered the presence of the mountain maiden to overcome me for a time, and we actually walked towards her for a quarter of an hour with the intention of making the attempt. I paused and pondered, and finally resolved to give the maiden up. I turned my face in the opposite direction and went over the snow towards the Finsteraarhorn. What a glorious sunrise—I wish! We fed ourselves at the foot of the mountain, and faced it. At half past 10. A.M. we were on the top. A world of indescribable magnificence lay before & around us. Water boils there at 187° Faht. We swept down the slopes of the mountain at a terrific speed checking each other at intervals at the approach of crevasses—We simply sat upon the snow & let gravity do the rest. Bennen proved himself worthy of the character given of him. I never had a guide like him. We reached this hotel at 7½ P.M. after 16½ hours of as hard work as I ever went through. We were both fresh. Eight hours sound sleep have given me back my thoughts.

To day I am resting, and simply feel a pleasant drowsiness, which you now feel by infection—dont you?

In a few days I shall be at Zermatt[32]

Affectionately Yours, | <u>J. Tyndall</u>

I have never once shaved since I came to Switzerland. my beard is quite furious looking!

RI MS JT/1/T/258
RI MS JT/1/TYP/7/2365–66
RI MS JT/5/11/1330–31

---

1. *My route*: Tyndall's journey from Paris to Rosenlaui occurred between 19–22 July 1858 (Journal, RI MS JT/2/13c/1077–79).

2. *Zürich*: see letter 1333, n. 11.

3. *Lucerne*: a city in central Switzerland south of Zurich.

4. *Meyringen*: Meiringen, a town in the Bern canton of Switzerland.

5. *Brüning*: a mountain pass connecting Meiringen to Rosenlaui to the south.

6. *Rosenlaui*: a small town in the Rosenlaui Valley, near the Rosenlaui glacier, which Tyndall examined on 23 July 1858 (Journal, RI MS JT/2/13c/1079)

7. *souliers*: boots (French).

8. *Scheideck*: Gross Scheidegg, a mountain pass between the Schwarzhorn and the Wetterhorn mountains in the Bernese Alps of Switzerland. One would cross this pass when traveling between Meiringen and Grindelwald.

9. *deprived of his booty*: Tyndall wrote about this incident in his journal: 'While I was thus employed a servant at the hotel of Rosenlaui was in chase of a Frenchman who had stolen my boots. I say stolen for he left a pair behind for which it was impossible to mistake mine. He went deliberately into the kitchen and chose my thorough-bred English out of the assembled host of boots. He was overtaken and dismantled, but the man who gave chase took the trouble of carrying the vile boots which really belonged to the Varlet with him. It would have been a fit punishment to have left him barefooted upon the mountains' (Journal, RI MS JT/2/13c/1079–80).

10. *Grindelwald*: a town in the Bern canton of Switzerland.

11. *two days upon the glaciers*: see letter 1312, n. 3; observed on 24–25 July 1858 (Journal, RI MS JT/2/13c/1080–83).

12. *Strahleck*: Strahlegg, a mountain pass between Grindelwald and the Grimsel, crossed on 26 July 1858.

13. *two travellers to one guide*: Tyndall and geologist Andrew Ramsay, and guide Christian Lauener (he would guide them throughout summer 1858, except for when Tyndall made the ascent of the Finsteraarhorn with Johann Bennen as guide, while Lauener went with Ramsay to the Rhone).

14. *two nights at the Grimsel*: 26–27 July 1858; see letter 1517, n. 2.

15. *a day upon the Rhone glacier*: 28 July 1858; see letter 1297, n. 12.

16. *Viesch*: Fiesch, a town in the Rhone Valley.

17. *ascended to a hotel*: the Hotel Jungfrau halfway up the Eggishorn, opened in 1841. Tyndall stayed there often in his alpine travels.

18. *Eggishorn*: see letter 1297, n. 6.

19. *Aletsch glacier*: the longest glacier in the Alps, which flows past the Eggishorn.

20. *Jungfrau*: a mountain of the Bernese Alps.

21. *Monk*: Mönch, a mountain of the Bernese Alps.

22. *Eiger*: a mountain of the Bernese Alps.

23. *Aletsch horn*: Aletschorn, the second highest mountain of the Bernese Alps.

24. *My guide*: Christian Lauener.

25. *brother*: Johann Lauener (1777–1853), who died on the Jungfrau.

26. *'the celebrated Bennen.'*: Johann Bennen.

27. *Finsteraarhorn*: the highest mountain of the Bernese Alps, which Tyndall climbed with Bennen on 2–3 August 1858.

28. *Oberland*: see letter 1527, n. 13.

29. *xxxxxx—*: transcribed exactly as in the typescript of this letter.

30. *The night before last*: the night of 2 August 1858.

31. *mountaineer*: see n. 26.

32. *Zermatt*: see letter 1297, n. 7.

To Michael Faraday   4 August 1858   1537

Eggischhorn[1] | 4 Aug 1858

My dear Mr. Faraday,

I wrote a letter to Mrs. Faraday five or six days ago but it was written during bad weather, and after a succession of rainy days, which probably cast their dull influence into the writing—so that on the whole the letter is not worth sending. My course hetherto has been thus. (1) To Paris where I saw nobody except Wertheim. Duboscq I learned was suffering from inflammation of the brain. (2) To Zurich[2] where I saw Clausius and bought a watch. (3) To Lucerne,[3] making three <u>voyages</u> on the way. (4) to Meyringen[4] over the Brünnig.[5] (5) to Grindelwald.[6] Spent two days there upon the glaciers. At Rosenlaui[7] we slept a night, and next morning to my consternation found that a French tourist had taken away my boots. They were quite new and first rate. He left behind him a pair of vile Continental <u>souliers</u>[8] in their place. A strong man followed him, overtook him at the Scheideck[9] and deprived him of his illgotten booty. The subject of the structure of glacier ice has long been a source of discomfort to me. I had offered an opinion upon the subject, but still I was not <u>quite sure,</u> and the different opinions entertained by many intelligent glacier observers increased my hesitation. On this point I think I am now at peace, and it is the principal point which I had to settle during the present excursion. From Grindelwald we crossed the Strahleck[10] and came down along the Finsteraar glacier to the Grimsel.[11] Here we spent a day examining the traces of ancient glacier action. These are perfectly astounding: To a height of at least 2000 feet above the valley of Hasli[12] this action can be traced with perfect distinctness. We[13] crossed the Grimsel pass to the Rhone glacier[14] and spent some time upon it—it was very instructive. How different things appear when the mental eye is cleared for their proper apprehension! The weather was here dismal, we spent a night at the little Auberge[15] at the foot of the glacier and started down the valley in the rain next day. From Viesch[16] we ascended to the Hotel Jungfrau[17] situated half way up the Eggishorn, round which runs the great glacier of the Aletsch.[18] From the summit of the Egg-ishorn the view is perhaps the finest I ever saw. But the great object of interest is the glacier: it is a most noble stream, and its origin is the grandest conceiv-able. Five valleys converge upon a single point, each pouring down a massive névé:[19] all unite to form the trunk of the Aletsch. The Jungfrau,[20] the Monk,[21] the Eiger,[22] the Trugberg,[23] the Aletsch Horn,[24] and other mighty masses are the collectors of the material. The mountain forms are beautiful, and laden with their snows smooth and shining in the sunlight appear lovely beyond description. I was anxious to make some observations on the diathermancy[25] of the lower atmospheric strata, and hence wished to make simultaneous observations upon a high summit and in a low valley. With this object in view the night before last I took lodgings in a wild mountain cavern, with a single hardy mountaineer at my side purposing to start early to make an attempt upon the highest mountain of the Oberland,[26]—the Finsteraarhorn.[27] At 3 o'clock yesterday morning we were on foot, and at half past 10 we were on the summit of the mountain. I had sent Ramsay[28] to the valley of the Rhone[29] with a black bulb thermometer, but unfortunately the day was not sufficiently serene to give us information on the principal point in question. Water boils at 187° farht, on the top of the mountain. I left a minimum thermometer there, which some future tourist may read, and give us some notion of the cold attained in these high regions. I reached this hotel[30] at half past seven yesterday evening; as fresh and well as could be expected after 16 ½ hours as hard work as I ever went through. I am resting to day. Eight hours unbroken sleep have done much to restore my forces. The only effect I feel is a kind of pleasant drowsiness, which to day's rest will wear away, so that I shall be fit for a similar excursion tomorrow.

Will you kindly remember me to Mrs. Faraday and Miss Barnard. I have learned from a visitor here that Mr. Geo. Barnard[31] is at Zermatt.[32] I shall feel much obliged if you would ask Anderson to direct any letters that may be for me to Saas,[33] Canton Valais,[34] Switzerland.

Good bye | Ever Yours most truly | John Tyndall

RI MS JT/1/TYP/12/4084–86
Faraday Correspondence, 5:3495
Typed Transcript Only

1. *Eggischhorn*: Eggishorn, see letter 1297, n. 6.
2. *Zurich*: see letter 1333, n. 11.
3. *Lucerne*: see letter 1536, n. 3.
4. *Meyringen*: see letter 1536, n. 4.
5. *Brünnig*: see letter 1536, n. 5.
6. *Grindelwald*: see letter 1536, n. 10.
7. *Rosenlaui*: see letter 1536, n. 6.
8. *souliers*: boots (French).
9. *Scheideck*: see letter 1536, n. 8.
10. *Strahleck*: see letter 1536, n. 12.
11. *Grimsel*: see letter 1517, n. 2.
12. *valley of Hasli*: the major valley of the river Aare.
13. *We*: Tyndall, geologist Andrew Ramsay, and guide Christian Lauener.
14. *Rhone glacier*: see letter 1297, n. 12.
15. *Auberge*: inn or hotel in French-speaking regions.
16. *Viesch*: see letter 1536, n. 16.
17. *Hotel Jungfrau*: see letter 1536, n. 17.
18. *Aletsch*: see letter 1536, n. 19.
19. *névé*: see letter 1436, n. 9.
20. *Jungfrau*: see letter 1536, n. 20.
21. *Monk*: see letter 1536, n. 21.
22. *Eiger*: see letter 1536, n. 22.
23. *Trugberg*: a mountain of the Bernese Alps, located south of the Mönch.
24. *Aletsch Horn*: see letter 1536, n. 23.
25. *diathermancy*: see letter 1436, n. 13.
26. *Oberland*: see letter 1527, n. 13.
27. *Finsteraarhorn*: see letter 1536, n. 27. LT transcribed this as 'Pinsteraarhorn', and it was copied the same in the *Faraday Correspondence*, although it was clearly meant by Tyndall to be 'Finsteraarhorn'.
28. *Ramsay*: Andrew Ramsay.
29. *valley of the Rhone*: the major valley of the river Rhône.
30. *this hotel*: see n. 17.
31. *Mr. Geo. Barnard*: George Barnard (1807–90), a painter, founding member of the alpine club, and brother of Sarah Faraday. See P. H. Hansen, 'Founders of the Alpine Club (act. 1857–1863)', in *ODNB*.
32. *Zermatt*: see letter 1297, n. 7.
33. *Saas*: Saas Fee, the main village of the Saas Valley in the Swiss canton of Valais.
34. *Canton Valais*: a southwestern canton of Switzerland, which includes portions of the Bernese Alps.

To Thomas Archer Hirst   4 August 1858   1538

4th. <u>Aug. 1858</u>

My dear Tom.

I ought to have written to you sooner—but the weather has delayed my expedition to Monte Rosa.[1] My course thus far has been 1st to Zurich,[2] thence to Lucerne[3] across the three intervening lakes. Thence over the Brunning[4] to Meyringen.[5] Thence to Grindelwald.[6] Thence across the Strahleck[7] pass to the Grimsel[8] thence to the Rhone glacier[9] & down to the Rhone Valley to Viesch.[10] Thence to the hotel Jungfrau[11] on the Eggishorn[12] where I now sit. For the last few days the weather has been fine and a wholesome north wind gives promise of its continuing so. We[13] have had some capital excursions. The Aletsch glacier[14] is the finest I ever saw: and round about its origin a series of mountain splendours congregate which I do not think I have seen equalled. The night before last I lay in a cave in the wildest spot imaginable. A simple hardy mountaineer[15] was at my side who snored vociferously—you may judge that I heard it all. Next morning 3 oC. and a bright moon saw us on foot; and yesterday I boiled water and planted a minimum thermometer at the summit of the Finsteraarhorn.[16] It is a splendid mountain. Water boils there at 1870 Faht.! The mountain is the highest in the Oberland.[17]

My guide[18] was a splendid fellow. Prudent, strong, but with a courage and hard head which were quite inspiring. We were at the top of the mountain at 10½ oC. A.M. descended and reached the hotel, fresh and strong at 7½ P.M. having had 16½ hours as hard work as I ever went through.

Today I am resting—but I feel wonderfully well. I had eight hours sound sleep last night—which have gone a wonderful way in restoring my expended force. Friday morning next I purpose leaving this place for Zermatt[19] or that region, I will leave word at the hotels in the neighbourhood where I am to be found so that you may easily find me when you come. We shall ascend Monte Rosa; and I hope you will be along with us. I have half a dozen more letters to write so will conclude this

most affectionately | <u>John</u>

BL 63902–874E-10

---

1.   *Monte Rosa*: see letter 1478, n. 22.
2.   *Zurich*: see letter 1333, n. 11.
3.   *Lucerne*: see letter 1536, n. 3.
4.   *Brunning*: see letter 1536, n. 5.
5.   *Meyringen*: see letter 1536, n. 4.

6.   *Grindelwald*: see letter 1536, n. 10.

7.   *Strahleck*: see letter 1536, n. 12.

8.   *Grimsel*: see letter 1517, n. 2.

9.   *Rhone glacier*: see letter 1297, n. 12.

10.   *Viesch*: see letter 1536, n. 16.

11.   *hotel Jungfrau*: see letter 1536, n. 17.

12.   *Eggishorn*: see letter 1297, n. 6.

13.   *We*: Tyndall, geologist Andrew Ramsay, and guide Christian Lauener.

14.   *Aletsch glacier*: see letter 1536, n. 19.

15.   *hardy mountaineer*: Johann Bennen.

16.   *Finsteraarhorn*: see letter 1536, n. 27.

17.   *Oberland*: see letter 1527, n. 13.

18.   *My guide*: Johann Bennen.

19.   *Zermatt*: see letter 1297, n. 7.

To Harriet Jane Moore or Julia Moore   4 August 1858   1539

4th Aug 1858.

My dear Miss Moore

I am writing to Mr Faraday[1] and seize the occasion to send you a very little note. Since I saw you I have scampered over many of the glaciers and mountains of the Oberland[2] including Rosenlaui,[3] the two glaciers of Grindelwald,[4] the Aar[5] and the Rhone.[6] I plucked some flowers from the top of the Scheideck[7] pass intending to enclose them in this note for your sister,[8] I put them in my pocket where they were ground to pieces by my keys! I am now upon the Eggishorn[9] round which the finest glacier in Switzerland winds. I have already been several times upon this glacier, and the night before last I lay under a roof of rock near the origin of the glacier. The Jungfrau[10] was quite at hand, with the Monk[11] and Eiger[12] & other great mountains of the Oberland. My object in lying there was to start early on an expedition to the top of the Finsteraarhorn[13] which is the highest of the Bernese Alps.[14] We[15] started at 3 o'clock in the morning and succeeded in reaching the summit at half past 10. The climb was a very formidable one and I carried an apparatus for boiling water on my back which greatly increased the difficulty. My guide[16]—I had only one—wished to carry it, but the poor fellow had enough without it, and I insisted on bearing my own share. Water boils at the top at a temperature of 187° Fah^t. We descended the greater part of the mountain after the fashion of avalanches, sitting on the snow and sliding down with extraordinary rapidity. After 16½ hours of great toil we reached this hotel.[17] I am resting today, and feel quite fresh and well. Tomorrow I expect to be sufficiently strong to undertake another journey of the kind.

Pray give my kindest remembrances to all my friends at Peters.[18]—In a few days I expect to be at Zermatt.[19]

Goodbye Ever Yours. | J. Tyndall.

RI MS JT/1/TYP/3/863–64

1. *writing to M͟r Faraday*: letter 1537.

2. *Oberland*: see letter 1527, n. 13.

3. *Rosenlaui*: the Rosenlaui glacier in the Swiss Bernese Alps.

4. *two glaciers of Grindelwald*: see letter 1312, n. 3.

5. *Aar*: see letter 1305, n. 2.

6. *Rhone*: see letter 1297, n. 12.

7. *Scheideck*: see letter 1536, n. 8.

8. *your sister*: either Julia Moore or Harriet Moore.

9. *Eggishorn*: see letter 1297, n. 6.

10. *Jungfrau*: see letter 1536, n. 20.

11. *Monk*: see letter 1536, n. 21.

12. *Eiger*: see letter 1536, n. 22.

13. *Finsteraarhorn*: see letter 1536, n. 27.

14. *Bernese Alps*: the range of the Swiss Alps located within the Swiss canton of Bern and neighboring cantons of Valais, Fribourg, and Vaud.

15. *We*: Tyndall and Johann Bennen.

16. *My guide*: Bennen.

17. *this hotel*: see letter 1536, n. 17.

18. *Peters*: Petersham, see letter 1515, n. 3.

19. *Zermatt*: see letter 1297, n. 7.

To Juliet Pollock   4 August 1858   1540

4th. <u>August</u> 1858

My dear M[rs]. Pollock.

I have just made a little excursion which I must describe to you, more perhaps for the purpose of proving to you that amid the mountains I have not forgotten my friends of Montagu Square,[1] than for any other purpose. I now sit in a hotel[2] situated half way up the Eggishorn.[3] It has been built for the accommodation of those who ascend the 'horn', from which the view is magnificent. Round the base of the horn winds the finest glacier I ever saw, which takes its origin from the snows of the Junfrau[4] and other mountains of similar magnitude. Five valleys contribute their contents to form the trunk of the great Aletsch glacier.[5] A nobler or more beautiful scene cannot be imagined than that presented from the common point of convergence of these five valleys. The mountain forms are most beautiful, and the snows shine upon them as whitely and smoothly as polished Carrara marble.[6] I have been amongst them at sunrise and sunset, at midday and at midnight when the moon renders the scene magical,—But I am forgetting my story. I wanted to make some observations upon the effect of the lower layers of the atmosphere on the sun's rays, and for this purpose wished to make simultaneous observations at the top of a high mountain & in a valley.[7] The highest mountain in the Oberland[8] is the Finsteraarhorn,[9] and it had been ascended twice. Attached to this hotel is a remarkable looking man,[10] between 30 & 40 years old, round shouldered, and with a trunk of the same thickness from hips to shoulders; a thick neck with a massive head set upon it. His legs, set rather widely apart, are bent so as to bring his feet sufficiently near each other. The figure of the man is expressive of great strength and his countenance is full of energy & decision, brightened up, however, by a gleam of good nature. This man was mentioned to me as the most daring climber in this canton.[11] He had been up the Finsteraarhorn, and I thought that in his company I might possibly carry out my intention. I asked him whether he would undertake to come along with me, engaging to follow where he felt inclined to lead, and not to weary him by calling upon him to help me. He agreed to make the trial, and the night before last we lay together in a cavern purposing to start upon our expedition before the dawn.

At 3 o'clock we were on foot. A cloudless moon was in heaven and the stars twinkled all colours. Between two high mountains which are the first sentinels of one of the tributary valleys to which I have referred stretches a 'saddle' of pure white snow with a graceful chainlike curvature. We first ascended to this saddle, and looking back the valley through which we had passed lay before us, and beyond it another magnificent valley—another of the tributaries—terminated by a similar saddle. The valley lay east and west, and each carved a space from the firmament towards one of which the rising sun was climbing, while the western one received his rays. Had Tennyson[12] been there he would have seen his 'daffodil sky'[13] illustrated as it never has been illustrated in England. —But I am again forgetting my expedition—We reached the foot of the Finsteraarhorn at 6. AM Clouds were still upon his head, but the wind was north, and we had sufficient faith in Nature's constancy to believe that the day would be a fair one. We fed at the base of the mountain—my guide[14] rejoiced in wine, but I had a bottle of cold tea and choice cream in equal volumes, which I had before proved to be a good antidote for thirst. Leaving everything not absolutely needed here behind us we commenced our task. The mountain sends down a number of parallel spurs, formed of loose & spiky gneiss,[15] and between these were great couloirs[16] filled with snow and ice. Our way lay sometimes along the spine of rock, sometimes upon the snow. I once paused upon a little ledge of ice which jutted from one of the slopes to take the angle downwards: it was 45°, and about 50 yards below the point where this observation was made yawned a deep crevasse. The slope was as steep as that of the roof of a gothic Cathedral. We climbed several slopes of this kind, and there was a certain melody in our ascent of some of them in which steps were to be cut. The guide was in front and swung his mattock once, cutting to the necessary depth with a single stroke. The raising of his foot kept time with the swing of his implement and gave a kind of rhythm to his operations. The last 1000 feet of the mountain formed by far the most formidable piece of climbing that I had ever encountered—My guide, strong as he was, laid his head upon his hatchet sometimes and panted like a chased deer. We reached the top at half past 10 A.M. We made all the observations we could, and I left a thermometer behind me which will enable some future tourist to tell us the greatest degree of cold attained in these high regions. The boiling point of water is 25 degrees lower than at the sea level, which explains the fact of my having poured a quantity of boiling water over my guide's hand without scalding him. We came down the greater part of the mountain in avalanche fashion, sitting upon the snow and sliding with extraordinary speed. After 16½ hours of hard walking & climbing we reached the hotel. I am fresh and well and after this days rest shall be ready for a similar excursion tomorrow.

Ever yours | <u>John Tyndall</u>

<u>Best love to Walter</u>[17] | In a few days I hope to be at | <u>Zermatt</u>[18] Canton Valais[19]

BL 63902–874E–10–51
RI MS JT/1/TYP/6/1934–36

---

1. *Montagu Square*: see letter 1433, n. 7.

2. *a hotel*: see letter 1536, n. 17.

3. *Eggishorn*: see letter 1297, n. 6.

4. *Junfrau*: Jungfrau, see letter 1536, n. 20.

5. *Aletsch glacier*: see letter 1536, n. 19.

6. *Carrara marble*: marble quarried from the town of Carrara in northern Italy.

7. *in a valley*: Andrew Ramsay was to make the observations in the valley. See *Glaciers of the Alps*, pp. 104–5.

8. *Oberland*: see letter 1527, n. 13.

9. *Finsteraarhorn*: see letter 1536, n. 27.

10. *a remarkable looking man*: Johann Bennen.

11. *canton*: a subdivision of a country (*OED*); in this case, Bern.

12. *Tennyson*: Alfred Tennyson.

13. *'daffodil sky'*: a reference to Alfred Tennyson's 1857 poem 'Come into the garden Maud', at the end of the first part of *Maud,* XXII.ii.856–59: 'For a breeze of morning moves | And the planet of Love is on high, | Beginning to faint in the light that she loves | On a bed of daffodil sky'.

14. *my guide*: Johann Bennen.

15. *gneiss*: a metamorphic rock, composed, like granite, of quartz, feldspar or orthoclase, and mica, but distinguished from it by its foliated or laminated structure (*OED*).

16. *couloirs*: a steep gorge or gully on a mountain side (*OED*).

17. <u>*Walter*</u>: Walter Pollock.

18. <u>*Zermatt*</u>: see letter 1297, n. 7.

19. *Canton Valais*: see letter 1537, n. 34.

From George Biddell Airy   10 August 1858   1541

Royal Observatory Greenwich | S.E. | 1858 August 10

My dear Sir

I was at the Hospice of the Grimsel[1] on July 28, and found your name entered on July 27. I wish that fortune had so far helped me as to give me a walk on the Unter Aar Gletscher[2] with you: my solitary walk (i.e. with my two sons,[3] like the spectator's 'enter a king and two fiddlers solus'[4]) was however a very interesting one, and I would give much for a repetition of it.

I had in my hand the copy of Agassiz' map[5] (made at the Royal Institution) with the accuracy and value of which I was delighted. I do not know a more striking sight than that of the subordinate moraines[6] from the Studer Horns[7] and other places sweeping round the promontory in the south area of the glacier opposite the Abschwung;[8] or than that of the two moraines from the two sides of the Abschwung, a few yards apart, one entirely of granite and the other entirely of schist.[9]

I looked with some care for the critical phenomena, especially those of the bands of ice, and confess myself entirely puzzled. I had been upon the lower Glacier of Grindelwald[10] (a very interesting one) and there I saw very numerous instances of the lenticular form which you have pointed out. On the Lower Aar Glacier[11] I did not see one: I only saw the interminable long lines of band: I call them interminable because in no case did I see a termination except where cut off by a crevasse.

The terminal face of the Rhone[12] shews well the bands transversal to the crevasses (they look like the fan-groining[13] in the roof of King's College Chapel)[14], and in passing in front of it I remarked that in one part the upper beds of ice, as divided by the horizontal bands, were pushed over the lower beds, so as to form what carpenters call a rabbet.[15] This looks very much like formation of the bands by bed sliding on bed.

There is a meteorological phenomenon which perplexes me. When the lower
end of the valley, below the glacier front, is choked with mist and rain-storm, the upper end of the glacier is perfectly free, and even the mountain tops are sometimes
seen. I remarked this in two instances.
I am, my dear Sir, | Yours very truly | <u>G. B. Airy</u>
Professor Tyndall

RI MS JT/1/A/35
RI MS JT/1/TYP/1/25

---

1. *Grimsel*: see letter 1517, n. 2.

2. *Unter Aar Gletscher*: see letter 1305, n. 2.

3. *my two sons*: Wilfrid and Hubert Airy. See letter 1518, n. 3.

4. *'enter a king and two fiddlers solus'*: quoted by English essayist Joseph Addison (1672–1719) in his daily newspaper *The Spectator* (in its first year, issue 29 of 3 April 1711), which he founded in March 1711 with Richard Steele. The quote refers to, in theater, instrument players providing music during an important character's solo speech.

5. *Agassiz' map*: presumably a copy of the map made by Johannes Wild for Agassiz's *Nouvelles études et expériences sur les glaciers actuels* (1847). See letter 1302, n. 11.

6. *moraines*: see letter 1415, n. 17.

7. *Studer Horns*: a smaller alpine peak on the eastern side of the Finsteraarhorn.

8. *Abschwung*: see letter 1436, n. 11.

9. *schist*: a crystalline rock whose component minerals are arranged in a more or less parallel manner (*OED*).

10. *lower Glacier of Grindelwald*: see letter 1312, n. 3.

11. *Lower Aar Glacier*: see letter 1305, n. 2.

12. *Rhone*: see letter 1297, n. 12.

13. *fan-groining*: vaulting composed of pendant semi-cones covered with foliated panel-work (*OED*).

14. *King's College Chapel*: the gothic chapel of King's College, Cambridge, took over a century to complete after construction began in 1446. It is noted for its fan-vaulted ceiling.

15. *rabbet*: a channel, groove, or slot, cut along the edge or face of a piece of wood or other material and intended to receive the edge or end of another piece or a tongue made specially to fit (*OED*).

From Michael Faraday   10 August 1858   1542

Royal Institution | 10 Aug 1858

My dear Tyndall,

We had your letter[1] this morning, and was very glad to hear of you, especially as we had some reason to think that Ramsay's[2] circumstances and trouble would derange you. I find you have both been together, but I suppose you will know before you receive this that he has been informed of the death of his mother,[3] i.e. if Anderson's answers to the enquiries here have caused the letter sent to him to arrive. It must be a great grief and disturbance to him. I have given Anderson your message, and I take two letters from the hall today, that I may send them to you with this. We arrived at home from Eastbourne[4] two days ago, all well, and all desiring to be remembered to you kindly. My wife does not like losing the letter you had written.[5] We have no news like yours to send you in return for your descriptions, but we can enjoy your letter very much, for we have been at so many of the places—not the summits that you talk of, but the bases of the elevations. I remember how weariness and illness laid hold of us at the little inn[6] at the foot of the Rhone glacier.[7] I shall be very glad if you meet Mr. Barnard,[8] but I doubt he will have left. Dr. Bence Jones is well and active. He had two of the Niger electrical fishes,[9] and we had the luck of electrical shocks from them; since then they have gone off to Dubois Reymond[10] and arrived safe. They must have been very hungry, nevertheless a couple of minnows that went in company with them arrived <u>alive and well</u> in Berlin. Mr. and Mrs. Barlow[11] are off, and I suppose by this time near to Bad Homburg.[12] We have been a good deal disturbed, since we came on, with the accounts of Mrs. Edwd. Forbes.[13] You have probably heard that she married in Edinburgh a while ago, but another lady claims the new husband as her husband,[14] and I understand he admits a simulation of marriage with that lady, but denies the legality. The affair must be a sad one for all concerned. Anderson says he sent off a letter[15] to you at Zermatt,[16] and hopes you have received it.

Ever, my dear Tyndall | Yours most truly | M. Faraday

I sent off Miss Moore's letter.[17]

Dr. J. T. | Poste Restante[18] | Saas[19] | Canton Valais[20] | Switz[21]

RI MS JT/1/TYP/12/4147
Faraday Correspondence, 5:3497
Typed Transcript Only

---

1.   *your letter*: letter 1537.
2.   *Ramsay's*: Andrew Ramsay.

3. *death of his mother*: Elizabeth Ramsay. When Tyndall and Ramsay arrived in Zermatt, Ramsay learned that his mother had died unexpectedly and returned immediately to England. Tyndall thus climbed Monte Rosa without him.

4. *Eastbourne*: a seaside resort town on the southern coast of England.

5. *losing the letter you had written*: a reference to a letter Tyndall wrote to Sarah Faraday and then did not send, mentioned in letter 1537.

6. *little inn*: probably Hôtel du Glacier du Rhône, managed by Joseph Seiler of Zermatt, who managed other hotels in the Swiss Alps as well.

7. *Rhone glacier*: see letter 1297, n. 12.

8. *Mr. Barnard*: George Barnard, a painter, alpinist, and brother of Sarah Faraday. See letter 1537, n. 31.

9. *Niger electrical fishes*: an electric fish of the Niger and other rivers in Africa, *Nile silurus*. Expeditions into Africa in the mid-nineteenth century had reported coming across fish with electric properties, and specimens were brought back to Europe for investigation. See M. Piccolino, 'Electric fishes research in the nineteenth century, following the steps of Carlo Matteucci and Giuseppe Moruzzi', *Archives Italiennes de Biologie*, 149 (2011, Suppl.), pp. 10–17.

10. *Dubois Reymond*: Emil Dubois-Reymond (1818–96), a German anatomist and physiologist from Berlin who focused on the emerging field of electrophysiology and challenged views propounded by Matteucci (*CDSB*). He delivered a course of lectures on electrophysiology at the RI in the spring of 1855. Dubois-Reymond published 'Note sur le Malaptérure électrique', *Annal. Chim. et Phys.*, 52 (1858), pp. 124–25.

11. *Mr. and Mrs. Barlow*: John Barlow, and Cecilia Anne Barlow (née Law, d. 1868), the daughter of a wealthy member of Parliament, were married in 1824.

12. *Bad Homburg*: a popular spa town in central Germany.

13. *Mrs. Edwd. Forbes*: Emily Marianne Forbes (née Ashworth, 1825–1909), the daughter of Sir Charles Ashworth and widow of naturalist Edward Forbes (who died in 1854; see letter 1493, n. 9).

14. *new husband as her husband*: Emily Forbes married William Charles Yelverton (1824–83), who had kept secret his marriage with Theresa Longworth (1833–81). A marriage dispute resulted in the 1861 case *Thelwall v. Yelverton,* which ruled in favor of Yelverton's marriage with Longworth. An appeal in 1864 reversed the decision and favored the marriage with Forbes (*ODNB*). See C. Schama, *Wild Romance: A Victorian Story of a Marriage, a Trial, and a Self-Made Woman* (New York: Walker & Company, 2010).

15. *a letter*: letter missing.

16. *Zermatt*: see letter 1297, n. 7.

17. *Miss Moore's letter*: letter 1539 to Harriet Jane Moore or Julia Moore which Tyndall possibly sent via Faraday with letter 1537.

18. *Poste Restante*: see letter 1293, n. 5.

19. *Saas*: see letter 1537, n. 33.

20. *Canton Valais*: see letter 1537, n. 34.

21. *Dr. J. T. | Poste Restante | Saas | Canton Valais | Switz*: written on the envelope.

To Auguste de la Rive   12 August 1858   1543

Zermatt.[1] Thursday. | 12th. Aug. 1858

My dear Sir.

I am anxious to sink a number of minimum thermometers in the snow at the summit of Mont Blanc,[2] in order to ascertain the minimum winter temperature at the top of the mountain. I have two thermometers with me, but for the sake of certainty I should like to have two or three more. Mine are of the ordinary construction; spirit[3] with a moveable index within, which is drawn back on the contraction of the liquid. You would do me a very great favor if you would send to me to Chamouni[4] two or three minimum thermometers of any construction which you might deem suitable for the observations; and if you would send them to me as soon as you can you would add to the obligation. I expect to be in Chamouni in about 10 days from the present time, and shall lose no time in ascending the mountain. I placed a thermometer at the summit of the Finsteraarhorn[5] a few days ago, and the day before yesterday I boiled water at the top of Monte Rosa.[6]

Would you kindly favour me with a line addressed to me Poste Restante[7] Visp[8] Canton Valais,[9] letting me know whether it is possible to obtain the thermometers in Geneva.[10] They might be graduated between +40° and -40° Centigrade.

Believe me dear Sir | Ever yours most truly | John Tyndall

If the maker of the thermometers would enclose his account I would send him the amount immediately

RI MS JT/1/T/373
RI MS JT/1/TYP/1/357

---

1. *Zermatt*: see letter 1297, n. 7.

2. *Mont Blanc*: see letter 1320, n. 6.

3. *spirit*: an alcohol thermometer, an alternative to mercury-in-glass thermometers.

4. *Chamouni*: see letter 1353, n. 9.

5. *Finsteraarhorn*: see letter 1536, n. 27.

6. *Monte Rosa*: see letter 1478, n. 22.

7. *Poste Restante*: see letter 1293, n. 5.

8. *Visp*: a town in the Rhône valley in the Swiss canton of Valais.

9. *Canton Valais*: see letter 1537, n. 34.

10. *Geneva*: see letter 1293, n. 3.

To Thomas Archer Hirst   [12]¹ August 1858   1544

13th. Aug. 1858.

My dear Tom

Your note[2] has just reached me. I wrote to you[3] from the Eggishorn,[4] I hope you received my note. I have been here for two or three days and two days ago made the ascent of Monte Rosa.[5] I shall wander through this neighbourhood for 6 or eight days more, and at the end of eight days shall probably be at Saas.[6] I wrote yesterday to Balmat[7] asking him to accompany me to the top of Mont Blanc[8] to set some thermometers there. I said to them that I should be at Chamouni[9] in about 10 days from the present time. I shall try and wait in this neighbourhood until you come. Enquire at the Hotel de la Poste as you pass Visp.[10] Enquire at Saas. I am almost sure to be at Saas at the time you mention. There is a little hotel[11] in the Distel Alp[12] near Saas, that will probably be my resting place when you come. Ramsay[13] suddenly went away from me just as I came here. He received the unexpected intelligence of his mother's[14] death.[15] A misfortune appears to hang over my company in this respect.

Poinsot is undoubtedly elected a Fellow of the R.S. you could not see his name in the list of candidates. The election of foreign members is a special honour, and is done in its own way. I don't quite recollect; but the formality of the general meeting on the 30th. of Nov. may have to be gone through. But there is not a doubt that he is elected. Cast every shade of doubt out of your mind.[16]

I am on the point of starting on an expedition and must therefore cut myself short

Ever yours affectionately | John Tyndall

RI MS JT/1/HTYP/514
LT Typescript Only

---

1. *[12]*: although 13 August 1858 is given on LT's typescript, this letter more likely dates to 12 August as that day Tyndall was at rest. As he noted in his journal, on 13 August he headed out very early for climbing, and the letter likely was postmarked that day (Journal, RI MS JT/2/13c/1111). In this letter he also noted, 'two days ago made the ascent of Monte Rosa'. This occurred on 10 August 1858.
2. *Your note*: letter missing.
3. *I wrote to you*: letter 1538.
4. *Eggishorn*: see letter 1297, n. 6.

5.   *Monte Rosa*: see letter 1478, n. 22. Tyndall made the ascent of Monte Rosa on 10 August 1858.

6.   *Saas*: see letter 1537, n. 33.

7.   *I wrote yesterday to Balmat*: Auguste Balmat. Letter missing.

8.   *Mont Blanc*: see letter 1320, n. 6.

9.   *Chamouni*: see letter 1353, n. 9.

10. *Visp*: see letter 1543, n. 8.

11. *a little hotel*: Mattmark Hotel on the Distel Alp.

12. *Distel Alp*: see letter 1532, n. 3.

13. *Ramsay*: Andrew Ramsay.

14. *his mother's*: Elizabeth Ramsay

15. *death*: see letter 1542.

16. *Poinsot is undoubtedly elected a Fellow . . . out of your mind*: Poinsot was indeed elected a Foreign Member of the RS in 1858. While it appears that his name was not included in the list of Foreign Member candidates prior to the anniversary meeting on 30 November 1858, it is included in the list given in President Lord Wrottesley's address for that meeting. See *Roy. Soc. Proc.*, 9 (1857–59), pp. 497–522, on p. 498.

From Juliet Pollock   13 August 1858   1545

59 Montagu Square W. August 13th 1858

My dear M[r]. Tyndall

Your letter[1] found me so full of the thoughts of my beloved Walter,[2] that a communication from his especial friend had a peculiar value for me; I thought how pleased Walter would have been to hear me read it out, and then I sat down to write, to tell that sweet little man all about it. The truth is, it was the second day of our parting from him, and I have found this parting so trying that except to tell you it has taken place I won't say much about it— only I must tell you that when his trunk was being packed (the school trunk) and it was proposed to him to take some of his playthings—he said 'no—I want only a few books, a little glass dog, my last present from Mama, for a remembrance of <u>her</u> and the portrait of M[r]. Tyndall.'—This portrait I was to be very careful in packing, and I unpacked all his things for him at school to ascertain its perfect safety!—By this, you see that we went to Brighton[3] with him—and we spent the afternoon there,—feeling sad, but still sadder the next morning in waking, to miss his little tap at our door and his sweet cheerful tones. But he writes that he is happy and likes cricket though he is a great muff at it—And now I must tell you something pleasanter. Our eldest boy[4] has given us very great satisfaction by the position he has taken at Eton[5] where he has come in, triumphant, for the College—the topmost boy competing with

45 picked boys. —It is considered a high distinction—and I am sure you will be glad to hear of it—.

Your letter interested me deeply and I fancied I could see what you described so powerfully, and that I could have liked the climbing but for the coming down again and the following the lead of that <u>unheimlich</u>[6] looking guide,[7]—I can feel glad that you have made this ascent, but I must wish that you were not about to do another—

Beware the pine tree's withered branch,
Beware the awful avalanche—
This was the peasant's last good night
A voice replied far up the height
<u>Excelsior</u>—[8]

Avalanche in this case refers to your own mode of descent (an avalanche of yourself) and the peasant must stand for myself and you persist—as in the song. —but we won't quote the last verse.[9]—Only as it dwells in my remembrance, I must just beg you to do nothing rashly, and not to contemplate any climb to which the making of a will is the first preparatory step.

We spent last Tuesday at Petersham[10] with the Moores[11] very pleasantly— Only M[r]. Byam[12] (clergyman) M[rs]. Robinson,[13] and M[r]. Spedding[14] there besides ourselves.—We walked in the garden between showers, and sat outside Gay's summerhouse,[15] contemplating the unfortunate Thames[16]—and so on till dinner. Poor old M[r]. Moore was pretty well till the close of the evening, when he began badly to wish that 'all the company was gone,' and Julia M.[17] then had a hard task with him. It is delightful to me to watch her in her true unobtrusive filial devotion. We returned home by the 9.15 train. Tomorrow we are going to Ely[18] to spend a few days with our friend M[r]. Thompson[19] who is a cannon of the cathedral—

On the 23.[d] I go with Fred.[20] and my baby[21] I hope to S[t]. Julians[22]—but alas—no Walter—and I am sorry to say, Frederick[23] is detained in London till the 2.[d] Sept[r].

I have been sitting with M[r]. Babbage to day and seeing his work shop. He is concentrating all his strength mental and bodily on the machine,[24] and I fear he's overtasking himself. I have urged him to sleep out of town for a fortnight. He is in an excitable highly wrought condition and clearly wants some relaxation. But I believe he is like the cork leg[25] & he <u>must</u> go on.

There is a very good review of Buckle in the last Quarterly.[26]

Yours alw[ays]. Most truly | Juliet Pollock

RI MS JT/1/P/169
RI MS JT/1/TYP/6/1938–39

1.   *Your letter*: letter 1540.

2.   *Walter*: Walter Pollock.

3.   *Brighton*: see letter 1365, n. 7. Where Walter was in school, possibly at Hurstpierpoint or Brighton College.

4.   *eldest boy*: Frederick Pollock.

5.   *Eton*: see letter 1296, n. 8.

6.   *unheimlich*: uncanny, weird (*OED*).

7.   *guide*: Johann Bennen.

8.   *Beware the pine tree's withered branch . . . Excelsior—*: H. Longfellow, 'Excelsior' (1841), vi.1–5. Longfellow's poem was about a young man's passage through a mountain village.

9.   *last verse*: Longfellow, 'Excelsior', ix.1–5: 'There in the twilight cold and gray, | Lifeless, but beautiful, he lay, | And from the sky, serene and far, | A voice fell like a falling star, | Excelsior!'.

10. *Petersham*: see letter 1515, n. 3.

11. *the Moores*: James and Harriet Moore and their family Harriet Jane, Julia, John, and Graham.

12. *M*ʳ. *Byam*: Richard Burgh Byam (1785–1867), Vicar of Kew and Petersham. See 'Obituary Memoirs', *The Gentleman's Magazine and Historical Review*, 3 (1867), pp. 672–73.

13. *M*ʳˢ. *Robinson*: not identified.

14. *M*ʳ. *Spedding*: probably James Spedding (1808–81), an English literary editor and biographer known especially for editing the works of Francis Bacon (*ODNB*). He dined often with the Pollocks, and on at least one occasion with Tyndall as well, later in 1864. See W. Pollock, *Personal Remembrances of Sir Frederick Pollock: Second Baronet, Sometime Queen's Remembrancer*, 2 vols (London: MacMillan and Co., 1887), p. 117.

15. *Gay's summerhouse*: a pavilion on the bank of the Thames at Petersham, built by the Duke and Duchess of Queensberry in the early eighteenth century. The reference is to the poet John Gay, author of *The Beggar's Opera* (1728). The Queensberrys were Gay's long-term patrons.

16. *the unfortunate Thames*: a reference to 'The Great Stink' of July and August 1858, where hot weather created unfavorable stenches from the untreated human and industrial waste throughout London and along the banks of the Thames. Newspapers christened it 'the great stink'. Cholera outbreaks before this time were blamed on longterm problems with sewage disposal in the river and some major engineering works were already in hand. The smell brought renewed attention to inadequate sewer systems and prompted action to fix the problem (which occurred over the next several decades). See S. Halliday, *The Great Stink of London: Sir Joseph Bazalgette and the Cleansing of the Victorian Metropolis* (Stroud: History Press Limited, 2001).

17. *Julia M.*: Julia Moore.

18. *Ely*: a cathedral city north-northeast of Cambridge.

19. *M*ʳ. *Thompson*: William Hepworth Thompson (1810–86), an English classical scholar, canon of Ely Cathedral, who in 1866 succeeded William Whewell as the Master of Trinity College at the University of Cambridge (*ODNB*).

20. *Fred*: Frederick Pollock.

21. *my baby*: Maurice Pollock.

22. *S^t. Julians*: see letter 1470, n. 4.

23. *Frederick*: William Frederick Pollock.

24. *machine*: see letter 1341, n. 3.

25. *cork leg*: an artificial leg. A reference to the short story by Henry Glassford Bell, 'The marvellous history of Mynheer von Wodenblock', (1832), about a man with a magic cork leg who could not stop walking, subsequently popularized in a doggerel song, 'The Cork Leg' (*ODNB*).

26. *review of Buckle in the last Quarterly*: a review of the first volume of Henry Buckle's *History of Civilization in England* (1857), in *The Quarterly Review*, 104:207 (1858), pp. 38–74.

To Thomas Archer Hirst   [22 August 1858][1]   1546

Sunday

My dear Tom.

Tho' hardly hoping that this will reach you, I write it. Your letter[2] reached me a short time ago, and I send this on by a guide. I am at the Matmark Hotel,[3] about 3 hours above Saas[4] & I will remain here till you come. Come quickly![5]

Yours affectionately | John.

RI MS JT/1/T/646
RI MS JT/1/HTYP/514

---

1. *[22 August 1858]*: the date is given by the addition of 'Aug 22nd 1858' on the letter, presumably written by Hirst.

2. *Your letter*: letter missing.

3. *Matmark Hotel*: see letter 1544, n. 11.

4. *Saas*: see letter 1537, n. 33.

5. *Come quickly!*: Tyndall and Hirst met in Saas on 26 August 1858 (Journal, RI MS JT/ 2/13c/1031–32).

To Juliet Pollock   [23 August 1858][1]   1547

My dear M^{rs}. Pollock

Excuse this crumpled paper. I carried it for 3 hours in my coat pocket this morning for the purpose of giving myself the pleasure of writing to you this afternoon, and it has got crushed in the passage. 100 feet above me is a gray impenetrable cloud, out of which heavy rain gushes, splashing dismally upon the rocks. A boulder I suppose a thousand tons in weight, the unshipped

cargo of a glacier rises before my window. I am on the brink of a mountain torrent, which foams and roars in response to the falling rain. A little below me
a huge glacier[2] coming from a lateral valley throws itself athwart the stream,
dams it up and forms the so called Matmark see.[3] In short if you look upon
a map of Switzerland, and find the valley of Saas,[4] high up under the ridge
of the Monte Moro,[5] you will see the Distel Alp[6]—this is now my dwelling
place. I am the only traveller here—alone in a room which could hold 100
with a pine fire sparkling, and gleaming, and crackling behind me. Next to
the bodily presence of a friend is the pleasure of being able to write to one, so
I shake away my solitude, and render my happiness for a time independent of
wind and weather by writing to you. I met John Clerk[7]—son of Sir George
Clerk[8] at the Riffel,[9] accompanied him to Zermatt,[10] met him again at Saas[11]
and parted from him here this morning: he is gone over the Monte Moro,
and I remain among my glaciers. Your letter reached me just as I was leaving
Zermatt, and accompanied me as a pleasant companion all along the valley
of the Visp,[12] which without it would have been very dull. Indeed without it
I think I should have fallen back on the Society of M^r. Barnett,[13] the banker
and his family, among whom were some exceedingly nice young ladies whose
society I am sure would have been very pleasant to me. But instead of thinking of them I thought of you and Walter.[14] It is an old theory, perhaps an old
weakness, of mine, that a mother's love is the most beautiful love of all. All
suffering, all enduring, it burns ever brightly, and trials which would quench
all other passions pass over this unharming. I read that portion of your letter
which referred to Walter two or three times over, and by a kind of magnifying
process applied to my own feelings tried to realize yours. But this little man
'muss wirken und streben'[15] and he must be prepared to do this. You must
seek for joy in other phases of his development, not only in the budding and
blossoming but also in the slow formation of sound and solid timber which
marks the progress of the child to the man. It gave me pleasure to find that
he thought of me, and it has only been a fair return, for since I came to Switzerland I have very often thought of him. I therefore <u>owe</u> him nothing, but
present to him as a free gift my best love. With regard to Bobo[16] I can only say
that had I a son who could accomplish such a triumph[17] I should be exceedingly proud of him.

Of mountain climbing since I wrote to you last I could talk a good deal
for I have made some very unusual marches. I have explored all the grand glacier region of Monte Rosa[18] and the neighbouring mountains. To this work
I devoted ten days, during which I had my quarters upon the Riffel. These
mountain excursions teach a man many things—they teach what can be
accomplished physically by persistent exertion, and as types obtain an intellectual value also. Not until the 10th day of my sojourn at the Riffel, did I solve
in a manner completely satisfactory to my own mind, and I trust convincing

to the minds of others, the principal problem which brought me to Switzerland this year. The day after I ascended the Riffel, and at 3 oC. in the morning the candle of my guide[19] gleamed into my bedroom, and he announced to me that the weather was good. We had made an arrangement to ascend Monte Rosa together, an achievement which neither of us had ever before accomplished. At 4 oC. we were on our way, and half an hour revealed to us the mountain from head to foot with the beams of the rising sun upon its rock pinnacle and slopes of snow. From the brother of my guide[20] we obtained some general information as to the route we ought to follow, and guided by this and trusting to our own heads and feet we faced the mountain. First across a large glacier to a boss of rocks at the foot of the mountain. Then up a steep snow slope to ascend mass of rock which rose precipitously out of the snow and round which we had to wind to reach its summit. Thence our way lay for hours over the unsullied snow, climbing slopes, rounding vast bosses[21] crossing curves and sweeps of extreme beauty, winding round the rifted ridges and pyramids of névé[22] and going zigzag up the steeper inclinations. For some hours it was merely childs play to a mountaineer, but afterwards it became stern and earnest work. The mountain at its upper portion forms what the Germans call a 'Kamm'—a comb; a word suggested I should think by the toothed edges which some mountain ridges exhibit, but now more widely applied to snow on rocks which forms a thin edge. Monte Rosa contracted her snowy shoulders to such an edge, and up this edge our way lay for hours. The incessant admonition of my guide was to fix my staff well into the snow, which admonition was a mere work of supererogation on his part. The wind had so acted upon this 'Kamm' that the snow was folded over so as to form an overhanging cornice, and for a good while this cornice was our only support. Once in endeavouring to fix my staff securely it went right through this cornice and I could see through the hole that I had made into the gulf below. On that side the mountain was a sheer precipice, along the edge of which we crawled. I am wrong here. The word crawl is not correct; we walked erect, for any attempt at crawling, or assisting the feet by the hands or knees would assuredly have been fatal. At the other side also the slope was most perilously steep, and intersected by precipices. I looked towards the summit of the mountain—A heavy cloud stuck to it, but like a royal mountain queen she dashed it aside and allowed the sun once more to shine upon her head. Her triumph was short.—dusky masses again gathered round her, and her efforts to get rid of them became ever feebler. They reached towards us. The valley underneath filled their torn and dislocated névés seethed like boiling cauldrons, and sent up their vapours to meet those descending from the summit. We were soon in the midst of them, the partial triumphs of the sun & wind in clearing the glaciers and lighting them up at intervals with a supernatural glare, served merely to render the scene more awful. The darkness thickened, relieved more and more rarely by those partial illuminations: and once in the thickest gloom an avalanche was discharged from the slopes of a neighbouring mountain the sound of which had an element of horror in it. We could not see it:—could not judge of its distance; could only hear its roar, and check the imaginings which it naturally excited. We came to a row of crags[23] which jutted above the Kamm; breathed a moment, and crossed them. During our breathing time I asked my guide how he would have acted had my foot slipped in crossing up the 'Kamm'—He seemed not to like the question, and to wish to drive all thought of it from my mind. Doubtless these men have learned the philosophy of not permitting the mind to dwell on danger. We could hardly see 20 yards in advance; the mountain still turned its icy edge towards us and up we went. 'From description' said my guide 'I should judge this to be the last comb of the mountain, but in this darkness we cannot be sure'. After a time we reached some rocks, spikes and prisms of weathered granite, which we hoped was the crest of the mountain. Among these we climbed for some time: often walking erect over edges of rock with terrible depths at each side. We rounded the larger masses, crept up and crept down, forced ourselves through fissures. My guide once dropped his pocket book, containing something of value to him. It was caught upon a little ledge some distance down. He descended & I went on. I could soon hear him clattering behind me.—the summit came fairly into view—down one cliff up another; and your friend was safely planted on the highest point of Monte Rosa. My guide was in a few minutes along side.

Since I wrote these words 12 hours have past. I have often wished to see winter in these mountain wilds and here I have it to my heart's content. There is a foot of snow on the steps leading up to the house door. It has heaped itself against the windows where they face the wind. A delta of gray stones through which a many armed river ran last evening is now a plain of snow. The mountains are all covered, save where the sharp hewn crags, too steep to hold the snow peer through it. It shines in upon me with a curious dazzling gleam producing a sensation akin to giddiness. It will stop my excursions upon the glacier for a little time; for such a traitor among the crevasses would be very dangerous. He sheets them over smoothly, but woe to the foot which trusts his apparent stability. I will now quit one snow scene for another slightly more perilous as far as I am concerned.

Snow began to fall ere we reached the summit of Monte Rosa and thickened as we remained there. I tried to ascertain the boiling point of water but was met by difficulty. My lamp would not unscrew. I had to remove the wick but it became sprinkled with snow, and afterwards was ignited with difficulty. I placed my apparatus on a ledge where it was sheltered laterally, but not above. I put my hat over it to protect it, and thus by patient care boiled my water. 184.92 Faht. is the boiling point. Clouds thickened and darkened and the snow fell with greater resolution. It was very thick but extremely fine, and dear old mother nature, who seemed to scowl on us so pitilessly cheered us with a gleam of her beauty, which we interpreted as an act of kindness on her part. The snow was all <u>flowers</u>, the most lovely that human eye ever gazed upon. When it fell upon the rocks it melted instantly and I noticed it first upon my felt hat which I had put to shelter my machine. There in the highest atmospheric regions this symmetry of force manifested itself and built up those blossoms of the frost. There was no deviation from the 6 leaved type, but numberless variations: any thing so beautiful I had never before beheld. Had some spirit of the mountain looked out upon me from the cloud and demanded my choice between the snow flowers and the prospect from the mountain top, I should have hesitated before parting with my exquisite vegetation. Still our position was an anxious one. We feared the loss of our track and the consequent possibility of taking a false route. But we dashed the gloom of thought aside more successfully than the Monte Rosa had dashed her vapours, and with the muscles braced, and a certain grim feeling about the heart we commenced our retreat. We were soon clear of the rocks which constitute the summit of the mountain, and saw our 'Ice Kamm' stretching a little way and losing itself in the gloom. Coming up each had to take care of himself, but coming down a rope was thought advisable. It was tied securely round my waist, but I noticed that my guide did not tie it around his: he slipped a loose noose over his arm. This being altogether unusual I asked him why he did not attach himself in the customary way, but he said that he should have more power with his own arrangement.[24] I am not yet quite convinced of this, but I knew full well that he would do all in his power to save me if my footing gave way, and I thought the exertion of that power would be in no way diminished by the consciousness on his part that he could as a last resource release himself and leave me to my fate. Still though I did not expostulate I did not like the arrangement: but it did me good. It cannot be called anger, it cannot be called pride, but a slight flush ran through me as I muttered 'I shall take care that you shall have no opportunity of proving your strength upon me'. I went first; steady and without fear, but with every energy awake to meet any possible chance that might arise. The staff on such occasions is the veritable staff of life. It would be perfectly impossible to get on without it: driven into the snow you can hold on by it or by an anchor. Twice our steps instead of being on snow were upon solid ice, which increased the danger immeasurably but these places once past, though the snow continued to fall heavily, we felt that our further progress was secure. There was great pleasure in feeling this. It was a most agreeable variation of that grim mental tension which had hitherto beset me, but which in itself was by no means unpleasant. The colour of the newly fallen snow was wonderful. Not only were the holes formed by our batons filled with the delicate blue light, but the feet of my companion, who now took the lead, seemed enveloped in a blue gleam as he lifted them out of the snow. This, and the other observation to which I have referred were sufficient to repay me for the journey. The people at Zermatt first refused to believe in the ascent and afterwards condemned my companion for undertaking what had never been accomplished, even in fine weather, with less than two guides. He has soon an answer which relieved him of all importunity.

Precisely a week after the above ascent took place the weather was indescribably glorious: not a trace of cloud to sully the deep blue of heaven. On that Tuesday morning I awoke at 5½ oC. and felt a strength in my muscles capable of encountering any fatigue. I had lent my guide to a party of gentlemen who proposed to ascend Monte Rosa. They had started from the hotel at 3½ oC. I breakfasted swiftly. I strapped half a bottle of tea and a ham sandwich on my back and at 6½ oC. faced the ridge which rises over the Riffel. From this ridge the mountains looked inexpressibly grand and Monte Rosa promised a glorious prospect. I shook my limbs to try whether they were in good order. They were. I bore swiftly down upon the Görner Glacier,[25] crossed it. Came fairly to the base of Monte Rosa and faced the mountain. I could see the party of 8[26] who had left the hotel nearly 3 hours before me high up the mountain. They reached the summit. I could hear their reverberated hurrah while I was alone with Nature below. They returned—I met them, passed them, reached the summit—saw Italy on one side and Switzerland on the other: returned, overtook the party, and thus with two mouthfuls of sandwich and a draught of tea accomplished the ascent of Monte Rosa alone.[27] I will <u>talk</u> to you about this ascent—I have already written enough to tire you.

My guide did good service on this day, for he saved the life of a fine young man. This young gentleman[28] complained of a swelled knee, before he was half way up the mountain, but with true English pluck[29] he persisted in the ascent. My guide had charge of him all through, holding him firmly by the arm, and helping him as much as possible. On returning they came to one of the icy places to which I have referred and here the young man's footing gave way. He slid, carrying my guide along with him. The direction of his motion led him fair to the brink of the precipice which forms the Lysskamm[30] side of the mountain, and over which he would have gone. The other side was also terribly steep, but there was at least a hope of rescue here. With a sudden[31] wrench my guide changed the direction of motion: but at the same moment the young gentleman's baton became entangled in his legs and both fell. They rolled over each other towards the brink of another precipice; but by a sudden effort my guide threw himself in advance dug his armed staff into the ice and stopped the motion. Both were saved, but the escape I am told was next to miraculous.

Now I wish you a patient head and good eyes to read this letter: The sun shines again, and a patch of blue overarches the Monte Rosa. If this continues the snow will soon disappear and I shall be on the ice once more. I have just learned that old age is coming upon me, not through any consciousness of failing vigour, but the intelligence is telegraphed through a tooth. I never had a tooth drawn, I never had toothache, I never till now had a hollow tooth. I have just made the discovery that I have one: but it is not a case of fair decay: at least I will thus comfort myself. I broke the enamel off it two years ago, and this gave space for the tooth of time to work upon. But I am degenerating into gossip. Write to me to Chamouni[32] and confess that you are wearied. I had no idea of stretching this letter to such an extent when I commenced. Best regards to M[r] Pollock and love to all the boys.[33]

Believe me ever | most Sincerely yours | J. Tyndall

You must dot the i's and cross the t's of this letter. I have not time to read it through. | I was glad to hear a word of M[r] Babbage. Glad to hear of our friends at Petersham.[34] All those little things have a value in the mountains.

BL 63902–874E–10–51
RI MS JT/1/TYP/6/1940–46

---

1. *[23 August 1858]*: the date is given by the addition of '22nd Aug. 1858.' on the letter, presumably written by Hirst. Tyndall began writing this letter on 22 August and finished it the following day.

2. *a huge glacier*: the Allalin, see letter 1436, n. 28.

3. *Matmark see*: Mattmark See, a lake in the valley of Saas formed by the damming of a river by the moraines of the Allalin glacier.

4. *valley of Saas*: a valley in the Swiss canton of Valais.

5. *Monte Moro*: the key pass between Saas and Macugnaga, below the mountain of the same name.

6. *Distel Alp*: see letter 1532, n. 3.

7. *John Clerk*: John Clerk (1816–1900), a lawyer and third son of George Clerk. See P. Townend (ed.), *Burke's Peerage, Baronetage & Knightage,* 105th edn (London: Burke's Peerage Ltd., 1970), p. 572.

8. *Sir George Clerk*: George Clerk (1787–1867), sixth baronet, a Scottish politician, FRS, and president of the Zoological Society in the 1860s (*ODNB*).

9. *Riffel*: the Riffel hotel on the Riffelalp, underneath the Riffelhorn, a mountain in the Pennine Alps, located south of Zermatt in the Swiss canton of Valais. The hotel was managed by a brother of Joseph Seiler of Zermatt who managed many alpine hotels, including the Hôtel du Glacier du Rhône (see letter 1542, n. 6).

10. *Zermatt*: see letter 1297, n. 7.

11. *Saas*: see letter 1537, n. 33.

12. *Visp*: the River Visp, which flows down the Mattertal (from Zermatt) and joins the Rhone at Visp.

13. *M*ͬ. *Barnett*: Henry Barnett (1815–96), an English banker, landowner, and politician. 'Obituary', *Standard*, 8 May 1896, p. 3.

14. *Walter*: Walter Pollock.

15. *'muss wirken und streben'*: must work and strive (German), from the 1798 poem 'Das Lied von der Glocke' (The Song of the Bell) by Friedrich von Schiller.

16. *With regard to Bobo*: a nickname for Frederick Pollock, the Pollock's eldest son.

17. *accomplish such a triumph*: Frederick Pollock was named first in his class at Eton College (see letter 1545).

18. *Monte Rosa*: see letter 1478, n. 22.

19. *my guide*: Christian Lauener.

20. *brother of my guide*: Ulrich Lauener.

21. *bosses*: a knoll or mass of rock (*OED*).

22. *névé*: see letter 1436, n. 9.

23. *crags*: a steep or precipitous rugged rock (*OED*).

24. *but he said that he should have more power with his own arrangement*: Tyndall apparently had reason to be unconvinced about how his guide held the rope. Two years later, on 18 August 1858 while Tyndall was in Breuil, he learned of a tragic climbing accident on the Col du Géant, in which three Englishmen and a guide died from a fall down a slope. Two other guides, who held the ropes rather than tying them around their waists, survived. On 24 August, Tyndall visited the site of the accident himself, and on 8 September wrote to the *Times* about his observations and conclusions regarding the accident. In the letter, Tyndall argued that for a guide to not have the rope around their waist was to ensure that they would not have their hands free to strike an axe or baton into the surface of the slope in an effort to stop a fall. He also commented that guides should be in a position with their climbers so as remain inseparable, writing that 'no member of their fraternity is fit for his vocation who is unwilling to share the fate of those whose lives are committed to his charge' (see J. Tyndall, 'The Accident on the Col-du-Geant', *Times*, 8 September 1860, p. 8).

25. *Görner Glacier*: a glacier located on the west side of the Monte Rosa close to Zermatt in the Swiss canton of Valais.

26. *the party of 8*: while Tyndall mentioned encountering this party in *Glaciers of the Alps*, he does not identify them beyond the guides (Lauener and Peter Bohren) and Lane Fox (p. 156). Bohren (1822–82) was a Swiss mountain guide, see C. D. Cunningham and W. Abney, *The Pioneers of the Alps*, 2nd edn (London: Low, Marston, Searle and Rivington, 1888), pp. 143–45. Lane Fox is possibly George Sackville Frederick Lane Fox (1838–1918), son of the English landowner and High Sheriff, George Lane-Fox (1816–96). In 1859, he was the first traveller to ascend the Aiguille du Midi. See Cunningham and Abney, *Pioneers of the Alps*, p. 113; and *Burke's Peerage* (2003), vol. 3, p. 3646.

27. *ascent of Monte Rosa alone*: this was the first solo ascent of the mountain.

28. *This young gentleman*: identified by Tyndall as 'Mr. F.' in *Glaciers of the Alps*, p. 159. Presumably the young Lane Fox. See letter 1554.

29. *pluck*: viscera; courage (*OED*).

30. *Lysskamm*: Lyskamm, a mountain in the Pennine Alps lying on the border between Switzerland and Italy, to the southwest of Monte Rosa, and separated from it by the Lysjoch.

31. *steep, but there was at least a hope of rescue here. With a sudden*: only in LT's typescript.

32. *Chamouni*: see letter 1353, n. 9.

33. *the boys*: Frederick, Walter, and Maurice Pollock.

34. *our friends at Petersham*: James and Harriet Moore and their family Harriet Jane, Julia, John, and Graham; for Petersham, see letter 1515, n. 3.

To Sarah Faraday   23 August 1858   1548

23 Aug 1858

My dear—,

—I now sit down to wipe away the reproach of having written a letter to you and not sent it. I reached this mountain wild the day before yesterday. Soon after my arrival it commenced snowing, and yesterday morning the mountains were all covered by a deep layer. It heaped itself up against the windows of this room, obscuring half the light. To-day the sun shines, and I hope he[1] will soon banish the snow, for the snow is a great traitor on the glacier, and often covers smoothly chasms which it would not be at all comfortable to get into. I am here in a lonely house, the only traveller. If you cast your eye on a map of Switzerland you will find the Valley of Saas[2] not far from Visp.[3] High up this valley, and three hours above Saas[4] itself, is the Distil Alp,[5] and on this Alp I now reside. Close beside the house, a many-armed mountain torrent rushes; and a little way down a huge glacier, corning down one of the side valleys, throws itself across the torrent, dams it up, and forms the so-called "Matmark see."[6] Looking out of another window I have before me an immense stone, the unshipped cargo of a glacier, weighing at least 1,000 tons. It is the largest boulder I have ever seen, is composed of serpentine, and measures 216,000 cubic feet. Previous to coming here I spent ten days at the Riffel Hotel,[7] above Zermatt,[8] and explored almost the whole of that glorious glacier region. One morning the candle of my guide[9] gleamed into my room at 3 o'clock and he announced to me that the weather was good. I rose, and at 4 o'clock was on my way to the summit of Monte Rosa.[10] My guide had never been there, but he had some general directions from a brother guide,[11] and we hoped to be able to find our way to the top. We first reached the ridge above the Riffel,[12] then droped down upon the Görner glacier,[13] crossed it, reached the base of the mountain, then up a boss of rock, over which the glacier of former days had flowed and left its marks behind. Then, up a slope of ice to the base of a precipice of brown crags;[14] round this we wormed till we found a place where we could assail it and get to the top. Then up the slopes and round the huge bosses of the mountain, avoiding the rifted portions, and going zigzag up the steeper inclinations. For some hours this was mere child's play to a mountaineer,—no more than an agreeable walk on a sunny morning round Kensington-gardens.[15] But, at length the mountain contracted her snowy shoulders to what Germans call a kamus[16]—a comb; suggested, I should say, by the toothed edges which some mountain ridges exhibit, but now applied to any mountain edge, whether of rock or snow. Well, the mountain formed such an edge. On that side of the edge which turns towards the Lyskamm[17] there was a very terrible precipice, leading straight down to the torn and fissured <u>névé</u>[18] of the Monte Rosa glaciers. On the other side the slope was less steep, but exceedingly perilous-looking, and intersected here and there by precipices. Our way lay along the edge, and we faced it with steady caution and deliberation. The wind had so acted upon the snow as to fold it over, forming a kind of cornice, which overhung the first precipice to which I have alluded. Our track for some time was upon this cornice. The incessant admonition of my guide was to fix my staff securely into the snow at each step,[19] the necessity of which I had already learned. Once, however, while doing this, my staff went right through the cornice, and I could see through the hole that I had made into the terrible gulf below. The morning was clear when we started, and we saw the first sunbeams as they lit the pinnacles of Monte Rosa, and caused the surrounding snow summits to flush up. The mountain remained clear for some hours, but I now looked upwards and saw a dense mass of cloud stuck against the summit. She dashed it gallantly away, like a mountain queen; but her triumph was short. Dusky masses again assailed her, and she could not shake them off. They stretched down towards us; and now the ice valley beneath us commenced to seethe like a boiling cauldron and to send up vapour masses to meet those descending from the summit. We were soon in the midst of them, and the darkness thickened; sometimes, as if by magic, the clouds partially cleared away, and through the thin pale residue the sunbeams penetrated, lighting up the glacier with a kind of supernatural glare. But these partial illuminations became rarer as we ascended. We finally reached the weathered rocks which form the crest of the mountain, and through these we now clambered up cliffs and down cliffs, walking erect along edges of granite with terrible depths at each side, squeezing ourselves through fissures, and thus by jumping, swinging, squeezing, and climbing we reached the highest peak of Monte Rosa.

Snow had commenced to fall before we reached the top, and it now thickened darkly. I boiled water, and found the temperature 184.92 deg. Fahrenheit. But the snow was wonderful snow. It was all flower; the most lovely that ever eye gazed upon. There, high up in the atmosphere, this symmetry of form manifested itself, and built up the exquisite blossoms of the frost. There was no deviation from the six-leaved type, but any number of variations. I should hardly have exchanged this dark snowfall for the best view the mountain could afford me. Still, our position was an anxious one. We could only see a few yards in advance of us, and we feared the loss of our track. We retreated, and found the comb more awkward to descend than to ascend. However, the fact of my being here to tell you all about it proved that we did our work successfully. And now I have a secret to tell you regarding Monte Rosa. I had no view during the above ascent, but precisely a week afterwards the weather was glorious beyond description. I had lent my guide to a party of gentlemen, so I strapped half a bottle of tea and a ham sandwich on my back, left my coat and neck-cloth[20] behind me, and in my shirt-sleeves climbed to the top of Monte Rosa alone. When I see you I will tell you all about this ascent, which was a very instructive one. I expect to remain here a week. The house is cold, and at present the wet comes through the ceiling. I have caught a slight cold, which I hope will soon pass away, as I want all my vigour upon the ice. When I quit this place I shall make my way to Chamouni,[21] where I expect to be in[22] eight or nine days. With kindest, &c.

"Most sincerely yours | John Tyndall."

M. Faraday and J. Tyndall, 'The Glaciers of Switzerland', *Times*, 3 September 1858, p. 10.
Faraday Correspondence, 5:3506
Typed Transcript Only

---

1. *he*: 'he' was incorrectly transcribed as 'it' in the *Faraday Correspondence*.

2. *Valley of Saas*: see letter 1547, n. 4.

3. *Visp*: see letter 1543, n. 8.

4. *Saas*: see letter 1537, n. 33.

5. *Distil Alp*: see letter 1532, n. 3.

6. *"Matmark see"*: see letter 1547, n. 3.

7. *Riffel Hotel*: see letter 1547, n. 9.

8. *Zermatt*: see letter 1297, n. 7.

9. *my guide*: Christian Lauener.

10. *Monte Rosa*: see letter 1478, n. 22.

11. *brother guide*: Ulrich Lauener.

12. *Riffel*: Riffelalp, see letter 1547, n. 9.

13. *Görner glacier*: see letter 1547, n. 25.

14. *crags*: see letter 1547, n. 23.

15. *Kensington-gardens*: at first the private gardens of Kensington Palace in London, the grounds are now public.

16. *kamus*: in letter 1547, Tyndall referred to this as a 'kamm', a comb. 'Kamus' appears to be an editorial misspelling on the part of *The Times*.

17. *Lyskamm*: see letter 1547, n. 30.

18. *névé*: see letter 1436, n. 9.

19. *step*: 'step' was incorrectly transcribed as 'side' in the *Faraday Correspondence*.

20. *neck-cloth*: see letter 1311, n. 25.

21. *Chamouni*: see letter 1353, n. 9.

22. *in*: 'in' is missing from the *Faraday Correspondence*.

From Charles Anderson and Michael Faraday   31 August 1858   1549

Royal Institution | 31 Aug 1858

Sir,

A letter[1] was left at the Institution[2] by post for you on Saturday morning[3] on Her Majestys Service, I opened it and found a cheque[4] in it and sent it in the afternoon to Mr. Francis, all here are very glad to hear that you are so strong and in good health. We hope you will long continue so. Every thing is very quiet here at present and we are getting on with the repairs.[5]

Your Humble Servant at all times, | C. Anderson

My dear Tyndall, as this is all the paper that can go, I take[6] one side from Anderson just to shew myself. We all think of you and are very glad you are so strong.

Ever yours, | M. Faraday

RI MS JT/1/TYP/12/4148
Faraday Correspondence, 5:3511
Typed Transcript Only

---

1. *A letter*: not identified.

2. *the Institution*: RI.

3. *Saturday morning*: 28 August 1858.

4. *a cheque*: possibly payment for Tyndall's military examination duties.

5. *the repairs*: not identified.

6. *take*: this word was given as 'that' in LT's typescript with a note that reads '[sic—sh^d be take]'; 'that' was retained in the *Faraday Correspondence* followed by '[sic]'.

From Juliet Pollock   31 August [1858][1]   1550

S^t. Julians. Sevenoaks. August 31.st

My dear M^r. Tyndall

Here comes one of my maternal lectures for you: I am dismayed at your solitary ascent of Monte Rosa,[2] and I call upon you to remember that however muscularly strong you may be, you must, in common with the rest of humanity, be subject to casualties, and that such a casualty as a sprained ankle a swelled knee, a sudden faintness, or a sudden fog, would place you (alone on a steep glacier) in considerable jeopardy; and to my thinking it is really not right to risk a valuable life in such a way. Now I have done. I have administered the draught of physic, and I must see if I can find some plums to take out the taste but here, I am not in contact with any of your particular friends, though nature the common friend of us all, is showing her face to me, full of sweetness and beauty. The distant landscape is much of the same character that you have at Tunbridge Wells[3] & especially like that you can perhaps recall, that opens out upon you from the churchyard of Rusthall,[4] but here, we have a greener and fuller foliage in the foreground—more variety of wood, and a smoother surface of grass. The present richness of the hop gardens is an additional charm in our walks. The poles bend beneath their weight and bowing towards each other make sometimes Gothic and sometimes Norman arches and in one instance the hop plant climbing above its own pole has attached itself to the lower branches of an old oak, hanging upon them caressingly and gracefully the clusters of its fruit. If you have never been in a hop country there is something beautiful that you have not yet seen. —

My husband[5] is in town at present but he will I trust join me on Thursday. I have Fred[6] and Maurice[7] with me—There is my Aunt,[8] I grieve to say still an invalid, to endeavour to cheer, for her spirits are tried, and there is Edward Herries[9] who has been to me always a very dear Brother. Charles[10] and Isey[11] are both absent at present. —

I feel the truth of all you say about my darling Walter,[12] but I confess that having had him to spend two days with us before coming here I felt the 2.^d parting quite as grievously as the first, for he appeared not able yet to reconcile himself to our separation. —

He is anxious to write to you but I have advised his waiting till you return to England as his hand is so very large. The time that children are kept at large text seems to me a mistake in the system of education.

The quiet intervals that I have here, I give to the reading of Carlyle's Cromwell[13] a book full of deep interest and of great entertainment. Do you know it? You probably do. —

I have read a good deal also of Hogg's life of Shelley,[14] and whatever Keats may say of Adonais,[15] it is evident to me from his own letters, I take no other testimony, that he was a man of an essentially bad mind, and that his total denial of all authority, beginning from the highest, his total abnegation of all restraints led to such a course of vice as ought to be the consequence of such a theory of life. When one man writes to another

I shall hope to hear from you again. —

Believe me always, my dear M^r. Tyndall | Yours most truly | Juliet Pollock | I am sorry for the tooth of time

If you write to me after the 16th address | Mrs J. Pollock | Greta Bank | Keswick | Cumberland[16]

RI MS JT/1/P/193

1. *[1858]*: the year is given by the reference to Tyndall's solitary ascent of Monte Rosa on 17 August 1858.

2. *your solitary ascent of Monte Rosa*: see letter 1547.

3. *Tunbridge Wells*: see letter 1423, n. 1.

4. *Rusthall*: St Paul's Church in Rusthall, Kent, England.

5. *My husband*: William Frederick Pollock.

6. *Fred*: Frederick Pollock.

7. *Maurice*: Maurice Pollock.

8. *my Aunt*: Isabella Maria Herries. See letters 1418 and 1420.

9. *Edward Herries*: Edward Herries (1821–1911), a British diplomat and cousin of Juliet Pollock. See 'Mr. Edward Herries', *Times,* 18 November 1911, p. 13.

10. *Charles*: Charles Herries (1815–83), a financier and cousin of Juliet Pollock (*ODNB*).

11. *Isey*: Isabella Herries (d. 1897), a cousin of Juliet Pollock. See *Faraday Correspondence,* 5:3225, n. 2.

12. *Walter*: Walter Pollock.

13. *Carlyle's Cromwell*: T. Carlyle, *Oliver Cromwell's Letters and Speeches: with elucidations,* 4 vols (London: Chapman and Hall, 1845).

14. *Hogg's life of Shelley*: T. J. Hogg, *The Life of Percy Bysshe Shelley,* 4 vols (London: Edward Moxon, 1858).

15. *whatever Keats may say of Adonais*: a reference to Percy Bysshe Shelley's *Adonaïs: An Elegy on the Death of John Keats,* written in 1821 following the death of the English poet John Keats.

16. *Greta Bank | Keswick | Cumberland*: Greta Bank is an estate located in the town of Keswick in the county of Cumberland in northwestern England, a Spedding family home. The Pollocks were good friends with James Spedding (see letter 1545, n. 14), and were likely visiting him at Greta Bank.

To Félix Bergoin   1 September 1858   1551

A Monsieur le Président de la Commission des Guides à <u>Chamonix</u>.

Monsieur,

Les phénomènes des glaciers ont toujours, comme vous savez, un grand intérêt pour des hommes scientifiques. En ce moment cet intérêt pour des raisons que je ne puis pas expliquer ici est beaucoup augmenté. Voilà ce qui m'a causé de commencer une série d'observations sur ce sujet, et de dédier pendant trois années une portion de mon temps à l'investigation de cette question. J'ai passé, comme vous savez, l'année dernière 6 semaines en observant les phénomènes de la Mer de Glace; J'avais reçu permission d'aller partout avec mon guide, et de prendre des garçons pour porter mes petits instruments. J'ai cité ceci en Angleterre, comme une preuve de la modération, et du bon sens de la Commission de Guides à Chamonix, qui ne jetterait pas des obstacles dans le chemin d'un homme de science.

Cette année je suis venu à Chamonix pour compléter mes observations. J'en ai déjà présenté une portion à la Société Royale de Londres, dont je suis membre, et j'ai aussi signalé le fait que M. Auguste Balmat avait l'intention de placer des thermomètres sur le sommet du Mont Blanc. J'ai recommandé à la notice formelle du Conseil de la Société Royale l'intention de M. Balmat, comme une entreprise digne d'assistance et d'encouragement. Le Conseil tout a pensé comme moi; et me voilà maintenant à Chamonix prêt de me *[1 word illeg]* à M. Balmat, et d'assister dans cette expédition, soit par l'argent, soit par conseil.

Mais Monsieur je ne trouve cette année que des difficultés. Je ne pouvais pas avoir hier un garçon pour porter un petit instrument—il faut avoir un <u>guide</u>. Le Guide Chef m'a dit aussi que si je monterait le Mont Blanc il faut avoir quatre guides. Enfin, Monsieur, il me forcera d'observer rigoureusement les règlements qui sont faits pour les <u>touristes</u>. Quand je pense à ce que les hommes de science ont fait pour Chamonix, il me semble Monsieur que celle-ci est une mauvaise récompense. Les hommes de science ont fait la découverte de Chamonix, et les ouvrages de telles hommes l'ont invité d'un intérêt qui l'a fait plus de service que des centaines des voyageurs ordinaires. Et voilà la récompense! A lieu d'assister, et de supporter un homme scientifique dans ses observations, Monsieur le Guide Chef, qui ne comprend pas la science, ne jette que des difficultés dans son chemin.

Le Guide Chef m'a dit hier que pour résoudre cette question il faut rassembler la Commission des Guides; et il m'a proposé d'assister à une séance à 7 heures du soir. Je faisait une excursion pénible sur la Mer de Glace, et sans cette convention entre le Guide Chef et moi j'aurais resté hier au Montanvert. Mais je suis descendu dans la pluie, pour me rendre au temps spécifié au bureau du Guide Chef. Je l'ai vu, et il m'a courtement dit qu'il a consulté la Commission, qu'elle a répondu "qu'elle ne pouvait se rassembler pour une chose comme ça; que les règlements étaient fixes, et il faut les observer." Vous Monsieur qui est Président de cette Commission—vous savez si Monsieur le Guide Chef m'a dit la vérité

Voici donc Monsieur que je demande respectueusement:—Je demande permission de faire mes observations dans la manière qui me paraît convenable; de prendre des garçons avec moi quand jes les considère nécessaires; de prendre aux grandes élévations l'assistance qui me semble d'être suffisante. de prendre par exemple au sommet du Mont Blanc <u>un</u> guide expérimenté comme Balmat, et des porteurs. de prendre un ami avec moi qui peut faire des observations simultanées. Vous savez que j'ai déjà monté le Mont Blanc avec un seul guide, et pendant, cette ascension j'ai coupé moi-même des escaliers dans les endroits les plus difficiles. Je ne tenterai rien qui est imprudent. Je prendrai conseils des guides expérimentés, et tout ce qu'ils considèrent imprudent je ne tenterait pas. Mais comme un homme qui a la force, qui pendant plusieurs années a été accoutumé aux montagnes et aux glaciers, et qui peut monter presque comme un guide même, surtout comme un homme qui fait un travail purement scientifique je demande permission de continuer mes recherches, sans l'embarras de ces règlements qui sont peut-être bons pour les voyageurs ordinaires.

J. T.

To the President of the Chamonix Guide Company.[1]

Sir,

As you know, glacier phenomena are always of great interest to scientists. At this moment for reasons that I cannot explain here that interest has greatly increased. This caused me to start a series of observations on the subject and dedicate for three years part of my time to investigating this matter. As you know, I spent six weeks last year observing the phenomena of the Mer de Glace;[2] I had received permission to go anywhere with my guide, and take boys to carry my small instruments. I have mentioned this in England as a sign of the moderation and common sense of the Chamonix Guide Company,[3] which would not throw obstacles in the path of a man of science.

I came to Chamonix[4] this year to complete my observations. I already presented a portion of it to the Royal Society of London, of which I am a member, and I also mentioned the fact that Mr. Auguste Balmat intended to place thermometers on the summit of Mont Blanc.[5] I recommended Mr. Balmat's intention as a formal notice to the Royal Society Council as an endeavour worthy of support and encouragement.[6] The Council agreed entirely with me: and here I am now at Chamonix, ready to *[1 word illeg]* Mr. Balmat, and to support him in this expedition either with money or with advice.

But Sir I found nothing but difficulties this year. Yesterday I could not take a boy to carry a small instrument—it is compulsory to have a <u>guide</u>. The Guide Chef[7] also told me that if I were to go up Mont Blanc, I should have four guides. Finally, Sir, he would force me to strictly observe rules that are made for <u>tourists</u>. When I think about what men of science have done for Chamonix, it seems to me Sir that this is a poor recompense. Men of science made the discovery of Chamonix, and the works of such men brought on much interest which did more service to it than hundreds of ordinary travelers. And here is the reward! Instead of assisting and supporting a scientist in his observations, Monsieur the Guide Chef, who does not understand science, only throws obstacles in his path.

The Guide Chef told me yesterday that in order to resolve this issue the Guide Company must assemble; and he proposed that I should attend a meeting at 7 o'clock in the evening. I did a laborious excursion on the Mer de Glace, and without this agreement between the Guide Chef and myself I would have stayed yesterday at Montanvert.[8] But I climbed down in the rain to make it to the specified time to the Guide Chef's office. I saw him, and he told me shortly that he consulted the Company, and that they replied "that they could not assemble for a thing like this; that the rules were fixed, and they must be observed". You Sir who is the President of this Company—you know if the Guide Chef told me the truth

Here I am Sir making a respectful request:—I ask for permission to make my observations in the manner that best suits me; to take boys with me when I consider it necessary, to take to the great heights the assistance that seems sufficient to me. to take for instance to the summit of Mont Blanc <u>one</u> experienced guide like Balmat, and boys. to take a friend with me who can make simultaneous observations. You know that I have already ascended Mont Blanc with a single guide, and during that ascent I cut steps with my own hands in the most difficult of places. I will not try anything unwise. I will take the advice of experienced guides, and I will not try anything that they consider to be unwise. But as a man who has strength, who became accustomed to mountains and glaciers over many years, and who can climb almost as a guide himself, above all as a man who does purely scientific work I ask for permission to continue my research, without the constraint of these regulations which may be good for ordinary travelers.

J. T.

RI MS JT/2/13c/1136–38
RI MS JT/2/9/524–29

1.	*To the President of the Chamonix Guide Company.*: The letter is transcribed from a copy in Tyndall's typescript journal, 1 September 1857. The journal entry preceding the letter reads: 'Called at the bureau of the Syndic—he takes the fares of passengers by the diligence to Geneva. I asked him whether he could grant me 5 minutes conversation; expecting of course that he had already made up his mind in my case. There was a certain severity in the old man's countenance, but still it seemed intelligent and sincere. I proceeded to talk to him under the impression that he was already acquainted with the case. But he appeared to be wholly ignorant of it. He said the Guide Chef had power to arrange the whole matter; and that no subject could be more proper for an assembling of the Commission. He advised me to wait for Balmat, but in the mean time considered it well for me to write him a letter stating the case, and making my request. This he said he would send to the Intendant of Bonneville, stating at the same time his own conviction that the conduct of the Guide Chef was "ridiculous". I was very glad to find things in this state; returned to my hotel and wrote the following letter, which I afterwards handed to the Syndic' (Journal, RI MS JT/2/13c/1136–38). For the response from the Intendent of Bonneville Félix Bergoin's secretary to Tyndall, see letter 1556. For more on the placing of the thermometers and difficulties with the Chamonix Guide Company, see letters 1566 and 1590, and *Glaciers of the Alps,* pp. 169–76, 178, 192–94.
2.	*Mer de Glace*: see letter 1306, n. 12.
3.	*Chamonix Guide Company*: a Chamonix-based alpine guide company founded in 1821, with climbing regulations set by the government, and overseen by a guide chef.
4.	*Chamonix*: see letter 1353, n. 9.
5.	*Mont Blanc*: see letter 1320, n. 6.
6.	*I recommended Mr. Balmat's intention . . . of support and encouragement*: Tyndall asked Balmat to climb Mont Blanc with him to place the thermometers (see letter 1544). The Council of the RI did grant Balmat funds for equipment.
7.	*Guide Chef*: see letter 1419, n. 17.
8.	*Montanvert*: see letter 1415, n. 15.

## To Sarah Tyndall				2 September 1858				1552

My dear mother,

It is a long time since I wrote to you,[1] and no doubt you will be desirous to hear something of me. For several weeks I have been wandering among the mountains and the glaciers of Switzerland, and I am at the present moment seated near the base of the highest mountain in Europe.[2] My health is very good, and my general strength considerable. In fact the people here consider me a first rate mountaineer. On the 20th of this month I must be in London, so that my tour is drawing to a close. I do not think you would in the least recognise me if I were to appear before you at present. My whiskers have grown to an enormous bulk, and for the last 6 weeks I have not applied a razor to my face. The consequence is that it is half hidden by a great beard and moustache. They are very ugly, but very comfortable, and save a great deal of trouble. They are also good as a protection against the sun which sometimes glares with immense power from the mountain snow. I have also a green veil round my hat to protect me from the same influence, and in addition to this to save my eyes am often compelled to wear dark spectacles. Imagine me then with my beard, and my veil and my spectacles and I think you will agree with me that you could hardly recognise me. I am also turned very brown by the sun; and appear a totally different person from the pale faced individual who came here six weeks ago. I hope the strength which I have acquired will help me over the duties of the coming winter.

When I return to London I will write to you again—until then goodbye—Your affectionate son | John

Chamouni Savoy.[3] 2nd, Sep, 1858

I shall leave this place in a few days, so it would not be safe to send a letter to me here.

RI MS JT/5/16/-

---

1. *long time since I wrote to you*: the last extant letter from Tyndall to his mother is letter 1270, John Tyndall to Sarah Tyndall, 8 August 1856, *Tyndall Correspondence*, vol. 5.

2. *highest mountain in Europe*: Mont Blanc. See letter 1320, n. 6.

3. *Chamouni Savoy*: see letter 1353, n. 9.

From Michael Faraday   2 September 1858   1553

Royal Institution | 2 Sep 1858

My dear Tyndall,

I found Anderson a day or two ago about to write to you on a little bit of paper just passable by the post, and I made free with one side of it.[1] In the uncertainty of knowing where you might be found, perhaps I might not have written to you again, but for the receipt of your letter[2] by my wife, detailing the ascent of Monte Rosa,[3] and the enormous indiscretion I have committed thereupon. What shall I say? I have sent it to the Times.[4] There, the whole is out. I do not know whether to wish it may appear tomorrow or next day or not. If you should dislike it, I shall ever regret the liberty I have taken. But it was so interesting in every point of view, shewing the life and spirit of a philosopher engaged in his cause: shewing not merely the results of the man's exertions, but his motives and his nature:—the philosophy of his calling and vocation, as well as the philosophy of his subject; that I could not resist, and I was the more encouraged to do so because, from the whole character and appearance of the letter, it shewed it was an unpremeditated relation and that you had nothing to do with its appearance, i.e. it will shew that if it should appear. Now I hope you forgive me. Nobody will find fault with me but you. It came too late for the Phil. Mag., but if the Times does not put it in, I shall send it to the Phil. Mag. However as this is only the 3rd of the month, there is time enough for that.

I won't give you any scolding. I dare say my wife will, when you see her: 'êtes-vous marié',[5] indeed! I cannot but feel glad you have done it now it is done, but I would not have taken the least portion of responsibility in advising you to such a thing.

I have no philosophy and no news for you. I feel just out of the world— forgetful, and dull headed in respect of science and of many other things— but well and content, as I have great reason to be. My wife and Jane[6] are pretty well: the latter absent, or she would send her remembrances with ours. I shall send this to Chamouni[7] on the chance of catching you.

Good bye, my dear friend | Ever truly yours | M. Faraday

Friday morning, 3rd. The letter is there.

Address: Dr. Tyndall | Chamounix | <u>Savoie</u>[8] | Switz[erland]

RI MS JT/1/TYP/12/4149
Faraday Correspondence, 5:3512
Typed Transcript Only

---

1. *one side of it*: letter 1549.

2. *your letter*: letter 1548.

3. *ascent of Monte Rosa*: see letter 1547.

4. *I have sent it to the Times*: M. Faraday and J. Tyndall, 'The Glaciers of Switzerland', *Times,* 3 September 1858, p. 10. See letter 1548.

5. *'êtes-vous marié'*: are you married (French).

6. *Jane*: Jane Barnard.

7. *Chamounix*: see letter 1353, n. 9.

8. *Savoie*: the Duchy of Savoy, a region of the Kingdom of Sardinia (now southeastern France) where Chamonix is located.

To Thomas Henry Huxley   3 September 1858   1554

Chamonix, 3rd, Sept. 1858.

My dear Huxley,

A dim memory has crossed my soul like a shadow several times during the last 3 or 4 days—namely that I promised to write to you and had not yet performed that promise. It is not the time necessary to write a letter that makes me shun the process, but somehow or other one's mind gets out of joint—but I won't explain. Ramsay[1] has told you doubtless of our sojourn at the Eggishorn.[2] This place I found all you described it to be. We spent 8 days there. The Mergilen See[3] was very lovely with its blue waters covered by its white icebergs. I have a spite against the latter, for I thought of using them as rafts, but they turned basely round and popped me into the water. That Aletsch Glacier[4] is the noblest stream that I have ever seen. We spent many hours upon it. At its source, where it teems through grand corridors formed by the Jungfrau,[5] Mönch,[6] Eiger,[7] Trugberg,[8] Aletschhorn,[9] and others of the same kin, its magnificence is indescribable. I walked to the top of the Finisteraarhorn[10] and returned along the Viesch glacier[11] to the Jungfrau hotel.[12] I had a most daring devil of a guide.[13] 'I am like the Tyroler'[14] he said to me once at a difficult spot on the side of the Finisteraarhorn 'who went to his priest to confession'—'How so?' I asked. 'Why,' he replied, 'he was obliged to confess that he loved women <u>ausserordentlich</u>[15]—'Oh!' said the priest, 'mein Freund, Frauen zu lieben und in den Himmel zu kommen, <u>das geht nicht</u>!'[16] 'Bei Gott es <u>muss</u> gehen'[17] was the Tyroler's reply—and I say now 'Bei Gott es <u>muss</u> gehen', we must get to the top!' Each took care of himself going up, but in coming down we tied ourselves together. When fastened he exhorted me thus, 'Slip if you like, fall if you like, throw yourself where you like, I hold you!' A quieter guide[18] but one equally strong to whom I told this afterwards, shook his head and said it was an 'unwahrheit'[19] that he could not hold me. At all events I did not give him the chance. In fact I held him once or twice as he was sweeping down <u>en avalanche</u> towards some crevasses. Along the Viesch glacier, which you know is terribly cut up, he led me with wonderful promptitude and hardihood. The crevasses at its upper portion were concealed. He was often half submerged, but always managed to claw himself out. Once he heard the icicles ring in the depth of a fissure into which my legs had gone. It was the only moment in which I saw concern in his face. 'Gott's Donner!'[20] he exclaimed 'You have not followed my steps.' The monosyllable 'Doch'![21] was my only response. He was a gallant devil, and if ever I go to that quarter again I shall be sure to foregather with him.

From the Eggischhorn[22] to Zermatt;[23] from Zermatt to the Riffel;[24] where Ramsay just looked in and vanished: a fate seems to hang over my companions: an exactly similar, and equally sudden event separated Hirst (he is now in Italy) from me 9 years ago. The morning after my arrival at the Riffel, the candle of my guide[25] gleamed in upon me. He pronounced the weather weatherable. I rose, and in an hour we were on our way to Monte Rosa.[26] He had never been there but he got some general directions from his brother[27] who had been, and we faced the mountain cheerily together. The morning was like a woman—<u>beautiful, but treacherous</u>, and when we had committed ourselves utterly to our enterprise the sunny smile was changed to a frown. The mountain, you know, contracts itself to an edge at its upper portion and along this edge you have to march for hours. The wind had so acted on the snow as to fold it over like an overhanging cornice: underneath was a most awful precipice, and once or twice while endeavouring to secure anchorage with my staff it went right through this cornice, and I could see through the hole the depths below. While crawling up the 'Kamm'[28] the valleys filled with fog, clouds wrapped themselves round the head of the mountain. The former seethed upwards and the latter seethed downwards, and soon your beloved Tyndall was in the midst of the atmospheric embroilment. On we went; reached the crags[29] which form the crest of the mountain; balancing, squeezing, climbing up ledge and down rock. We reached the summit. Snow fell thickly, and we feared the obliteration of our footmarks. Nevertheless we boiled water and found its temperature to be 184.92. The snow was lovely, it was all flowers, there was not a breath of wind to destroy their forms, any thing so beautiful I had never seen. I caught them by millions upon my felt hat. They were all of the six-leaved type, but of endless varieties. Down thro' the gloom, an avalanche was let loose; we could not see it, could not gauge its distance, but the sound reaching us thro' the obscurity had an element of horror in it. We descended the 'Kamm', I first, with a rope round my waist, and my companion afterwards with the said rope round his arm. I did not quite like the arrangement, for I did not see in it the resolution that we were to live or die together. He could readily have slipped me off in case I had been in peril, but with a feeling of grim indifference I said to him I shall give you no chance of doing so. This was not my Eggischhorn guide[30] but he proved an excellent one: and I dare say I did him here wrong. He declared to me afterwards that it was because he felt sure that I was perfectly competent to take care of myself that he adopted the arrangement. We made the journey to the summit and back to the Riffel in 11½ hours.

A week afterwards the weather was glorious. I rose at 6 o'clock and, seeing the weather, the thought of the prospect from Monte Rosa rushed upon me. I had lent my guide for the day to a party[31] who purposed to ascend the mountain. They started at 3 o'clock. I had a swift breakfast, left my coat and neckcloth[32] behind me, strapped half a bottle of tea and a ham sandwich on my back. Faced the mountain, met the party in their descent, went to the summit alone, enjoyed the prospect, down the kamm, overtook the party in advance of me and returned in the jolliest manner possible. I had to hang leaden[33] weights on the wings of my imagination when I stood alone upon the top, for if let loose she would have made wild work of it. My guide did good service on that day. At the risk of his own life he saved that of a fine young fellow named Fox—Lane Fox,[34] who slipped on the kamm and would have gone over the precipice had he not been prevented.

I came here a few days ago. I have been twice up the Mer de Glace,[35] but the fresh snow obscures the surface and defeats me But I have done an immense amount of work. The glaciers of Monte Rosa have taught me a great deal, and the question of structure and stratification is finally settled in my mind.[36] But I did not intend to talk science when I commenced, nor shall I do so now. If the weather clears and the snow melts I shall walk with Balmat[37] to the top of Mont Blanc[38] to sink thermometers there: but I have little hope of this; the quantity of fresh snow is so great that I fear it will not be melted this year. I have met Wills[39] and like him, he is an earnest truthful fellow. I have seen the photograph of his wife[40] and am quite in love with her. The lady I have not seen. Kindest regards to Mrs Huxley.[41] Goodbye old Hal.[42] I shall be in town on the 20th, my examinations commence then.

Ever and always | Tyndall | I am living with Simond.[43]

RI MS JT/1/TYP/9/2890–92
IC HP 8:36–38
Typed Transcript Only

---

1. *Ramsay*: Andrew Ramsay.
2. *Eggishorn*: see letter 1297, n. 6.
3. *Mergilen See*: Marjelen See, a lake under the Eggishorn and next to the Aletsch glacier.
4. *Aletsch Glacier*: see letter 1536, n. 19.
5. *Jungfrau*: see letter 1536, n. 20.
6. *Mönch*: see letter 1536, n. 21.
7. *Eiger*: see letter 1536, n. 22.
8. *Trugberg*: see letter 1537, n. 23.
9. *Aletschhorn*: see letter 1536, n. 23.
10. *Finisteraarhorn*: see letter 1536, n. 27.
11. *Viesch glacier*: the Fiescher Glacier, on the south side of the Bernese Alps in the Swiss canton of Valais.
12. *Jungfrau hotel*: see letter 1536, n. 17.
13. *daring devil of a guide*: Johann Bennen.
14. *Tyroler*: a person of the Tyrol region of central Europe, located within the Eastern Alps.

15. *ausserordentlich*: extraordinarily (German).

16. *'mein Freund, Frauen zu lieben und in den Himmel zu kommen, das geht nicht!'*: 'my Friend, to love women and to go to Heaven, that will not do' (German).

17. *'Bei Gott es muss gehen'*: 'By God it must go' (German).

18. *A quieter guide*: possibly Christian Lauener.

19. *'unwahrheit'*: 'untruth' (German).

20. *'Gott's Donner!'*: 'God's thunder!' (German).

21. *'Doch'*: 'But' (German).

22. *Eggischhorn*: see n. 2.

23. *Zermatt*: see letter 1297, n. 7.

24. *Riffel*: see letter 1547, n. 9.

25. *my guide*: Christian Lauener.

26. *Monte Rosa*: see letter 1478, n. 22.

27. *his brother*: Ulrich Lauener.

28. *'Kamm'*: a comb (German), meaning a crest or ridge. See letter 1547.

29. *crags*: see letter 1547, n. 23.

30. *not my Eggischhorn guide*: Bennen, the guide attached to the Hotel Jungfrau under the Eggishorn, mentioned above (n. 13), who worked with Tyndall on the mountains while Ramsay was accompanied by Christian Lauener in the valley below. Christian Lauener, however, appears to have accompanied Tyndall and Ramsay on their ascent of the Eggishorn on 31 July, while they only engaged Bennen as guide on 2 August. See *Glaciers of the Alps*, pp. 92–119. The summit of the Eggishorn is mentioned on p. 100 and the ascent of Finsterarhorn on pp. 104–19.

31. *a party*: see letter 1547, n. 26.

32. *neckcloth*: see letter 1311, n. 25.

33. *leaden*: heavy as if made of lead (*OED*).

34. *Lane Fox*: possibly George Sackville Frederick Lane Fox (1838–1918), son of the English landowner and High Sheriff, George Lane-Fox (1816–96). See letter 1547, n. 26.

35. *Mer de Glace*: see letter 1306, n. 12.

36. *and the question of structure and stratification is finally settled in my mind*: the issue was whether the stratification in the névé was preserved in the structure lower down. On this trip with Ramsay, Tyndall resolved that it was not, and that the structure was due to pressure (see *Glaciers of the Alps*, p. 120). In his journal entry of 18 August 1858, he wrote 'Here upon the Furgge glacier; with the solemn Matterhorn as witness and earnest Nature looking one everywhere in the face, no man could resist the evidence that the structure and the stratification of glacier ice were things as distinct as the cleavage and bedding of rocks' (Journal, RI MS JT/2/13c/1128).

37. *Balmat*: Auguste Balmat.

38. *Mont Blanc*: see letter 1320, n. 6.

39. *Wills*: Alfred Wills (1828–1912), an English judge and mountaineer who served as the third President of the Alpine Club from 1863–65 (*ODNB*).

40. *his wife*: Lucy Wills (née Martineau, d. 1860).
41. *Mrs. Huxley*: Henrietta Huxley.
42. *Hal*: nickname for Thomas Huxley.
43. *Simond*: Edouard Simond.

To Elizabeth Dawson Steuart[1]   3 September 1858   1555

3rd Sep. 1858.

My dear Mrs Steuart,

The thought of you crossed my mind a few days ago upon the glaciers—I know not how it came, but I asked myself at the time 'how is it that I never receive a letter from my old friend!' I laid the blame upon myself, and formed on the spot the laudable resolution to write to you as soon as I should find a peaceful resting place. Chamouni[2] has furnished me with such a pause and from Chamouni I write to you.

For the last three years I have devoted a portion of my time to the investigation of glaciers, and hence I have been brought to the Alps every year. Last year I lived for six weeks in a little hut beside the Mer de Glace.[3] Every day I was upon the ice observing its motion, which is exactly like that of a river, and examining its properties. On the 13th of August I walked in company with a friend[4] and a single guide[5] to the summit of Mont Blanc.[6]

This year my ramble among the glaciers has been very extensive. I have visited most of the glaciers on the Bernese Alps;[7] and ascended to the summit of the Finsteraarhorn,[8] which is the highest mountain in the Oberland.[9] Thence to the district about Monte Rosa[10] where I spent a fortnight, and learned a great deal concerning its magnificent glaciers. I ascended the mountain twice during my stay in the neighbourhood. Such ascents are not without peril and in my case a little more perilous than usual for I do not take with me the retinue of guides which generally accompany tourists who make such excursions.

When I came here I hoped to be able to climb to the summit of Mont Blanc to sink some thermometers in the ice at the top. But so much fresh snow has fallen, and the weather is so unsettled, that I almost fear I shall not be able to accomplish the ascent this year. I will wait here for a week or so and then act as circumstances dictate.

When I return I shall try to get into some quiet nook, throw my observations together and publish them in the form of a book.[11] If this intention should ever be fulfilled I shall take care to send you a copy, and from it you will be able to judge of the nature of my work and the labour which it has rendered necessary.

I had the pleasure of seeing Mr Duckett[12] in London some weeks before I came away, and I think I must ask you to apologise to him for the shortness of the time which I was able to devote to him when he had the kindness to call upon me last. It is always a delight to me to see him, but I am sometimes in the midst of experiments which demand incessant attention and hence may appear neglectful to my friends on such occasions. I hoped to see Mr Duckett afterwards and to explain this to him myself, but unfortunately I had not the opportunity. Will you kindly remember me to him?

Some military examinations,[13] in which I take a part, commence on the 21st of this month, so I hope to be in London by the 20th. I should be very glad to hear from you once more if you have a minute's time to write to me. The Royal Institution | Albemarle Street | London | is my permanent address.

And now I wish you goodbye and believe me | Ever yours sincerely | John Tyndall

Would you kindly have the enclosed[14] forwarded to my mother?[15]

RI MS JT/1/TYP/10/3338–39
LT Typescript Only

1. *Mrs. Steuart*: Elizabeth Dawson Steuart (née Duckett, 1802–93), was, like her husband, born into a wealthy family. She married William Richard Steuart on 18 August 1820, and they lived at Steuart's Lodge, Leighlinbridge in County Carlow, Ireland. She took a keen interest in Tyndall's career and was probably his most consistent and long-lived patron. She continued to support Tyndall after her husband died in 1852. See G. Cantor and G. Dawson (eds), *The Correspondence of John Tyndall, Volume 1: The Correspondence, May 1840–August 1843* (Pittsburgh: University of Pittsburgh Press, 2016), p. 444.

2. *Chamouni*: see letter 1353, n. 9.

3. *Mer de Glace*: see letter 1306, n. 12.

4. *a friend*: Thomas Hirst.

5. *a single guide*: Edouard Simond.

6. *Mont Blanc*: see letter 1320, n. 6. They left Chamonix on 12 August and reached the summit on 13 August.

7. *Bernese Alps*: see letter 1539, n. 14.

8. *Finsteraarhorn*: see letter 1536, n. 27.

9. *Oberland*: see letter 1527, n. 13.

10. *Monte Rosa*: see letter 1478, n. 22.

11. *a book*: *Glaciers of the Alps*.

12. *Mr Duckett*: probably one of Elizabeth Steuart's brothers, John Dawson Duckett (1791–1866), William Duckett (1796–1868), or Joseph Fade Duckett (1796–1875). Most likely Joseph, since in her reply she mentioned that he would be making a visit to London and 'no doubt will call upon you while there' (see letter 1561). Elizabeth's father, William Duckett, was born in 1761.

13. *military examinations*: beginning on 21 September 1858, Tyndall conducted exams in the Experimental Sciences. See *Examination Papers used at the Examinations for Direct Commissions, in September, 1858* (London: Harrison, 1858), pp. 13–16.
14. *the enclosed*: letter 1552.
15. *my mother*: Sarah Tyndall.

From the Intendant of Faucigny[1]   11 September 1858   1556

Intendance Royale de la Province de Faucigny,
| Bonneville, 11 Septembre 1858.

Monsieur,—

J'apprends avec une véritable peine des difficultés que vous rencontrez de la part de M. le Guide Chef pour l'effectuation de votre périlleuse entreprise scientifique, mais je dois vous dire aussi avec regret que ces difficultés résident dans un règlement fait en vue de la sécurité des voyageurs, quelque puisse être le but de leurs excursions.

Désireux néanmoins de vous être utile notamment en la circonstance, j'invite aujourd'hui même M. le Guide Chef à avoir égard à votre projet, à faire en sa faveur une exception au règlement ci devant eu, tant qu'il n'y aura aucun danger pour votre sûreté et celles des personnes qui vous accompagneront, et enfin de se prêter dans les limites de ses moyens et attributions pour l'heureux succès de l'expédition, dont les conséquences et résultats n'intéressent pas seulement la science, mais encore la vallée de Chamounix en particulier.

Agréez, Monsieur, | l'assurance de ma considération très-distinguée. | Pour l'Intendant en congé, | Le Secrétaire, | Deleglise.

Royal Intendent of the Province of Faucigny,
| Bonneville, 11 September 1858.

Sir,—

It truly pains me to learn of the difficulties that you are experiencing[2] from the Guide Chef[3] regarding the accomplishment of your perilous scientific endeavour, but I must also say with regret that these difficulties stem from regulations made to oversee the safety of travellers, regardless of the purpose of their trips.

Wishing nevertheless to be of service to you in this instance, I invite M. le Guide Chef today to consider your proposal, to make an exception to the regulation in his favour, as long as there is no danger to your safety and to that of those individuals in your company, and finally to be at your service within his means and responsibilities for the happy success of the expedition, the consequences and results of which do not only interest science but also the valley of Chamounix[4] in particular.

Please accept, Sir, the assurance of my highest regards. | For the Intendant[5] on leave, | The Secretary, | Deleglise.

Glaciers of the Alps, p. 172

---

1. *Indendant of Faucigny*: the writer signed as the secretary for the Intendant who was on holiday; Secretary Deleglise is not identified. Tyndall's description of his interaction with the president of the Commission of Guides is related in *Glaciers of the Alps,* pp. 171–72.
2. *difficulties that you are experiencing*: see letter 1551.
3. *Guide Chef*: see letter 1419, n. 17.
4. *Chamounix*: see letter 1353, n. 9.
5. *Intendant*: Félix Bergoin.

To Michael Faraday   11 September 1858   1557

Chamouni | 11 Sep 1858

My dear Mr. Faraday,

I see 'it is all out', for I have just read it.[1] There is nothing <u>very</u> stupid in it, and it gives me pleasure to think that you considered it sufficiently interesting to be made public. For the little bit at the commencement I thank you much. I was puzzled two or three days ago on opening a letter[2] from General Portlock,[3] to find the first words of it referring to my letter to you published in the Times. Beside his letter came your own explaining all. The only difference it makes is that instead of the single lecturing of Mrs. Faraday, I shall have half a dozen ladies solemnly admonishing me. But I can patiently bear any amount of lecturing from ladies, and so on this score I am not very much disturbed.

I came here soon after I wrote to Mrs. Faraday,[4] but the weather for a long time proved obstinately bad. Heavy rain in the valleys and heavy snow on the mountains. I came to look at the Mer de Glace[5] once more, and to see whether hints obtained upon other glaciers this year were illustrated upon it. The fresh snow, however, on the upper portions of the glacier disguises the structure of the ice, and renders observations difficult. It was also my intention to assist the eminent guide Auguste Balmat in placing some thermometers at the summit of Mont Blanc.[6] We have not a single observation to show either the minimum winter temperature, or the depths to which the cold of winter penetrates the ice at the summit. Saussure[7] has some <u>conjectures</u> upon this latter point, but we have no direct observations. The weather however opposed itself to this expedition—aided and abetted, I am sorry to say, by the <u>guide chef</u>[8] at Chamouni,[9] who attempted to impose on me, in all their rigour, the regulations which have been made for <u>tourists</u>. He would not permit me to have a boy to carry a little instrument up the Mer de Glace—I must take <u>a guide</u>. He also opposed himself to my concerted ascent with Balmat, and declared that I must conform to the rules and take <u>four</u> guides. I vainly endeavoured to shew him the difference between my position and that of a tourist; or to make him understand that the works of scientific men had done more for Chamouni than hundreds of ordinary travellers. To their credit, however, be it spoken, his superiors think a little differently upon the subject from the <u>Guide Chef</u>; The 'Intendent'[10] of the Province[11] has told him that in the case of a man of science he must interpret the laws widely and liberally, and must not attach to them a 'Judaical' signification. Having come to this conclusion that the bad weather offered an insuperable barrier to the ascent of Mont Blanc I went the day before yesterday with Balmat to the glacier du Taléfre,[12] and at a height of about 10,000 above the sea we sank a thermometer in the ice. We found it excessively hard and difficult to pierce. A second thermometer was placed beside a rock which forms the summit of the Jardin,[13] so as to give the minimum temperature of the <u>air</u>. Next year Balmat will ascend and read the result. I had never seen the wonderful circus of the Taléfre so wonderful, rendered so by the glorious weather, which has suddenly changed, and the heavy fresh snow which covered the surface of the glacier and rolled incessantly in avalanches from the surrounding mountains. During portions of our little expedition we had to plod through snow nearly three feet deep. (I had a most intelligent companion in Mr. Wills,[14] who as you know has written an interesting little book upon the Alps).[15] The weather at present is magnificent, but our thermometers are disposed of. Balmat posseses[16] one of his own, which, though not graduated low enough to give us the temperature of the air, might tell us something regarding the depth to which the winter cold penetrates the ice. Balmat himself conceives the idea of ascending Mont Blanc for the sole purpose of making his observation. I learned this last year, and made it known to the Council of the Royal Society, recommending the enterprise as one worthy of assistance and encouragement. The Council promptly voted me a small sum[17] out of the government grant;[18] but as for personal remuneration Balmat steadily refused it. He affirms that he was actuated by no hope of pecuniary reward when he conceived the idea of placing the thermometer at the summit, and that he will not now accept such recompense. It gives me pleasure to make known to you the spirit which actuates[19] this brave, gentle, and independent Chamouni guide, who never once shunned[20] fatigue or danger if a scientific object was to be gained by encountering it. He has ascended in winter through the snow to the Mer de Glace, and observed the motion of the boulders upon the glacier. These observations, which are recorded in the excellent papers of Professor Forbes,[21] are the most important, if not the only ones, that we possess, as to the influence of the seasons upon glacier motion.

I think we must help him to carry out his idea. The observation will not be a complete one, but it will teach us something, and others may be associated with it to repay the ascent. Thus matters stand at present, and a day or two will decide whether the mountain is to feel this year the shock of a crowbar upon his head.

Remember me most kindly to Mrs. Faraday and Miss Barnard[22]

And believe me always | Most sincerely Yours | John Tyndall

Would you have the goodness to have the enclosed[23] posted for me?

Like many other apparently 'impractical' things I think the climbing tendency of Englishmen might be turned to profitable account. There are many men of intelligence and culture among these mountain climbers who would be rejoiced to lend a hand in making scientific observations.

Water boils here at 194.6 Faht.

Chamouni, 11th, Sep, 1858.

RI MS JT/1/TYP/12/4087–90
Faraday Correspondence, 5:3514
Typed Transcript Only

1. *I see 'it is all out', for I have just read it*: two weeks previously Faraday had informed Tyndall that he had sent his letter (letter 1546) to *The Times* (published on 3 September 1858, p. 10). See letters 1548, 1549, and 1553.

2. *a letter*: letter missing.

3. *General Portlock*: see letter 1478, n. 17.

4. *wrote to Mrs. Faraday*: letter 1548.

5. *Mer de Glace*: see letter 1306, n. 12.

6. *Mont Blanc*: see letter 1320, n. 6.

7. *Saussure*: Horace-Bénédict de Saussure.

8. *guide chef*: see letter 1419, n. 17.

9. *Chamouni*: see letter 1353, n. 9.

10. *'Intendent'*: see letter 1556, n. 5.

11. *Province*: the Province of Faucigny at Bonneville.

12. *glacier du Taléfre*: see letter 1416, n. 6.

13. *Jardin*: a rocky spot in the middle of the Talèfre glacier frequently visited by tourists to the glacier, containing a profusion of plants at high altitude.

14. *Mr. Wills*: see letter 1554, n. 39.

15. *interesting little book upon the Alps*: A. Wills, *Wanderings among the High Alps* (London: Richard Bentley, 1856).

16. *possesses*: 'possess' in LT's typescript and in the *Faraday Correspondence.*

17. *The Council promptly voted me a small sum*: Tyndall described this action as well in *Glacier of the Alps*, p. 169.

18. *government grant*: see letter 1451, n. 2.

19. *actuates*: 'actuate' in LT's typescript and in the *Faraday Correspondence*.

20. *shunned*: 'shuned' in LT's typescript and in the *Faraday Correspondence*.

21. *excellent papers of Professor Forbes*: For example, J. Forbes, 'Sixteenth Letter on Glaciers,– (1) Observations on the Movement of the Mer de Glace down to 1850. (2) Observations by Balmat, in continuation of those detailed in the Fourteenth Letter. (3) On the gradual passage of Ice into the Fluid State. (4) Notice of an undescribed Pass of the Alps', *Edinburgh New Philosophical Journal*, 50 (1851), pp. 167–74.

22. *Miss Barnard*: Jane Barnard.

23. *the enclosed*: not identified.

To Auguste de la Rive [16 September][1] 1858 1558

My dear Sir

Your note[2] reached me last night at Chamouni:[3] and about an hour ago I reached Geneva.[4] It is late and I am loth to intrude upon you, otherwise I would call. I <u>must</u> start tomorrow morning very early for London, to attend an examination of military candidates.[5] Had I been aware of your being at home I might have arranged to come a day earlier and spent that day in Geneva. But this I am sorry to say is now impossible.

Ever Yours | John Tyndall | Hotel de la Couronne[6] | Thursday 7 oC. P.M.

RI MS JT/1/TYP/1/357/1
LT Typescript Only

1. *[16 September]*: the month and day are given by Tyndall's reference to needing to head toward London for a military examination on 21 September (see n. 5). The Thursday before that date was 16 September.

2. *Your note*: letter missing. See also letters 1531 and 1534.

3. *Chamouni*: see letter 1353, n. 9.

4. *Geneva*: see letter 1293, n. 3.

5. *examination of military candidates*: beginning on 21 September 1858 at the Royal Military Academy at Sandhurst, Tyndall conducted exams in the Experimental Sciences. See *Examination Papers used at the Examinations for Direct Commissions, in September, 1858.* (London: Harrison, 1858), pp. 13–16.

6. *Hotel de la Couronne*: a hotel in Geneva.

From Millicent Bence Jones[1]   24 September 1858   1559

Folkestone[2] | Sep. 24th. 1858

My dear Dr Tyndal[3]

Thank you very much for the pretty crochet-needle you sent me. I like it so very much. It will be very useful. Edy[4] and Baby[5] both like their things very much. Edy does worsted[6] work so it holds her needles and is very useful as well as pretty.

Papa[7] told us you had a great many adventures in the mountains; I hope you will tell us about them when we see you in London. It was very kind of you to think of us all so far off. Master Archie[8] would if he had any manners at all write you a letter of thanks but he has none as yet so mean time I send his thanks for him.

With very many more from | Your's affectionately and very much obliged | Millicent M. Bence Jones.

RI MS JT/1/J/119
RI MS JT/1/TYP/3/748

---

1. *Millicent Bence Jones*: Millicent Mary Bence Jones (1843–1933), eldest child of Henry and Millicent Bence Jones.
2. *Folkestone*: see letter 1447, n. 3.
3. *Tyndal*: misspelled in the original letter.
4. *Edy*: Edith Mary Bence Jones (1853–1919), daughter of Henry Bence Jones and his wife Millicent (née Acheson, 1817–87). She would marry Edward John Stapleton. See 'Deaths', *Times,* 23 July 1919, p. 1.
5. *Baby*: Archibald Bence Jones (1856–1937), son of Henry and Millicent Bence Jones. See 'Deaths', *Times,* 25 February 1937, p.1.
6. *worsted work*: embroidery done with worsted yarn on canvas (*OED*).
7. *Papa*: Henry Bence Jones.
8. *Master Archie*: see n. 5.

From Juliet Pollock   24 September [1858][1]   1560

Greta Bank | September 24.

My dear Mr Tyndall,

I am amazed at your retort[2] courteous and admit that your rebuke may be just but I grieve to say that Walter's[3] letters continue to be desponding.

Our baby[4] is recovering and we are able to enjoy the agreeable society of the family here at present without a drawback.

We have had such a variety of weather as exhibits the mountains under all their different aspects, and yesterday they were very grand under a storm clearing off. I wish I had a portion of your climbing powers but my ambition is greater than my strength and though for a woman I can accomplish a fair extent of walking I cannot achieve any thing brilliant in that way.

The comet[5] has during the last two nights been concealed in stormy skies but I have some hopes of seeing him to night and I think I shall die the happier for having seen a party of that description superior to the famous one of 1811[6] hitherto held over me triumphantly by my elders. Is that a sentiment worthy of a philosopher's friend?

I suppose you and Anderson are laboring together now at the R.I. in the accustomed manner, and that you have seen Mr Faraday and our friends the Moores.[7]

If you are so kind as to think of writing to me again my present address and during the next week is Greta Bank | Keswick, Cumberland

Fred's[8] best regards | Yours ever most truly | Juliet Pollock.

RI MS JT/1/TYP/6/2050
LT Typescript Only

---

1.   *[1858]*: the year is given by the reference to Donati's comet. See n. 5.

2.   *your retort*: letter missing.

3.   *Walter's letters continue to be desponding*: Walter Pollock, second son of Juliet and William Frederick Pollock, was in school in Brighton (see letter 1545).

4.   *Our baby*: Maurice Pollock.

5.   *comet*: Donati's comet, a celestial spectacle of the nineteenth century, was first observed by Italian astronomer Giovanni Battista Donati (1826–73) on 2 June 1858. The comet was visible to observers in both hemispheres between September 1858 and March 1859. See A. Gasperini, D. Galli, and L. Nenzi, 'The worldwide impact of Donati's comet on art and society in the mid-19th century', in D. Valls-Gabaud and A. Boksenberg (eds), *The Role of Astronomy in Society and Culture: Proceedings IAU Symposium No. 260* (Cambridge:

Cambridge University Press, 2011), pp. 340–45. The painting of a coastal scene in south-eastern England by William Dyce (1806–64), *Pegwell Bay, Kent—a Recollection of October 5th 1858*, depicts the comet in daytime above people exploring the beach at low tide.

6. *famous one of 1811*: the Great Comet of 1811, visible for 260 days, was discovered by French astronomers Honoré Flaugergues (1755–1835) on 25 March 1811 and Jean-Louis Pons (1761–1831) on 11 April 1858. This comet held the record for longest period of visibility until the discovery of Hale-Bopp in 1997.

7. *the Moores*: James and Harriet Moore and their family Harriet Jane, Julia, John, and Graham.

8. *Fred's*: William Frederick Pollock.

From Elizabeth Dawson Steuart[1]   27 September 1858   1561

Sept. 27. 58

My dear John,

It had indeed appeared long to me since I had heard from you, until the receipt of your last letter,[2] but I <u>know</u> how much your valuable time is occupied in scientific pursuits, and feel that I should not expect it to be devoted to me, warm as my interest always is in your welfare and happiness. The day before your note reached me, I received a copy of 'The Times', wherein a letter of yours was published.[3] My Brother William,[4] knowing the gratification I should feel on reading it, sent me the paper: You have indeed encountered perils and dangers, and been mercifully protected through all. The Almighty arm of Omnipotence guided your steps, and gave you success in your undertakings, for without this aid you must have perished. My Brother Joseph[5] is gone on a tour to the Channel Islands[6] and some parts of France he has not yet visited—such as Avranches,[7] Caen[8] &c. He is sure to end with London, and no doubt will call upon you while there, and I hope may be fortunate enough to find you at home. This place is so much changed, I do not think it now contains any you would care to hear of: All the former kind people are gone, and certainly have not been replaced by <u>better</u>: it is miserable to see such a thin congregation in the Church, which used to be so well and respectably filled in days of yore. But so it is in this ever varying world, where nothing stands still. I suppose you have found the Comet[9] an object of great interest: I have had fine views of it from my bed-room window about 4 o'clock in the morning.

Believe me with kindest wishes, | ever most sincerely | your friend | E. D. Steuart

RI MS JT/1/TYP/10/3340
LT Typescript Only

1. *Elizabeth Dawson Steuart*: see letter 1555, n. 1.

2. *your last letter*: letter 1555.

3. *a copy of 'The Times', wherein a letter of yours was published*: see letter 1548.

4. *My Brother William*: William Duckett (1796–1868). See B. Burke, *A Genealogical and Heraldic History of the Landed Gentry of Great Britain & Ireland,* 8th edn, 2 vols (London: Harrison and Sons, 1894), vol. 1, p. 546.

5. *My Brother Joseph*: Joseph Fade Duckett (1796–1875). See Burke, *A Genealogical and Heraldic History of the Landed Gentry of Great Britain & Ireland,* vol. 1, p. 546.

6. *Channel Islands*: a group of islands in the English Channel, off the French coast of Normandy.

7. *Avranches*: a town in the Normandy region of northwestern France.

8. *Caen*: another town in the Normandy region of northwestern France.

9. *Comet*: Donati's comet. See letter 1560, n. 5.

To Heinrich Gustav Magnus   [late Sept or early Oct 1858][1] 1562

My dear Friend

When we parted company at Chamouni[2] I hoped to have the pleasure of meeting you again at Carlsruhe,[3] but on counting up my time I found this impossible—In fact I was compelled to be in London on the 20th of September.[4] Prof. Delarive wrote to me[5] saying that you had been at Geneva,[6] and asking me to halt there: but his note arrived too late. When I go to Berlin I shall undoubtedly scold Miss Magnus[7] for returning the little telescope—it would have been a great pleasure to me if she had kept it in memory of my achievement. Well you saw us at the Grand Mulets[8]—next morning at 1½ oC. we were on the glacier, and no moon being in heaven we were compelled to carry lanterns. From the Grand Mulets I first saw the Comet,[9] having heard nothing about it previously. The wind had been high during the night, and though it afterward lulled, it rushed violently down upon us at intervals during our ascent; Auguste Balmat, who led us, once or twice expressed his fears that we should not reach the top. We advanced however, and finally found ourselves upon the calotte, or last slope of the mountain: here a thick cloud enveloped us and snow commenced to fall. We reached the summit[10] where I attempted to make some observations, but I was compelled to relinquish them fearing the man who accompanied me might be frost bitten. We sank a thermometer at the top, and in digging the hole for it Balmat used his hand incautiously— He lost sensation, and for a long time he despaired of recovering it. We stood up on the mountain, beating his hands and rubbing them with snow & hands for a long time.[11]

RI MS JT/1/T/1061

1. *[late September or early October 1858]*: the date range is given based on the references to Tyndall's recent travels in the Alps. This letter is at least after 22 September and probably after Tyndall returned from the BAAS meeting in Leeds. In letter 1558 (16 September), he mentioned needing to leave Paris toward London because he had to be at the Royal Military Academy at Sandhurst on 21 September for a military examination. Presumably he left London on 20 September (to already be in Sandhurst the next morning), and after his examination duties were complete, then travelled to Leeds to attend part of the twenty-eighth meeting of the BAAS from 22–29 September 1858 (see letter 1566, n. 17 and n. 18).

2. *Chamouni*: see letter 1353, n. 9.

3. *Carlsruhe*: a city in southwestern Germany, now Karlsruhe.

4. *be in London on the 20ᵗʰ of September*: see n. 1.

5. *wrote to me*: letter missing.

6. *Geneva*: see letter 1293, n. 3.

7. *Miss Magnus*: either Anna, or Christine Magnus, one of Gustav Magnus's two daughters.

8. *Well you saw us at the Grand Mulets*: on 11 September 1858, see letter 1419, n. 9.

9. *Comet*: Donati's comet. See letter 1558, n. 5. In *Glacier of the Alps*, Tyndall described first seeing the comet on 11 September 1858: 'Immediately before lying down on the previous evening I had opened the little window of the cabin [on the Grand Mulets] to admit some air. In the sky in front of me shone a curious nodule of misty light with a pale train attached to it' (p. 164).

10. *the summit*: of Mont Blanc (see letter 1320, n. 6), reached on 13 September 1858 after resting a night at the cabin near the Grand Mulets.

11. *for a long time*: if this is the whole of the letter, then Tyndall did not close it with an ending salutation. It could possibly also be a letter which he started writing and never finished nor sent.

To John Murray   [September]¹ 1858   1563

My dear Mʳ. Murray

I return you my best thanks for the book,² which I shall no doubt read with great interest—I have already dipped a little into it & have had occasion to admire the beauty and force of the author's³ style.

With regard to Monte Rosa⁴ it shall all be laid before you in due time. I am preparing a work, the title of which is to be 'The Glaciers,'⁵ and in which I purpose giving a clear exposition off all their phenomena. The book will also contain an account of the ascents that I have made and also a description & explanation of various interesting physical phenomena. Such descriptions will form breathing places—little islets thrown in amid the scientific flow of the work on which the general reader may find rest & refreshment. I am persuaded that such a work will be well received, at all events it is much needed, as no existing work at all comes up to our wants upon the subject.

I have no difficulty in finding a publisher for this book, but the publisher I should like would be yourself—For once in your hand I need not trouble myself further about it, feeling assured that you would do all things for the best.

Ever yours | <u>J. Tyndall</u>
Tyndall J | 1858.[6]

NLS

---

1. *[September]*: the month is given by relation to letter 1530. This letter appears to have been written after Tyndall had returned from Switzerland to London in September 1858. Murray's reply to letter 1530 is missing. Tyndall's account of his solo ascent of Monte Rosa was published in *The Times;* see letter 1548.
2. *the book*: not identified.
3. *the author's*: not identified.
4. *Monte Rosa*: see letter 1478, n. 22. Tyndall's solo ascent of the Monte Rosa occurred on 17 August 1858. See *Glaciers of the Alps,* pp. 151–60, and n. 1 above.
5. *preparing a work, the title of which is to be 'The Glaciers'*: *Glaciers of the Alps.*
6. *Tyndall J | 1858.*: written at the end of the letter, presumably by Murray.

To George Gabriel Stokes   9 October 1858   1564

<u>9th Oct.</u> 1858

My dear Sir

I have read Sir. Cha. Lyell's paper,[1] and recommend it for publication in the Philosophical Transactions.[2] Its style is clear, its facts are valuable, and its descriptions often highly interesting.

Judged of with strict reference to its title, the paper would, I think, bear consideration. The facts & arguments bearing upon the principal point, which is a physical one would, if drawn more closely together, be more impressive and convincing.

I am not sufficiently acquainted with the literature of geology to know the grounds of the assumption that continuous sheets of lava cannot be formed on slopes exceeding two or three degrees of inclination. But whatever these grounds may be I have no hesitation in holding with Sir Chas. Lyell, that they are untenable.

I may be perhaps allowed to remark that the observation of the <u>changes</u>

of inclination, which modify the pressures and tensions of a lava stream in a known manner, are likely to be as instructive as the observation of the absolute slope of any part of the stream.

A summary of the results of the paper, stating what is new, and what corroboration would, I imagine, materially assist the reader in remembering what it contains.

Believe me dear Sir | Very truly yours | <u>John Tyndall</u>
Prof. Stokes | <u>Sec. R.S.</u>

RS RR/3/185 Tyndall 9 October 1858

1.	*Sir. Cha. Lyell's paper*: C. Lyell, 'On the Structure of Lavas Which Have Consolidated on Steep Slopes; With Remarks on the Mode of Origin of Mount Etna, and on the Theory of "Craters of Elevation"', *Phil. Trans.*, 148 (1858), pp. 703–86; 904–6 (maps and figure).

2.	*and recommend it for publication in the Philosophical Transactions*: Tyndall acted as a referee for the *Phil. Trans.*, reviewing submissions and sending reports to Stokes, the journal's editor.

## From Thomas Archer Hirst				10 October 1858				1565

address | Signor Battelle | 113 Via Felice
| Rome | <u>via Marseilles</u> | Oct^r <u>10th</u> 1858

My dear John

I arrived at Rome about ten days ago[1] thus concluding as interesting a tour as I ever made in my life. Besides natural beauties, which Italy shares with other countries without surpassing them, Italy as every body knows is peculiarly distinguished by its richness in historical associations and for its fine art collections. As to the latter I got my own peculiar pleasure and instruction from them. I did not look upon them with an artists eye I did not study their execution or details, but I looked upon them as I would upon a poem solely to appropriate as far as possible the conception. I have no doubt that in my journal I have pronounced to be beautiful what an artist would consider inferior, and vice versa. But in picture galleries as well as in the streets and country I was continually made aware that I was extracting a very small portion only of the interest latent in every object around me. I was made to regret daily my want of familiarity with classical history, for here more than in any other country every Town, River Hill and valley has a representative as well as an intrinsic interest. I resolved to remedy this grave <u>defect</u> in my education as much as my leisure time in Italy would permit. For this purpose I have devoted myself almost exclusively to the first centuries of the Christian Era during the time

I have been in Rome I have procured a concise well written history of Rome[2] which I read in the Evening whilst in the day time I stroll through the scenes of the actions described and surround myself with the likenesses in Marble of the actors themselves. In this manner Rome possesses an interest for me greater than I anticipated, and the most wonderful of all Histories becomes illustrated and in a certain sense realized.

After leaving you in that sunny valley of Saas[3] I strode away and reached Mattmark[4] at about the same time as you would reach Visp.[5] Next morning we started at four O clock and were on the Summit of the Pass before seven. The air was clear and fresh Monte Rosa[6] was close to us and visible from top to base. Its massive form and steep rugged rocky slopes contrasting with the whitest purest snow formed as magnificent a sight as I have ever seen. On the other side the whole panoramic of the Oberland[7] was revealed to us, the Jungfrau[8] in the background. We reached Macugnaga[9] at 9 A.M. and after a hearty breakfast I set off as fresh as ever down the lovely Italian valley of Anzasca[10] and reached Vogogna[11] on the Simplon Road[12] by night fall. There was a good days work! All this time my heavy luggage had been carried by porters at a great expence to me. I saw clearly enough that I must use diligences[13] and railways for the rest of my journey—in fact the only fault I can find with the tour just completed is that it has furnished me with too few opportunities for physical exercise. Nothing agrees worse with my constitution than coaches, Railways and Steamers. I spent a couple of days on the lakes of Maggiore and Como,[14] and in the town of Como[15] I did not fail to visit with a students veneration, the statue of the great Volta.[16] He stands there a noble figure at the head of the loveliest lake, his left hand resting on the memorable 'pile'.[17] I spent three or four days in Milan,[18] where I made the acquaintance of one of the ablest and most promising of Italian mathematicians Brioschi.[19] I confess I was a little pleased to find myself at once known & welcomed in that strange land on presenting my card. My next halting place, for a day only, was Verona the City of the Capulets and Montagues.[20] It was out of respect to Shakespear[21] that I visited the quaint old town and I <u>tried</u> my best to believe that the washing trough they shewed me had in reality contained Juliets fair form.[22] Venice[23] more than fulfilled my expectations which had been accumulating since childhood. Amongst all cities I have seen this one, with canals for streets, destitute of wheeled carriages and quadrupeds larger than dogs, is unique. It was like realizing a page of the Arabian nights[24] to find myself on my arrival at night put into a black gondola and made to glide noiselessly and swiftly along the dark canal on whose banks I could just discern the marble palaces of the Doges.[25] With wonderful grace and precision the gondolier with his single oar caused the little bark[26] to glide along dark narrow canals where there appeared to be no room for his oar. At length we landed and following him along narrow passages darker even than the canal we had left, he ushered me suddenly into the magnificent square of S$^{\underline{t}}$ Mark's,[27] brilliantly lighted and full of gaily dressed people promenading to the music of a fine band. I was at first bewildered and it was only by degrees that the beauty of the scene revealed itself. The magnificent square with marble pavement, as large as the Palais Royal[28] and infinitely more picturesque, the rich oriental looking church of S$^{\underline{t}}$ Marks[29] at one end, the graceful towering Campanile,[30] and above all the black blue sky studded with brilliant stars formed a concert room such as I had never seen before. From Venice I returned to Padua[31] and there struck off southwards. Between Padua and Ferrara[32] I crossed the Po[33] which forms the division between Austrian and Papal territory. I shall not soon forget it for having neglected to get a visé[34] from his ambassador his sacred majesty the Pope[35] caused me to be arrested, to descend from the Diligence forfeit my fare, and remain his prisoner for 24 hours at a dirty village on the River's bank. After undergoing this chastisement and <u>paying sufficiently</u>, a permission to enter his Holiness' dominions was granted to me. At Bologna[36] I only stopped half a day but I saw Galvani's[37] house on which there is this inscription (in Italian of course) | 'It was in this house; his temporary residence, that Louis Galvani on the 1$^{\underline{st}}$ of September 1786 discovered in dead frogs animal electricity—the source of wonder to all centuries.'

I passed over the Appenines,[38] I am sorry to say, cooped up in the interior of a diligence the rain falling in torrents most of the time and clouds hiding every vestige of scenery. Next Florence[39] where I next stopped I was greatly pleased, there and in its lovely neighbourhood I spent more than a week. I will not attempt to extract from my journal any of the numerous accounts of the effects its pictures and statues had upon me suffice it to say that I seriously studied many of them & not without profit. Often at moonlight I have gazed at the beautiful cathedral[40] standing on the spot, now called 'Sasso di Danti',[41] where habitually the Divine Poet[42] used to sit and contemplate. Guide books even go so far as to say that the architectural beauties of the cathedral formed the subject of his contemplations it is more probable, however, that he found its vicinity appropriate to thoughts that roamed far beyond the walls of his native town into Hell itself.[43] Galileo[44] too lived here, the Grand Duke of Tuscany has erected a beautiful little tribune[45]—a kind of Chapel—to him. Inside there is a worthy statue of the fine old fellow around are preserved religiously many relics of him compasses, telescopes lenses &c &c and above all the <u>forefinger of his right hand</u> The flesh has long since departed but the skeleton and even the nail is still there.

From Venice I went to Pisa[46] where I remained another day. I expected to meet there with more reminiscences of Galileo than I found I booked my full, however, at the leaning tower[47] and a curious object it is: The italian leaning towers, and there are many, embody of course some thought as do all other artificial works. I should like to know more of the thought or motive which lies at the bottom of these towers built intentionally non-vertical. This tower at Pisa gives a tumble-down look to everything around it, the Cathedral and Babtistery[48] close to it appear to be leaning also. In this cathedral which is a very fine one I sat for a long time watching the oscillations of the lamps suspended from the ceiling The same lamps were swinging to and fro on that memorable morning when Galileo came to pray and when one of them, the guides pretend to know which one, revealed to the young philosopher the secret of its isochronism.[49] I embarked at Leghorn[50] for Civita Vecchia[51] and thus made my first voyage on the classic, tideless sea[52] about which I have dreamed as a boy. I should be very glad to be able to enter into its beauties but to tell the truth I saw little of them for I was not quite well.

I have now brought this skeleton of my tour up to my arrival in Rome what I have been doing since I have already told you. I know well than when once my Equally Attracting Bodies[53] have laid hold of me there will be small chance more for Roman History and Antiquities so I am making the most of the last few days. I have got pleasant lodgings and I have furnished them with a Piano for I anticipate being thrown on my own resources for relaxation more here than I was in Paris. The comet[54] has been and still is a magnificent object here, every night I watch it from the balcony before my window which is on the third story. A few nights ago whilst gazing at it I saw another sight entirely new to me, I do not know what value others more accustomed to search the heavens may attach to my observation but at any rate I will describe it on a separate sheet[55] for you to make use of as you think fit.[56] Write to me soon and tell me what other hazardous things you have done since we parted. Did you go up M. Blanc?[57]

The story you told me in Saas as you might expect interested me deeply and had formed the subject of many a long meditation since; you promised me to keep me informed with the sequel whatever might be its nature[58]

Yours affectionately | <u>Tom</u>

RI MS JT/1/H/242

---

1. *I arrived at Rome about ten days ago*: from his journal, it appears that Hirst arrived in Rome on 27 September 1858. See Brock & MacLeod, *Hirst Journals,* p. 1401.

2. *a concise well written history of Rome*: not identified. Hirst mentioned this book in his journal as well (for 29 September 1858), while also carrying with him John Murray's *A Handbook for Travellers in Central Italy,* 4th edn (London, 1857). See Brock & MacLeod, *Hirst Journals,* p. 1402.

3. *valley of Saas*: see letter 1547, n. 4.

4. *Mattmark*: see letter 1547, n. 3.

5. *Visp*: see letter 1543, n. 8.

6. *Monte Rosa*: see letter 1478, n. 22.

7. *Oberland*: see letter 1527, n. 13.

8. *Jungfrau*: see letter 1536, n. 20.

9. *Macugnaga*: a village in northern Italy, located at the eastern base of Monte Rosa.

10. *Anzasca*: a valley in northern Italy, of which Macugnaga is at its western end at the eastern base of Monte Rosa.

11. *Vogogna*: a village in northern Italy, east of Macugnaga toward the eastern end of the valley of Anzasca.

12. *Simplon Road*: a mountain road from northern Italy into the Rhone valley of Switzerland, passing through Vogogna.

13. *diligences*: a public stage-coach (*OED*).

14. *lakes of Maggiore and Como*: thin lakes on the southern side of the Alps in northern Italy.

15. *Como*: a city in northern Italy, at the southern-western most point of lake Como.

16. *Volta*: Alessandro Giuseppe Antonio Anastasio Volta (1784–1827), an Italian physicist and chemist credited with the invention of the electrical battery (in 1799). Volta retired to Como in 1819, and is buried there on his estate (*CDSB*).

17. *'pile'*: a voltaic pile, Volta's electrical battery which consisted of stacked copper and zinc discs.

18. *Milan*: a large city in northern Italy.

19. *Brioschi*: Francesco Brioschi (1824–97), an Italian mathematician (*CDSB*).

20. *Capulets and Montagues*: a reference to the dueling families of Shakespeare's tragedy *Romeo and Juliet*.

21. *Shakespear*: William Shakespeare (1564–1616), the English poet and playwright (*ODNB*).

22. *Juliets fair form*: Juliet of the Capulet family.

23. *Venice*: see letter 1293, n. 1.

24. *Arabian nights*: *One Thousand and One Nights,* a collection of Middle Eastern and southern Asian stories put together in between the eighth and thirteenth centuries. Its first English translation (1706) was titled *Arabian Nights*.

25. *Doges*: Doge's Palace, the residence of the Doge of Venice (or chief magistrate) from 697–1797, next to Piazza San Marco.

26. *bark*: a rowing boat (*OED*).

27. *square of S͟t Mark's*: Piazza San Marco, a public square in Venice.

28. *Palais Royal*: a palace in Paris.

29. *church of S͟t Marks*: Saint Mark's Basilica, a Catholic church in Venice, adjacent to Piazza San Marco.

30. *Campanile*: a bell-tower (*OED*).

31. *Padua*: a city in northern Italy just west of Venice.

32. *Ferrara*: a city in northern Italy to the south of Padua.

33. *Po*: the Po River, Italy's longest, which flows from east to west through its northern region.

34. *visé*: an entry or note on a passport, certificate, or other official document signifying that it has been examined and found correct; a formal official signature or entry of this nature (*OED*).

35. *Pope*: Giovanni Maria Mastai-Ferretti (1792–1878), Pope Pius IX from 1846–78. See F. Gallo (ed.), *Dizionario Biografico Degli Italiani* (Rome: Istituto della Enciclopedia Italiana, 2015), vol. 84, pp. 29–40.

36. *Bologna*: a large city in northern Italy, south of Ferrara.

37. *Galvani's*: Luigi Galvani (1737–98), an Italian physicist and biologist. He discovered animal electricity when he found that a frog's legs reacted to an electric current when connected to muscles nerves (*CDSB*). The term *galvanism*—electricity developed by chemical action (*OED*)—is named after him.

38. *Appenines*: the northern chain of a range of mountains extending along the peninsular part of Italy.

39. *Florence*: see letter 1293, n. 4.

40. *cathedral*: Cathedral of Saint Mary of the Flower in Florence.

41. *'Sasso di Danti'*: Dante's rock, or stone, in the Piazza del Duomo where apparently Dante sat and found inspiration for writing. In the nineteenth century, English travellers on a Grand Tour would seek out and sit in the same spot.

42. *Divine Poet*: a reference to Dante.

43. *into Hell itself*: a reference to Dante's *Inferno* (see letter 1322, n. 3).

44. *Galileo*: Galileo Galilei (1564–1642), an Italian astronomer and natural philosopher known for his support of heliocentrism (*CDSB*).

45. *Grand Duke of Tuscany has erected a beautiful little tribune*: Leopold II (1797–1870), Grand Duke of Tuscany from 1824–59, ordered the erection of the Tribune of Galileo in La Specola, the observatory and natural history museum in Florence, completed in 1841. See M. Caravale (ed.), *Dizionario Biografico Degli Italiani* (Rome: Istituto della Enciclopedia Italiana, 2005), vol. 64, p. 664–67.

46. *Pisa*: see letter 1300, n. 5.

47. *leaning tower*: the famed Leaning Tower of Pisa was the bell tower of Pisa's cathedral; its soft foundation caused leaning to occur before completion of its construction in 1319.

48. *Cathedral and Babtistery*: the Pisa Cathedral and the Pisa Baptistery of St. John, together in the Piazza del Duomo (now Piazza dei Miracoli).

49. *isochronism*: the character or property of oscillating or taking place in equal spaces of time (*OED*).

50. *Leghorn*: the English name for Livorno, a port city in central Italy to the south of Pisa.

51. *Civita Vecchia*: Civitavecchia, a port town in central Italy to the south of Livorno and to the northwest of Rome.

52. *tideless sea*: the Tyrrhenian Sea, part of the Mediterranean Sea, which has limited tides because of the narrow connection it has with the Atlantic Ocean (the Strait of Gibraltar).

53. *my Equally Attracting Bodies*: see letter 1319, n. 8.

54. *comet*: Donati's comet. See letter 1558, n. 5.

55. *I will describe it on a separate sheet*: this sheet is missing, but Tyndall referred to it in his reply (letter 1566).
56. *A few nights ago whilst gazing . . . as you think fit*: these lines are crossed out but were still read by Tyndall as he replied to this in letter 1566.
57. *M. Blanc*: Mont Blanc, see letter 1320, n. 6. Tyndall summited the mountain on 13 September 1858 in order to place thermometers there with his guide Auguste Balmat.
58. *The story you told me . . . might be its nature*: these lines are crossed out but were still read by Tyndall as he replied to this in letter 1566.

To Thomas Archer Hirst   19 October 1858   1566

Queenwood 19ᵗʰ Oct. 1858.

My dear Tom

I received your long and truly interesting letter this day[1] and have read it with great delight. The picture of your journey is so beautiful that I am half tempted to send it off to those dear friends of mine whom I mentioned to you at Saas,[2] and to whom you refer in the last paragraph of your letter:[3] for well I know it would interest them. I hope you will give yourself a good holiday before you commence your 'equally attracting bodies.'[4] I mentioned the title of your investigation to Miss Edmondson[5] today, and explained it by stating that two young ladies were violently in love with you, but their attractions were so equal that you did not know how to choose between them. Your meteor account[6] was so interesting that I transcribed it and sent it this evening to the Times: they may print it or they may not—if they dont I will send it to the Philosophical Magazine.[7] Again I say give yourself a good holiday and carve a good picture of the 'Eternal City'[8] upon your brain before you commence your mathematics.

Now I do not intend to refer to all the points in your letter which gave me delight, but I will just repay you by giving you a sketch of what has occurred to myself since we parted.

I went to Chamounix[9] but the weather was very bad, and it was aided by a stupid guide Chef[10] who affirmed that I must conform to the ordinary regulations. I addressed the Syndic of the Commune,[11] the Intendent of the province[12] & secured their sympathies. Armed thus I was led to the top of Mont Blanc[13] by Auguste Balmat. We were driven down by a tourmente,[14] and Balmat very nearly lost his hands. He was prosecuted afterwards by the Commission of guides,[15] and I dont know the result.[16] But I went down to Leeds and brought the case before the British Association,[17] where a resolution was passed[18] that application shᵈ. be made to the Sardinian authorities for increased facilities in making observations upon the summits of the Alps.

I doubt not that before next year all those fetters will be broken and it will be no longer in the power of an ignorant guide chef to thwart a man of science.

While at Leeds I was the guest of Mr Beckett of Kirkstall Grange.[19] Lord Carlisle[20] was there at the time and some other great people. One evening at Mrs Beckett's[21] request I gave them a simple account of my ascent of Monte Rosa.[22] Lord Carlisle came to me afterwards apparently much interested, and declared that we must always be friends. He said more pleasant things to me than I have room or time to repeat.

You know I intended to throw the Alpine observations into the form of a book.[23] To promote this idea I came down to Queenwood.[24] I have got on but slowly, still I have accomplished something & the book when finished will, I hope, have many readers.

But now you ask me to let you know something of that matter regarding which I spoke to you at Saas—well here goes to gratify you.

M^rs. D.[25] writes a beautiful letter[26]—She has a great love for the country and describes it with the warmth which that love prompts. Well I had one of her letters at Chamounix,[27] dwelling upon the beauties of the little paradise which they occupy in Surrey.[28] 'How I should should[29] like to shew it all to you' She said 'Will you not come to see us when you return from the Alps?' I could not say 'no' so I promised, and <u>went</u>.

One evening I reached a beautiful little country hotel near to the place where they live. I wrote a note[30] saying that I had arrived there and would join them next morning at breakfast. My messenger brought a reply[31]—'Come here this minute you shall not sleep in any house but mine'. I waited to finish a little sherry, took my bag in my hand and went towards their house. The evening was raw & foggy: on the road I thought I saw some ladies figures and soon heard a pleasant voice crying '<u>well</u>!' They had all come out to meet me & surely no man could have received a warmer welcome. I stayed with them 3 day: the forenoon of each day was spent rambling through the woods and along the downs of the most beautiful country that my eyes ever rested on. We usually dined at 2 oC. and the carriage was invoked in the afternoon to take us to some distant scene of loveliness. We returned to tea at 7 oC. and after tea had another excursion. There is a dense wood of box[32] at the summit of a neighbouring hill, and thither we climbed by moonlight and wandered along its dark corridors listening to the owl hooting forth its rich melancholy notes. We had our little frights and surprises, for I was the only man in the company and I sometimes disappeared mysteriously leaving my fair friends alone. The moon whitened the boles[33] of the trees, and the light was sometimes sent in silver spangles from the glassy surface of the box leaves. Two of the girls who are tall and straight were provided with black mantles having hoods: these they drew over their heads and walked silently on in front of us; exactly like meditating monks in the gloom of a cathedral. In the open moonlight spaces they sometimes turned round and let the light shine upon their white dresses. —the whole thing was more like a romance than a reality. In short I never spent three such days, if the value of days be measured by social enjoyment. Before I came away I had engaged to eat my Christmas dinner with them: this is taking time by the forelock with a vengeance.

We have written to each other since I came here that is M^rs. D & I. I left my axe behind me with them when I came away, and a few days ago I had a note[34] informing me that the said axe was reposing in dignified idleness in the chimney corner of their drawing room, waiting for its master to come and take it away. I was reminded of the cosiness of the said corner and requested to come before the conclusion of the month, when they depart for Torquay.[35] Next Friday I go to Abbotts Ann[36] to spend a couple of days with Mr Best[37]— on Monday the 25th I hope to reach my friends: I purpose staying with them until the 28th and then finally return to London.

Thus stands the matter my dear Tom. I thought it all past, but I sometimes think it is not over yet. If I could find any flaw in these people it would perhaps be well for me; but they are altogether beautiful. These girls[38] have been brought up in opulence and among fashionable people, still there is not a trace of artificiality among them. A noble truthfulness of demeanour, a direct reliance upon high natural qualities, and a lovingness of disposition which warms and beautifies all, are the characteristics which I have ever observed among them. I sometimes ask myself, is it manly to permit them to take such a hold of me? would it not be manlier to say sternly I will see them no more, and get on with my work? but then my reason tells me that these people are admirable. It is not blind feeling which produces this conviction but the open eye of my intellect and this 'manly' part may therefore be the part of a fool. I know that others like them as much as I do, that many a fine fellow has laid seige[39] to that fair citadel in vain. When I think of this and of my own position, the whole thing appears so wild to me, that I shake it suddenly from my thoughts as an utter absurdity. Still I doubt if ever she was kinder to any man than she has been, & continues to be, to me. Nay if I dare trust my instincts in the matter I should dare to hope. But the odds against me are too patent to be overlooked.

Again I sometimes reflect on her mother's knowledge of the world, and of the workings of the human heart. I think of her practical wisdom and of her truth of character, and I ask myself what am I to conclude from the liberties which she allows me? Do I commit the mistake of an ignoble nature and put a false interpretation upon her great kindness? Still she knows what I have felt towards her child.[40] She knows quite well that I have gone through a fiery struggle to obtain any freedom; and still, with this knowledge before her eyes she puts me again in peril. I know she likes me; I believe I am not presumptuous in thinking this, for the proofs of her affection to me have been numberless. but is it either kind or wise to throw me into such danger? Perhaps she deems me old enough to protect myself—and I hope I am so, even should this dream go to pieces.

Sometimes I doubt not the counsel of a friend like you Tom would be worth something to me; for I fear my own pride in this matter; I mean that pride which has clung to me through life, and which causes my soul to revolt from any compromise of independence. This girl has been brought up in opulence: Even if I had never heard it mentioned, the manner in which they live is an evidence of their wealth. I confess this is one of the grave thoughts that beset me. I would not drag her down, I am unable to maintain her at her present level, and the thought of her dragging me <u>up</u> is so terribly repugnant to me that even supposing what is far too daring to suppose, I could not bear it. Thus on all sides this question is surrounded with difficulties. But I will not seek them. I will resign myself calmly to the guidance of that spirit which has hitherto shaped my almost miraculous career, and wait for the solutions which time must bring. God bless you Tom

always yours affectionately | <u>John</u>.

While writing this I have had a note from my friend[41]—kind, good and affectionate as ever.[42] Atkinson[43] is here: he will write to you in a few days. This is a terribly long letter all about myself, still I do not think it will weary you.

RI MS JT/1/T/647
RI MS JT/1/HTYP/515–16

---

1. *I received your long and truly interesting letter this day*: letter 1565. Tyndall would have received Hirst's letter of 10 October several days prior, suggesting that this letter probably took several days to write and completed on 19 October.
2. *Saas*: see letter 1537, n. 33, and at the end of letter 1565.
3. *to whom you refer in the last paragraph of your letter*: not identified, but presumably someone within the 'story you told me in Saas' (see end of letter 1565).
4. *'equally attracting bodies.'*: see letter 1319, n. 8.
5. *Miss Edmondson*: Jane Edmondson (1823–1906), daughter of Anne (née Singleton) and George Edmondson. She married Davis Benson in 1859 and wrote about her family's experience in Russia in *Quaker Pioneers in Russia* (London, Headley Brothers, 1902), and *From the Lune to the Neva Sixty Years Ago* (London: Samuel Harris & Co., 1879).
6. *meteor account*: see letter 1565, where Hirst gives an account of observing Donati's comet and another celestial 'sight'.
7. *sent it this evening to the Times . . . I will send it to the Philosophical Magazine*: it does not appear that it was published in either the *Times* or the *Phil. Mag.*
8. *'Eternal City'*: Rome.

9. *Chamounix*: see letter 1353, n. 9.

10. *stupid guide Chef*: see letter 1419, n. 17.

11. *Syndic of the Commune*: Michel Favret (1796–1870), Syndic of the Commune of Chamonix in 1858, previously Guide Chef, and Mayor of Chamonix in 1861 (Journal, RI MS JT/2/13c/1150).

12. *Intendent of the province*: see letter 1556, n. 5.

13. *Mont Blanc*: see letter 1320, n. 6.

14. *tourmente*: a whirling storm or eddy of snow (*OED*).

15. *Commission of guides*: a body of persons that decided matters regarding climbing in the Chamonix region of the Alps.

16. *He was prosecuted . . . I dont know the result*: Tyndall did not mention the result in *Glaciers of Alps* (see *Glaciers of the Alps*, p. 192).

17. *I went down to Leeds and brought the case before the British Association*: the twenty-eighth meeting of the BAAS was held in Leeds from 22–29 September 1858, but Tyndall did not attend the entirety of it (see letter 1562, n. 1).

18. *resolution was passed*: see the summary of Tyndall's testimony, 'Particulars of an Ascent of Mont Blanc', *Brit. Assoc. Rep. 1858,* pp. 39–40.

19. *Mʳ Beckett of Kirkstall Grange*: William Beckett (1784–1863), banker and Member of Parliament for Leeds (1841–52) and Ripon (1852–57). R. V. Taylor, *Biographia Leodiensis: Or, Biographical Sketches of the Worthies of Leeds and Neighbourhood* (London: Simpkin, Marshall, & co., 1865), pp. 506–9.

20. *Lord Carlisle*: George William Frederick Howard, seventh earl of Carlisle (1802–64), Lord Carlisle from 1848–64 (*ODNB*)

21. *Mʳˢ Beckett's*: Frances Adelina Beckett (née Ingram), married to William Beckett in 1841.

22. *my ascent of Monte Rosa*: see letter 1478, n. 22, and letter 1547.

23. *book*: *Glaciers of the Alps*.

24. *Queenwood*: see letter 1298, n. 13.

25. *Mʳˢ. D.*: Maria Drummond.

26. *a beautiful letter*: letter missing.

27. *one of her letters at Chamounix*: letter missing.

28. *Surrey*: a county in England, to the south and southwest of London, where the Drummonds had an estate.

29. *should*: Tyndall repeated the word 'should', but the first ended a page of the letter while the second began the next page.

30. *a note*: letter missing.

31. *a reply*: letter missing.

32. *box*: a genus (*Buxus*) of small evergreen trees or shrubs of the family Euphorbiaceæ; specially *B. sempervirens,* the Common or Evergreen Box-tree, a native of Europe and Asia (*OED*).

33. *boles*: stem or trunk of a tree (*OED*).

34. *a note*: letter missing.

35. *Torquay*: a seaside resort town in southeast England, on the English Channel.

36. *Abbotts Ann*: a village near Andover, England, west of London.

37. *Mr Best*: Samuel Best (1802–73), a reverend in charge of a primary school in Abbotts Ann. Personally interested in geology, he authored *After Thoughts on Reading Dr. Buckland's Bridgewater Treatise* (London: J. Hatchard and Son, 1837), and made science a focus of his school's curriculum. See 'Deaths', *Hamphire Advertiser*, 25 January 1873, p. 4. For a short biography of Best, see A. Geddes, *Samuel Best & the Hampshire labourer* (Andover: Andover Local History Society, 1981).

38. *These girls*: Emily, Mary, and Fanny Drummond.

39. *seige*: Tyndall misspelled 'siege'.

40. *her child*: Mary Drummond.

41. *a note from my friend*: not identified, possibly Maria Drummond.

42. *affectionate as ever*: here several lines, presumably about Maria Drummond, are scratched out and left unreadable, implying Tyndall did not want Hirst to read them.

43. *Atkinson*: Edmund Atkinson (1831–1900), a chemist and physicist that taught at Queenwood College. 'Obituary Notices', *Journal of the Chemical Society: Transactions*, 79 (1901), pp. 888–89.

From Mary Anne Coxe [19] October [1858][1] 1567

9 Athol Crescent | Tuesday 9th October

My dear Dr Tyndall

I was so glad to see your well known hand writing[2] once more—and to be assured, that though we have unfortunately not met this year, our interest in each other has suffered no diminution.

I cant tell you <u>how</u> disappointed I was to find you at Queenwood[3] as we went to the Continent; on our return we tried again, but with the ill success we almost <*1 word missing*> counted on.

Yes, James[4] told me of his meeting you, and that 'out of goodness' he had told you not to call. (we left no address because the man said you would be absent a longer time than we were to be in London) our stay <u>was</u> very short, and the chances were many—against our being in, unless an appointment were made, and <u>that</u> with a man of your much occupation would have been unfair—(and we <u>were</u> a long way off—Cambridge Terrace[5]) This last time we were in Old Quebec Street[6] and I should have so enjoyed a 'craze'[7] with you!

I wanted to go back again to you, when I heard of your return, but James threw cold water upon me[8] too, declaring it would only bore you—and so I very unwillingly relinquished the idea then, expecting that by the time <u>we</u> were to return, <u>you</u> would also. But we had to return a month sooner than we had intended, (the middle of September) and you were still absent. We often wondered where you were, and if we had any chance to meet—(Wherever I fell in with a 'Book' I hunted over the names with you in my mind's eye.

And yet we might have foregathered, for we must have been, indeed were in many of the same places. Paris and all over Switzerland with a great deal of Germany. We found the heat intolerable—(seldom under 80. in the night) at Geneva[9] we spent nearly ½ an hour every morning and evening in the Rhone,[10] and this toned us down for the rest of the day and night.

And now how shall I thank you for having <u>thought</u> of me in your travels.[11] I assure you I feel it no small compliment, and to prove it in a way that I know to be so troublesome—for I well know that it is—therefore can thoroughly appreciate the kindness. I do hope you say truly that the 'little trinket of a box' is 'worthless intrinsically'[12]—I dont know how it is, but I really <u>never</u> enjoy the receiving of a costly present; where as some small thing that bears on its face that it was got for one individual, and therefore a proof of kindly remembrance, gives me unmixed pleasure—and such I do hope will be the case now—but how is it that I have got the letter and <u>not it</u>. I <u>hope</u> it is merely stupidity on the part of the Post office gentry, but cant help <u>fearing</u> it may be their honesty is at fault.

Tomorrow morning will I trust make all right.

Here I am interrupted.

We were at Frankfort[13] during the 'Fair',[14] and of course for luck, brought away some 'fairings'[15]—here was yours—

'The gift is small | But give to all'![16]

I am writing against time—and have only moments to say that I am always dear Dr Tyndall

Yours very much | & most sincerely | Mary Anne Coxe

RI MS JT/1/TYP/1/276
LT Typescript Only

———————

1. *[19] October [1858]*: the year is given by the reference to Queenwood (see n. 3) and the day by relation to letter 1568. While LT's typescript shows 9 October as the day, this was a Saturday. However, 19 October was a Tuesday and closer to Coxe's next letter to Tyndall (1568), and thus LT probably transcribed the date incorrectly.

2. *to see your well known hand writing*: letter missing.

3. *Queenwood*: Tyndall travelled to Queenwood College (see letter 1298, n. 13) in mid-October 1858 to work on the manuscript of his *Glaciers of the Alps*.

4. *James*: James Coxe.

5. *Cambridge Terrace*: Cambridge Terrace is a row of mansions overlooking Regent's Park in London, some two miles from the RI.

6. *Old Quebec Street*: a street in Westminster, London.

7. *'craze'*: not identified but presumably dialect or slang for eager conversation.

8. *threw cold water upon me*: to heap discouragement on, disparage, 'damp' (*OED*).

9. *Geneva*: see letter 1293, n. 3.

10. *Rhone*: see letter 1293, n. 6.

11. *And now how shall I thank you for having thought of me in your travels*: possibly in Tyndall's letter that this replies to (which is missing), he shared that he toasted her while atop Monte Rosa in August, like he did when atop Mont Blanc in August 1857 (see letter 1445).

12. *I do hope you say truly that the 'little trinket of a box' is 'worthless intrinsically'*: from a letter from Tyndall to Coxe, which is missing.

13. *Frankfort*: alternative spelling of Frankfurt, a large city in Germany that serves as a significant transportation hub.

14. *'Fair'*: probably the Dippemess, a traditional fair held annually in Frankfurt that dates back to the 14th century. Pottery goods were usually for sale.

15. *'fairings'*: a present, souvenir, or other item given at or brought back from a fair (*OED*).

16. *'The gift is small | But give to all'*: not identified. Most likely a line of poetry from a book Coxe obtained at the fair.

From Mary Anne Coxe 30 October [1858][1] 1568

Edinbro. Saturday | October 30th. (1858)

You worry me Brutus![2]

I <u>did</u> know that an address was useless—We were but three days in London—and your ancient Cerberus[3] told us <u>you</u> would be 10 at Queenwood.[4] So cui bono[5] letting you hunt after us at 39 Jermyn Street—as my vanity whispered you would do—when the birds were flown?[6] Methinks I have thee on the hip Master Tyndall!

I knew of your previous where abouts by the very interesting letter in the Times[7]—(to Professor Faraday) and could have better forgiven your being in <u>foreign parts</u> when we were in London, than to have missed you so narrowly. I feel quite jealous of Queenwood you go there perpetually <u>and never come to us.</u> (there must be metal more attractive there than we wot[8] of)

We were three weeks in Paris—our ultima thule[9]—our chief enjoyment was the dolce far niente[10] of lolling back in open carriages driving in the Bois de Boulogne[11]—No bad pastime I can assure you—What a glorious place Paris is! and what exquisite weather we had. I must tell you a capital joke at the Commissioner's[12] expense—he complained to the Landlord of the 'Windsor'[13] on wishing to change our rooms 'que les lits étaient trop petits'[14] 'Pardon', smilingly replied mine Host—'ce n'est pas les lits qui sont trop petits, c'est monsieur qui est trop grand!!!'[15]

A clincher, wasn't it? which made the Commissioner for the nonce[16] look <u>very</u> <u>small</u>!

Uncle Willie[17] <u>will</u> persist in calling his affection paralysis, though medical men say 'were it <u>really</u> so he would not be so ready to give it that name'! His <u>brain</u> and <u>spine are utterly untouched</u>, but after sitting 4 hours in a very cramped position—in a small boat fishing—he found on coming to rise, that <u>one leg</u> was completely powerless—at first he laughed and thought it great fun, but when hours and days elapsed with little improvement, he set it down as paralysis—and so he still persists in calling it—though <u>all</u> the medical men—my husband[18] among them—declare it to be 'the rupture of a small blood vessel at the sheath of the nerve' and that gradually as the blood was absorbed the use of his limb would be restored—though probably not before the expiry of a year would he be altogether right. (his general health of mind and body—absolutely unimpaired) this opinion is quite being borne out by the fact—that whereas he at first could not move without two crutches—he now walks about the room with <u>one stick</u>—and on the street with two. Yet strange to say he himself will scarcely admit that there is any improvement, and still clings to the theory of paralysis—till James and I could sometimes shake him! it is a sort of monomania[19] with him—<u>I</u> allege to make himself interesting—And such levee's of visitors as he used to hold—grave judges of the land, Duchesses—(even the magnificent Grace of Sutherland[20]) and pretty damsels by the dozen. Could you not make a run down here at Xmas? I do <u>so</u> wish you would. 'Where there's a will there's a way',[21] says the proverb, and if you dont manage it, I shall know that all the love you profess for me is 'words—words—mere words'.[22]

Voila! | Yours as you behave yourself! | M. A. C.

I say Xmas as most likely to be your free time—but any time that the Com.[23] is not absent on duty would suit us equally well.

RI MS JT/1/TYP/1/277
LT Typescript Only

---

1. *[1858]*: the year is given by the reference to Tyndall's letter to Faraday being published in the *Times* (see letter 1548).

2. *Brutus*: Marcus Junius Brutus, a politician of the late Roman Republic involved in the assassination of Julius Caesar, an act made well-known through William Shakespeare's play *Julius Caesar*.

3. *your ancient Cerberus*: in Greek and Latin mythology the proper name of the watch-dog which guarded the entrance of the infernal regions, represented as having three heads (*OED*). Probably a reference to Charles Anderson at the RI.

4. *Queenwood*: see letter 1567, n. 3.

5. *cui bono*: 'for whose benefit?' (Latin), taken popularly but erroneously in English as 'to what use or good purpose?' (*OED*).

6. *the birds were flown*: a phrase meaning to have escaped or disappeared.

7. *letter in the Times*: see letter 1548.

8. *wot*: to know (*OED*).

9. *ultima thule*: the highest or uttermost point or degree attained or attainable, the acme, limit (Latin, *OED*).

10. *dolce far niente*: delightful idleness (Italian, *OED*).

11. *Bois de Boulogne*: a large public park in Paris.

12. *Commissioner's*: Mary's husband James Coxe was appointed Commissioner of the General Board of Lunacy for Scotland in 1857.

13. *Landlord of the 'Windsor'*: presumably the hotel where the Coxes stayed while in Paris.

14. *'que les lits étaient trop petits'*: 'because the beds were too small' (French).

15. *'ce n'est pas les lits qui sont trop petits, c'est monsieur qui est trop grand!!!'*: 'it is not that the beds are too small, it is the gentleman who is too big!!!' (French).

16. *for the nonce*: for the particular occasion (*OED*).

17. *Uncle Willie*: not identified.

18. *my husband*: James Coxe.

19. *monomania*: a Victorian category of mental illness characterized by a single pattern of repetitive and intrusive thoughts or actions (*OED*).

20. *Grace of Sutherland*: probably a mistranscription for Her Grace, the Duchess of Sutherland, Harriet Sutherland-Leveson-Gower (1806–68), a member of Queen Victoria's household.

21. *'Where there's a will there's a way'*: a nineteenth-century revision of the proverb 'To him that will, ways are not wanting', meaning that someone who is determined to achieve something can overcome obstacles to that end.

22. *'words—words—mere words'*: W. Shakespeare, *Troilus and Cressida*, V.iii.108, Troilus, 'Words, words, mere words, no matter from the heart: | The effect doth operate another way'.

23. *Com.*: see n. 12.

To Sarah Tyndall [October 1858][1] 1569

London. Sunday morning

My dear mother

I have at length a quiet hour to write to you. Since I last wrote to you[2] I have been very busy, but the store of strength which I obtained during my visit to the Isle of Wight[3] enables me to get through my work cheerfully and with satisfaction. It is the greatest pleasure of my life to work when I am quite well. In fact I only value life for the purposes of real earnest work.

The weather has been extremely dry in England, so dry indeed as to injure some parts of the country. This morning the clouds look heavy, but the wind is still east and as long as this continues—as long as the wind blows over us from Asia and Africa instead of from the Atlantic ocean we shall have but little rain. Still the country looks lovely and in an hour or two I take the train, and roll into one of the most lovely parts of England. There is one beautiful family[4] which I often visit on Sundays and with them I intend to spend the afternoon and evening of today, returning by a night train so as to commence my work in good time tomorrow.

I invited a son of Dr Tyndall's[5] to come to see me, but as he was getting ready he received an appointment as surgeon on board an emigrant ship.[6] He has a younger son named John[7] and he is coming over to see me in June. Were it not that I have such a load of work pressing upon me I would make a trip to Ireland and pay you a visit.

I think you told me the name of your grandmother's father was Malone[8] and that he cut her off with a shilling for marrying out of connexion. Your great grandfather Malone was, I think you told me a man of property, and that he divided his property among his sons. I think you told me that Janeville[9] belonged to him: is this Janeville near Ballybromwell?[10] Excuse me for bothering you with these questions. They may be of interest at some future day. Did not my grandfather Tyndall[11] own some property at Coolcullen,[12] and did he not change his will on his deathbed taking his property from my father;[13] and did not Lord Frankfort[14] afterwards obtain the property by lawsuit. All this I remember hearing when a boy. I suppose I am quite right.

Your affectionate son

RI MS JT/5/16a/-
LT Typescript Only

---

1. *[October 1858]*: the year and month are given by a note LT wrote on the typescript: 'Oct. 1858 is the date of the Mother's answer, pinned on to last page of Journal VIII a'. Tyndall recorded an extract of his mother's reply (included in n. 8 below) at the end of his manuscript journal which begins in September 1858 and ends in June 1871 (Journal, RI MS JT/2/10/66 (manuscript) and RI MS JT/2/13c/1383 (transcript)). This letter and his mother's response probably relate to Tyndall's courting Mary Drummond. He often felt he was not in the same social class as the Drummond family, and was likely seeking information from his own mother about his family's heritage to garner the favour of Mary and her mother Maria. See letter 1578 in which Tyndall shares with Hirst about his relationship with the Drummond family.

2. *last wrote to you*: letter 1552.

3. *my visit to the Isle of Wight*: Tyndall frequently visited the Isle of Wight, the largest island in England (located in the English Channel) which was a popular destination for holidays

4. *one beautiful family*: probably Maria Drummond and her daughters Mary, Emily, and Fanny.

5. *a son of Dr Tyndall's*: not identified. Perhaps the son of John Tyndall (d. 17 April 1875), a doctor in Gorey, County Wexford, Ireland. See 'Obituaries', *Lancet,* 1 May 1875, p. 633. The obituary states he was 67 when he died, so in 1858 he would have been around 50 years old, old enough to have two older sons that could potentially travel from Ireland to London to visit with Tyndall.

6. *an emigrant ship*: not identified.

7. *a younger son named John*: not identified. Perhaps another son of John Tyndall (d. 17 April 1875), a doctor in Gorey, County Wexford, Ireland (see n. 5).

8. *the name of your grandmother's father was Malone*: William Malone. Sarah Tyndall's reply letter is missing, but an extract can be found in Tyndall's journal for 1871, with the date October 1858: 'My grandmother's family were people of wealth & property. their name was Malone, but my grandmother marrying out of meeting she was denied by them. Her father left three townlands one to each of her brothers—Ballybromwell to William, Ballintrain to Thomas, and Killkay to Pim. I believe that is Nehemiah, and to my grandmother one angry shilling This however never altered the affections of her husband towards her, and they had abundance while they lived. there were no such farms in the country as he kept. They had one son & that was my father and she reared him as she was reared herself as a Quaker, and I went along with him to meeting where M^rs. Lecky, first saw me. Oct. 1858' (Journal, RI MS JT/2/13c/1383 (transcript) and RI MS JT/2/10/661 (manuscript)).

9. *Janeville*: a townland in County Carlow, Ireland.

10. *Ballybromwell*: Ballybrommell, a townland in County Carlow.

11. *my grandfather Tyndall*: William Tyndall (1766–1823), of Coolcullen, County Kilkenny, Ireland, was the father of Tyndall's father John Tyndall (1792–1847). He married Emelia Byrne in 1790. John Tyndall moved to County Carlow, settling in Leighlinbridge. He worked for William Steuart, who owned many of the properties in Leighlinbridge. John Tyndall and his father looked to the Steuarts for patronage and approached them for support when Tyndall was considering various forms of employment as a young man. For letters between Tyndall and Elizabeth Steuart, see letters 1555 and 1561.

12. *Coolcullen*: a townland in County Kilkenny.

13. *my father*: John Tyndall (1792–1847), grew up in Leighlinbridge and married Sarah McAssey in 1813. They had two children that are definitely known: John and his older sister Emma (b. 1817). He probably served in the Royal Irish Constabulary at the police station in Nurney from 1828–33. The family then moved to Castlebellingham, County Louth for the next three years before returning to Leighlin Bridge, where he worked as a boot and shoemaker. He may also have served as a land agent for William Steuart, one of the principal landlords in Leighlin Bridge and also the main patron of the Tyndall family. His 1847 death was a blow to his son, who relied on his father for advice.

14. *Lord Frankfort*: the Viscount Frankfort, a landowner in County Kilkenny and County Carlow.

To William Thomson   6 November 1858   1570

Dear M[r]. Thomson
  I have now the pleasure of transmitting the syllabus[1] to you.
  & remain | very sincerely yours | <u>John Tyndall</u>
  <u>6th. Nov. 1858</u>

RI MS JT/1/T/1439

---

1. *syllabus*: not identified.

To George Gabriel Stokes   23 November 1858   1571

23rd. Nov. 1858.

My dear Sir.
  I do not think that it would be very good taste in me to offer an opinion[1]
upon those parts of M[r] Moore's[2] paper[3] where he as a mathematician differs
from other mathematicians. He undertakes to shew the incompleteness of
the equation hitherto regarded as containing the expression of the motion
of sound in narrow tubes, and proposes one containing an additional plus
or minus quantity as a substitute. From his theory he deduces that a wave
of rarefaction is more rapidly propagated than a wave of condensation; that
hence it is natural that the sensation of sound should be produced, not by
condensation, but by the nine rapid signals given by rarefactions; and finally
he adverts to the circumstance that these considerations explain an acknowledged difficulty in the ordinary theory of sound—the difficulty namely, that
the calculated velocity falls short of the velocity derived from experiment.
  The most decisive argument in favour of the idea that sound is produced
by rarefactions is drawn by M[r] Moore from the construction of the human
ear. His observations here are of a highly ingenious and interesting character,
but it appears to me that it would be extremely perilous to admit this argument <u>from design</u>[4] as capable of deciding a physical problem. The author also
refers, though he does not insist on the point, to the double report heard in
certain states of the atmosphere, where a gun is fired, as an illustration that
sounds are transmitted in duplicate; but it appears to me that in the experiments referred to the double sound was due to echoes from the clouds, and
does not favour M[r] Moore's notion.
  With regard to the acknowledged difficulty referred to by the author, it
is only fair to him to state that other mathematicians also have experienced
a similar difficulty in accounting for the discrepancy between the observed

& calculated velocities of sound.[5] But this never has been an acknowledged difficulty with any one who really comprehended the explanation of Laplace;[6] and I think it hardly excusable to call this a difficulty after all that has been recently written upon the subject & more especially after the lucid physical exposition of the theory which we owe to one of the secretaries of the Royal Society. (Phil Mag, 1851. page 309.)[7] A paper therefore which, in part at least, rests its merit on the removal of a difficulty which has no real existence is at least of doubtful value to me. But while on physical grounds I am unable to recommend the paper of M[r]. Moore for publication in the Phil. Transactions I should be glad to see it reported on by a mathematician. For devotion to other modes of enquiry has so far weakened the portion of mathematical power which I once possessed as to render me doubtful in some degree of my own competence, without great expenditure of time, to offer an opinion upon purely mathematical questions.

I am dear Sir | Yours very sincerely | <u>John Tyndall</u>
Prof. Stokes | &c &c &c | Sec. R.S.

RS RR/3/203 Tyndall 23 November 1858

1. *to offer an opinion*: Tyndall acted as a referee for the *Phil. Trans.*, reviewing submissions and sending reports to Stokes, the journal's editor.

2. *M[r] Moore's*: not identified. Perhaps John Carrick Moore.

3. *paper*: not identified.

4. *argument <u>from design</u>*: one year later, Tyndall's friend Charles Darwin would publish *On the Origin of Species* (Murray, 1859), his theory of evolution by natural selection countering the widely held 'argument from design' presented in clergyman William Paley's *Natural Theology* (1802), which maintained that the complexity of nature showed evidence of design by a creator. Tyndall owned an 1845 edition of Paley's works (which he dated October 1848), and made annotations throughout, including marking passages in *Natural Theology*. In his journal entry of 2 January 1849, he wrote, 'I am willing enough to be convinced upon the subject of which Paley writes, but what he says is insufficient to this end. The Great Spirit is not to be come at in this way; if so, his cognition would only be accessible to the scientific and to very little purpose even here' (Journal, RI MS JT/2/13b/413).

5. *difficulty in accounting for the discrepancy between the observed & calculated velocities of sound*: Tyndall later published on this topic. See J. Tyndall, 'Note on Laplace's correction for the velocity of sound', *Phil. Mag.*, 26 (1863), pp. 384–87; and *Phil. Mag.*, 27 (1864), p. 41.

6. *Laplace*: Pierre Simon Laplace (1749–1827), a pioneering French mathematician, physicist, and astronomer (*CDSB*).

7. *(Phil Mag, 1851. page 309.)*: G. G. Stokes, 'An Examination of the possible effect of the Radiation of Heat on the Propagation of Sound', *Phil. Mag.*, 1 (1851), pp. 305–17.

From David Brewster   25 November 1858   1572

My Dear Mr Tyndall,

In consequence of the promotion of Mr Adams[1] to the Lowndes Chair of Astronomy in Cambridge,[2] it is more than probable that our Chair of Natural Philosophy,[3] which is in the gift of the College[4] will be vacant. Its value will doubtless be raised to £500 per annum from £350. Would you accept of it? I am

My Dear Mr Tyndall | Ever Most Truly y[ou]rs | D. Brewster. | S^t. Leonard's College | S^t. Andrews | Nov. 25th 1858

RI MS JT/1/B/127
RI MS JT/1/TYP/1/158

---

1. *Mr Adams*: John Couch Adams (1819–92), a British mathematician and astronomer, most well-known for his prediction of the existence and position of the planet Neptune (*ODNB*).
2. *Lowndes Chair of Astronomy in Cambridge*: the Lowndean Chair of Astronomy at the University of Cambridge in England, established in 1749 by the astronomer Thomas Lowndes. Adams held this professorial chair from 1859 until his death in 1892.
3. *Chair of Natural Philosophy*: the Chair of Natural Philosophy at St. Andrews University in Scotland was vacated by John Couch Adams in 1859 (he had only served since 1858) and filled by William Swan (1818–94), a Scottish mathematician and physicist, who served until 1880.
4. *the College*: St. Leonard's College.

To Rudolf Clausius   27 November 1858   1573

London 27th Nov. 1858

My dear Clausius

Many times since my return to England has the thought of writing to you occurred to me. I wished among other things to tell you the great pleasure which my short visit to Zürich[1] gave me, and how gladly I would have returned there in accordance with your kind information if my engagements had permitted me. After parting with you upon the <u>Lauzplatz</u>[2] we[3] made the best of our way to Rosenlaui,[4] and spent a morning upon the glacier. Here a Frenchman deliberately <u>stole my boots</u>; he was followed, overtaken on the Sheideck[5] where they were taken off his feet. He had left me a miserable old pair in their place. We stayed two days at <u>Grindelwald</u>, and examined both the glaciers.[6] Then we crossed the <u>Strahleck</u>[7] to the Grimsel;[8] visited the Rhone glacier[9] and afterward went down the valley to Viesch.[10] We ascended the Eggishorn,[11] and remained at the hotel there 8 days. During this time I examined most of that grand Aletsch glacier,[12] from its source downwards, and finding a hardy guide[13] at the hotel he and I ascended to the summit of the Finsteraarhorn.[14]

We afterwards went <u>to Zermatt</u>,[15] where my friend Ramsay[16] receiving intelligence of the death of his mother, left me and returned to England. I had a guide[17] from Lauterbrunnen,[18] a strong fellow & good glacier man, he had never been up Monte Rosa,[19] but I thought as both of us were good climbers we might try it together. We did so on the day after our arrival at the Riffel[20] and succeeded in reaching the top. We saw nothing: for hours we toiled amid impenetrable gloom, dark, dense clouds and falling snow. I made a few observations on the top, but the most wonderful thing I have ever seen was the form of the snow crystals. There was not a breath of air stirring to injure them, and they descended as a shower of <u>frozen flowers</u>. Each flower had 6 leaves: there were endless variations, but no deviation from the 6 leaved type. any thing so beautiful I had never before seen. But what will interest you most was the following observation. In the earlier portion of the day we had sunshine, and not until we came near the summit did the snow begin to fall, though we were enveloped in clouds long before. I tried the colour of the snow <u>as we ascended</u>, driving the Alpenstock[21] into it, but the blue was hardly perceptible. On coming down however after a thick layer of <u>fresh snow</u> had fallen the colour was magnificent, on drawing the Alpenstock from the snow, a most exquisite blue would fill the hole made by the stock, and as my guide walked in front of me, where he lifted his feet, and shook the snow from them, gleams of blue light seemed to flash from it. At this time the darkness was so great that we could not see more than 20 yards in any direction—<u>there was not a trace of the blue heaven to be seen</u>, and this observation agrees with all that I made before and since, that this exquisite blue color is a property of the newly fallen, and finely granular snow. A week afterwards the weather was magnificent. One glorious day I had lent my guide to a party of gentlemen; so wishing to take advantage of the weather I started in my shirt sleeves, carrying with me about three ounces of food, and half a pint of cold tea, and without wine, or brandy, coat, or guide, I climbed to the Höchste Spitze[22] of Monte Rosa.

I had often seen interference colours on the clouds, but had never observed any thing to equal the magnificence of those I saw during this second ascent of Monte Rosa. The day was glorious, and generally without a trace of vapour. Three times however during my ascent a light veil of cloud drew itself across the sun, and at each time the veil was flooded with the most gorgeous colours. Red, orange, green, yellow, blue—all nearly as vivid as the hues of the spectrum. This was by <u>transmission</u>—but I have seen similar colours by reflective light among the clouds. They must be excessively thin, and I think the colours must be those of <u>mixed plates</u>—do you not think so?

I waited 11 days at the Riffel, and explored the whole system of glaciers from the Cima de Jazzi[23] to the Zmutt glacier.[24] My main object in going to Switzerland was to learn something more about the <u>structure</u> of the ice. M[r]. Mousson[25] (to whom I beg of you to remember me most kindly) was quite right in regarding the question unsolved. I had a strong opinion on the subject it is true, but as your friend Herr Moleschott[26] says it is easy to have opinions—but I wanted <u>proofs</u>. These proofs I am happy to say I obtained this year, and I think I shall be able to establish a perfect parallel between the structure of glacier ice & the cleavage of rocks.[27]

I passed from the Riffel to the Allalein glacier,[28] then to the Fee gl.[29] in company with our friend Hirst, who was on his way over the Moro[30] to Rome. Afterwards to Chamouni[31]—I met Magnus[32] & his family on the slope of the Montanvert,[33] and also saw Ettingshausen.[34] I reached the top of Mont Blanc[35] a second time & planted a thermometer there. This was the end of my Campaign—I wish you would give me your opinion about the snow & the <u>clouds</u>. When you see the professors Fick[36] remember me kindly to them— Believe me dear Clausius

Ever most sincerely yours | <u>John Tyndall</u>

I wish you a happy Christmas.

RI MS JT/1/T/175

<hr>

1.   *Zürich*: see letter 1333, n. 11.

2.   *Lauzplatz*: unidentified. A location in Zurich, possibly a public square.

3.   *we*: Tyndall, geologist Andrew Ramsay, and guide Christian Lauener. See *Glaciers of the Alps,* pp. 92–93.

4.   *Rosenlaui*: see letter 1536, n. 6.

5.   *Sheideck*: see letter 1536, n. 8.

6.   *at <u>Grindelwald, and examined both the glaciers</u>*: see letter 1536, n. 10, and letter 1312, n. 3.

7.   *<u>Strahleck</u>*: see letter 1536, n. 12.

8.   *Grimsel*: see letter 1517, n. 2.

9.   *Rhone glacier*: see letter 1297, n. 12.

10.   *Viesch*: see letter 1536, n. 16.

11.   *Eggishorn*: see letter 1297, n. 6.

12.   *Aletsch glacier*: see letter 1536, n. 19.

13.   *hardy guide*: Johann Bennen.

14.   *Finsteraarhorn*: see letter 1536, n. 27.

15.   *<u>Zermatt</u>*: see letter 1297, n. 7.

16.   *Ramsay*: Andrew Ramsay.

17. *guide*: Christian Lauener.

18. *Lauterbrunnen*: see letter 1527, n. 12.

19. *Monte Rosa*: see letter 1478, n. 22.

20. *Riffel*: see letter 1547, n. 9.

21. *Alpenstock*: see letter 1424, n. 6.

22. *Höchste Spitze*: the highest summit on Monte Rosa, renamed in 1863 to Dufourspitze.

23. *Cima de Jazzi*: a mountain of the Pennine Alps north of Monte Rosa.

24. *Zmutt glacier*: a glacier in the Pennine Alps in the Swiss canton of Valais, near the Matterhorn).

25. *M^r. Mousson*: Albert Mousson.

26. *Herr Moleschott*: possibly Jacob Moleschott (1822–93), a Dutch physiologist and scientific materialist who taught in Switzerland and Italy (*CDSB*).

27. *a perfect parallel between the structure of glacier ice & the cleavage of rocks*: see letter 1295, n. 3 and n. 4, and letter 1554, n. 36.

28. *Allalein glacier*: see letter 1436, n. 28.

29. *Fee gl.*: a glacier in the Pennine Alps in the Swiss canton of Valais, near the Allalinhorn.

30. *Moro*: see letter 1547, n. 5.

31. *Chamouni*: see letter 1353, n. 9.

32. *Magnus*: Heinrich Magnus.

33. *Montanvert*: see letter 1415, n. 15.

34. *Ettingshausen*: Andreas Freiherr von Ettingshausen (1796–1878), an Austrian physicist who held positions in mathematics and physics at the University of Vienna since 1817 and was appointed director of its physics institute in 1852 (*NDB*).

35. *Mont Blanc*: see letter 1320, n. 6.

36. *professors Fick*: possibly the German brothers Adolf Eugen Fick (1829–1901), a physiologist who developed laws of diffusion (of gas through membranes) in 1855 (*NDB*), and Franz Ludwig Fick (1813–58), a professor of anatomy at the University of Marburg who studied bone growth and vision. See O. Gerland (ed.), *Grundlage zu einer Hessischen Gelehrten-, und Schriftsteller-Geschichte, von 1831 bis auf die neueste Zeit,* 2 vols (Kassel: August Freuschmidt, 1868), vol. 2, pp. 48–50.

From John Douglas Cook   30 November 1858   1574

My dear Tyndall

Although, having been in the country, I have been long in saying it, you really are a very good Boy, and I am very much in love both with you and your Article.[1] But, pray, increase my affection for you, by giving me another very, very soon, and be sure to repeat the operation as often as you can.

Ever most truly yours | John D. Cook. | Tuesday Novr 30/58

RI MS JT/1/C/47
RI MS JT/1/TYP/1/226

---

1. *your Article*: John Cook (*c.* 1808–68) started the *Saturday Review* in 1855, and Tyndall
    and some of his fellow scientists offered to contribute science articles for it (see letters
    1509 and 1528). This particular article mentioned is not identified for certain, but could
    be 'The Clouds', *Saturday Review*, 28 August 1858, p. 207; 'The English in Switzerland',
    Saturday Review, 4 September 1858, p. 228, or 'Colour of Water', *Saturday Review*, 16
    October 1858, p. 374.

From Christian Gerling   7 December 1858   1575

Marburg den 7 Decbr 1858.

Theurer Freund!

Seit lange habe ich Ihnen nicht geschrieben; ja ich glaube beinahe daß ich
Ihnen noch nicht einmal für Ihre letzten gütigen Zuschriften und Zusendun-
gen gedankt habe. Ich war in den beiden letzten Jahren von allerhand Abhaltungen sehr in meiner Correspondenz gestört. Ueber den wissenschaftlichen
Theil dieser Abhaltungen gebe ich Ihnen in der Abhandlung Rechenschaft,
welche ich gleichzeitig mit diesem Briefe unter Streifband absenden werde.
Ich sende Ihnen gleich zwei Abdrücke indem ich Sie bitte einen davon bei
Gelegenheit unserm Freunde Dr. Hirst zukommen zu lassen, von welchem
ich nicht weiß ob er wieder in England oder in Frankreich oder Italien noch
ist.

Der Gegenstand hat mich lange und mühselig beschäftigt, macht mir aber
jetzt da er fertig ist recht viel Vergnügen und wenn Sie einmal den Apparat
bei mir sehen, so hoffe ich wird er Ihnen gefallen. Vorzüglich aber empfehle
ich Ihrer Aufmerksamkeit die, meines Wissens, neue Wellenbewegung und
Wellenfläche welche ich damals in Wien nur noch ganz vorläufig kannte,
und welche mir von einiger Bedeutung zu seyn scheint. Ich halte es nämlich
für <u>möglich</u> (mehr wage ich bisjetzt nicht zu sagen) daß dies die Welle des
natürlichen homogenen unpolarisirten Lichts sey. Wir würden dann nur zu
definiren haben: Stehende Welle in welcher <u>innerhalb einer Wellenlänge</u> alle-
möglichen Polarisationsbewegungen vorkommen.

Wie dem aber auch sey, so liegt die Frage nahe, ob sich nicht diese Art
von Bewegung schon irgendwo sonst in der Natur <u>nachweisen</u> lässt. Ich habe
vorläufig an die Schwingungen bei Wheatstons Kaleidophon gedacht, oder
an gewisse Erscheinungen beim Ausfliessen von Wasserstrahlen.—Bitte
belehren Sie mich wenn Sie etwas der Art wissen.

Nun will ich Ihnen doch aber auch etwas aus unserer Häuslichkeit mittheilen, und damit anfangen Ihnen recht freundliche Grüsse aus der ganzen Familie zu bestellen und Sie zu versichern, daß wir uns Ihrer stets in altbekannter Gesinnung erinnern.

Zuerst kann ich Ihnen gottlob melden daß wir alle Vier gesund sind. Meine Frau und ich fangen freilich an zuweilen daran erinnert zu werden daß wir lange schon aufgehört haben jung zu seyn; indessen lässt sich das doch gut ertragen. Ich habe in letztem Sommer mein 70 stes Jahr zurückgelegt, und lese in diesem Winter meine Vorlesung über Physik mit derselben Lust wie vor 10 Jahren. Auch von unsern auswärtigen Kindern und Enkeln haben wir gottlob gute Nachrichten.—Ein sehr freudiges Familien-Ereigniß haben wir aber im Oktober erlebt. Meine Marie hat sich nämlich verlobt, mit dem Universitäts-Syndicus Platner. Da dieses ein sehr achbarer Mann ist, welchen wir von seiner Kindheit an als solchen kennen (obwohl wir nie anders dachten als er habe beschlossen unverheirathet zu bleiben) und da derselbe mit Marie in so vielen Dingen gleiche Interessen und gleiche Ansichten hat; so hoffen wir dass ihre Verbindung, welche im Frühjahr statt finden wird, eine recht glückliche seyn wird. Ueberdies ist es uns eine große Freude daß wir sie hier in Marburg bei uns behalten, da die andern Kinder fern sind.

Sonst kann ich Ihnen von hier eben nichts Neues melden. Est geht alles so ziemlich denselben Weg wie damals als Sie hier waren.—Diesen Sommer hatten wir einen kurzen Besuch von Dr. Wrightson mit seiner liebenswürdigen jungen Frau. Mit Simpsons wechseln wir von Zeit zu Zeit einen Brief.— Frankland und Frau besuchten uns auch im Sommer.—Unser Freund Tyndall wird ja hoffentlich auch einmal kommen.

Zu der Versammlung in Carlsruhe bin ich nicht gekommen weil ich damals mich geistig und körperlich etwas angegriffen fühlte, und fürchtete daß die mit solchen Versammlungen verbundene Unruhe mich wieder, wie in Wien, krank machen könne. Doch habe ich meinen Apparat hingeschickt, und derselbe scheint auch den Sachkennern gut gefallen zu haben.

Nun für heute leben Sie wohl mein theurer Freund und behalten mich ferner in wohlwollendem Andenken.

Der Ihrige | Gerling

Marburg 7 Decbr 1858.

Dear friend!

I have not written to you for a long time; indeed, I almost believe that I still have not even thanked you for your last kind letters and mailings.[1] The last two years I was disrupted very much in my correspondence by all kinds of hindrances. I am giving you an account of the scientific part of these hindrances in the paper which I shall be sending[2] at the same time as this letter under postal wrapper. I am sending you two offprints right away, with the request to send one of them when you have the opportunity to our friend Dr. Hirst,[3] of whom I do not know whether he is back in England or still in France or Italy.

The item has occupied me long and toilsomely, but gives me very much pleasure now that it is finished, and if you see the apparatus[4] here at my place some time, then I hope you will like it. Above all, however, I recommend to your attention the, to my knowledge, new wave movement and wave surface which I knew only still quite tentatively back then in Vienna, and which seems to me to be of some significance. You see, I think it <u>possible</u> (I dare not say more as yet) that this is the wave of natural homogenous unpolarised light. We would then only have to define: a standing wave in which all possible polarisation movements take place <u>within one wavelength</u>.

However, the obvious question is, whether this kind of movement can be shown to exist somewhere else in nature. I have tentatively thought of the vibrations in Wheatstone's Kaleidophone,[5] or of certain phenomena in the outflow of jets of water. —Please inform me if you know anything of the sort.

But now I shall also tell you something from our domestic life, and start by offering you very kind greetings·from the whole family and by assuring you that we always remember you in our same old cast of mind.

First of all, I can bring you word, thank goodness, that all four of us are healthy.[6] My wife[7] and I admittedly are beginning to be reminded every now and then that we have long since ceased to be young; however, that is still quite bearable. Last summer I completed my 70th year, and this winter I am giving my lecture on physics with the same pleasure as 10 years ago. We also have good news, thank goodness, from our children and grandchildren[8] who are now living elsewhere. —We experienced a very happy family event in October, however. My Marie, you see, has got engaged to the university advocate Platner.[9] As he is a very respectable[10] man whom we have known as such since he was a child (although we never thought other than that he had decided to remain unmarried) and as he has the same interests and same views as Marie in so many things, we hope that their marriage, which will take place in the spring, will be a very happy one. Furthermore, it is a great joy for us that we shall keep her here in Marburg[11] with us, as the other children are far away.

Apart from that, I cannot give you word of anything new from here. Everything is going[12] in more or less the same way as at that time when you were here. —This summer we had a brief visit from Dr. Wrightson[13] with his amiable young wife.[14] From time to time we exchange a letter with the Simpsons.[15]—Frankland and his wife[16] also visited us in the summer.—Our friend Tyndall will of course hopefully come too some time.

I did not go to the meeting in Carlsruhe[17] because I was feeling somewhat worn out in mind and body at the time, and feared that the upheaval that is connected with such meetings could make me ill again, as it did in Vienna.[18]

However, I did send my apparatus[19] there, and all the experts seems to have liked it too.

Well, for today, fare well, my dear friend, and keep[20] me further in benevolent memory.

Yours | Gerling

RI MS JT/1/G/3

1. *your last kind letters and mailings*: the previous letter from Tyndall to Gerling is 1399 of this volume (dated 12 June 1857). There are none in *Tyndall Correspondence,* volume 5.

2. *in the paper which I shall be sending*: probably C. L. Gerling, 'Darstellung aller Polarisationsbewegungen und einer zweiten verwandten Wellenbewegung durch Zusammensetzung zweier Schraubenbewegungen, nebst Nachricht von einem Apparat dazu', *Poggend. Annal.,* 105 (1858), pp. 175–210.

3. *Dr. Hirst*: Thomas Hirst.

4. *the apparatus*: described as an apparatus for visualizing the ether oscillations of polarized light. 'Ein neuer Apparat zur Versinnlichung von Aetherschwingungen polarisirten Litches'. See *Tagblatt der 34 Versammlung Deutscher und Ärtze* (Karlsruhe: Chr. Fr. Müller, 1858), p. 24.

5. *Wheatstone's Kaleidophone*: a device made of a ball or disk attached to the end of a vibrating rod. Light reflected against the ball produces an optical illustration of the vibration. See C. Wheatstone, 'Description of the Kaleidophone, or Phonic Kaleidoscope', *Quarterly Journal of Science, Literature, and Art* (January–June, 1827), pp. 344–51.

6. *all four of us are healthy*: see letter 1399, n. 9.

7. *My wife*: see letter 1399, n. 8.

8. *grandchildren*: not identified.

9. *Marie, you see, has got engaged to the university advocate Platner*: Marie Platner (née Gerling), married Hermann Platner (1814–93), a Syndic of the University of Marburg.

10. *respectable*: in the German, Gerling spelled 'achtbarer' incorrectly.

11. *Marburg*: Marburg, Germany, where Gerling taught at the University of Marburg. Tyndall was a student of Gerling's.

12. *Everything is going*: in the German 'Est geht alles', Gerling spelled 'Es' incorrectly.

13. *Dr. Wrightson*: Francis Trippe Wrightson (1817–93), son of a Birmingham manufacturer, was a student who struggled in his studies in Marburg. He obtained a PhD in 1853 under the direction of Kolbe, and was a professor of chemistry at Sydenham College, Birmingham. See W. H. Brock 'Bunsen's British Students', *Ambix,* 60 (2013), pp. 203–33.

14. *his amiable young wife*: not identified.

15. *the Simpsons*: Maxwell Simpson and his wife Mary Simpson (née Martin), see letter 1372, n. 8.

16. *Frankland and his wife*: See 1308, n. 8.

17. *the meeting in Carlsruhe*: the thirty-fourth meeting of the Gesellschaft Deutscher Naturforscher und Ärtze took place in Carlsruhe (now Karlsruhe) in September 1858.

18. *in Vienna*: the thirty-second meeting of the Gesellschaft Deutscher Naturforscher und
    Ärzte, which Tyndall attended, took place in Vienna in September 1856.
19. *my apparatus*: see n. 4.
20. *me*: in the German, Gerling did not include 'sie' ('further') after 'behalten'.

To Christian Gerling   10 December 1858   1576

10$\underline{^{th}}$ Dec. 1858

My dear Friend.

Your letter,[1] and your two memoirs[2] reached me this day and I hasten to thank you for both and to assure you that it gave me great pleasure to hear from you. It is a long time since I received any intelligence from dear old Marburg.[3] It gives me sincere satisfaction to learn that you and your family are well and happy; with regard to your being no longer young you are only about 2 years older than Faraday, and I sometimes say to him that <u>he</u> is the young man and <u>I</u> the old one.[4] I saw Mr Biot last year in Paris, and when I mentioned Faraday's age to him he exclaimed 'Il est un jeune homme moi, J'ai quatre vingt trois!'[5]

I hoped at one time to be able to go to Carlsruhe[6] this year, but I found myself unable to spare the time. I spent the summer among the Alps, and one day while sitting in the hotel at the Montenvert[7] Professor v. Ettingshausen[8] and his daughter[9] came in—It gave me great pleasure to see them. Another day as I was ascending towards the Mer de Glace[10] a gentleman and three ladies were riding on before me. As he turned a corner I caught sight of the gentleman's face—it was my friend Prof. Magnus who was travelling with his wife and two daughters.[11] I was extremely pleased to see him, though at first he did not at all remember me on account of the huge beard which I had permitted to grow among the mountains.

My attention was drawn three years ago towards the glaciers, and until this year I was not able to finish the subject. I spent a glorious time in Switzerland and became an expert mountaineer—I ascended the Finsteraarhorn[12] this year with a single guide[13] and placed a <u>minimum</u> thermometer on the summit—I ascended Monte Rosa[14] twice—once with a single guide,[15] and once <u>alone</u>. Afterwards I came to Chamouni[16] and waited there for some time for good weather so that I might reach the top of Mont Blanc[17] & plant thermometers there. I did reach the summit but we were attacked by a kind of <u>tourmente</u>,[18] and it was so terribly cold, that my guide,[19] the best man in Chamouni, very nearly lost his two hands. They were quite <u>gelées</u>,[20] and it was only by the greatest efforts that sensation was restored to them.

Our friend Hirst is now in Rome—I had the happiness of meeting him in the valley of Saas,[21] as he was about crossing the Monte Moro.[22] We spent two delightful sunny days upon the glaciers together—I hope he will return to England in the Spring.

I have not yet had time to read your memoir, but shall do so in my first moments of leisure—you shew a good example of activity to younger men, and I hope you will long continue to do so—I wish you many happy Christmasses I offer Miss Gerling[23] my felicitations on the approaching happy occasion, and with best and kindest wishes to you all

Believe me ever | Most sincerely yours | John Tyndall

UB Mbg 319:740

---

1. *Your letter*: letter 1575.
2. *your two memoirs*: possibly C. L. Gerling, 'Darstellung aller Polarisationsbewegungen und einer zweiten verwandten Wellenbewegung durch Zusammensetzung zweier Schraubenbewegungen, nebst Nachricht von einem Apparat dazu', *Poggend. Annal.*, 181 (1858), pp. 175–210; and C. Gerling, 'Bemerkungen über das indirecte Eliminiren beigeodätischen Arbeiten', *Zeitschrift für Mathematik und Physik*, 3 (1858), pp. 377–82.
3. *dear old Marburg*: see letter 1399, n. 10.
4. *he is the young man and I the old one*: Tyndall here repeated a phrase he used in another letter to Gerling (1399).
5. *'Il est un jeune homme moi, J'ai quatre vingt trois!'*: loosely translated as 'He is a younger man than I, I'm eighty three!'
6. *Carlsruhe*: see letter 1562, n. 3.
7. *Montenvert*: see letter 1415, n. 15.
8. *Professor v. Ettingshausen*: Andreas Freiherr von Ettingshausen (1796–1878), see letter 1573, n. 34.
9. *his daughter*: either Sophia Karoline (1835–1913) who married Joseph Grailich in 1857; or, Antonie (1828–1916), a women's rights activist, who had married Anton Schrotter von Kristelli sometime before 1851.
10. *Mer de Glace*: see letter 1306, n. 12.
11. *wife and two daughters*: Bertha, Anna, and Christine Magnus.
12. *Finsteraarhorn*: see letter 1536, n. 27.
13. *a single guide*: Johann Bennen.
14. *Monte Rosa*: see letter 1478, n. 22.
15. *a single guide*: Christian Lauener.
16. *Chamouni*: see letter 1353, n. 9.
17. *Mont Blanc*: see letter 1320, n. 6.
18. *tourmente*: see letter 1566, n. 14.

19. *my guide*: Auguste Balmat.
20. *gelées*: ice-cold or frost (French); Tyndall likely used this to mean 'frozen' or 'frost-bitten'.
21. *Saas*: see letter 1537, n. 33.
22. *Monte Moro*: see letter 1547, n. 5.
23. *Miss Gerling*: Marie Platner (née Gerling).

From George Gore   25 December 1858   1577

8 Broad St. Birmingham. | Decr. 25/1858

Dear D^r. Tyndall,

Immediately upon my return from London I succeeded in obtaining a 3rd variety of electro-deposited antimony, apparently possessing the same heating power as the others, and containing Iodide of antimony instead of Bromide or Chloride; and I have sent you a small specimen separately by Post. It requires a higher temperature than either of the others for its discharge; a piece 5/32^{nds} of an inch thick placed upon melted fusible metal suddenly discharged when the latter acquired a temperature of 358 Fah^t, and lost a little of its contained Iodide. It may be safely broken at 60 Fah^t. If you should consider it advisable to mention this variety at the reading of my paper[1] pray do so.

I hope to be able to make comparisons of the specific quantities, specific heats, electro-chemical equivalents, chemical analyses, amounts of molecular heat, temperatures of sudden discharge, chemico-electric and thermo-electric relations, etc of the 3 varieties, by experiment; from which some further information will probably be gained.

By enquiring I have met with a second and independent observer of the supposed molecular discharge in electro-deposited zinc, and am making further trials to verify it.

On electrolysing a hot solution of arsenite of copper in cyanide of potassium with a copper anode, singularly to relate <u>arsenic</u> alone appeared to be deposited instead of the copper!.

Believe me to remain | Yours very truly | <u>George Gore</u>
Dr John Tyndall. FRS. | &c. &c.

CUL SC—add 7656

---

1. *my paper*: G. Gore, 'On the Movements of Liquid Metals and Electrolytes in the Voltaic Circuit', *Roy. Soc. Proc.*, 10 (1859), pp. 235–55.

To Thomas Archer Hirst   [31 December 1858][1]   1578

My dear Tom

It is 10 oC. the last day of the old year, but so dense and dark in London streets that my candle is hardly able to enlighten the gloom. I received your paper[2] with your kindly greeting—a greeting which it is needless for me to reciprocate in words, inasmuch as you know it is responded to by my heart. You are gradually making your way Tom, slowly, surely, and I doubt not deeply writing your name among the world's workers. Your papers I am told have attracted the attention of mathematicians here, Sylvester for example, deems you a very able fellow—Go on and prosper my son, and fulfil all the hopes which your father[3] entertained regarding you.

I should have written to you long since, but that affair[4] in which you take so great an interest was in a doubtful & unsatisfactory condition. They[5] all went to Torquay[6] on the coast of Devon early in November and remained there until the 22nd. of December. Every thing looked as prosperous as possible; but my head began to get bad, and I wrote some letters which I think must have disturbed my friends a little. I once told her mother for example that I hated her opulence because it threw an impassable barrier in my way; I had long thought of sending them my autobiography, and your note strengthened me in this idea. In fact if I could even gain the girl's heart I would not do it while she laboured under the slightest misapprehension regarding me.[7] Well I sent them the sketch. A prompt reply came back, which though kind & full of noble sentiments contained a passage or two which stirred me. For instance she (Mrs D.) said that 'though "May" (we call her <u>May</u>) was not to be won her friendship was mine, and the affectionate esteem of a highminded woman was not to be lightly thrown away.' My first impulse was when I read the letter to tear it up and have done with the matter. —but on reflection I saw that other passages in the letter quite neutralized the above, so I did not know what to think. I wrote asking to be releieved[8] from my engagement on X mas day. I had a prompt reply[9]—'it might do me good to have a run in the country, but then I <u>must</u> come to them <u>on X mas eve</u>'. This note contained no trace of coolness, I said in reply to it, that I had looked into it searching for a trace of coldness, but finding none it would be stupid, if not cowardly, to back out of my engagement: 'If then said I you really wish me to dine with you on X mas day tell me the hour & I will be there'. Throughout the week I had no reply to this—it was written in a very independent style & I thought I had offended them. Well X mas eve came, I had just said to Faraday, who was in the laboratory with me, that it was my intention to have a solitary dinner in the country on X mas day, when a messenger came to say that M^rs^. D. was up stairs & wished to see me. Of course I was with her in a moment & I was so glad to see her that I could not help kissing her tender little hand. She came to tell me their arrangements for X mas day. I was to go at 10C. and have luncheon, then to Westminster Abbey to hear the chanting—then home again to dinner. 'Are you alone at present' I asked—'no she said' <u>May</u> is in the carriage at the door'. I was soon at the carriage—and stood at the door talking—'get in!' said Mrs D. You can't talk there—so I did get in and we had such a jolly ten minutes in the carriage—

Next day I was with them of course, and a most genial happy day it was. I remained with them till 11. P.M. One other gentleman,[10] brother of the Education man Sir James Kay Shuttleworth[11] was with us & that was all. He has known them for longer than I have and is as fine & lovable a fellow as ever I met: full of truth, thought & feeling—he is a barrister and loves poetry— and possibly may love somebody else <u>not May</u>[12]—but I like him very much. He read some poetry for me, and I thought & told him of your reading the Raven[13] and <u>In Memoriam</u>[14] for me, and the furze bushes[15] of the General's field.[16] I have seen my girl[17] many times since—I saw her last night. She is just turned 22, and I believe a dozen fellows at least have proposed to her within the last 4 years. She is a very true and noble girl, and I think—though I am almost afraid to express it—I think that she loves me—It is <u>certain</u> that I love her. I have known her for 3 years & a half, so that she has had no time to love any body else. Of course when I think of it dear Tom the whole thing looks more like a romance than a reality. I have seen that girl in her mother's receptions, surrounded by fine young fellows, with bishops, lords, and marquises her mothers guests, and to think of her choosing me almost bewilders me. Mind she has not told me that she chooses me, and even two days ago I received a peremptory order from her mother not to write to May—telling me that she has resisted many an agonized entreaty for permission to write to her, saying that she was very particular & bidding me to take care—I do not care a straw for the order. For now that my sight is clear I see plainly enough that May has felt kindly towards me for a long long time—but my sceptical, proud heart, kept me from believing this. I am sure I have afflicted her greatly, and I must try to make amends for it in the future—

Goodbye my dear Tom, & God bless you | Ever affectionately yours | <u>John Tyndall</u>

Of course this is all <u>entre nous</u>.[18]

RI MS JT/1/T/648

―――――――――

1. *[31 December 1858]*: the year is given by the reference to Hirst's paper and that Maria Drummond and her three daughters travelled to Torquay (Tyndall had mentioned their plans to do this in letter 1566, also to Hirst). The month and day are given by Tyndall stating that he is writing on 'the last day of the old year'.

2. *your paper*: see letter 1317, n. 8.

3. *father*: Tyndall often referred to Hirst as his son and himself as a father figure. Hirst's father, also Thomas, died in an accident when Hirst was 14 years old.

4. *that affair*: Tyndall was courting Mary Drummond.

5. *They*: Maria Drummond and her daughters Mary, Emily, and Fanny.

6. *Torquay*: see letter 1566, n. 35.

7. *I had long thought of sending them my autobiography . . . the slightest misapprehension regarding me*: see letter 1569.

8. *relieved*: misspelled by Tyndall.

9. *Every thing looked as prosperous as possible . . . I had a prompt reply*: in this passage, Tyndall referred to 'some letters which I think must have disturbed my friends a little', 'your note', 'the sketch', 'A prompt reply', and his own 'prompt reply'. These are all missing.

10. *gentleman*: Joseph Kay (1821–78), an English economist, lawyer, and author of several books on the social condition of the poor. He married Mary Drummond in 1863 (*ODNB*).

11. *Sir James Kay Shuttleworth*: James Kay Shuttleworth (1804–77), an English civil servant and educationalist. Along with Joseph Kay, Edward Ebenezer Kay was another brother (*ODNB*).

12. *May*: nickname for Mary Drummond.

13. *Raven*: Edgar Allen Poe's poem 'The Raven' (1845).

14. *In Memoriam*: Alfred Tennyson's poem 'In Memorium' (1839).

15. *furze bushes*: popular name of *Ulex europæus*, a spiny evergreen shrub with yellow flowers, growing abundantly on waste lands throughout Europe (*OED*).

16. *General's field*: likely a field at or near Queenwood College (see letter 1298, n. 13).

17. *my girl*: Mary Drummond.

18. *entre nous*: between us (French).

To George Gabriel Stokes   31 December 1858   1579

My dear Stokes

I send you a note[1] from Mr Gore which bears upon his paper.[2]

I am much obliged to you for your last note[3]—and most sincerely reciprocate your kind wishes.—A happy new year to you & yours.

Very sincerely yours | John Tyndall

31st. Dec. 1858

CUL SC—add 7656

1. *a note*: letter 1577.
2. *his paper*: see letter 1577, n. 1.
3. *your last note*: letter missing.

From Rudolf Clausius   31 December 1858   1580

Zürich, 31. Dec. 58.

Lieber Tyndall,

Sie haben mich durch Ihren Brief und die vollständige Erzählung von Ihrer Reise sehr erfreut. Ich hatte im Herbst immer auf Nachricht von Ihnen oder Ihre eigene Ankunft gewartet, indessen konnte ich mir wohl denken, dass es bei einer solchen Reise immer möglich ist, seinen ursprünglichen Plan festzuhalten. Die Reichhaltigkeit Ihrer Unternehmungen hat nicht blos mich in Erstaunen gesetzt, sondern auch die naturforschende Gesellschaft, der ich sie mitgetheilt habe. Ein Unternehmen hat man sogar als zu verwegen getadelt, dass Sie es gewagt haben <u>allein</u> den Monte-Rosa zu besteigen.

Durch Ihre Beobachtungen über die blaue Farbe des Schnees ist meine frühere Idee, dass der blaue Himmel dabei von Einfluss sei, widerlegt. Ein Punct, der mir bei Ihrer Beschreibung aufgefallen ist, ist der, dass die Schneedecke in ihrem natürlichen Zustande nicht blau erscheint, dagegen die von einem Alpenstock gemachte Höhlung und der Schnee, welchen der Führer von seinen Füssen schüttelte. Das scheint dahin zu deuten, dass ein gewisser Druck auf den Schnee ausgeübt werden muss um die Farbe hervorzurufen. Dieses könnte man vielleicht so erklären. In dem natürlichen Schnee haben die kleinen Crystalle aus welchen die Blumen bestehen, alle möglichen Lagen, und es finden daher so mannigfaltige Brechungen und Reflexionen statt, dass daraus nur Weiss entstehen kann. Durch einen gelinden Druck aber, werden die Crystalle, welche die Form von dünnen Blättchen haben, in eine gemeinsame Richtung gebracht, und nun können sie alle in gleicher Weise zur Hervorbringung einer Interferenzfarbe wirken, welche bei sehr dünnen Blättchen das Blau erster Ordnung sein würde. Indessen ist dieses nur eine unbestimmte Idee, von der ich nicht weiss ob sie nicht durch andere bei dieser Erscheinung vorkommende Umstände widerlegt wird.

Die Farbenerscheinung in der Wolken ist mir aus Ihrer Beschreibung nicht ganz klar geworden. Wenn eine und dieselbe Wolke an verschiedenen Stellen verschiedene Farbe gezeigt hat, so glaube ich nicht dass das Farben dünner Blättchen sein konnten, denn sonst müsste man annehmen, dass an den verschiedenen Stellen die Blättchen verschiedene Dicke hätten und an jeder Stelle eine bestimmte Dicke vorwaltend war. Ich würde eher an Beugungsfarben denken, wie man sie sieht wenn man durch ein Glas, was mit

Lykopodiumsaamen bestreut ist, ein Licht betrachtet. Ich weiss übrigens nicht genau was Sie unter colours of mixed plates verstanden haben.

Auf Ihre weitere Entwickelung der Gletschertheorie bin ich sehr gespannt. Dass Ihre Hauptidee, <u>dass der Druck bei der Bildung der Gletscherstructur von wesentlichem Einfluss ist</u>, die richtige ist, bezweifele ich gar nicht, und es handelt sich nur darum, ins Einzelne zu verfolgen, in welcher Weise die Wirkung des Druckes stattfindet. Die im Dec. Heft des Phil. Mag. ausgesprochene Ansicht von Thomson, dass der verschiedene Druck an den verschiedenen Stellen eines Bläschens an gewissen Stellen Schmelzung und an anderen Gefrieren bewirkt, ist zwar sinnreich aber etwas complicirt, und ich sollte denken dass das Plattwerden der Bläschen sich einfacher erklären lässt. Auch spricht dagegen der Umstand, dass viele Bläschen diese Gestaltänderung nicht zeigen, während doch die Thomson'sche Wirkung sich auf alle erstrecken müsste.

Ich sende mit diesem Briefe zugleich eine kleine Abhandlung an Sie ab. Den ersten Theil, bis S. 253, wo der Strich ist, betrachte ich als eine wesentliche Vervollständigung meines Aufsatzes 'über die Art der Bewegung etc–' der letzte Theil dagegen enthält nur einige Bemerkungen von untergeordnetem Interesse.

Mit freundlichen Grüssen und herzlichen Gluckwünschen zum neuen Jahre, | Ihr | Clausius.

Zürich, 31. Dec. 58.

Dear Tyndall,

You have given me much pleasure through your letter[1] and the full account of your journey. I had been waiting all along through the autumn for news from you or for you to arrive yourself, however, I was able to well imagine that it is not always possible to keep to one's original plan with such a journey. The wide variety of your activities astonished not only me, but also the natural history society,[2] to which I communicated them. One undertaking was even criticised for being too foolhardy, that you dared to climb the Monte Rosa <u>alone</u>.[3]

Through your observations on the blue colour of snow,[4] my earlier idea that the blue sky may be an influence here is now disproved.[5] One point that struck me with your description is the one that the snow cover in its natural state does not appear blue, in contrast to the cavity made by an alpenstock[6] and the snow which the guide shook off his feet. That seems to point toward the fact that a certain pressure must be exerted on the snow to give rise to the colour. Perhaps one could explain this thus. In natural snow, the small crystals of which the flowers consist have all possible positions, and there are therefore such diverse refractions and reflections taking place that from this only white can emerge. By means of gentle pressure, however, the crystals, which have the form of thin lamellae,[7] are brought into a common direction, and now they can all work in an identical way to produce an interference colour, which, with very thin lamellae, would be first order blue. However, this is only an indeterminate idea, about which I do not know whether it would be disproved by the other circumstances occurring with this phenomenon.

The appearance of colour in clouds did not become entirely clear to me from your description. If one and the same cloud has displayed different colour in different places, then I do not believe that that could be the colouring of thin lamellae, for otherwise one would have to assume that in different places the lamellae had different thickness, and that in every place a particular thickness prevailed. I would rather think of diffraction colours, as one sees them when one looks at a light through a glass which is sprinkled with lycopodium seeds.[8] At any rate, I do not know exactly what you have understood by 'colours of mixed plates'.

I am looking forward very much to your further development of the theory of glaciers. I do not doubt at all that your main idea, <u>that pressure is of fundamental influence with the formation of glacier structure</u>, is the correct one, and it is only a matter of tracing into specific detail in which way the effect of the pressure occurs. The view expressed by Thomson in the Dec. issue of the Phil. Mag.[9] that differing pressure on different areas of a vesicle[10] causes melting on some areas and freezing on others is indeed ingenious, but somewhat complicated, and I should think that the flattening of the vesicle can be explained more simply. What also speaks against it is the fact that many vesicles do not show this change in shape, while Thomson's effect would really have to extend to all of them.

I am sending a short paper[11] to you with this letter. The first part, up to p. 253, where the line is, I consider to be a fundamental improvement of my essay 'über die Art der Bewegung etc–',[12] however, the last part only contains some remarks of secondary interest.

With kind regards and warm good wishes for the New Year, | Your | Clausius.

RI MS JT/1/TYP/7/2232–33
LT Typescript Only

1.	*your letter*: see letter 1573.

2.	*natural history society*: see letter 1461, n. 3.

3.	*you dared to climb the Monte Rosa <u>alone</u>*: on 17 August 1858, Tyndall made a solo summit of Monte Rosa, the second highest mountain in the Alps. For an account of this ascent, see letter 1547.

4.	*your observations on the blue colour of snow*: see letter 1573.

5.	*my earlier idea that the blue sky may be an influence here is now disproved*: see letter 1461, n. 30.

6.	*alpenstock*: see letter 1424, n. 6.

7. *lamellae*: plural of lamella, a thin plate, scale, layer, or film, especially of bone or tissue (*OED*).

8. *lycopodium seeds*: a fine flammable powder made of the spores of a club-moss, or Lycopodium (*OED*).

9. *Thomson in the Dec. issue of the Phil. Mag.*: W. Thomson, 'On the Stratification of Vesicular Ice by Pressure', *Phil. Mag.*, 16 (1858), pp. 463–66.

10. *vesicle*: a minute bubble or spherule of liquid or vapour (*OED*).

11. *a short paper*: R. Clausius, 'Ueber die mittlere Länge der Wege, welche bei der Molecular-bewegung gasförmiger Körper von den einzelnen Molecülen zurückgelegt werden, nebst einigen anderen Bemerkungen über die mechanische Wärmetheorie', *Poggend. Annal.*, 105 (1858), pp. 239–58, which appeared in English in February 1859 as R. Clausius, 'On the Mean Length of the Paths described by the separate Molecules of Gaseous Bodies on the occurence of Molecular Motion; together with some other Remarks upon the Mechanical Theory of Heat', *Phil. Mag.*, 17 (1859), pp. 81–91.

12. *my essay 'über die Art der Bewegung etc–'*: 'on the kind of motion etc–' (German). See R. Clausius, 'Ueber die Art der Bewegung, welche wir Wärme nenne', *Poggend. Annal.*, 100 (1857), pp. 353–80.

1859

From George Day[1]   11 January 1859   1581

United College | St. Andrews. | Jan. 11th 1859.

My dear Dr. Tyndall,

You are, I believe, aware that our Chair of Natural Philosophy at St. Andrews[2] is now vacant.

The Professors are the Electors. Several of us feeling that we could hardly find better judges, asked Stokes and Adams[3] to suggest the best man for us, and they both agree in strongly recommending Mr Liveing[4] Fellow of St. John's College, Cambridge,

<Handwritten letter begins>

as the best Candidate. You may perhaps have read his paper on the Transmutation of Matter in the Cambridge Essays.[5]

Now we should, I have no doubt, succeed in carry Liveing, if it were not for a document published by another Candidate, a M[r] Swan,[6] a private teacher of Mathematics in Edinburgh, emanating from M[r] Faraday whose name very justly carries with it so much weight that the testimonial (which I enclose)[7] will in all probability overbalance the strong evidence of Stokes and Adams in favour of Liveing.

Would you so far oblige us as to ask M[r] Faraday whether he intended his note to be used as a general Testimonial for a Natural Philosophy Chair, or whether he merely meant it as a kind encouragement to a young and unknown investigator?

If M[r] Faraday really does think that M[r] Swan's Papers entitle him to aspire to our Chair, we shall probably all (or almost all) vote for him; if however this is not the case we should act upon the recommendation of Stokes and Adams.

Your answer will of course only be confidentially shown to two or three of my Colleagues.

The matter is one of so much importance that I feel we need offer no apology either to M[r] Faraday or yourself for the trouble we are imposing on you.

An early reply will oblige | Dear D[r] Tyndall | yours ever truly | Geo. E. Day.

We all very much regret that we could not persuade you to accept our Chair.[8] Wont six months holiday still tempt you?

RI MS JT/1/D/12
RI MS JT/1/TYP/1/317

---

1. *George Day*: George Edward Day (1815–72), a Welsh physician who was Chandos Professor of Anatomy and Medicine at the University of St Andrews from 1849 until his retirement in 1863 (*ODNB*).
2. *St. Andrews*: St Andrews University in Scotland.
3. *Adams*: John Couch Adams (1819–92), a British mathematician and astronomer, most well-known for his prediction of the existence and position of the planet Neptune (*ODNB*).
4. *Mr Liveing*: George Downing Liveing (1827–1924), an English chemist at the University of Cambridge. From 1860–80, he served as professor of chemistry at the Royal Military College at Sandhurst (*ODNB*).
5. *his paper on the Transmutation of Matter in the Cambridge Essays*: G. D. Liveing, 'On the Transmutation of Matter', *Cambridge Essays, contributed by Members of the University*, 1 (1855), pp. 123–47.
6. *Mr Swan's*: William Swan (1818–94), a Scottish mathematician and physicist. He was appointed as Chair of Natural Philosophy at St Andrews in 1859 and served until 1880 (*ODNB*). See letter 1572, n. 3.
7. *the testimonial (which I enclose)*: this is not found in the *Faraday Correspondence*.
8. *could not persuade you to accept our Chair*: see letter 1572, where David Brewster asked Tyndall if he wanted the position; Tyndall's reply is missing.

To George Biddell Airy   17 January 1859   1582

17th Jan. 1859

My dear Sir

I have been reading with pleasure and interest your lectures at Ipswich[1]—would you kindly inform me whether any models similar to those you used in the lectures are to [be][2] borrowed, or then loan purchased, at any place in London?

In one of your lectures at the R.I.[3] you had a model of the experiment of Cavendish[4]—is it to be found in London?

I have a course of lectures[5] which commence on the 27th. and I must talk a little of the gravity & motion of the planets—as you know very well our audience requires illustration.

Have you ever tried the experiment of Cavendish with a fine torsion balance and Gauss's reflector?[6] I almost think the attraction would manifest only with much smaller masses than those used by Cavendish.

Most truly yours | John Tyndall

I am hard at work upon a glacier paper[7] which I hope will interest you.

RGO MS.RGO 6/408.378–79

1. *lectures at Ipswich*: G. B. Airy, *Six lectures on astronomy: delivered at the meetings of the friends of the Ipswich Museum, at the Temperance Hall, Ipswich, in the month of March 1848* (London, 1849). The Ipswich Museum in the English town of Ipswich in the county of Suffolk, was founded in 1846 and devoted largely to natural history collections.

2. *[be]*: the word 'be' was missing from the letter.

3. *one of your lectures at the R.I.*: on 2 February 1855, Airy delivered a lecture at the RI entitled 'On the Pendulum-experiments lately made in the Harton Colliery, for ascertaining the mean Density of the Earth' (for an abstract, see *Roy. Inst. Proc.*, 2 (1854–58), pp. 17–22).

4. *experiment of Cavendish*: Henry Cavendish's (1731–1810) torsion balance experiment in the late nineteenth century that measured the force of gravity between two masses, and yielded an estimated mass density of the Earth.

5. *course of lectures*: Tyndall delivered 'Twelve Lectures on the Force of Gravity' at the RI in 1859, on Thursday afternoons from 27 January to 14 April (*Roy. Inst. Proc.,* 2 (1854–58), p. 4). See RI Guard Book, vol. 2, p. 110–11.

6. *Gauss's reflector*: heliotrope, invented by Carl Friedrich Gauss (1777–1855), a German astronomer and mathematician (*CDSB*).

7. *glacier paper*: J. Tyndall, 'On the Veined Structure of Glaciers; with observations upon White Ice-seams, Air-bubbles and Dirt-bands, and remarks upon Glacier Theories', *Phil. Trans.,* 149 (1859), pp. 279–307. Read for the RS on 24 February 1859.

To George Gabriel Stokes   17 January 1859   1583

17th Jan 1859

My dear Sir.

I should not infer from the paper of M[r]. Thomas[1] 'On the nature and action of gunpowder'[2] that he is acquainted with all that is known at the present day of the constitution of elastic fluids; In my opinion he treats the fundamental point of his paper in what may be termed an unphysical spirit—masking indeed the physics of the process which he undertakes to examine under the vague term 'impulsion'. Neither do I think that his experiments in all cases justify his conclusions. And therefore I cannot recommend the paper, in its present state, for publication in the Philosophical Transactions.[3]

I am not sufficiently acquainted with the facts of gunnery to say whether under the supposition of Mr Thomas anything of practical value lies concealed; but I think a paper to be accepted by the Royal Society ought to exhibit more of scientific knowledge & precision than are to be found in the present paper.

I am dear sir | very truly yours | <u>John Tyndall</u>
Prof. Stokes | &c. &c. &c. | Sec. R.S.

RS RR/4/264 Tyndall 17 January 1859

1.　*M^r. Thomas*: William Lynall Thomas, of Anderton in southwest England, filed for a patent in 1855 (and sealed in 1856) 'for an improvement in projectiles and in gun wads'. In 1877, he went to trial against the Queen over a patent for heavy ordnance. The trial went in his favour, and he was awarded by a special jury the sum of £8,790 11s. 6d., but the case remained under litigation.

2.　*'On the nature and action of gunpowder'*: L. Thomas, 'On the Nature of the Action of Gunpowder', *Roy. Soc. Proc.*, 9 (1859), pp. 586–88. Read to the RS on 16 December 1858 by a 'Dr. Gray', presumably Henry Gray (1827–61), FRS, whose *Gray's Anatomy* was published in 1858.

3.　*I cannot recommend the paper . . . the Philosophical Transactions*: Tyndall acted as a referee for the *Phil. Trans.*, reviewing submissions and sending reports to Stokes, the journal's editor.

From Robert Main[1]　　　　18 January 1859　　　　1584

1859, January 18.

Dear Sir

The Astronomer Royal[2] is not at home, but I expect his return in a day or two, when I have no doubt, he will send you an answer to your inquiry[3] respecting models for illustration of lectures on Astronomy.

I am, | dear Sir, | Your faithful servant, | Robert Main
John Tyndall Esq[ui]^r[e]·

RGO MS.RGO 6/408.380

1.　*Robert Main*: Robert Main (1808–78), First Assistant at the Royal Observatory, Greenwich from 1835–60 (*ODNB*).
2.　*The Astronomer Royal*: George Airy.
3.　*your inquiry*: letter 1582.

From George Biddell Airy 26 January 1859 1585

1859 January 26

My dear Sir

Most unhappily your letter of 17th.[1] arrived just when my return from the country was doubtful for a day or two, and was not forwarded: and since returning I have had to go through a vast mass of papers, and reached it only last night. I fear that my answer now is useless.

1. The apparatus for my lectures at Ipswich[2] was cobbled up here as I wanted it. It is the property of the Ipswich Museum. Lately I borrowed some of it for a lecture at Bury:[3] it required considerable repair. I suppose that all the rest, so far as it exists, would require repair. I do not imagine that any thing of the sort is to be found at shops.

2. <u>Part</u> of the Cavendish model[4] is here: it must have come by some mistake. You shall have it tomorrow (unless you countermand it). I enclose a note[5] to my friends Mess[rs]. Ransomes Sims[6] of the Orwell Works[7] Ipswich, who have so much influence in the Museum that they can do anything to serve you there. You can forward the note, or not, as you think best.

I am, my dear Sir, | Yours very truly | <u>G B Airy</u>

I have no doubt that the use of the mirror would have sensibly increased the delicacy of the observation.

Professor Tyndall

RGO MS.RGO 6/408.381–82

---

1. *your letter of 17th*: letter 1582.

2. *my lectures at Ipswich*: see letter 1582, n. 1.

3. *a lecture at Bury*: in October 1858, Airy gave a lecture on astronomy at the Athenaeum in Bury St Edmunds, Suffolk. Airy's talk, during which the spectacular Donati's comet was on display (see letter 1560, n. 5), inspired the construction of an astronomical observatory. See M. P. Mobberley and K. J. Goward, 'The Bury St Edmunds Athenaeum Observatory', *Journal of the British Astronomical Association,* 115 (2005), pp. 251–60, on p. 254.

4. *Cavendish model*: see letter 1582, n. 4.

5. *a note*: letter missing.

6. *Mess[rs]. Ransomes Sims*: Robert Ransome (1795–1864), James Allen Ransome (1806–75), and William Dillwyn Sims (1825–95) operated an agricultural machinery production company in Ipswich which also produced iron works for astronomical observatories. The company, founded in the late eighteenth century by Robert Ransome (1753–1830), changed ownership between many members of the Ransome family and others. See B. Bell, *Ransomes Sims & Jefferies: Agricultural Engineers* (Sheffield: Old Pond Publishing, 2001).

7. *Orwell Works*: the name of the Ransomes and Sims' production plant.

To George Biddell Airy   28 January 1859   1586

28th. Jan. 1859

My dear Sir.

I am extremely obliged to you for both your notes.[1]

The <u>case</u>[2] has arrived & I hope its contents will be turned to good account.

I <u>have</u> forwarded your note[3] to Ipswich[4]—if the rotating frame reach me any time before Wednesday next it will be quite early enough.

Thank you very much for your kindness

believe me | most sincerely yours | John Tyndall

RGO MS.RGO 6/408.386

———————

1. *both your notes*: letter 1585, and possibly 1584 (from Robert Main).

2. *case*: presumably this case contained part of Airy's Cavendish model (see letter 1585).

3. *your note*: a note from Airy to 'Mess[rs]. Ransomes Sims' (see letter 1585, n. 5).

4. *Ipswich*: see letter 1582, n. 1.

From Thomas Archer Hirst   [5 February][1] 1859   1587

Rome. Jan 30th 1859

My dear John.

Your last letter[2] arrived just at a time when a kindly word was very acceptable to me—it was at one of those transient moments when a slight indisposition or mental fatigue begets a tendency to take a gloomy view of things in general. I learnt from your letter that to you at least the horizon was brightening and thus some reflected light could not fail to reach me too. I shut up my books and set out for my daily walk the better to chew the cud of the good news your letter contained and thus ruminating my spirits rose even to cheerfulness. I recalled with a smile our last words at Saas[3] when I asked to be kept informed of certain affairs and you with resolution apparently fixed replied that 'all was finished' to which I being very incredulous on that point pleaded to be informed of the end of it. That same end will very probably be what I always suspected and what I ardently hope for. At all events matters are proceeding slowly and well to my hearts content if my sincere wishes can help the 'finis'[4] to which I alluded at Saas will not tarry long.

I look every week at the 'Atheneum'[5] for some account of proceedings at the Royal Institution. Your long vacation is just finished and you are about commencing to amuse and instruct once more your fashionable audience. My solitary life has of course a great influence on my imagination but I must say

I feel an internal shudder when I read that you were to give twelve lectures on the Law of Gravitation[6] I imagined myself sentenced to talk for twelve successive nights to <u>your</u> audience on the Law of Gravitation and the task appears so formidable that to me transportation for twelve years would be preferable. At the same time I have no fear whatever of your passing even successfully through the terrible ordeal and I should very much like to be one of your hearers You have certainly deviated widely from the ordinary routine of Physical lectures wherein the said Law of Gravitation usually presents little interest and is regarded as one of the branches most barren in interesting experiments. I am sure you will prove that it is not so and I should greatly like to see how you will do this. Forbes I see has spoken once more on the subject of regelation[7] I found his article[8] little instructive, it bears the traces of a man labouring under hopeless indisposition. Poor man the vigor which animates his former memoirs seem all departed, all replaced by discontent and lassitude—such at least was the impression produced upon me by his short article. Tell me in your next what you have been working upon, principally since you left Switzerland; how does the book on Switzerland progress?[9]

And now I must say a word about myself and first about my health. You will be glad to hear that it is long since I enjoyed such good health with such an amount of sedentary work. I do not know whether it is that I am learning to treat myself more prudently or whether it is that the climate and diet of Rome are suitable to me, but such is the fact—it is long since I passed through a winter so well. The bowels do their part very creditably and—though they threaten to rebel occasionally I have hitherto succeeded by natural means, without medicine of any description—to wax them again into good humour. Now and then I have to put down my pen and walk for a couple of days much to the astonishment of my Italian friends, whose temperaments differ so widely from mine and from those of all other northern barbarians. I will venture to say there are few Romans who, during their whole lives, have scampered so much over the wide Campagna[10] as I have done during the last three months, and my scamperings are of infinite service to me. As is natural my work succeeds in the same proportion as my health is good. I am just putting the finishing strokes to a memoir[11] (in French) which will appear shortly in Tortolini's journal[12] If I can judge it will be as solid a piece of work as I have yet produced. The subject is purely geometrical and not extensive, but well laboured. Mathematicians will find nothing brilliant in it—no wide flashes illuminating vast fields, such as they continually meet with in the productions of Cayley,[13] Sylvester and others. Such brilliances are not in me and I must content myself with more modest gifts, with less extensive fields in fact with a little garden which I can well dig and in my plodding way render capable of producing some fruit though only a few flowers.

I know very few people here either English or Italian. The uniformity of my week is broken only by a visit which good Tortolini always pays me on a Thursday Evening and which I return generally on a Sunday He is a simple kindly soul who ardently loves his mathematics and never tires talking about them. There is little depth or 'resistance' in him in this respect he reminds me greatly of Knoblauch[14] In mathematics he flies like a busy bee from flower to flower and buzzes contentedly if he finds even a minimum of honey in them. A man destitute of guile, incapable of inveterate hate or of deep love and as often happens generous enough of the love and friendship of which he is capable. Above all he is incapable of silently manifesting respect and esteem, for his part he is convinced that the expression of these feelings is inseparable from their existence This being the man you will at once know how to interpret the following lines I found written on the fly leaf of a volume of his researches which he presented to me the other day. You will easily be able to translate 'Al mio carissimo (superlative) Amico Sig^r Dottor, T. A. Hirst in occasione del suo viaggio in Roma—in attestato di sincera Stima ed Affetto[15]—Barnaba Tortolini' It is needless to say that the lines taken even at their precise value gave me pleasure

A week has passed by since I wrote the above. I thought by keeping my letter a while to increase its length but I find I must let it go. I enclose an account sent to me by Williams and Norgate[16] which I will ask you to pay for me and in so doing, please draw their attention to the question they put at the foot of the note and to my answer there written. It is Saturday night and I am a little weary of holding the pen; tomorrow I must devote religiously to exercise in the open air hence the necessity for finishing abruptly my letter.

You know of course that the Prince of Wales[17] is here, he lives very near to me but I have never seen nor probably ever shall. The knowledge of his movements will reach me through English papers.

The old king of Prussia[18] is here too. I <u>did</u> see him today, poor old fellow he is a sad wreck of a man. This reminds me to tell you that I have read Carlyles life of Frederick the Great[19] i.e. as much of it as is at present published It is an admirable piece of work and has been a source of great pleasure and relaxation to me.

With best wishes Dear John Yours affectionately <u>Tom</u>

RI MS JT/1/H/243

---

1. *[5 February]*: while the letter is dated as 30 January 1859, in the letter Hirst wrote on a 'Saturday night' that 'A week has passed by since I wrote the above', making the date of completion 5 February.
2. *Your last letter*: letter 1578.
3. *Saas*: see letter 1537, n. 33.
4. *'finis'*: the conclusion, end, finish (Latin, *OED*).

5. *'Atheneum'*: *The Athenaeum,* a British literary magazine published from 1828–1921.

6. *twelve lectures on the Law of Gravitation*: see letter 1582, n. 5.

7. *regelation*: see letter 1306, n. 10.

8. *his article*: possibly J. D. Forbes, 'On some Properties of Ice near Its Melting Point', *Proceedings of the Royal Society of Edinburgh,* 4 (1857–58), pp. 103–7. Read 19 April 1858.

9. *the book on Switzerland progress*: *Glaciers of the Alps.*

10. *Campagna*: Campania, a mountainous region in southern Italy which includes Mount Vesuvius (Campagna is a town within Campania).

11. *memoir*: 'Sur la courbure d'une série de surfaces et de lignes', *Annali di matematica pura ed applicata,* 2 (1859), pp. 95–112, 148–67.

12. *Tortolini's journal*: *Annali di matematica pura ed applicata,* edited by Barnaba Tortolini (1808–74), an Italian priest and mathematician. From 1850–57, the journal was devoted to broader exact sciences topics, and in 1858 narrowed to a focus on pure mathematics. See 'Barnaba Tortolini', *Annali di Matematica Pura ed Applicata,* 7 (1875–76), pp. 63–64.

13. *Cayley*: Arthur Cayley (1821–95), a British mathematician who focused on pure mathematics (*ODNB*).

14. *Knoblauch*: Hermann Knoblauch, see letter 1414, n. 1.

15. *Al mio carissimo (superlative) Amico Sigr Dottor, T.A. Hirst in occasione del suo viaggio in Roma—in attestato di sincera Stima ed Affetto*: To my dearest Friend Mr. Doctor, T.A. Hirst in the circumstance of his travel to Rome—as a token of my respect and affection (Italian).

16. *Williams and Norgate*: a London and Edinburgh publisher. The enclosure is missing.

17. *Prince of Wales*: see letter 1478, n. 5.

18. *old king of Prussia*: Frederick William IV (1795–1861), who reigned as King of Prussia from 1840–61 (*ADB*).

19. *Carlyles life of Frederick the Great*: the first two volumes of Thomas Carlyle's six-volume *History of Friedrich the Second of Prussia, Called Frederick the Great* (London: Chapman and Hall) were published in 1858 (and the entire work completed by 1865).

From James David Forbes　　　　12 February 1859　　　　1588

D[r] Tyndall | R. Institution | London | Edinburgh | 12th Feb. 1859

My dear Sir

I should like to address to you as one of the Editors of the Phil. Magazine a very few remarks[1] on the little paper on Ice[2] which you sent me. I will endeavour to let you have them in the course of next week as I should like them to appear in the Magazine for March. My letter will be short so I hope you will try to keep a place for it. I remain

dear Sir Yours faithfully | James D. Forbes

StA JDF Incoming letters 1859, no. 532

---

1. *few remarks*: 'Remarks on a paper "On Ice and Glaciers" in the last Number of the Philosophical Magazine. In a letter to Prof. Tyndall. By Prof. J.D. Forbes', *Phil. Mag.*, 17 (1859), pp. 197–201.

2. *little paper on Ice*: J. Tyndall, 'Remarks on Ice and Glaciers', *Phil. Mag.*, 17 (1859), pp. 91–96.

To Julius Plücker   13 February 1859   1589

13th Feb. 1859

My dear Sir

It is with mixed feelings of regret and doubt that I commence this letter—regret that I should have left your last letter[1] to me so long unanswered, and doubt as to whether you will accept either my reasons for doing so, or my tardily expressed repentance.

I received your letter in the midst of the mountains of Switzerland when every day was more or less filled with excitement. I <u>lost</u> it there, and hence had it not before me on my return to remind me of my duty—Add to this that my health has been often so delicate as to take away from me sometimes the power of writing to my friends, and you have the chief reasons why this letter has been deferred until this time.

I hope however that your beautiful researches have not been neglected by the Philosophical Magazine, but that they have had in England the publicity which is so justly due to them.

M[r]. Gassiot still continues to look at the subject. M[r]. Grove gave a lecture[2] upon it sometime ago at the Royal Institution, in which he advanced a new, but to me not very clear, nor yet satisfactory theory of the stratifications. He supposes them to be due to the conflict of opposing currents, and assumes the excitement and play of <u>tertiary</u> and other currents in the circuit which produces the phenomena of the striae. By interposing a break in the circuit he supposes that he can cut off the feebler of the two currents which produce the effect—and such a gap destroyed the stratification in an experiment which he made before our members. As I have stated already neither his experiments nor his reasonings are to me clear or conclusive.

M[r]. Faraday continues very well—he has lately collected all his minor and earlier researches into a single volume[3] which has been published by Taylor & Francis. I ought, perhaps, to say Francis only, for old M[r]. Taylor[4] died some weeks ago.

I am still at work upon the glaciers, but hope to finish them this year—I had no idea when I commenced the enquiry that it would occupy me so long—I have now I think reduced their principal phenomena to their <u>physical causes.</u>

Need I say that it will give me pleasure to hear from you whenever you are at leisure to write to me?

Believe me dear Sir | very sincerely yours | <u>John Tyndall</u>

NRCC Plücker, item #70 (Box 2/6)

———

1. *your last letter*: letter missing; as Tyndall mentioned, he lost it while in Switzerland the previous summer.
2. *a lecture*: on 28 January 1859, Grove presented at the RI 'On the Electrical Discharge, and it Stratified Appearance in Rarefied Media'. For an abstract, see *Roy. Inst. Proc.,* 2 (1858–62), pp. 5–10.
3. *a single volume*: M. Faraday, *Experimental Researches in Chemistry and Physics* (London: Taylor and Francis, 1859).
4. *M*. *Taylor*: Richard Taylor (1781–1858), a prominent English publisher, especially of scientific journals (including the *Phil. Mag.*). He went into partnership with his son William Francis following a nervous breakdown in 1852, establishing the Taylor & Francis publishing firm. Taylor died on 1 December 1858 (*ODNB*).

From Richard Owen[1]   18 February 1859   1590

12, Hertford-street, Mayfair, W., | February 18th, 1859.

My dear Sir,—

Having, as I informed you in my last note,[2] communicated with the Sardinian Minister Plenipotentiary[3] the day after receiving your statement relative to the guides at Chamouni,[4] I have been favoured by replies from the Minister, of the 4th and 17th February. In the first the Marquis d'Azeglio assures me that he will bring the subject before the competent authorities at Turin,[5] accompanying the transmission 'd'une récommendation toute spécial.'[6] In the second letter the Marquis informs me that 'the preparation of new regulations for the guides at Chamouni had for some time occupied the attention of the Minister of the Interior,[7] and that these regulations[8] will be in rigorous operation, in all probability, at the commencement of the approaching summer.' The Marquis adds that, 'as the regulations will be based upon a principle of much greater liberty, he has every reason to believe that they will satisfy all the desires of travellers in the interests of science.'

With much pleasure at the opportunity of having been in any degree able to bring about the fulfilment of your wishes on the subject,

I remain, my dear Sir, | Faithfully yours, | RICHARD OWEN. | Pres. Brit. Association.

Prof. Tyndall, F.R.S.

Glaciers of the Alps, pp. 193–94

1. *Richard Owen*: see letter 1330, n. 5.
2. *my last note*: letter missing.
3. *Sardinian Minister Plenipotentiary*: Massimo Taparelli d'Azeglio (1798–1866), an Italian statesman and novelist. He served as Prime Minister of Sardinia from 1849–52. See A. Ferrabino (ed.), *Dizionario Biografico Degli Italiani* (Rome: Istituto della Enciclopedia Italiana, 1962), vol. 4, pp. 746–52.
4. *Chamouni*: see letter 1551, and 1566, n. 18.
5. *Turin*: a city in northern Italy.
6. *'d'une récommendation toute spécial'*: of a special recommendation (French).
7. *Minister of the Interior*: Urbano Rattazzi (1808–73) served as minister of the interior for the Kingdom of Sardinia-Piedmont from 1854–58 and 1859–60. See F. Gallo (ed.), *Dizionario Biografico Degli Italiani* (Rome: Istituto della Enciclopedia Italiana, 2005), vol. 86, pp. 558–62.
8. *these regulations*: various editions of John Murray's *Handbook for Travellers in Switzerland and the Alps of Savoy and Piedmont* show the changes to the regulations of Chamouni guides. The 8th edition (1858) stated that the guide chef assigned guides based on a rotation called the 'tour de rôle', and that guides who went out of turn were fined. 'This arrangement is certainly calculated to repress emulation amongst the guides, and great complaints are made of it by those who wish to make expeditions into the High Alps' (*A Handbook for Travellers in Switzerland and the Alps of Savoy and Piedmont,* 8th edn (London: John Murray, 1858), p. 347). The 9th edition (1861) stated that new regulations removing these restrictions were established in 1859 by the King of Sardinia 'on the application of the Alpine Club, and some eminent scientific men'. The regulations were, however, re-established after Savoy was annexed to France in 1860 (*A Handbook for Travellers in Switzerland and the Alps of Savoy and Piedmont,* 9th edn (London: John Murray, 1861), p. 386). The 11th edition (1865) stated that travellers were again permitted to select 'their own guides if | 1. They desire to undertake "extraordinary" expeditions. | 2. If engaged in scientific pursuits. | 3. If they do not know French, and require a guide to speak a language they know. | 4. If they desire to re-engage a guide they have had in former years. | 5. Ladies, unaccompanied by a gentleman. | Members of the Alpine Club, and those who can give evidence of having frequently made expeditions in the high Alps, are exempt from restrictions of the tour de rôle, and also the number of guides they take with them' (*A Handbook for Travellers in Switzerland and the Alps of Savoy and Piedmont,* 11th edn (London: John Murray, 1865), p. 396).

To George Biddell Airy   26 February 1859   1591

26th <u>Feb. 1859</u>

My dear Sir

Would you have the kindness to say to me in a simple line whether I am to send the whole of the apparatus[1]—including that which you were good enough to send to me from Greenwich[2]—to Ipswich?[3] I hope you will excuse the troubles I give you.

Most truly yours | John Tyndall

RGO MS.RGO 6/408.385

---

1. *apparatus*: see letter 1585.
2. *Greenwich*: see letter 1374, n. 2.
3. *Ipswich*: see letter 1582, n. 1.

From George Biddell Airy   26 February 1859   1592

1859 Feb^y.26

My dear Sir

The whole of the Apparatus[1] ought to go to Ipswich.[2] The part which was sent from this place[3] had come by some accident which I do not understand.

I am my dear Sir | Yours very truly | G B Airy
Professor Tyndall

RGO MS.RGO 6/408.387

---

1. *Apparatus*: see letter 1585.
2. *Ipswich*: see letter 1582, n. 1
3. *this place*: the Royal Observatory, Greenwich, where Airy served as Astronomer Royal from 1835–81.

From Herbert Spencer[1]　　　[February 1859][2]　　　1593

13 Loudoun Road | St John's Wood

My dear Tyndall,

Had not the announcement of coffee prevented, I had hoped to carry much further the discussion we commenced on Saturday evening.[3] Lest you should misunderstand me, let me briefly say now what I wished to say.

In the first place I fully recognize, and have all along recognized, the tendency to ultimate equilibrium; and have, after sundry other chapters on the general laws of change, a final one entitled 'The Equilibration of Force'.[4] Indeed of the general views which I have of late years been working out this was oddly enough the first reached. Among memoranda jotted down for a second edition of 'Social Statics',[5]—memoranda written toward the close of 51 or early in 52—I have some bearing on this law in its application to society. Seeking, as I had been doing, for a deeper basis for the principles of equity set forth in that book, I had arrived at this as the ultimate. The universal tendency towards equilibrium must apply to human actions as well as to others. All our hourly doings are endeavours to equilibrate our acts and wants. Those modifications of faculty which we are ever undergoing are a continuous approximation to equilibrium between constitution and conditions. So long as desires and their satisfactions are not balanced there must be change either by the abatement of the desire or the obtainment of the satisfactions. That is, there must be change until equilibration is reached. This applies to man in his social relations. Society must go on changing until all men's constitutions and their circumstances are in correspondence—<u>until the desires of each are such only as may be gratified without thwarting the desire of others.</u> That is, change must continue until all men are organically just—until none have any desires at variance with the welfare of others. And thus the moral law that 'Every man is free to do whatever he will provided he does not trench upon the equal freedom of others' (which it is the aim of 'Social Statics' to establish) is necessarily the law of the ultimate social state. Thus you see that my views commit me most fully to the doctrine of ultimate equilibration.

That which was new to me in your position enunciated last June, and again on Saturday, was that equilibration was death. Regarding as I had done, equilibration as the ultimate and <u>highest</u> state of society, I had assumed it to be not only the ultimate but also the highest state of the universe. And your assertion that when equilibrium was reached life must cease, staggered me. Indeed, not seeing my way out of the conclusion, I remember being out of spirits for some days afterwards. I still feel unsettled about the matter and should like some day to discuss it with you.

Let me further say that you must not take the views put forth in the articles on Progress and Physiology[6] as anything more than rude, misshapen germs. During the past year they have been undergoing a development which I never anticipated; and the ideas at present published seem to me to stand towards a true theory much as an egg does to a bird, or an Amphioxus[7] to a man. Indeed it is the wide and rapid evolution which these (and some connate ideas) have been undergoing, and the power which I see they have of absorbing and organizing all I had before thought, which renders me so anxious to get the opportunity of working them out.

ever yours truly | Herbert Spencer

RI MS JT/1/S/191
RI MS JT/1/TYP/3/1175

---

1. *Herbert Spencer*: Herbert Spencer (1820–1903), an English philosopher, sociologist, biologist, and anthropologist. A prolific writer on diverse subjects, one of Spencer's most lasting contributions was his coining the phrase 'survival of the fittest' in his *Principles of Biology* (1864) (*ODNB*).

2. *[February 1859]*: the year and month are based on the suggestion that this letter is from late 1858 or early 1859 in D. Duncan, *Life and Letters of Herbert Spencer* (London: Williams and Norgate, 1911), p. 103. February 1859 is the last date that it could have been written since Spencer moved from Loudoun Road that month, see H. Spencer, *Autobiography*, 2 vols (London: Williams and Norgate, 1904), vol. 2, p. 3, pp. 30–31.

3. *the discussion we commenced on Saturday evening*: not identified.

4. *other chapters on the general laws of change, a final one entitled 'The Equilibration of Force'*: although Spencer wrote much about the topic of equilibrium in *First Principles of a New System of Philosophy* (London: Williams and Norgate, 1862), he did not include a chapter titled 'The Equilibration of Force'.

5. *'Social Statics'*: H. Spencer, *Social Statics: or, the Conditions Essential to Human Happiness Specified, and the First of Them Developed* (London: John Chapman, 1851). An abridged and revised edition was published in 1892.

6. *articles on Progress and Physiology*: H. Spencer, 'Progress: its Law and Cause', *Westminster Review*, 67 (1857), pp. 244–67; and H. Spencer, 'The Ultimate Laws of Physiology', *National Review*, 5 (1857), pp. 332–55. The essay on progress, in which Spencer addressed the recently published ninth edition of Lyell's *Principles of Geology*, Humboldt's *A Sketch of a Physical Description of the Universe*, and William B. Carpenter's *Principles of Comparative Physiology*, formed the basis for his *First Principles* (1862).

7. *Amphioxus*: an order of fish-like marine vertebrates, also called the Lancelet, which is classified at the bottom of the vertebrate series (*OED*).

BIOGRAPHICAL REGISTER

This register contains the names and biographical details of people who are mentioned three or more times in the letters included in this volume. Biographical information about people mentioned only once or twice is contained in the notes appended to the appropriate letter(s).

Conventions

The sources for the biographical entries are standard biographical dictionaries, such as the *Complete Dictionary of Scientific Biography*, the *Oxford Dictionary of National Biography*, Poggendorff's *Biographisch-literarisches Handwörterbuch der exacten Wissenschaften*, *Historishes Lexikon der Schweiz*, *Neue Deutsche Biographie*, and the *Dictionary of Nineteenth-Century British Scientists*. These have not been specified in the entries unless necessary. Additionally, the editors have found material from the Tyndall correspondence and from the journals of Tyndall and T. A. Hirst very useful. Where additional sources add important material, they are given in "( )" at the end of the entry.

Agassiz, Louis (1807–73), was a Swiss naturalist. He studied medicine, botany, geology, and zoology in Germany and France before being appointed to a position at the University of Neuchâtel. In 1847 he was appointed professor of zoology and geology at Harvard in the United States and founded the Museum of Comparative Zoology. He was a key figure in developing the study of the Earth's history, proposed the theory of ice ages, and made a close study of glaciers in the Alps. In 1840 he published his important *Études sur les Glaciers* (2 vols., Neuchâtel: Gent et Gassmann), which helped establish the scientific study of glaciers.

Airy, George Biddell (1801–92), was educated at Cambridge University, graduating senior wrangler in 1823. In 1826 he was appointed Lucasian professor of mathematics, and in 1828 to the Plumian Chair of astronomy and directorship of the Cambridge University observatory. In 1835 he was appointed Astronomer Royal, a position he held until 1881, and set about improving the instrumentation and research activities of the Royal Observatory, Greenwich.

He vigorously pursued research on a wide range of topics, including optical diffraction, calculating the Earth's mean density, and analyzing the inequalities in the motions of the Earth and Venus. During the period covered by this volume Airy was interested in electricity and magnetism, as shown by his correspondence with Tyndall about experiments with induction coils.

Anderson, Charles (1790–1866), served in the Royal Artillery, achieving the rank of sergeant. In 1827 he was first employed at the RI to assist Faraday in the project (on behalf of the Board of Longitude) of improving optical glass. With the termination of that project in 1830 Anderson became Faraday's laboratory assistant and also assisted him in his RI lectures.

Asher, George Michael (1827–1900), was the second and only surviving son of Berlin bookseller Adolphus Asher (1800–53), who established the firm Asher and Company. Asher and Co. had shops in Berlin and London and was a major supplier to the British Library. George Asher initially trained as a bookseller but later became a jurist after a dispute with Anthony Panizzi, the chief librarian of the British Museum from 1856 to 1866. (David Paisey, "Adolphus Asher (1800–1853): Berlin Bookseller, Anglophile, and Friend to Panizzi," *British Library Journal* 23 [1997]: 131–53.)

Babbage, Charles (1791–1871), was a Cambridge-educated mathematician who held the Lucasian Professorship in mathematics at Cambridge after George Biddell Airy (1828–39), but never delivered a lecture. A controversial figure, he contributed to a wide range of subjects, including mathematics, astronomy, cryptography, the productivity of manufacturing, and, most famously, the theory of computing. He designed and tried to implement the first automatic calculating engines—the difference engine and the analytical engine.

Ball, John (1818–89), an Irish politician and glaciologist, was admitted to Cambridge University in 1835, but, as a Catholic, and unwilling to subscribe to the Protestant Articles, did not have his degree conferred. During the 1840s he began to study the glaciers of the Alps, returning to Ireland to work for the poor law commission during the Irish famine (1845–52). In Ireland he noted that the geological formations of the Dingle peninsula resembled those of the Alps and proposed that the landscape must have been formed by ancient glaciers. He became Member of Parliament for Carlow County in 1852 and in 1855 was named assistant undersecretary of state in the colonial department by Lord Palmerston. He was the first president of the Alpine Club (1858–60) and edited the club's first publication, *Peaks, Passes, and Glaciers* (London: Longman, Green, Longman, and Roberts, 1859).

Balmat, Auguste (1808–62), was a mountaineer and guide born in Chamonix, Savoy, who at the age of twenty-two made his first ascent of Mont Blanc. He met J. D. Forbes in 1842 and assisted him during his studies of the movement of glaciers and survey work leading to the important map of the Mer de Glace published in Forbes's *Travels through the Alps of Savoy* (Edinburgh: A. and C. Black, 1843). In 1854 he guided the much-celebrated ascent of the Wetterhorn by Alfred Wills (1828–1912). Wills later dedicated his *Wandering among the High Alps* (London: Bentley, 1856) to Balmat. In 1858, while guiding Tyndall and Wills's ascent of Mont Blanc, he nearly lost his hands in the process of placing thermometers at the summit. (C. D. Cunningham and W. Abney, *The Pioneers of the Alps*, 2nd ed. [London: Low, Marston, Searle and Rivington, 1888].)

Balmat, Edouard (b. 1846), was the son of Chamonix porter and guide Alexandre Balmat (1815–75). Rather than hiring a guide, in the summer of 1857 Tyndall explored the Mer de Glace and climbed the Col du Géant with Edouard Balmat, "a little boy, who can climb well" (letter 1419, and *Glaciers of the Alps*, 47, 67). Tyndall referred to Balmat as his "little demon." Known as "le petit Balmat," he was also among the porters who assisted Tyndall in his research on the Mer de Glace in December 1859 (*Glaciers of the Alps*, 195–219).

Barlow, John (1798–1869), was an Anglican clergyman and secretary of the RI. Barlow studied at Trinity College, Cambridge, and was ordained in 1823. He was subsequently appointed curate at Uckfield (Sussex), rector of Little Bowden (Northamptonshire), and from 1854 until his death was chaplain-in-ordinary at Kensington Palace. From 1842 to 1860 he was the secretary of the RI and was close to both Michael Faraday and Tyndall. Together with Faraday, he undertook the day-to-day running of the institution. After marrying, in 1824, Cecilia Anne Barlow (née Law, d. 1868), the daughter of an MP who made his fortune in India, Barlow lived near the RI and entertained many of its wealthy and aristocratic supporters as well as fellow scientists, including Tyndall, who was a frequent guest.

Barnard, Jane (1832–1911), a niece of Michael and Sarah Faraday, lived at the RI and assisted the Faradays. She was the daughter of Sarah's brother, John Barnard, and Michael's younger sister Margaret. Like her aunt and uncle, she was a member of the Sandemanian church.

Bennen [Benet], Johann Joseph (1824–64), was a Swiss mountain guide. Born Johann Joseph Benet, he was nicknamed "Bennen" by English alpinists. Bennen made the first ascent on the Aletschorn in 1859, and the first ascent on the Weisshorn with Tyndall in 1861. In 1862 they attempted the Lion ridge of the Matterhorn. In the summer of 1858 he was Tyndall's guide while ascending the Finsteraarhorn. Bennen died in an avalanche on Haut de Cry. (C. D. Cunningham and W. Abney, *The Pioneers of the Alps*, 2nd ed. [London: Low, Marston, Searle and Rivington, 1888], 148–53.)

Bergoin [Bergoën], Félix (1806–77), was a Savoyard lawyer, judge, and civil servant. In December 1857 he took up the post of intendant of the province of Faucigny at Bonneville. As intendant, he was also the president of the Compagnie de Guides de Chamonix. Tyndall addressed Bergoin seeking exemptions for "men of science" from the rules of the *compagnie*. Following the Treaty of Turin (24 March 1860), which annexed Savoy and Nice to France, Bergoin was replaced by Joseph Guy. (Christian Sorrel, "Bergoin Félix 1806–1877," in *La Savoie et l'Europe, 1860–2010: Dictionnaire historique de l'annexion*, ed. Christian Sorrel et al. [Montmélian: La Fontaine de Siloé, 2009], 168–69.)

Biot, Jean-Baptiste (1774–1862), was educated at the École Polytechnique, Paris, and later professor of physics at the Collège de France. Biot conducted experiments on balloon flights and also made a close study of meteorites. His main research was on the polarization of light, for which he was awarded the RS's Rumford Medal in 1840. Tyndall met Biot in Paris in July 1857 while attending Anna Hirst's funeral.

Booth, Sarah (1791–1856), was the widow of John Booth and mother of Tyndall and Hirst's friend Francis Booth (d. 1853). According to Tyndall she lived in a cottage known as "The Birdcage" near Skirtcoat Moor in Halifax, where she sold sweets to children visiting the moor. When her son Francis contracted tuberculosis, and died in 1853, Hirst agreed to support Sarah financially. Tyndall acted as an intermediary while Hirst was in France seeking a cure for his wife, Anna's illness. (John Tyndall, "Memoranda Concerning Dr. Hirst," *Roy. Soc. Proc.* 52 [1892–93]: xiv–xviii.)

Bossoney, Michel (1814–83), "guide chef" of the Compagnie des Guides de Chamonix, was responsible for assigning guides to travelers. In the summer of 1857 Bossoney permitted Tyndall to ascend the Col du Géant without a guide. The following year Bossoney required that Tyndall hire four guides to ascend Mont Blanc. Tyndall's response was to address the president of the

Commission of the Guides of Chamonix, and the intendant of the Province of Faucigny, who instructed Bossoney to make an exception to the rules. In the summer of 1857 Bossoney was also the antagonist to T. W. Hinchliff and a party of climbers, as recounted in his "Poaching on Mont Blanc A Dozen Years Ago," *Fraser's Magazine* 80, no. 475 (1869): 97–110.

Brande, William Thomas (1788–1866), was an English chemist and professor at the RI. While studying chemistry at Saint George's Hospital he attended Davy's lectures at the RI. A year after Davy resigned as professor of chemistry at the RI in 1812, Brande succeeded him. That year he was awarded the RS's Copley Medal. Brande was an active lecturer and prolific writer. His titles include *Outlines of Geology* (1817), *Manual of Chemistry* (1817), *Manual of Pharmacy* (1825), *Dictionary of Materia Medica* (1839), and *Dictionary of Science, Literature, and Art* (1842). In 1852 Brande resigned his professorship of chemistry at the RI in order to become superintendent of the Royal Mint.

Brewster, David (1781–1868), was a Scottish mathematician, astronomer, editor, and educator, known for his work in optics and his role in founding the BAAS. He was a regular contributor to numerous periodicals, by which he mostly earned his living. He was one of three founding editors of the *Edinburgh Magazine.* He was also known as a controversialist. A bitter and public dispute with Charles Wheatstone over the invention of the stereoscope erupted in the pages of the *Times* in October and November 1856. He was appointed principal of the United College of Saint Salvator and Saint Leonard at the University of Saint Andrews in 1838, and principal of the University of Edinburgh in 1859.

Brücke, Ernst Wilhelm (1819–92), was a German physiologist trained at the University of Berlin, graduating as a doctor of medicine and surgery in 1842. He became professor of physiology at the University of Königsberg in 1848 and the University of Vienna the following year. Brücke's research interests were wide ranging, contributing to the physiology of the eye and digestion, as well as language and speech. In 1845 he founded, with Emil du Bois-Reymond and Hermann von Helmholtz, the Physical Society of Berlin. He was given the noble title von Brücke in 1873 by Emperor Franz Joseph I.

Buckle, Henry Thomas (1821–62), was an English historian best known for his unfinished *History of Civilization in England* (2 vols., London: John W. Parker and Son, 1857–61). When his father, a wealthy London merchant, died in 1840, Buckle inherited a significant fortune, which allowed him to live comfortably. He was educated at home by his mother except for a single year at Gordon House School when he was fourteen. In 1858 Buckle delivered at the RI his only public lecture, titled "The Influence of Women on the Progress of Knowledge," which was later printed in *Fraser's Magazine*. Tyndall related to Hirst how in 1858 he engaged in an intense debate with Buckle at a dinner party where he also met David Livingstone. He shared with Huxley that Carlyle thought Buckle "a weak watery unfruitful creature out of whom no good can come."

Bunsen, Robert (1811–99), was a German chemist who received his PhD from the University of Göttingen, where he then lectured (1833–36). Bunsen subsequently held positions at the Polytechnikum at Kassel, the University of Marburg (1839–51), the University of Breslau, and, from 1852, the University of Heidelberg. He made many contributions to chemistry, including the Bunsen burner and the discovery of cesium and rubidium, and developed a battery that bears his name. He made extensive use of both electrolysis and spectrum analysis to analyze chemical substances. A renowned researcher, he was also an extremely popular teacher, mentoring both Tyndall and Edward Frankland at Marburg. He received the RS's Copley Medal in 1860, and its Davy Medal, with Gustav Robert Kirchhoff, in 1877.

Carlyle, Thomas (1795–1881), was a Scottish essayist, novelist, and historian, and noted intellectual figure in the Victorian era. He was a major literary influence on Tyndall, who avidly read his publications. After reading *Past and Present* (1843) Tyndall commented that Carlyle "must be a true hero. My feelings towards him are those of worship 'transcendental wonder' as he defines it" (Journal, 18 July 1847, RI MS JT/2/13a/231). Tyndall met Carlyle and his wife, Jane Baille Carlyle (née Welsh), in the 1850s and wrote to Hirst about their meetings.

Chasles, Michel Floréal (1793–1880), was a French mathematician. He entered the École Polytechnique in 1812 and participated in the defense of Paris in 1814. After giving up his position at the École Polytechnique and working briefly at a Paris stockbroker, he moved to Chartres, where he continued his work in history and mathematics. From 1841 until 1851 he taught at the École Polytechnique, and held a chair of geometry at the Sorbonne from 1846 until his death. As a keen collector of manuscripts and autographs, he became embroiled in a public controversy after he defended the claim that Pascal had discovered the law of gravitation before Newton. The letters he had purchased turned out to be forged by Denis Vrain-Lucas, at whose trial Chasles testified in 1869.

Clausius, Rudolf (1822–88), was a renowned German scientist and one of the founders of modern thermodynamics. His seminal paper on the theory of heat, "Ueber die bewegende Kraft der Warme," was published in the *Annalen der Physik* in 1850. He taught physics at the Royal Artillery and Engineering School in Berlin in 1850 before moving to a professorship in mathematical physics at the Polytechnicum in Zurich in 1855. During the 1850s Tyndall arranged publication of English translations (by Hirst and himself) of Clausius's works in *Phil. Mag.* In 1858 Clausius nominated Tyndall to the Naturforschende Gesellschaft in Zürich for his work on Swiss glaciers.

Cook, John Douglas (c. 1808–68), was a journalist and newspaper editor born in Scotland. Cook worked as a reporter in London for the *Times*, and served as editor of the *Morning Chronicle* from 1848 until the paper was sold in 1854. In 1855 Cook and Beresford Hope launched the *Saturday Review of Politics, Literature, Sciences and Art* (1855–68), to which Tyndall became a contributor.

Coxe, James (1811–78), was a Scottish physician who studied at Göttingen, Heidelberg, Paris, and received his MD in 1835 from the University of Edinburgh, where he then practiced medicine. In 1841 he married Mary Anne Cumming. In 1855 he was appointed to the royal commission that examined the conditions of the lunatic asylums in Scotland and, following the 1857 act that established the General Board of Lunacy for Scotland, served as one of its two paid commissioners. He was knighted in 1863 for his services. ("Sir James Coxe," *Illustrated London News*, 18 May 1878, 471.)

Coxe, Mary Anne (née Cumming, 1811–c. 1875), was a daughter of Lesley Baillie and Robert Cumming of Logie, Co Moray. Her mother was the subject of Robert Burns's "Bonnie Lesley," as noted in the novelist Charles Gibbon's dedication of *In Honour Bound*, vol. 1 (London: Richard Bentley and Son, 1874) to Lady Coxe. In 1841 she married the Scottish physician James Coxe, and became Lady Coxe after he was knighted in 1863. She was a correspondent of Tyndall's and he claimed to have toasted her on the summit of Mont Blanc in 1857.

Dante (c. 1265–1321), born Durante degli Alighieri in Florence, but known simply as Dante. He is best known for his *Divine Comedy*, which he completed in 1320.

Darwin, Charles (1809–82), was an English naturalist and geologist, best known for his book *On the Origin of Species* (London: John Murray, 1859). Darwin encouraged Tyndall's research on glaciers and supported him in his disagreement with Forbes, writing, "I wish you all sorts of good fortune in your most interesting investigations; and the Lord have mercy on you, when Forbes answers you is my prayer." Darwin had attended part of Tyndall's lecture at the RS of his and Huxley's joint paper in January 1857. See Charles Darwin to Thomas Henry Huxley, 17 January [1857], in F. Burkhardt et al., eds., *The Correspondence of Charles Darwin*, vol. 6, 1856–1857 (New York: Cambridge University Press, 1991), 322–23.

Davy, Humphry (1778–1829), worked at the Pneumatic Institution in Bristol before moving to the RI in 1801. He was successively director of the laboratory, 1801–25; professor of chemistry, 1802–12; and honorary professor, 1813–23. There he achieved fame as a lecturer and for his chemical research that isolated such elements as sodium, potassium, and chlorine. He is often attributed—probably wrongly—with the invention of the miner's safety lamp. In addition to being the foremost scientist at the RI, Davy served as president of the RS from 1820 to 1827.

Debus, Heinrich (1824–1915), studied chemistry in Marburg (1845–48), where he worked as Bunsen's assistant and where he met Tyndall, Frankland, and Hirst. Moving to England in 1851, he taught chemistry at Queenwood College and was for a time a colleague of Tyndall's. After leaving Queenwood in 1867, he spent three years at Clifton College, Bristol, then moved to London as lecturer in chemistry at Guy's Hospital Medical School (1870–88), adding to this the position of professor of chemistry at the Royal Naval College, Greenwich (1873–88) under Hirst, who was founding director of studies at the college.

de la Rive, Auguste-Arthur (1801–73), was a Swiss physicist and one of the founders of the electrochemical theory of batteries. He served as professor of natural philosophy at the Academy of Geneva and worked principally on electricity, although he also investigated the specific heats of gases and calculated the temperature of the Earth's crust. His principal book was *Traité d'électricité théoretique et appliquée* (3 vols., Paris: J.-B. Baillière et H. Baillière, 1854–58). C. V. Walker was translating volume three into English when de la Rive asked Tyndall to read the proofs and correct what de la Rive called Walker's "Gallicisms."

Desor, Pierre Jean Édouard, (1811–82), was a German-Swiss geologist and mountaineer who studied law at Giessen and Heidelberg. After meeting Louis Agassiz in 1837 Desor became his chief collaborator and close friend, and followed him to America when Agassiz became professor of zoology and geology at Harvard. In the 1840s he made several ascents, including the Jungfrau, with James David Forbes in 1841. In 1852 he returned to Switzerland and was appointed professor of geology at the Academy of Neuchâtel. He later had a political career, which culminated in his being elected president of the Swiss Federal Assembly in 1874.

Despretz, César (1791–1863), was a Catholic chemist and physicist who taught at the Faculté des Sciences in Paris. His research covered a wide range of topics, but in the early 1850s he was primarily concerned with the conduction of heat and the functioning of the voltaic pile. He also published textbooks in both physics and chemistry.

Dove, Heinrich Wilhelm (1803–79), was a Prussian meteorologist and experimental physicist. He was educated at the University of Berlin and became an ordinary professor at Berlin's Friedrich-Wilhelms-Universität in 1845. In his research, he focused on meteorology and was awarded the RS's Copley Medal in 1853 for his contributions to understanding the distribution of heat in the atmosphere.

Drummond, Emily (1838–1930), was the second daughter of Thomas Drummond and Maria Drummond (née Kinnaird). In 1857 Tyndall became enamored with one of the Drummond sisters, possibly Emily, or Mary, her elder sister, whom Tyndall pursued in 1858.

Drummond, Fanny Eleanor (1840–71), was the youngest daughter of Thomas Drummond and Maria Drummond (née Kinnaird). In 1858 Tyndall told Hirst that the Drummonds were "a splendid family," "Rich, cultivated, beautiful & good—good to excellence.—a mother and three daughters, who rank among the choicest bits of humanity that I ever met, or hoped to meet" (letter 1510).

Drummond, Maria (née Kinnaird, 1810–91), was a literary hostess and the adopted daughter of the wealthy MP, banker, and merchant Richard "Conversation" Sharp. She was born on the Caribbean island of Saint Vincent and orphaned by a volcanic eruption in 1812. Through Sharp's social circle she was introduced to Faraday, Macaulay, Babbage, and Walter Scott, among others. When Sharp died in 1835 she inherited much of his £250,000 estate and married Thomas Drummond (1797–1840), who developed the Drummond light for the Irish Survey and was undersecretary for Ireland (1835–40). Together they had three daughters, Mary, Emily, and Fanny Eleanor, before Thomas died in 1840. Maria and her daughters continued to entertain a wide circle of London's elite, including Tyndall, at their home near Hyde Park, London, and their country house at Fredley Park, in Mickleham, Surrey. Around 1858 Tyndall fell in love with Mary Drummond, Maria's eldest daughter. (Charles Kegan Paul, *Maria Drummond: A Sketch* [London: Kegan Paul, Trench, Trübner, & Co., 1891].)

Drummond, Mary Elizabeth (1836–1924), "May" and probably "Δ" in Tyndall's journal, was the eldest daughter of Thomas Drummond and Maria Drummond (née Kinnaird). She would eventually marry Joseph Kay (1821–78) in 1863. In his journal and in his letters to Hirst, Tyndall revealed that he had fallen in love with Mary but had doubts over his own social position relative to the wealth in which she had been raised. "I have seen that girl in her mother's receptions, surrounded by fine young fellows, with bishops, lords, and marquises her mothers guests, and to think of her choosing me almost bewilders me" (letter 1578). Tyndall spent Christmas day 1858 with the Drummonds, and noted that Joseph Kay, Mary's future husband, was also present. ("Deaths," *Times*, 13 February 1924, 1; J. F. McLennan, *Memoir of Thomas Drummond* [Edinburgh: Edmonston and Douglas, 1867], 409–10.)

Duboscq, Louis Jules (1817–86), was a French inventor, instrument maker, and photographer known for producing high-quality optical instruments. Among his inventions was the electric arc lamp that bears his name. The lamp produced a very bright light useful for projecting images. On 5 June 1857 Jules Lissajous and Duboscq made the experiments while Tyndall gave his Friday Evening Discourse. Tyndall wrote that Duboscq "took charge of his own lamp, which however he did not manage with any thing like the promptness which I have attained myself. I was thus at times forced to extemporize, and fill up chasms which had not been anticipated. This was observed by many, and manifestly took their fancy" (letter 1404).

Edmondson, George (1798–1863), was headmaster of Queenwood College in Hampshire from 1847 until his death. Edmondson, a Quaker, had attended Ackworth School in Yorkshire. From 1817 to 1824 Edmondson lived in Russia, tutoring the children of his mentor, Daniel Wheeler. After returning to England he opened a school at Blackburn in 1825, then one near Preston, before becoming the head teacher at Queenwood College when it opened in 1847. Under his influence science and technological subjects were emphasized. Although he hired Tyndall, Hirst, Debus, and Frankland, they were critical of his management of Queenwood. Hirst wrote to Tyndall that he considered Edmondson "a mere zero or rather worse for he has a decidedly negative value. All your attempts at improvement are cramped first by his imbecility and again by all many of pecuniary embarrassments" (letter 1434).

Escher (von der Linth), Arnold (1807–72), was a Swiss geologist and son of Hans Conrad Escher von der Linth (1767–1823). A student of Horace-Bénédict Saussure in Geneva and Alexander von Humboldt in Berlin, he became professor of geology at the École Polytechnique in Zurich in 1856. In 1842 he was a member of the first party to ascend Mount Lauteraarhorn. He produced, in collaboration with Bernhard Studer, a detailed geological map of Switzerland in 1853. For his father's work diverting the Linth river the family was given the surname von der Linth.

Faraday, Michael (1791–1867), was a natural philosopher, Fullerian Professor of Chemistry at the RI, and a member of the Sandemanian church. His extensive research focused on electricity and encompassed electrochemistry, electromagnetism, diamagnetism, and magneto-optics (the Faraday effect). Faraday also held the positions of director of the laboratory (1825–67) and Superintendent of the House (1852–67) at the RI. He was Tyndall's patron and helped him obtain the position of professor of natural philosophy at the RI in 1853. Despite significant differences in their views on religion and on the nature of magnetism, they held the highest mutual regard for each other. Tyndall wrote to Hirst that "[i]t falls to the lot of few to have such a noble affectionate scientific father as your friend John possesses in Faraday. Indeed I know of no material advantage which I could set for an instant in competition with the friendship of that unspoiled and beautiful soul" (letter 1404). Tyndall's *Faraday as a Discoverer* (London: Longmans, Green & Co, 1868) was his tribute to his mentor. In 1867 Tyndall succeeded Faraday as Superintendent of the House and director of the laboratory at the RI.

Faraday, Sarah (née Barnard, 1800–79), was a daughter of Mary Boosey and Edward Barnard, a successful London silversmith. She was Sandemanian in her religion and likely met Michael Faraday at the Sandemanian meeting house in Paul's Alley, London. After they married in 1821 she moved into his rooms at the RI. In later years, she would take on many of Faraday's administrative responsibilities at the RI. Tyndall frequently dined and socialized with Sarah and Michael Faraday and their niece Jane Barnard. Tyndall wrote of Sarah Faraday: "She is very motherly to me, and Faraday says that were it not for her own feebleness she must become quite a mother to me, and take care of me" (Journal, 2 March 1855, RI MS JT/2/13c/730).

Forbes, James David (1809–68), was a Scottish physicist and glaciologist who held the chair of natural philosophy at the University of Edinburgh from 1833, until he became principal of the college of Saint Salvator and Saint Leonard (the United College), at Saint Andrews University in 1860. While much of his research focused on heat, he was famous for his studies of glaciers. An avid mountaineer, Forbes published *Travels through the Alps of Savoy* (Edinburgh: A. and C. Black) in 1843. His theory that a glacier is a viscous body brought him into conflict with Tyndall and Huxley in the late 1850s. Letters in this volume indicate both tensions and collegial relationships.

Francis, William (1817–1904), was a chemist, translator, and editor. He was the illegitimate son of the printer and publisher Richard Taylor. Francis learned the printing trade in Germany in the 1830s before training as a chemist with Justus von Liebig and Gustav Rose in Giessen, where he obtained his doctorate in 1842. He made many translations for his father's *Scientific Memoirs* and the *Philosophical Magazine* and became the managing editor of the latter in 1851. In 1852 he entered into partnership with his father, establishing the Taylor & Francis publishing firm. He first met Tyndall in 1850 and became one of Tyndall's close confidants. Tyndall joined Francis as an editor of the *Phil. Mag.* in 1854.

Frankland, Edward (1825–99), was a chemist and one of Tyndall's closest friends. He worked as an assistant to the chemist Lyon Playfair in 1845 before teaching at Queenwood College in 1847, where he met Tyndall. He and Tyndall studied together at the University of Marburg in 1848. After completing his degree in 1849, Frankland worked briefly in Liebig's laboratory at Giessen and taught at the Putney College of Engineering, near London, before assuming the chair of chemistry at Owens College, Manchester. In 1851, just before moving to Manchester, he married Sophie Fick (1821–74), whom he had met in Marburg. In 1857 Frankland returned to London as a lecturer at Saint Bartholomew's Hospital, and was awarded the RS's Royal Medal after Tyndall nominated him. In 1864, Frankland joined Tyndall, Thomas Hirst, Joseph Dalton Hooker, Thomas Henry Huxley, and three others as members of the X Club, a monthly dining club in which they promoted science in British society.

Gassiot, John Peter (1797–1877), was an electrician and wine merchant by trade. He was one of the founding members of the London Electrical Society in 1837, serving as the society's treasurer until 1841. He pursued research on electricity in the laboratory he outfitted at his home in Clapham Common, near London. During the 1860s James Clerk Maxwell would work there while establishing the unit for electrical resistance. Gassiot published papers on the functioning of the voltaic pile (battery) and showed that a spark could be produced before the circuit of a battery was closed.

Gerling, Christian Ludwig (1788–1864), was a German physicist and astronomer who obtained his PhD from Göttingen under the supervision of Carl Friedrich Gauss in 1812. He became professor of mathematics, astronomy, and physics at Marburg in 1817. He was known mostly for his work in geodesy, and was also a leading administrator of the University of Marburg during the several decades he taught there. Tyndall was one of his pupils while at Marburg, after which they maintained a friendly correspondence.

Gore, George (1826–1908), was an English electrochemist and son of a cooper (a maker or repairer of casks and barrels). While apprenticing as a cooper himself, he studied science privately and in 1851 moved to Birmingham to work in the electroplating industry. In 1853–65 he published dozens of papers on chemistry and electrometallurgy. Gore was a highly sought-after scientific consultant in Birmingham and wrote many important textbooks, including *The Art of Electro-Metallurgy* (London: Longmans, Green, and Co., 1877), *The Art of Scientific Discovery* (London: Longmans, Green, and Co., 1878), and *The Scientific Basis of National Progress* (London: Williams and Norgate, 1882).

Grove, William Robert (1811–96), was a Welsh judge and electrochemist. He invented a nitric acid voltaic cell, known as "Grove's cell," that caught the attention of Faraday and Tyndall. He also developed a gas battery (the first fuel cell), for which he was awarded the RS's Royal Medal in 1847. He held a professorship of experimental philosophy at the London Institution from 1841 to 1847, at which point financial constraints led him to practice law. He became a Queen's Counsel in 1853, was appointed a judge of the court of common pleas in 1871, and then the Queen's Bench in 1880. He remained a highly respected member of the scientific community and was elected president of the BAAS in 1866.

Harris, William Snow (1791–1867), was an Edinburgh-trained physician and natural philosopher. His marriage to Elizabeth Snow Thorne in 1824 provided him with sufficient financial means to give up his practice as a medical doctor and pursue his interests in electricity and magnetism. He invented a lightning conductor for ships' masts in 1820, which were later adopted by the Admiralty after years of his efforts to prove their efficacy. He was knighted for this work in 1847.

Haughton, Samuel (1821–97), was an Irish mathematician, geologist, anatomist, and physiologist who published a series of scientific and mathematical textbooks with his Dublin colleague James Allen Galbraith. He was professor of geology and mineralogy at Trinity College Dublin (1851–81), where he and Joseph Galbraith taught the growing number of students who were taking public examinations to enter the Indian Civil Service, the Royal Artillery, and Royal Engineers. Tyndall supported his admission to the RS in 1858 and advised Haughton about presenting research to the RS in 1857.

Hawtrey, Stephen Thomas (1808–86), was an English educator and mathematician who graduated eleventh wrangler in the Cambridge Tripos exams in 1832. He taught at Eton College from 1836 and was later mathematical master there (1851–71). In 1845 he founded and publicly promoted Saint Mark's School, Windsor (later Imperial Service College), and served as the school's warden (1871–86). Hawtrey wrote a number of books on education, including *A Narrative-Essay on a Liberal Education* (London: Hamilton, Adams, and Co., 1868), *The Aim, Duties, and Reward of a Schoolmaster* (London: Hamilton, Adams, and Co., 1870), and *An Introduction to the Elements of Euclid* (London: Longmans, Green, and Co., 1874). In 1857 Tyndall gave a series of lectures at Eton College at Hawtrey's request.

Henslow, John Stevens (1796–1861), was an English botanist, clergyman, and the father of Frances Hooker (née Henslow). After obtaining his BA at Cambridge University in 1818, he held the chair of mineralogy from 1822 until assuming the chair of botany in 1825. Henslow expanded the Cambridge Botanic Garden and planned its current forty-acre site. Henslow was an important mentor to the young Charles Darwin and ultimately a lifelong friend. After receiving a parish living in Hitcham, Suffolk, in 1837 he relocated there and returned to Cambridge only periodically to teach and attend his duties there. While at Hitcham, Henslow engaged in a number of efforts to support education for the working class, including schools, excursions, allotments of land for vegetable growing, and horticultural shows.

Herries, Isabella Maria (c. 1789–1870), was a daughter of Charles Herries and Mary Anne Johnson. She was a sister of Juliet Pollock's mother, Catherine Creed (née Herries). Isabella raised Juliet Pollock and her cousin Maria Julia Herries (c. 1820–1857) after the deaths of their mothers.

Herschel, John Frederick William (1792–1871), was an English mathematician and astronomer. He was the son of the famous astronomer William Herschel (1738–1822) and himself became one of the most eminent scientists in

Victorian Britain. He graduated from Cambridge University in 1813 as senior wrangler. He was awarded the RS's Copley Medal in 1821, and a second Copley Medal in 1847. His *Preliminary Discourse on the Study of Natural Philosophy* (London: Longmans, Rees, Orme, Brown & Green, 1830) was widely read and greatly admired. He was awarded a knighthood in 1831, followed by being created a baronet in 1838, on his return from mapping the southern skies in Cape Town, South Africa. In 1849 he published *Outlines of Astronomy* (London: Longman, Brown, Green, and Longmans), a popular presentation of astronomy that went through eleven editions by 1871. He made many contributions to the science of photography and also edited the Admiralty's *Manual of Scientific Enquiry* (London: John Murray, 1849), a science field guide for the use of the Royal Navy. From 1850 to 1855 he served as master of the mint, and on his death was buried next to Isaac Newton in Westminster Abbey.

Hirst, Anna (née Martin, 1831–57), born in Ireland, was the wife of Thomas Archer Hirst. They met in 1852 in Marburg, where she was a companion to her sister, Mary Simpson (née Martin), wife of the Irish chemist Maxwell Simpson, who was studying in Bunsen's laboratory. Anna's brother was the Irish nationalist John Martin. She married Hirst at her family home in County Down, Ireland, on 28 December 1854. She soon began to display symptoms of tuberculosis, leading Hirst to resign from Queenwood College in June 1856 to nurse her in the south of France. She died in Paris on 1 July 1857 and is buried in the cemetery at Montmartre. While Thomas Hirst remained on the Continent until 1860 for research and to visit with eminent mathematicians (in Paris and a year in Italy), he visited her grave often.

Hirst, Thomas Archer (1830–92), was an English mathematician, educator, and Tyndall's closest friend and correspondent. His friendship with Tyndall began in Halifax, Yorkshire, in 1845 when they were both working for the surveyor Richard Carter. Encouraged by Tyndall—who acted as a sort of mentor to him—Hirst studied at the universities of Marburg, Göttingen, and Berlin between 1849 and 1852. In 1853 he returned to England with a PhD and replaced Tyndall as mathematics teacher at Queenwood College. Like Tyndall, he did not get on with the headmaster, George Edmondson. He resigned in June 1856 to look after his sick wife, Anna, who died in Paris in 1857. Tyndall arrived in Paris for Anna's funeral while on his way to Chamonix. He convinced Hirst to accompany him to Chamonix, where they climbed Mont Blanc together and made observations on the Mer de Glace. In 1864 Hirst joined Tyndall, Edward Frankland, Joseph Dalton Hooker, Thomas Henry Huxley, and three others as members of the X Club, a monthly dining club in which they promoted science in British society.

Hooker, Frances Harriet (née Henslow, 1825–74), was a daughter of Harriet Jenyns and the botanist John Stevens Henslow, who had been Darwin's mentor at Cambridge University. She married Joseph Hooker in August 1851 and together they had four sons and two daughters. She was fluent in French and German and in addition to making private translations for her husband she published a translation of Emmanuel le Maout, *A General System of Botany* (London: Longman, 1873). She corresponded with Tyndall and shared literary references, poetry, and novels. She was an excellent editor and would occasionally look over manuscripts for him.

Hooker, Joseph Dalton (1817–1911), a British botanist who specialized in taxonomy and geographical distribution of plants, ultimately became the most famous British botanist of the Victorian era. He was the younger son of Sir William Jackson Hooker, the director of the Royal Botanic Gardens, Kew. After studying medicine at the University of Glasgow he became a naval surgeon and spent the years 1839–43 on James Clark Ross's geomagnetic survey of the southern oceans, where he collected plants as well as serving as surgeon. Subsequently he explored and collected plants in India (1847–51), achieving considerable mountaineering success, before joining his father as assistant director of Kew Gardens in 1855. Hooker was an intimate friend of Charles Darwin and Thomas Henry Huxley. He was first introduced to Tyndall by Huxley in November 1855 and the three became close friends. Together with Huxley, in August 1856 Hooker helped Tyndall with his first studies of Alpine glaciers. With Charles Lyell, Hooker was instrumental in bringing Darwin's theory of natural selection to public attention. In 1864 Hooker joined Tyndall, Edward Frankland, Thomas Hirst, Thomas Henry Huxley, and three others as members of the X Club, a monthly dining club in which they promoted science in British society.

Hooker, William ("Willy") Henslow (1853–1942), was the eldest child of Joseph and Frances Hooker. At age thirteen, for his health, he was sent to New Zealand, where he lived with the geologist James Hector and his wife, Georgina (1869–70). From 1877 he was employed in the India Office, later becoming a supervisor of the India Store Depot, London. In 1914 he married Sarah Ann Smith (1863–1952). Tyndall and Willy had a special affinity for each other, and a letter from Frances Hooker shows why Willy preferred Tyndall to Hooker's other friends, John Lindley and Charles Darwin—Tyndall gave him hugs (letter 1395).

Hopkins, William (1793–1866), was an English mathematician and geologist. He entered Cambridge University as an undergraduate at Peterhouse in 1822 after the death of his first wife. He graduated the Mathematical Tripos as seventh wrangler in 1827 and received his MA in 1830. Settling in Cambridge, Hopkins became a private mathematics tutor training undergraduates for the Tripos exams. He was one of the most successful tutors of his time and became known as the "senior wrangler maker." In his research he was one of the most significant proponents of the solidity of the Earth and came into conflict with J. D. Forbes over their opposing theories of glaciers. In 1857 Hopkins wrote to Tyndall supporting his theory of glaciers and warned that Forbes could not "carry on a controversy without a large spice of ill-humoured feeling" (letter 1359).

Huxley, Henrietta Anne (née Heathorn, 1825–1915), born in the West Indies to Henry Heathorn and Sarah Henrietta Richardson (née Harris), was schooled in Neuwied, Germany, for two years. In 1843 she, her mother, and half sister moved to Australia to join her father. She was keeping house for her sister Oriana and brother-in-law William Fanning in Sydney, Australia, when she met Thomas Henry Huxley in 1847, who was then a surgeon and zoologist on HMS *Rattlesnake*. After an eight-year engagement, she arrived in London in 1855 and married Huxley on 21 July. She assisted her husband in his work by translating German and drawing diagrams for his lectures. She wrote poetry and published a volume of her collected poems at the age of eighty-six, *Poems of Henrietta A. Huxley with Three of Thomas Henry Huxley* (London: Duckworth, 1913). ("Death of Mrs. Huxley," *Times*, 6 April 1915, 12.)

Huxley, Thomas Henry (1825–95), was a British zoologist, educator, science popularizer, and one of Tyndall's closest friends. Following medical training at Charing Cross Hospital in London, he was a ship's surgeon aboard the HMS *Rattlesnake* expedition to Australia (1846–50) under Captain Owen Stanley. After eking a living by writing for Chapman's *Westminster Review*, in 1854 he became professor of natural history at the Royal School of Mines, London. He held this position until 1872, and continued with the School of Mines when it then moved to South Kensington until 1885. While in Australia he met Henrietta (Nettie) Anne Heathorn, whom he married in July 1855 after an eight-year engagement. He joined Tyndall at the RI as Fullerian Professor of Natural History (1855–58). In the summer of 1856, and again in 1857, he and Tyndall studied the movement of glaciers and climbed in the Alps together. They published their research in January 1857 as "On the Structure and Motion of Glaciers," *Phil. Trans.* 147 (1857): 327–46. In 1864 Huxley joined Tyndall, Edward Frankland, Thomas Hirst, Joseph Dalton Hooker, and three others as members of the X Club, a monthly dining club in which they promoted science in British society.

Jones, Henry Bence (1813–73), was a wealthy English physician and medical chemist. He was educated at Harrow School and Trinity College, Cambridge, before studying medicine at Saint George's Hospital, London. Attracted by the potential applications of chemistry to clinical medicine, he studied organic chemistry with Liebig at Giessen (1841–43) before assuming a career as a successful physician in private practice and at Saint George's Hospital. In May 1842 he married a cousin, Lady Millicent Acheson, daughter of the second Earl of Gosford, with whom he had seven children. He published important research on protein chemistry and made an abstract in English of du Bois-Reymond's work on electrophysiology (1852) that annoyed Carlo Matteucci, the Italian pioneer of electrophysiology. In 1853 he became a manager (trustee) of the RI and was elected secretary in 1860. He frequently entertained Tyndall at both his London home and Folkestone, where he kept a summer house.

Lauener, Christian (1826–91), was a celebrated Alpine guide born in Lauterbrunnen, Switzerland, and younger brother of Ulrich Lauener, another Alpine guide. During the summer of 1858 he acted as Tyndall's guide during his exploration of the Aar, Rhone, and Aletsch glaciers and ascent of the Eggishorn. In 1855 he married Margaritha von Allmen. Lauener's first ascents included the Weisse Frau (1862) and the Gletscherhorn (1867). He was known as a leading guide into the late 1870s, retiring in the 1880s to operate an inn at Lauterbrunnen. (*Glaciers of the Alps*, 92–168; C. D. Cunningham and W. Abney, *The Pioneers of the Alps*, 2nd ed. [London: Low, Marston, Searle and Rivington, 1888], 81–83.)

Lauener, Ulrich (1821–1900), was an Alpine guide born in Lauterbrunnen, Switzerland, and the elder brother of Christian Lauener. He was a renowned Alpine guide for over fifty years, making several first ascents, including the Dufourspitze, the highest peak of Monte Rosa (1855), and the Tschingelhorn (1865), and he guided Leslie Stephen on the first crossing of the Eigerjoch (1859). He married Luzia Stäger in 1857. (C. D. Cunningham and W. Abney, *The Pioneers of the Alps*, 2nd ed. [London: Low, Marston, Searle and Rivington, 1888], 79–81.)

Lindley, John (1799–1865), was a English botanist and horticulturalist. As a young man he became a close friend of William Jackson Hooker, through whom he was introduced to Charles Lyell and Joseph Banks. In 1829 Lindley became the first professor of botany at the nondenominational University College, London and held the position until 1860. Lindley was a prolific writer, editor, and lecturer, with over two hundred publications to his name, many of which were written for a popular readership. He received the RS's Royal Medal in 1857.

Liouville, Joseph (1809–82), was a French mathematician educated at the École des Ponts et Chaussées, who worked in number theory and differential geometry. Liouville taught at the École Polytechnique from 1831 until 1851, when he was elected to the Collège de France in Paris, where he taught until 1879. In 1836 he founded the *Journal de mathématiques pures et appliquées.* The journal, which he edited until 1874, became known as the *Journal de Liouville.* At the Collège de France, Liouville gave eleven lecture courses on the theory of numbers. (Kenneth S. Williams, *Number Theory in the Spirit of Liouville* [New York: Cambridge University Press, 2011], 9.)

Lissajous, Jules Antoine (1822–80), was a French physicist who entered the École Normale Supérieure in 1841 and received his doctorate in 1850. He became professor of physics at the Lycée Saint-Louis in 1847, a position he held until 1874, when he became rector of the Academy Chambéry and later the rector of the Academy at Besançon. Lissajous is best known for developing an optical method of examining sound vibrations in 1855. The Lissajous apparatus reflects light on mirrors mounted on two tuning forks to produce an image of the vibration, known as the Lissajous curve.

Lyell, Charles (1797–1875), was a Scottish geologist and lawyer, whose *Principles of Geology* (3 vols., London: John Murray, 1830–33) advocated a "uniformitarian" view of the processes of geological change. Educated at Exeter College, Oxford, he studied law at Lincoln's Inn, London. He was briefly professor of geology at King's College, London (1831–33), after which he lived on private means. He married Mary Elizabeth Horner in July 1832; she assisted her husband in his scientific work. A prominent figure in the Geological Society of London, Lyell was twice elected as its president (1835–37; 1849–51). He was knighted in 1848. Lyell and Hooker were instrumental in bringing Darwin's theory of natural selection to public attention. The discovery in 1859 that humans had lived alongside the mammoths of the last ice age led to Lyell's last major publication, *Geological Evidences of the Antiquity of Man* (London: John Murray, 1863).

Magnus, Anna (1841–75), was the eldest child of Gustav and Bertha Magnus.

Magnus, Bertha (née Humblot, 1820–1910), was born to a French Huguenot family that settled in Berlin. She married Gustav Magnus in 1840. Together they had two daughters, Anna (1841–75) and Christine (1842–1936), and a son, Paul (1845–1930).

Magnus, Christine (1842–1936), was a daughter of Gustav and Bertha Magnus. In 1866 she married her cousin Victor von Magnus (1828–72), a baron and banker. She would later marry Ludwig Raschdau (1849–1943), a German diplomat.

Magnus, Heinrich Gustav (1802–70), was a German chemist and physicist. A pupil of the German chemist Eilhard Mitscherlich in Berlin and of J. J. Berzelius in Stockholm, he was professor of technology and physics at the University of Berlin from 1833 until his death. In Berlin, he developed a superbly equipped physics laboratory at his home, to which he invited young German physicists to work, including Tyndall in 1851. Although his early research was primarily in chemistry, by the 1850s he had begun to investigate a range of physical problems, as his letters to Tyndall demonstrate.

Magnus, Paul (1845–1930), was a son of Gustav and Bertha Magnus.

Martin, John (1812–75), was an Irish nationalist ("Honest John Martin") and brother of Thomas Hirst's wife, Anna. Educated at Trinity College Dublin, he became active in anti-British nationalist politics as a "Young Irelander." He was arrested and convicted of treason in 1848 and transported to Van Dieman's Land (Tasmania) in 1849. He was granted release in 1854 on condition that he live in exile, and chose to live in Paris. It was there that Hirst first met him when on his honeymoon. Following a full pardon in May 1856, he returned to Ireland, where he continued to engage in the cause for Home Rule. In 1856 and 1857 Martin lived with Anna and Thomas Hirst in Pau, France, where Anna sought treatment for tuberculosis. In later life he repatriated and became an MP for Meath in the British parliament, advocating Home Rule, 1871–75.

Matteucci, Carlo (1811–68), was an Italian chemist, physicist, physiologist, and politician who corresponded with Faraday from 1833 onward. After graduating in physics at the University of Bologna, in 1828 he undertook further studies in Paris before being appointed professor of physics at the University of Pisa in 1840. His pioneering researches in the field of electrophysiology led him into a vexatious dispute with du Bois-Reymond that is reflected in his correspondence with Tyndall. He married an Englishwoman, Robinia Elizabeth Young, in September 1846.

Maule-Ramsay, Fox (1801–74), second Baron Panmure and eleventh earl of Dalhousie, was a Scottish career politician. Born Fox Maule, he became the second Baron Panmure on the death of his father, William Ramsay Maule, in 1852. After serving in the army from 1820 to 1832, he was elected MP for various Perthshire constituencies (1835–52), and served as undersecretary of state (1835–41), and secretary at war (1846–52; 1855–58), following which he succeeded to the earldom of Dalhousie in Scotland.

Moore (Carrick-Moore), James (1762–1860), was a surgeon, biographer, and promoter of vaccination. Born in Glasgow, he studied medicine in Edinburgh and London. He served as a medical officer in the American War of Independence, and set up medical practice in London by 1784. Moore promoted vaccination in the 1800s and 1810s, and became director of the National Vaccine Establishment. In 1821 he inherited the significant estate of the Glasgow banker Robert Carrick, and adopted the name Carrick-Moore. The wealthy Moores had close connections with the RI and frequently entertained Tyndall at their Wimbledon home.

Moore, Graham Francis (Michell Esmeade) (1806–83), was a lawyer educated at Trinity College, Cambridge. He was the younger son of James Carrick-Moore and Harriet Moore (née Henderson). Born Graham Francis Moore, on inheriting the estate of his cousin Anne Michell in 1845 he took the name Michell Esmeade. He died without having married. The wealthy Moores had close connections with the RI and frequently entertained Tyndall at their Wimbledon home.

Moore, Harriet (née Henderson, 1779–1866), was a daughter of the actor John Henderson (c. 1747–85) and Jane Figgins (c. 1752–1819). Harriet married surgeon and vaccine promoter James Carrick-Moore. Together they had five children: Harriet, Louisa (1802–53), Julia, John, and Graham Francis. The wealthy Moores had close connections with the RI and frequently entertained Tyndall at their Wimbledon home.

Moore, Harriet Jane (1801–84), was a British watercolor artist who painted scenes of Faraday's apartment, study, and laboratory at the RI during the 1850s. Moore was the eldest child of surgeon James Carrick-Moore and Harriet Moore (née Henderson). She was elected a member of the RI in 1852. See Frank A. J. L. James, "Harriet Jane Moore, Michael Faraday, and Moore's Mid-Nineteenth-Century Watercolours of the Interior of the Royal Institution," in *Fields of Influence: Conjunctions of Artists and Scientists, 1815–1860*, ed. James Hamilton (University of Birmingham Press, 2001), 111–28.

Moore, John Carrick (1805–98), was a field geologist and the eldest child of surgeon James Carrick-Moore and Harriet Moore (née Henderson). He entered Queen's College, Cambridge, in 1823, receiving his BA in 1827 and MA in 1830. He was called to the bar in 1831, but became interested in geology and devoted much of his life to research. He was an active member of the Geological Society of London, becoming a fellow in 1838 and serving as secretary (1846–52). He was also a member of the RI from 1836, becoming a visitor in 1853 and a manager in 1857. Later in life he would write *Recollections of an Octogenarian* (London: Chiswick Press, 1888).

Moore, Julia (1803–1904), was the youngest of the three daughters of Harriet and James Carrick-Moore.

Moseley, Henry (1801–72), was an English mathematician and educationalist. He studied mathematics at Saint John's College, Cambridge, where he graduated seventh wrangler (1826) and later received his MA (1836). He took holy orders in 1827, and during a curacy in Somerset he published *A Treatise on Hydrostatics* (Cambridge, 1830), which led to his appointment as professor of natural and experimental philosophy at King's College, London (1831–44). He then became a school inspector for the Committee of Council of Education, where he did important work in introducing science teaching into elementary schools and teacher training colleges. He was closely involved in arranging examinations for admission to the Royal Military Academy at Woolwich in 1855 following the reform of military education at the time of the Crimean War.

Mousson, Albert (1805–90), was a Swiss physicist, glaciologist, and malacologist. He studied in Bern and at the Universities of Geneva, Göttingen, and Paris. In 1832 he married Barbara Seger. After the founding of the University of Zurich in 1833 he became a lecturer, then professor of physics there. He held the professorial chair of physics at the École Polytechnique Fédérale and the University of Zürich (1855–78). He contributed to the study of glaciers and the role of pressure on the melting of ice. In 1854 he published *Die Gletscher der Jetztzeit* (Zürich: Fr. Schulthess). He also studied meteorology and described 450 new species of molluscs.

Murray, John (1808–92), was an English publisher and the eldest child of Anne Elliot and publisher John Murray. This Murray was the third to carry that name. Murray was educated at Charterhouse School (1819–26) and for a year (1827) at Edinburgh University. He began his career in the family firm at 50 Albemarle Street, London, in 1828 and became head of the firm after his father's death in 1843. Murray published the *Quarterly Review*, a popular series of *Handbooks for Travellers*, and numerous scientific publications. On 24 November 1859 Murray published both Darwin's *On the Origin of Species* and Samuel Smiles's *Self-Help*. He became a member of the Alpine Club in 1858 and in 1860 published Tyndall's *Glaciers of the Alps*.

Newton, Isaac (1642–1727), was an English natural philosopher and mathematician educated at a grammar school in Grantham until entering Trinity College, Cambridge, in 1661. While an undergraduate the university was dispersed due to an outbreak of plague. It was during his time at home that he had his *anni mirabiles*, in which he laid the foundations of his later discoveries. In 1669 he was appointed Lucasian Professor at Cambridge University, which he held until 1702. His *Principia Mathematica* was first published in 1687. Later in his life he was master of the Royal Mint (1700–27) and was knighted by Queen Anne in 1705. He was buried in Westminster Abbey, London.

Platner, Marie (née Gerling), was a daughter of Christian Ludwig Gerling (1788–1864). She married Hermann Platner (1814–93).

Playfair, Lyon (1818–98), was a Scottish chemist who worked in Justus von Liebig's laboratory in Giessen (1839–41). In 1842 he became professor of chemistry at the Royal Manchester Institution. In 1845 he became chemist to the Geological Survey of Great Britain and professor of chemistry at the School of Mines, London, and in 1848 was elected FRS. In 1853 Playfair was appointed secretary for science in the Department of Science and Art. In this position, he attempted to establish a national system of science education. Playfair maintained a close relationship with the royal family and especially Albert, the Prince Consort, and for a time educated the queen's sons Albert and Alfred. In 1858 he became professor of chemistry at Edinburgh University.

Plücker, Julius (1801–68), was a German mathematician, experimental physicist, and professor of physics at the University of Bonn (1847–68). He first established a reputation for his work on analytical geometry, but under the influence of Faraday's research he began experimental work in electricity and magnetism in the 1840s. His ideas about diamagnetism and the behavior of crystals in magnetic fields were challenged by Tyndall, which led to ill feeling and controversy. They appear, however, to have resolved their differences by April 1858 (letter 1589; Journal, 18 April 1858, RI MS JT/2/13c/1069).

Poggendorff, Johann Christian (1796–1877), was a German physicist at the University of Berlin. His major contribution to science was as editor for fifty-two years (1824–76) of *Annalen der Physik und Chemie,* which became known as *Poggendorff's Annalen* during his editorship. He used his broad knowledge of science and scientific men in producing a *Biographisch-literarisches Hand-worterbuch zur Geschicte der exacten Wissenschaften* (Leipzig: Johann Ambrosius Barth), the first two volumes of which were published in 1863.

Poinsot, Louis (1777–1859), was a French mathematician and physicist who studied at the École Polytechnique and the École des Ponts et Chaussées, Paris. After three years' studying to become an engineer he decided to change to mathematics. He taught mathematics at the Lycée Bonaparte from 1804 to 1809, when he became an assistant professor at the École Polytechnique. His several publications, beginning with *Éléments de statique* (Paris: Calixte-Volland, 1803), secured his reputation as a mathematician and election to the Académie des Sciences. He became a member of the Paris Bureau des Longitudes in 1843, and the senate on its re-formation in 1852. He was elected a foreign fellow of the RS in 1858.

Pollock, Frederick, (1845–1937), was a third baronet, jurist, and the eldest son of Juliet and William Frederick Pollock. He was schooled at Eton College (1858–63) before entering Trinity College, Cambridge, where he obtained his BA (1867) and MA (1870). He was admitted to Lincoln's Inn in 1868, was called to the bar in 1871, and married Georgina Harriet Deffell in 1873. He was a distinguished jurist, author of many legal treatises, and became professor of jurisprudence at Oxford University (1883–1903).

Pollock, Juliet (née Creed, 1819–99), was a daughter of Catherine Herries and the Reverend Henry Creed, vicar of Corse, Gloucestershire. She married William Frederick Pollock, the barrister and author, in 1844, with whom she had three sons. She and her husband were loyal supporters of the RI, attended Tyndall's lectures, and frequently invited Tyndall to dinner, and were among Tyndall's closest friends during the period covered by this volume. Juliet Pollock was the author of several books, including *Julian and his Playfellows* (London: Grant and Griffith, 1852), *New Friends* (London: John W. Parker and Son, 1858), and *Macready As I Knew Him* (London: Remington and Co., 1885).

Pollock, Maurice Emilius (1857–1932), was the third son of Juliet and William Frederick Pollock. He became a sculptor and married Lydia Helen Roberts (1853–84) in 1880, and Mabel Mary McPherson (1860–1937) in 1889.

Pollock, Walter Herries (1850–1926), was the second son of Juliet and William Frederick Pollock. He was educated at Eton College (1862–65) and was admitted to Trinity College, Cambridge, where he obtained a BA (1871) and MA (1875). He was admitted to the Inner Temple in 1870 and called to the bar in 1873. Pollock became a writer, poet, and journalist, and was editor of the *Saturday Review* (1884–94). He married Emma Jane Pipon in 1876. Letters in this volume show a close affinity between Walter and Tyndall. Juliet referred to Tyndall as Walter's "especial friend," and related how he took a photo of Tyndall with him when leaving for school in Brighton at the age of eight.

Pollock, William Frederick (1815–88), was a second baronet, lawyer, and author. The eldest son of Jonathan Frederick Pollock, first baronet (1783–1870), and Frances Rivers (d. 1827), he married Juliet Creed in 1844. He was admitted to Trinity College, Cambridge, in 1832, obtaining his BA (1836) and MA (1840). Admitted to the Inner Temple in 1833, and called to the bar in 1838, he then became a master of the court of exchequer in 1846 and was appointed Queen's Remembrancer in 1874. Pollock's *Personal Remembrances* (2 vols., London: Macmillan and Co., 1887) illustrates how the family circulated among London's elite and entertained at their Montagu Square home. Both William and Juliet were loyal supporters of the RI, and Tyndall was frequently among their guests.

Pridie, William Roby (1825–74), was a Yorkshireman whom Hirst and Tyndall befriended when they joined the Mutual Improvement Society in Halifax. He was the son of the Reverend James Pridie (1786–1873), an independent minister in Halifax. Pridie and Francis Booth, another member of the Mutual Improvement Society in Halifax, were mentored by Hirst (just as Hirst was mentored by Tyndall). When Booth died of tuberculosis in March 1853, Hirst financially supported his mother until her death in 1856. Pridie, who was by then a bank clerk in Halifax, arranged the payments. Tyndall assisted with these payments while Hirst was in Europe in 1856.

Ramsay, Andrew Crombie (1814–91), was a Scottish geologist born in Glasgow. His study of geology was encouraged by John Nichol, and through this support Ramsay's abilities were recognized at the 1839 BAAS meeting in Glasgow. After working as a local director for the Geological Survey of Great Britain he became professor of geology at University College, London (1848). He held the chair of geology at the School of Mines from 1851 to 1872 (the Royal School of Mines from 1863), and in 1872 succeeded Roderick Murchison as director of the Geological Survey. He married Louisa Williams in 1852 and on their honeymoon in Europe became interested in glaciers. Quickly becoming a leading expert in the field, Ramsay argued that glaciers formed some lake basins. He accompanied Tyndall to the Alps in 1858.

Ramsay, Elizabeth (née Crombie, c. 1773–1858), was the mother of Andrew Crombie Ramsay and his three siblings John, Elizabeth, and William. Daughter of Andrew Crombie, born in Edinburgh, she married William Ramsay in 1809. After William's death in 1827 she took on boarders in the family home. Elizabeth died in 1858 while her son was exploring the Alps with Tyndall.

Regnault, Henri Victor (1810–78), was a French chemist, physicist, and photographer. A pupil of Liebig's at the University of Giessen, he was professor of physics at the Collège de France (1841–54), and director of the famous porcelain factory at Sèvres (1854–70), where he conducted many significant experiments in chemistry. In 1871 his laboratory at Sèvres was destroyed and his son Alex-Georges-Henri Regnault killed, both as a result of the Franco-Prussian War. He retired from science the next year, never recovering from these losses.

Rose, Heinrich (1795–1864), was a German analytical chemist and brother of the scientist Gustav Rose (1798–83). Heinrich Rose taught at the University of Berlin as *privatdozent* in 1822, extraordinary professor in 1823, and ordinary professor after 1835. Rose was known for his studies of minerals and his textbook *Handbuch der analytischen Chemie* (Berlin: Mittler, 1829), which he expanded in several editions over the course of his life.

Sabine, Edward (1788–1883), was an Anglo-Irish astronomer, geophysicist, and military officer who led the British effort to establish magnetic observatories all over the globe. A graduate of the Royal Military Academy at Greenwich, he reached the rank of general in the 1850s. He accompanied J. C. Ross's Arctic expedition (1818) and W. E. Parry's expedition (1819). He received the Copley Medal from the RS in 1821 for these magnetic observations in the Arctic. He was closely connected with the administration of the RS, and served as its treasurer (1850–61) and president (1861–71). He was an outstanding lobbyist for science, an effective administrator, and a valuable patron to Tyndall from 1851.

Sauerwald, E., was a scientific instrument maker whose workshop was located in Berlin. He constructed apparatuses for Gustav Magnus, John Tyndall, Emil du Bois-Reymond, and Hermann von Helmholtz, among others.

Saussure, Horace-Bénédict de (1740–99), was a Swiss geologist, botanist, and mountaineer, educated at the University of Geneva. He held the chair of philosophy at the Academy of Geneva in 1762 and married Albertine Amélie Boissier in 1765. After a number of trips to Chamonix his interest turned to geology. During the 1770s he began the research that would be published in his four-volume *Voyages dans les Alpes* (Neuchâtel: Samuel Fauche, 1779–96). Saussure made the third ascent of Mont Blanc in 1787, a year after Jacques Balmat and Michel Paccard. Spending four and a half hours on the summit, he conducted a number of scientific experiments. His expedition caught the popular imagination and earned him international fame.

Schulze, Oskar (1825–78), was the second son of Johann Friedrich Schulze and a member of the fifth generation of the Schulze organ builders. The family business was located in Paulinzella, Thuringia, Germany. Oskar studied acoustics and constructed an apparatus for demonstrating sound waves. He sent one of these instruments to Tyndall in 1857 seeking a recommendation. However, Tyndall appears to have refused the gift. (Bryan Hughes, *The Schulze Dynasty: Organbuilders 1688–1880* [St Leonard's on Sea: Musical Opinion Ltd., 2006].)

Simond, Edouard (d. 1870), was a Swiss Alpine guide. He guided Tyndall and Hirst on their ascent of Mont Blanc in 1857. According to Tyndall he kept a hotel in Chamonix, possibly the Hotel du Mont Blanc, discussed in D. Dunant's *Le Touriste à Chamounix en 1853* (Geneva, 1853), 181. Simond died in a storm on Mont Blanc in 1870 while guiding an American party of climbers. Tyndall wrote of him: "Simond had proved himself a very valuable assistant; he was intelligent and perfectly trustworthy; and though the peculiar nature of my work sometimes caused me to attempt things against which his prudence protested, he lacked neither strength nor courage." (*Glaciers of the Alps*, 86, 90–91; C. D. Cunningham and W. Abney, *The Pioneers of the Alps*, 2nd ed. [London: Low, Marston, Searle and Rivington, 1888], 33.)

Simpson, Maxwell (1815–1902), was an Irish chemist who had studied at Trinity College Dublin and University College, London. In 1847 he became a lecturer in chemistry at Park Street Medical School in Dublin (a renowned private medical school) and later at the Peter Street School of Medicine (another private medical school with an anatomy theater). He studied in Germany (1851–54) and Paris (c. 1858–60). His wife was Mary Martin, whose sister Anna accompanied them in 1851 to Marburg, where she would meet Thomas Hirst.

Sorby, Henry Clifton (1826–1908), was an English geologist who pioneered the field of petrography through his application of the microscope to measure and determine the composition of rocks. His research on sedimentary rock led him to conclude that pressure was the cause of slaty cleavage. He was elected to the RS in 1857 for this work. He also turned his microscopic studies to iron, steel, and crystals.

Stokes, George Gabriel (1819–1903), was an English mathematician and physicist. He graduated from Cambridge University as senior wrangler and Smith's Prizeman in 1841 and served as Lucasian Professor of mathematics at Cambridge University (1849–1903). He augmented his income as a lecturer at the Royal School of Mines (1849–56). As secretary of the RS (1854–85) he was responsible for refereeing papers for the *Phil. Trans.*, a task for which he frequently sought Tyndall's assistance.

Studer, Bernhard (1794–1887), was a Swiss geologist who, after earning a degree in theology, studied mathematics and science at Göttingen from 1816 to 1818. In 1834, after the establishment of the University of Bern, he became its first professor of geology, remaining there until his retirement in 1873. His research was largely concerned with the geology of the Alps. His first monograph examined the sandstones, shales, and conglomerates in the Alps known as Molasse (1825), and his second was a study of the geology of the western Swiss Alps (1834). He collaborated with Arnold Escher von der Linth on the first geological map of Switzerland (1853). His textbook, *Geologie der Schweiz* (Bern: Stämpflische Verlagshandlung, 1851–53), was written to accompany the map.

Sylvester, James Joseph (1814–97), was an English mathematician of Jewish descent. He was appointed to the professorship of natural philosophy at the nondenominational University College London in 1838, after which he accepted a professorship at the University of Virginia in 1841, although he stayed there only six months. After returning to England the following year, he met Arthur Cayley, a fellow mathematician, and together they made significant advances in the theory of invariants. He found work as an actuary before obtaining the professorship of mathematics at the Royal Military Academy, Woolwich (1855–70). Through the 1850s Sylvester was a prolific mathematical researcher, often publishing in the *Phil. Mag.*

Tennyson, Alfred (1809–92), was poet laureate from 1850 in succession to William Wordsworth. Born in Lincolnshire, he was educated at Trinity College, Cambridge. In 1850 he published *In Memoriam* (London: Edward Moxon). In 1853 he moved to Farringford on the Isle of Wight, where he published *Maud* (London: Edward Moxon, 1855). Tyndall greatly admired Tennyson's poetry and met him in 1858. Tennyson declined honors or titles from Queen Victoria four times between 1865 and 1880, but accepted a barony in 1883.

Tennyson, Emily Sarah (née Sellwood, 1813–96), married Alfred Tennyson in 1850, and was his manager and secretary. They had two sons, Hallam and Lionel, in 1852 and 1854. Emily has been recognized by her contemporaries and by historians for the important role she played in Tennyson's work. She answered his correspondence, edited and set his poetry to music, and read French, German, and Italian.

Thomson, James (1822–92), was a mechanical engineer and older brother of William Thomson, Lord Kelvin. As a teenager, he attended Glasgow University, receiving his BA (1839) and his MA (1840) in mathematics and natural philosophy. He worked as an engineer at firms in Glasgow, London, and Belfast before he became professor of engineering at Queen's University Belfast in 1854. Thomson applied his studies of pressure in steam engines to J. D. Forbes's theory of glacier motion. In 1857 he argued that liquefaction by pressure accounted for the plasticity of ice observed in glaciers, bringing him into disagreement with Tyndall's theory of regelation. (James Thomson, "On the Plasticity of Ice, as Manifested in Glaciers," *Roy. Soc. Proc.* 8 [1857]: 455–58; *Glaciers of the Alps*, 408.)

Thomson, William (1824–1907), ennobled in 1892 (and then known as Lord Kelvin), was a distinguished Irish-Scottish mathematician, physicist, and inventor. He was educated at the University of Glasgow and Peterhouse College, Cambridge University, where he was second wrangler and Smith's Prizeman in 1841. He was professor of natural philosophy in Glasgow from 1846, paying frequent visits to London. He first met Tyndall at the Edinburgh meeting of the BAAS in 1850, and from the start the two physicists had a strained relationship. For much of his career Tyndall regarded Thomson as a scientific rival.

Tyndall, Sarah (née Macassey [Macasey/McAssey], 1793–1867), was Tyndall's mother. She was widowed in March 1847. She remained in the family cottage at Leighlinbridge, Ireland, supported financially by Tyndall. Only two letters to her during the period covered by this volume survive (1552, 1569). Tyndall later told Hirst of his deep affection for her rare qualities.

Verdet, Marcel Émile (1824–66), was a French physicist who held professorships at the École Normale Supérieure, the École Polytechnique, and the

Faculté des Sciences, Paris. Verdet advocated for the acceptance of the law of conservation of energy, and worked to introduce foreign thermodynamics research to France. Between 1852 and 1864 he published abstracts of foreign research in every volume of the *Annales de chimie et de physique*. His own research focused on electromagnetic induction and the polarization of light in magnetic fields, known as the Faraday effect.

Walker, Charles Vincent (1812–82), was an electrical engineer, author, and technical editor. He was a founding member of the London Electrical Society and served as the society's secretary and treasurer, and edited the *Electrical Magazine* (1841–43). He also edited and wrote the second volume of Dionysius Larnder's *Manual of Electricity* (London: Longman, Brown, Green & Longmans, 1844). Walker's *Electrotype Manipulation* (London: George Knight and Sons, 1841) and *Electric Telegraph Manipulation* (London: George Knight and Sons, 1850) were printed in several editions and translated into both French and German. From 1845 he was the chief telegraph engineer of the South-Eastern Railway Company. Walker also translated many scientific works into English, including de la Rive's *Traité d'électricité théoretique et appliquée* (3 vols., 1854–58). Walker was translating volume three when de la Rive asked Tyndall to read the proofs and correct Walker's "Gallicisms."

Wertheim, Guillaume (1815–61), was a physician, mathematician, and physicist who studied the mechanical properties of wood, bones, and tissue. Born in Vienna, he attended the University of Vienna, receiving a doctorate of medicine in 1839. He became a naturalized French citizen in 1848 and received a doctorate in physics from the Faculty of Science of Paris (Sorbonne). He was a professor of physics at Montpellier in 1854, and became an admissions examiner at the École Polytechnique in 1855. (Marcel Verdet, "Notice sure les travaux scientifiques de M. Guillaume Wertheim," *Société philomathique de Paris: Extraits des procès-verbaux des séances* 26 [1861]: i–xvi.)

Wheatstone, Charles (1802–75), was an English experimental physicist, inventor, and businessman. He had been a musical instrument maker before becoming professor of experimental philosophy at King's College, London (1834–75), where he was a reluctant lecturer. There he developed the electric telegraph with William Fothergill Cooke as well as a number of optical instruments, including the stereoscope. The latter led him into controversy with David Brewster. He married Emma West in 1847; they had five children by 1855. The family settled in Hammersmith, where Wheatstone frequently entertained Tyndall, whom he first met in June 1852.

Wright, Fanny (née Bellamy), married Richard Pears Wright in July 1852 at Saint Martin-in-the-Fields, London. In 1855 she gave birth to Alice Maud Wright, who died in January 1857. Tyndall attended the child's funeral.

Wright, Richard Pears (1829–92), was an English mathematician and educator who taught engineering and surveying at Queenwood College from 1851. According to Hirst, who encouraged his development as a geometer, he suffered from epilepsy. He married Fanny Bellamy in July 1852. Tyndall went on holiday with the Wrights on the Isle of Wight in 1855 and attended the funeral of the Wrights' daughter, Alice Maud (1855–57), in January 1857. Wright left Queenwood in 1863. From 1870 until his death in 1892 he was a teacher of mathematics at University College School, London.

Wrottesley, John (1798–1867), second Baron Wrottesley, English astronomer, scientific administrator, and member of many parliamentary commissions. He was the son of John Wrottesley, a Staffordshire landowner, farmer, and MP who was made first Baron Wrottesley in 1838. He was educated at Westminster School and Christ College, Oxford (BA 1819) and trained as a lawyer. He was one of the founders of the Royal Astronomical Society in 1820 and served as its secretary (1831–41). On inheriting his father's title, he relocated his Blackheath observatory to Wrottesley Hall at Tettenhall. He succeeded Lord Rosse as president of the RS, serving from 1854 to 1858.

Wynne, Anne (née Osborne, 1808–64), was the daughter of Sir Daniel Toler Osborne of Beechwood, Tipperary. She married George Wynne in 1834. The Wynnes had three sons (Henry, Edward, and Frank), to whom Tyndall sometimes sent letters, and one daughter, Lucy. During the 1850s Tyndall was a frequent visitor to the Wynnes' home at Harrow, near London.

Wynne, George (1804–90), was an Irish surveyor, engineer, and military officer. Born in Dublin, he enlisted in the Royal Engineers (c. 1825) and was commissioned as lieutenant in 1826. Wynne met Tyndall while employed in the Irish Ordnance Survey (1835–40). When he and Tyndall met, Wynne was in charge of the division in which Tyndall worked (see vols. 1–2). Wynne later served with distinction in Greece, China, and the West Indies, rising to general by the time he retired in 1877. Over the period of this volume he had the rank of captain, and the position of inspector of railways (1847–58) under the Board of Trade, before being transferred to China. He married Augusta Osborne in 1834. Tyndall and Wynne had a high regard for each other. ("The Late General Wynne," *Times*, 12 July 1890, 11.)

INDEX

The index refers to the letters using their serial numbers. The letter 'n' denotes a footnote reference and appears if an item is not otherwise mentioned in the letter to which the footnote pertains. A number in **bold** signifies a letter to or from a correspondent.

Academia Caesarea Leopoldino Carolina Naturae Curiosorum, 1365
Academie der Naturforscher, **1427**
Academy of Sciences, 1367
Adams, John Couch, 1572, 1581
Addison, Joseph, 1541n
Agassiz, Louis, 1302, 1303, 1304, 1305, 1313, 1328, 1333, 1436, 1453, 1461, 1510, 1541
Airy, George Biddell, **1346, 1347, 1348, 1373, 1374, 1379, 1380, 1384, 1385, 1388, 1391, 1393, 1394, 1421, 1422, 1517, 1518, 1541**, 1582, 1584, **1585, 1586, 1591, 1592**
Airy, Hubert, 1518, 1541
Airy, Wilfrid, 1518, 1541
Algemeine Zeitung, 1514
Alighieri, Dante, 1322, 1323
Alpine Club, 1459, 1466, 1478
Alps, 1311, 1415, 1418, 1419, 1420, 1422, 1428, 1429, 1436, 1440, 1446, 1453, 1527, 1531, 1536, 1537, 1538, 1539, 1540, 1541, 1543, 1544, 1547, 1548, 1550, 1551, 1552, 1554, 1555, 1556, 1562, 1566, 1573, 1576, 1580, 1590. *See also* glaciers.
America, 1311
Anderson, Charles, 1542, **1549**, 1553, 1560
Annalen der Physik und Chemie, 1520
Antaeus, 1383
Arago, François, 1293
Asher, George Michael, 1309, 1327, 1408

Athenaeum, 1587
Atkinson, Edmund, 1566
aurora, 1501

Babbage, Charles, 1328, **1341**, 1485, 1545, 1547
Babinet, Jacques, 1356
Babington, Charles Cardale, 1364
Bacon, Francis, 1367, 1387
Bakerian Lecture, 1300
Ball, John, 1328, 1395, 1456
Balmat, Auguste, 1415, 1446, 1551, 1554, 1557, 1562, 1566
Balmat, Edouard, 1419, 1420, 1428n, 1576
Baring, Bingham, 2nd Baron Ashburton, 1510, 1511
Barlow, Cecilia Anne (née Law), 1452
Barlow, John, 1401, 1403, 1420, 1455, **1485**, 1490, 1542
Barnard, Anne Henslow, 1396, 1527
Barnard, George, 1537, 1542
Barnard, Jane, 1417, 1553
Barnett, Henry, 1547
Battersby, Eliza, 1489
Beckett, Frances Adelina (née Ingram), 1566
Beckett, William, 1566
Becquerel, Alexandre-Edmond, 1415
Becquerel, Antoine-César, 1415
Beetz, Wilhelm von, 1461
Bennen [Benet], Johann Joseph, 1536, 1538n, 1539n, 1540n, 1545n, 1554n, 1573, 1576
Bentley, Charles A., 1373
Bergoin [Bergoën], Félix, **1551,** 1556
Bernays, Adolphus, 1364n
Best, Samuel, 1566
Bevington, James Buckingham, 1298